Topics in Nonlinear Physics

Topics in Nonlinear Physics

Proceedings of the Physics Session
International School of
Nonlinear Mathematics and Physics

A NATO Advanced Study Institute
Max-Planck-Institute for Physics and Astrophysics
(Munich, 1966)

Edited by

Norman J. Zabusky

Springer-Verlag New York Inc. 1968

NORMAN J. ZABUSKY

Head, Plasma and Computational Physics Research Department

Bell Telephone Laboratories, Inc., Whippany, N. J. 07981

Library of Congress Catalog Card Number 67-30647

[redacted]

Title-No. 1468

Printed in the United States of America

Acknowledgements and Remarks

The International School of Nonlinear Mathematics and Physics was directed by Professor MARTIN D. KRUSKAL and Dr. N. J. ZABUSKY and was sponsored by the Advanced Study Institute Program of the North Atlantic Treaty Organization. Several travel grants were given to United States citizens by the National Science Foundation. Additional support was provided by the home institutions of the directors, Princeton University and Bell Telephone Laboratories, Incorporated.

The Institute of Astrophysics of the Max-Planck-Institute for Physics and Astrophysics was host to the school. We would like to thank Professor W. HEISENBERG and Professor L. BIERMANN for the hospitality and services they provided during our 8-week stay. We would also like to thank Dr. D. PFIRSCH who acted as liaison with the scientists at the Institute, and Fr. G. WESSLING for hotel arrangements and liaison with the service staff at the Institute. Drs. A. GIORGINI and M. PRASTEIN assisted Professors P. SAFFMAN and C. A. TRUESDELL, respectively, with the preparation of their lecture notes. Mrs. CHARLOTTE ZABUSKY ably served as office manager and editor during and after the school sessions.

The physics session ran 3 weeks and was opened on June 27, 1966, with some introductory remarks by N. J. ZABUSKY and a perspective lecture by W. HEISENBERG on "Nonlinear Problems in Physics" that is included in this volume.

The mathematics session was opened on July 18, with some introductory remarks by M. D. KRUSKAL and a perspective lecture by S. M. ULAM on "Nonlinear Problems in Mathematics". The proceedings of the mathematics session are being edited by Professor M. D. KRUSKAL and will be published by Springer-Verlag in a separate volume entitled "Topics in Nonlinear Mathematics". It will include the following:

P. GERMAIN (O.N.E.R.A. and University of Paris, France)
Shock Waves - Jump Relations and Structure

H. O. KREISS (Uppsala University, Sweden)
Numerical Solutions of Nonlinear Partial Differential Equations

M. D. KRUSKAL (Princeton University, U. S. A.)
Asymptotic Methods for Nonlinear Problems

P. D. LAX (Courant Institute of Mathematical Sciences, New York University, U. S. A.)
Shock Waves - General Theory
Properties of the Korteweg-deVries Equation

S. ORSZAG (Massachusetts Institute of Technology, U. S. A.)
Turbulence

S. M. ULAM and P. STEIN (Los Alamos Scientific Laboratory, U. S. A.)
Nonlinear Algebraic Transformations

N. J. ZABUSKY (Bell Telephone Laboratories, Inc., U. S. A.) and M. D. KRUSKAL (Princeton University, U. S. A.)
Nonlinear Dispersive Systems

Including faculty, there were 85 participants from 18 countries.

Introduction to Topics in Nonlinear Physics

Is the study of the physics of nonlinear phenomena like the study of the zoology of nonelephants? Or is there a rational structure to link our knowledge of eclectic nonlinear phenomena? CARL VON WEIZSÄCKER suggested the former; KRUSKAL and I believe the latter.

In 1964 we recognized that in recent years different areas of physics have been investigating phenomena in which nonlinear effects played an important if not essential role. Nonlinear ordinary differential equations first arose when scientists tried to describe the dynamics of interactions of one or more particles with each other and with an external field. However, the application of partial differential equations to the study of the dynamics of nonlinear *continua* did not begin in earnest until Riemann's classic paper appeared in 1860.

In the century following few general results were obtained for nonlinear systems, compared with the progress made in other (linear!) branches of mathematics and physics. The elucidation of nonlinear phenomena is difficult, because one often finds that the methods of solution are highly restricted or "rigid", in that they are rendered useless if the problem posed is modified only slightly.

During the past decade the study of the properties of nonlinear continua has been accelerated. This is a result of our improved capability to perform and interpret experiments which utilize controllable, intense incoherent and coherent sources, e.g. bombs, condenser banks, particle beams, and lasers. Furthermore, natural phenomena can be observed or monitored (e.g., via satellites, radio telescopes, etc.) for "long" times because of the precision and reliability of measuring and communication devices.

The success achieved by some past experiments (and probably all large future experiments) can be attributed in part to the high level of sophistication reached by high speed digital computers with associated graphical output devices. With them one can look for the *small* effect; experiments with many parameters can be optimized; large volumes of data can be processed and correlated conveniently and rapidly; and nonlinear partial differential equations that are analytically unmanageable can be solved.

In viewing this situation, we hoped that the Nonlinear School would present a broad view of several fields and focus attention on two questions. Are there similar concepts and techniques applicable to diverse areas of nonlinear physics? Can communication among scientific disciplines be improved by clearing the forest of jargon?

We were indeed fortunate in assembling a dynamic faculty. Except for Wheeler's fundamental quantum geometric approach to nature, our physics curriculum was chosen from areas that are mainly classical or that can be described classically. The course material presented at the physics school covered diverse phenomena in a variety of approaches.

HEISENBERG opened the physics school with a perspective lecture on "Nonlinear Problems in Physics", in which he touched on the questions posed above. His contributions to quantum theory are familiar to all of us; not so well known is his recurrent interest and experiences with nonlinear phenomena ranging from fluid flow and turbulence to nonlinear electrodynamics and the nonlinear spinor theory of elementary particles. His ideas may be summarized in four aphorisms: simplification by *symmetrizing*; initial progress by *linearizing*; general features by *statistical methods*; and phenomena with long-time *unpredictability*.

TRUESDELL gives a rigorous axiomatic development of the mechanics and thermodynamics of macroscopic classical nonrelativistic continua. This theory has been developed over the past twenty years. It attempts to unify and systematize the *macroscopic* treatment of solids and fluids including materials with unusual viscometric constitutive relations when they are subjected to small and large deformations. TRUESDELL remarks, "...it is pure illusion to think that learning all about tiny things is the path to knowledge about big things".

The theory begins with a careful restatement of macroscopic thermodynamics of homogeneous processes. These are combined with Euler's dynamical laws, and general thermomechanical "equations of balance", or conservation laws, are derived. Materials in nature are represented by a variety of constitutive relations, namely stress tensors related to the past history and present state of motion of the material. For example, a large part of the notes is devoted to studying "simple materials". These include many of the recent nonlinear theories. Simple materials can have long-range "memory" so that stress relaxation, creep, and fatigue can occur. Assuming motions with "constant stretch history" reduces the complexity of the general stress tensors. In essence, a particle of a continuum sees an unchanged sequence of past deformations relative to its present configuration. Viscometric flows of nonNewtonian fluids (like polymers) are included in this class; they have stress tensors with elements proportional to a power of the velocity shear. Examples illustrate unusual effects produced by viscometric properties in conventional arrangements: rectilinear flow; shear flow; helical flow; flow between rotating cylinders (Couette flow); flow in a circular pipe (Poiseuille flow); etc. Materials with "fading memories" are described by *thermomechanical* constitutive relations.

The lectures conclude with a general discussion of wave motion using Hadamard's ideas on the dynamics of singular surfaces. The results are applied to "acceleration" waves, that is, waves where the function (e.g., the density) and its first derivative are continuous but the second and higher derivatives may be discontinuous. These are "weak" nonlinear waves, unlike the strong discontinuities or shock waves that arise, for example, in lossless descriptions of fluids, like the Euler equations, and are treated at great length in the mathematics volume.

PRIGOGINE and BAUS give a formal, mathematical development of a *microscopic* theory of matter and radiation. Their approach relies heavily on standard perturbation methods, that is, nonuniform asymptotic expansions. They develop a statistical mechanics for Hamiltonian systems, and relate the results to macroscopic thermodynamics. The initial conditions and the nature of the Hamiltonian play a strong role in determining both the final state of a system and the time required to reach it. In fact, the long-range or short-range nature of the forces between the particles determines the rate of approach to thermodynamic equilibrium. Particular attention is given to two basic classical problems: that of reconciling microscopic reversibility with macroscopic irreversibility; and that of defining an H-function in the sense of BOLTZMANN. (TRUESDELL, on the other hand, starts with postulates governing irreversibility and entropy.)

The perturbation formalism is applied to the classical problems of Brownian motion and the dynamics of anharmonic solids. A "master" or evolution equation for the velocity distribution function is derived from the Liouville equation. One can identify a "destruction fragment", a term in the evolution equation, that describes the decay of dynamical correlations. The term depends on the dynamics of the system and the initial conditions.

For Hamiltonian systems, Boltzmann-like evolution equations are derived if one follows a BOGULIUBOV two timescale prescription. That is, if (1) we neglect "memory" effects or assume decay of initial "dynamical correlations" and (2) we retain only the asymptotic effects of the finite duration of a collision, then we may obtain a Fokker-Planck equation for weakly coupled systems, a Boltzmann equation for dilute gases, and a Balescu-Lenard equation for plasmas. The effect of nonMarkoffian terms can be easily evaluated by these methods.

The random-phase approximation is examined critically, particularly for systems having stable nonlinear states. Stable (organized or cellular) states may arise if there exist two competing processes. For example, Bénard cells arise in a fluid in a gravitational field heated from below because of competition between convection and dissipation. The "solitons" discussed by ZABUSKY and KRUSKAL in the mathematics volume are stable states resulting from the competition between the steepening effect of the nonlinear process and the radiating effect of the linear-dispersive process. They exist in a variety of lossless media, including the anharmonic lattice discussed by PRIGOGINE from another point of view.

The existence of an entropy depends on the loss of memory expressed by the destruction fragment. The problem of constructing a functional, which includes the effect of correlations, satisfies an H-theorem (in the Boltzmann sense) and gives at equilibrium the correct thermodynamic entropy, is considered. Only for weakly coupled or dilute systems is entropy to be associated with "disorder". A strongly interacting system is highly organized and may not exhibit thermalization. The problem of defining entropy is difficult because it is a nonlinear functional of the correlations, whereas energy and specific heat depend linearly on the correlations.

BAUS applies the methods and perturbation formalism of the Prigogine school to a plasma of relativistic charged particles interacting with an electromagnetic field. He derives a "master" equation for a homogeneous system, that is, a kinetic equation for the fluctuations of particle density and wave density. In this equation, the time derivative of the density is equal to the sum of two terms - a nonMarkoffian term depending upon the integral over the density at past times, and a term depending upon the initial condition. This master equation contrasts with other recently derived hierarchy equations, where the time derivative of the density depends only on binary correlations.

This master equation is analyzed by perturbation methods, with the expansion parameter being the "plasma parameter" (the inverse of the number of particles in a Debye sphere). In lowest order, the relativistic extension of the Balescu-Lenard equation is recovered. This equation describes the irreversible approach of a plasma to equilibrium because of Coulomb collisions.

For a stable plasma BAUS obtains a system of nonlinear integro-differential equations - nonlinear because the unknown (density) function appears in the denominator of the dielectric function. This nonlinearity manifests the continuing adaptation of the polarization cloud about a charged particle to the "mean" state of the system. Applications are made to the problems of normal and anomalous Bremsstrahlung, the relation between the absorption coefficient and electrical conductivity, etc.

Bloembergen's survey of nonlinear optics is guided by experiments. The description is mainly classical and macroscopic and uses Maxwell's equations for media whose electromagnetic polarization contains nonlinear terms in addition to the usual linear dispersive terms. The choice of terms included in any particular case is determined by the type of optical media (their symmetry properties, the nature of the nonlinearities, etc.), and the magnitudes of frequencies, velocities, relaxation times, and amplitudes (or coupling parameters) governing a particular phenomenon. All possible effects are tersely represented by a time-averaged "free enthalpy", a sum of terms formed from products of the electric and magnetic fields and their space and time derivatives.

Since most media are highly dispersive and would undergo changes of state, like rupture and ionization, before "long-time" phenomena could manifest themselves, nonlinear optical effects are weak. Thus, BLOEMBERGEN concludes that "shock waves" cannot form by the usual crossing of characteristics of the corresponding hyperbolic system (the well-known phenomena discussed by GERMAIN in the mathematics volume). Hence, first-order perturbation theory is adequate to describe most observations. If one goes to higher order, secular terms will arise, and he shows how to avoid this difficulty by using a Krylov-Bogoliubov averaging procedure. KRUSKAL's lectures on asymptotic methods at the mathematics school give a general and rigorous account of such procedures.

Harmonic generation and parametric processes are also discussed by BLOEMBERGEN. Usually there is a variety of wave propagation modes available for a spatially homogeneous physical system. However, geometrical considerations may restrict the number of strongly interacting modes to two or three. For example, stimulated Raman scattering is treated as a parametric process - a parametric "down process" between two modes, since energy in the electromagnetic modes is converted to energy of optical-phonon modes (at a lower frequency) in the medium. The important phenomenon of "self-focusing" of a narrow beam of radiation is studied by introducing an intensity-dependent index of refraction. Thus, waves of different frequency are simultaneously diffracted and dispersed.

SAFFMAN gives a lucid and stimulating review of the problem of homogeneous isotropic turbulence (HIT). He calls attention to those points where physical and mathematical assumptions are unsubstantiated, poorly correlated with experiment, and in need of further scrutiny. ORSZAG's lecture notes on turbulence (in the mathematics volume), on the other hand, are mathematically formal.

Since HIT is a *strongly* nonlinear process, attempts to apply standard perturbation theory have been fruitless. Unfortunately, there exists no small parameter to expand in, such that "solvable" equations are obtained, and the flow is characterized by being "intermittent" in space-time. Scattered throughout the chapters is a note of skepticism about the success of any approach involving interacting Fourier modes of the velocity field in wavenumber space. SAFFMAN feels that the phenomena of intermittency and "vortex-stretching" are obscured in wavenumber space treatments of turbulence.

The importance of the number of dimensions of the physical space is considered critically when he examines "Burgerlence", the one-dimensional turbulence of a fluid governed by Burgers' partial differential equation. In fact, he shows that the general arguments for the energy cascade associated with a universal equilibrium theory give the wrong answer when applied to Burgers' equation.

SAFFMAN introduces an heuristic theory treating turbulence as a random array of laminar flows in physical space. Two cascade processes are operative. In the primary (large-scale) process, vortex sheets and tubes are formed with a characteristic thickness proportional to the square root of the Reynolds number. The appearance of fluid-dynamic instabilities, e.g., of the Taylor-Görtler or Taylor-Couette type, gives rise to a secondary cascade process that results in fine-scale motions.

He concludes his discussion with a critique of several "quasi-analytical" theories of turbulence, including the Wiener-Hermite stochastic expansion and Kraichnan's direct-interaction approximation. SAFFMAN feels that the former is potentially "superior to the latter, which rests on a number of unjustified and probably unjustifiable assumptions of a mathematical form with no physical basis".

Wheeler's notes on a theory of superspace and quantum geometrodynamics differ in style and content from the lectures given in Munich. It is a closely reasoned, persuasive presentation. The arguments are drawn from the commonplace, the not-so-familiar, and the esoteric: breaking waves and the foamy sea surface; the electric contribution to chemical forces; and the transformation of equivalence classes of diffeomorphisms. Geometrodynamics is fundamentally a nonlinear theory in its infancy.

"Superspace" is the arena for geometrodynamics, just as Lorentz-Minkowski space-time is the arena for particle dynamics. Three-geometries (*not* 4-geometries) are the events in superspace. When no "real" mass energy sources are present, the dynamical evolution of geometry can be described in the context of classical geometrodynamics by the Einstein-Hamilton-Jacobi equation. The dependent variable is the "dynamical phase", a functional of the 3-geometries. Again and again it is stressed that in quantum geometrodynamics we must give up the concept of a 4-geometry.

Quantum effects are felt in at least two areas: gravitational collapse and microscopic quantum fluctuations in the geometry of space. Here the characteristic dimension is the very small Planck length $L = (\hbar G/c^3)^{1/2} = 1.6 \times 10^{-33}$ cm, and the characteristic energy density is $\hbar c/L^4 = c^5/\hbar G^2 \approx 10^{95}$ gm/cm^3. WHEELER maintains that geometrodynamics can be extrapolated to these small distances and violent energy fluctuations.

The "wormhole" concept is advanced and with this view the occurrence of electric charge provides WHEELER with the "evidence" for the existence of fluctuations in the topology of space over Planck distances.

WHEELER concludes with the vision that everything in physics will be constructed out of geometry, that particles will be viewed as geometrodynamic excitons.

The equations of geometrodynamics are certainly as nonlinear as the equations of classical continuum dynamics and richer in structure. Because the data base in geometrodynamics is sparse, progress will be made at first only by careful mathematical analysis and by carrying over relevant physical concepts from other fields.

In this introduction a feeling for the nonlinear way, as it is practiced in several branches of physics, is given. Many connections among the disciplines are established, some tenuous, some strong. Yet it seems clear, there will be benefits, at least in the techniques and concepts, as we step out in all directions to find a nonlinear Gestalt.

Whippany, New Jersey
December, 1967

NORMAN J. ZABUSKY

Faculty of the Nonlinear School

BAUS, M. Université Libre de Bruxelles, Brussels, Belgium

BLOEMBERGEN, NICOLAAS. Pierce Hall, Harvard University, Cambridge, Massachusetts 02138, USA

GERMAIN, P. Office National d'Étude et de Recherche Aérospatiales, 29 Avenue de la Division Leclerc, Chatillon s/Bagneux (Seine), France

KREISS, HEINZ-OTTO. Uppsala Universitet, Inst. för Informationsbehandling, Sturegatan 4b, Uppsala, Sweden

KRUSKAL, MARTIN D. Princeton University, Plasma Physics Laboratory, Box 451, Princeton, New Jersey 08541, USA

LAX, PETER. New York University, Courant Institute of Mathematical Sciences, 251 Mercer Street, New York, N.Y. 10012, USA

ORSZAG, STEVEN A. Department of Mathematics, Massachusetts Institute of Technology, Cambridge, Massachusetts, USA

PRIGOGINE, ILYA. Université Libre de Bruxelles, Faculté des Sciences Brussels, Belgium

SAFFMAN, PHILIP G. California Institute of Technology, Graduate Aeronautical Laboratories, Pasadena, California 91109, USA

STEIN, PAUL. University of California, Los Alamos Scientific Laboratory, P.O. Box 1663, Los Alamos, New Mexico 87544, USA

TRUESDELL, C. A. Johns Hopkins University, Department of Mechanics, Baltimore, Maryland 21218, USA

ULAM, S.M. University of California, Los Alamos Scientific Laboratory, P.O. Box 1663, Los Alamos, New Mexico 87544, USA

WHEELER, JOHN A. Princeton University, Palmer Physical Laboratory, Princeton, New Jersey 08540, USA

ZABUSKY, NORMAN J. Bell Telephone Laboratories, Inc., Whippany, New Jersey 07981, USA

Contents

Introduction to Nonequilibrium Statistical Physics

I. Prigogine

Interactions in a Classical Relativistic Plasma

M. Baus

Nonlinear Optics

N. Bloembergen

Lectures on Homogeneous Turbulence

P. G. Saffman

Superspace and the Nature of Quantum Geometrodynamics

J. A. Wheeler

Nonlinear Problems in Physics *

Werner Heisenberg

The task, to give a kind of introduction to this summer school on nonlinear problems in physics, seems to me especially difficult for several reasons. First, what are nonlinear problems? It has already been mentioned before that practically every problem in theoretical physics is governed by nonlinear mathematics except quantum theory, and even in quantum theory it is a rather controversial question whether it will finally be a linear or nonlinear theory. I will come back to this problem at the end of my talk. Besides that, it has been argued that every nonlinear problem is really individual; that is, it requires individual methods, usually very complicated and difficult methods, and it is rather improbable that one can learn from one nonlinear problem to solve another nonlinear problem. Finally, I have to emphasize that I am certainly not an expert in this field, neither in nonlinear problems in mathematics, nor in physics; but I have, as has been mentioned, actually come across a few nonlinear problems on my way through physics, so I at least know some of the horrible difficulties and troubles which one meets in these problems.

In this situation I feel that the only thing I can do is not to give a general survey of the field, but rather to tell about some of these problems which have passed my way in physics, and to see whether there are common features; and in fact I do think that all these problems have some features in common — difficulties which are similar in different problems and also methods which one may

* This lecture appeared in Physics Today 20, 27 (May, 1967).

use to solve the difficulties. So I hope you will allow me to go briefly through some of these problems and to see what are the common features.

I would like to start with a brief historical remark and to say a few words about point mechanics, and only then I will come back to the general problems of the School, which are mostly problems with continuous media, and finally, as I said before, I will discuss the problem to what extent quantum theory may be considered as a linear or nonlinear theory.

The course of history in physics usually follows the way from simple problems to more complicated problems, and for the questions to be discussed here one may consider as the simplest problem to solve a homogeneous linear equation, and then, slightly less simple, an inhomogeneous linear equation, and finally much more complicated is the solution of a nonlinear equation. Actually, going back to history, when mathematical physics started three hundred years ago, the description of the motion of inertia, simply the effect that bodies are either at rest or move with constant velocity, if no outer force acts upon them, may be considered as the solution of a homogeneous linear equation: the second derivative of the coordinate with respect to time equals zero. When one comes to the laws of free fall of GALILEO, we find that he actually has solved an inhomogeneous linear equation (the force being the inhomogeneous term), and finally NEWTON formulated the nonlinear equation; the equations of motion in Newton's mechanics definitely are nonlinear equations.

Of course NEWTON was not in a position to find general solutions of nonlinear equations; so from the very beginning he had to think about possible simplifications. Therefore, I might start by mentioning some of the most important simplifications which may be used not only in the nonlinear problems of point mechanics, but actually in all nonlinear problems which we have to deal with. There are mainly two different types of simplification which are important. The first comes from the symmetry of the problem. The physicists have learned from the mathematicians that the symmetry of a problem as a rule produces a conservation law. All the conservation laws which we know in physics — conservation of energy, momentum, angular momentum, and so on — rest upon fundamental symmetries in the underlying natural law. For instance, Newton's equations of motion are invariant under the Galilean group which is a continuous group of ten parameters. Therefore, one has ten conservation laws and these conservation laws may be used for the simplification of the problem. When NEWTON was able to solve the two-body problem in astronomy, this was due to the use of symmetry. With two bodies there are 12 degrees of freedom, because every body has 3 coordinates and 3 momenta; and of these 12 coordinates, or 12 degrees of freedom, he could actually eliminate 10 by means of conservation laws, and the rest then was rather simple. This simplification will be used in all problems.

The next kind of simplification which also has been very widely used and has actually meant much more for the discussion of nonlinear problems is what one may call the linearization of the problem. Let us assume that one has more or less by chance found simple solutions for the problem concerned, very special solutions — for instance, static solutions where all the bodies are at rest, or later in hydrodynamics, stationary solutions where the motion does not change with time, that is, the velocities at least are constant. In all these cases, one can study the general solutions of the nonlinear problem in the neighborhood of the special solution. That is more or less the main object of perturbation theory. For small perturbations around a known motion one has to solve linear equations. Thereby one comes back to simpler mathematical problems for which one may be able to find the complete system of solutions. Such solutions can be studied for the whole time interval from minus infinity to plus infinity, and the solutions may be characterized by their asymptotic behavior in space and time. If the special solution of the original system is a static solution, then we have the special advantage that the coefficients in these linear equations of perturbation theory are constant in time, and it will be possible to look for solutions which behave like an exponential function of time,

$$e^{i\omega t} \qquad \text{or} \qquad e^{\alpha t}.$$

If such a system of solutions has been found, one can at once answer questions of stability or instability. If only periodical solutions exist or only solutions which decrease in time, then the system is stable — a perturbed solution starting at a given time will in the course of time not deviate strongly from the original simple solution. If, however, there are solutions of the linear equation which increase exponentially with time, then the solution is not stable. This method of simplification can be applied even if the original solution is not a static solution, and you know that very many papers have been written on general perturbation theory, and much insight has been gained by such methods.

Beyond this point one comes very soon into the real difficulties of the nonlinear problems and I might first mention the difficulties in the case of point mechanics. For instance, let us think of the well known problem of the three bodies in astronomy or the more complicated problem of our planetary system. There, one has of course applied perturbation theory, but one has very soon found that it is extremely difficult to know how far such a theory can work. It is true that the motion of the planets can be followed rather easily by perturbational methods for a finite amount of time, say for a few thousand years. But when time increases then it may be that the perturbations are not small any more. Resonance effects have been observed on account of commensurabilities between the orbital frequencies of different planets, and in the course of time they may produce a big effect, so that in certain cases it may finally be impossible to say whether the orbit will be periodical or whether the planet will move out of the system. Such questions can be extremely difficult to answer. I might mention a most paradoxical result of this mathematical analysis — the theorem of Bruns. He proved that in an even infinitely close neighborhood of a point where the perturbation theory converges, there must always be other points where the perturbation theory diverges. So one can say that the points where the perturbation theory converges and those where it diverges form a dense manifold. This result suggests that after a very long time one can never know where the orbit finally will go. It is only for a finite time that we can really determine what will happen. This is not only a mathematical result interesting for the astronomers — it may have very important practical applications. I might just mention a problem which troubled us about 15 years ago when the CERN proton-synchrotron was constructed. For the big CERN-accelerator it was very important to know whether a proton running around the circle of the accelerator would be stable in the assumed orbit or not. One could select conditions such that there would be stable oscillations around the orbit. But when there is some small perturbing defect for the orbit in the tube then there may be resonances. The frequency of rotation of the proton around the machine can become commensurable to the frequency of oscillations around the orbit, and then similar difficulties could arise as in the astronomical problems. One would expect, that if there is resonance, the amplitude of the oscillations will increase. But then you come into the region where the problem is not linear any more, perturbation theory doesn't work. On account of the nonlinearity the frequency will be changed, and therefore the resonance will be stopped, and in this way the nonlinear terms should exert a stabilizing influence, until finally the motion has passed the critical region of resonance.

This was the hope; and when numerical calculations were carried out the result was that the particle may actually run around its orbit ten thousand times, that is, the stabilizing effect seems to work very well. First the amplitude increases, then the frequency gets out of resonance, then the amplitude decreases again, and so on. But after ten thousand revolutions it may happen that the particle runs out of its orbit just as well. So you can get this kind of surprise in nonlinear problems, and that, I think, is a very characteristic feature of nonlinear problems. Therefore one might say, to use a very simple term, that the nonlinear problems have a certain kind of unpredictability. One doesn't know how the solutions will behave after a very long time, and I think that this may be a very general feature of nonlinear problems.

What can one do if it is too difficult to study the single orbit, too difficult to get a kind of survey of all the possible solutions of such an equation? The next way out is to investigate not a single solution, but ensembles of solutions. One can argue that it may not be necessary to know every detail of the solution, one can be satisfied with an incomplete knowledge of the system. In this case one cannot ask: what will be the state of the system after a certain time; but one has to ask: what will be the probable state of the system after some time. This has been the line of research in statistical mechanics. If you follow it you may get into the difficult problem whether an average over the time variable is equivalent to an average over the ensemble. I just mention the problem of the ergodic hypothesis, but I cannot go into any details here.

From this brief review of point mechanics we may conclude that we have four characteristic features which we will probably find in many nonlinear problems. First, we can always use the *symmetries* to simplify our problems. The second, *linearization* of the problem, is possible around special solutions which are simple and may be found just by looking at nature. The third is that solutions have a kind of *unpredictability*—you never know what the solution will finally do and it may not be possible to extend the solution beyond a certain time. And finally, one possible way out are *statistical methods*; we can be interested not in special solutions but in ensembles of solutions or in systems which we do not know completely.

But let me now come to those problems in which you are really interested. They are the problems with infinitely many degrees of freedom or problems concerning continuous media. Such problems, of course, exist in very great number — fluid dynamics, gas dynamics, elasticity, electromagnetism; also electromagnetism is a nonlinear theory if the interactions with the bodies are taken into account— and finally gravitation. And let me mention at once one complication which will come up again and again in this field of physics: such equations where the field quantities depend on the continuous variables x, y, z and t, are almost necessarily incorrect for the following reason. In very small distances one has to take atomic structure into account. Usually in hydrodynamics we need not do it, and e.g. in the Navier-Stokes equation we do not speak about molecular structure of a liquid, we just introduce a viscosity term. The same is true for elastic bodies, for problems of gases, and so on. Still, we have to remember that at very small distances something new will happen, something that is not described by the equation, and this has to be studied if the equation does not determine the course of events within the bigger dimensions. It is only in quantum field theory or in electromagnetism that the continuum may be a fundamental continuum; but even there our doubts arise, and there we come into the general question of whether quantum theory is a linear or a nonlinear theory.

After these preliminary remarks, we can try to see whether in these many problems of fluid dynamics, elasticity, etc. one can use the same methods of simplification which I have mentioned in point mechanics. First of all, one will definitely always use the symmetry properties of the underlying system and the conservation laws. Then, one will use the method of linearization. How this linearization works I might just explain in one problem which I have studied myself some time ago and in which I was strongly interested for many years — the stability of laminar hydrodynamical motion. At that time one used to study the incompressible fluids, say, the flow of water, and so one started from the Navier-Stokes equation which introduces, besides inertia, also the viscosity of the liquid; and the question was whether a laminar flow is stable or unstable. One did know from the experiments that the laminar motions were only very special solutions of these equations. The general solutions were much more complicated; there we have to do with the big field of turbulent motions which one knows from daily experience. The laminar motions are known, one may say, from inspection. We see how a liquid can flow through a tube or between two parallel plates, and we can easily find the solutions of the Navier-Stokes equations corresponding to these simple laminar motions. They had been derived long ago and described in the textbooks. But such a solution is sometimes not stable. It is well known that when a liquid flows through a tube, then if it exceeds a certain speed, the type of motion changes completely. Instead of the smooth flow through the tube, we have all kinds of vortex motion appearing, eddies are formed and dispersed again, and one can ask why that happens at a certain speed. This was a typical problem which could be solved by the method of linearization. One could ask for motions in a very close neighborhood around the laminar motion, and since this latter motion was stationary, the coefficients in the linear equations for the perturbation were constant in time, so one could look for periodical solutions and then one had to solve an ordinary eigenvalue equation, as in quantum theory. One could use those mathematical methods which later on have proved useful in quantum theory, namely the investigation of asymptotic solutions, or one could introduce the modern techniques of applied mathematics including the use of the electronic computers. The result was that for small Reynolds numbers all deviations from the laminar motion would die down, but above a certain Reynolds number there are deviations where the amplitude increases, and then one gets into an entirely different type of motion, namely turbulent motion, which cannot be treated by perturbation theory.

But let me now come to really nonlinear problems in physics which especially in gas dynamics are interesting. While in linear motion, for instance in linear wave motions like sound waves, the velocity is essentially independent of the amplitude and independent of the frequency of the waves, the situation is entirely different for a nonlinear wave motion. If, for instance, we have an initial discontinuity, then in the linear wave equation we would expect that this initial discontinuity, say across a surface, will be propagated with the velocity of sound like any other wave, but in a nonlinear motion something different happens. This discontinuity may either disappear at once or may be propagated as a shock front — one or two shock fronts — not with the speed of sound but with supersonic speed. And not only that; it may be that out of a smooth motion where there is no discontinuity at the beginning a discontinuity will be formed later. I think that this fact corresponds in some way to the other fact mentioned before that nonlinear equations always have a tendency to diverge, that is, to get to points in time where one cannot continue the solution. One can say in a more mathematical language that nonlinear equations often do not admit solutions which can be continuously extended to any region where the differential equation remains regular. So the regularity of the equation does not guarantee the regularity of the solution, and you may after a finite time get to singularities in the solutions. If such a singularity occurs then the equation itself cannot determine what happens afterwards. At least the equation alone is not sufficient, since we come to a point where the behavior of the substance in very small dimensions becomes important. In hydrodynamics or gas dynamics the fundamental equation treats the gas or the fluid as a continuum, and this cannot be correct for very small distances; here we have to take the molecular structure into account. When our solutions become discontinuous, the molecular structure must play a role in the region of the discontinuity. Fortunately, in some cases in gas dynamics it is sufficient to know that in this region of discontinuity some irreversible processes must go on. The irreversibility will necessarily lead to an increase of entropy and this result alone is sometimes sufficient to determine the course of the shock wave. This seems to be a rather lucky incident in gasdynamics; at least in principle I don't see any reason why that should be so. Fundamentally, one could imagine that one must go into the microscopic details and study the behavior of the atoms in order to say what happens. But, as I said before, sometimes one can be lucky and things go better than one might expect.

But even when we come to problems in which the atomic structure plays no role — problems which are continuous by their very nature — even then we may be confronted with the fact that solutions cannot be continued over a certain time — that solutions can disappear into nothing or arise from nothing. I would like to mention one example which has worried me for some time in quantum field theory. Let us compare a linear relativistic wave equation, namely the well-known Dirac equation, with another equation which comes from the Dirac equation, if one replaces the mass term of the Dirac equation by a nonlinear term. These two equations are

$$\gamma_\nu \frac{\partial\psi}{\partial x_\nu} + \kappa\psi = 0, \tag{1}$$

$$\gamma_\nu \frac{\partial\psi}{\partial x_\nu} + \psi(\bar{\psi}\psi) = 0. \tag{2}$$

In a linear equation like (1) we know that when the solution has been zero everywhere for some time, it will never become different from zero. It is of course true that solutions of the inhomogeneous equation can have a different behavior; we can construct the Green's function to this wave equation. Such a Green's function could then be either a retarded or an advanced Green's function, and may be different from zero only in the future or only in the past. But for the homogeneous equation (1) the only solution of the Green's function type — namely zero for spacelike distances and different from zero in the future or the past — is a solution which is given by the commutator in the quantum field theory and is identical with the difference between the retarded and the advanced Green's function. In a certain sense this function may be called a relativistically invariant solution of the homogeneous equation (1).

For equation (2) however you have both types of solutions; that is, you can have solutions which are zero for spacelike distances and are different from zero for the past and the future, but you also can have solutions for the same equation which start, so to say, from nothing. The wave function may be zero for the past, and all of a sudden it starts to become nonzero at the singular point and develops like the retarded Green's function, in spite of the fact that there is no inhomogeneous term involved here. This emphasizes very strongly the strange behavior of nonlinear equations. There may always be situations in which the solutions of a nonlinear equation cannot be continued into the future or into the past. By the way, this problem plays an important role in the discussion of the later question whether quantum theory is linear or nonlinear.

Finally, we can treat nonlinear problems in hydrodynamics and gasdynamics with the statistical methods as we do in point mechanics. As I said before, it is as a rule not possible to follow the complete set of solutions of a nonlinear equation. Therefore it may be convenient to study ensembles of solutions, i.e. the statistical behavior of fluids. The general principle of such a statistical approach is that one is satisfied with an incomplete knowledge and incomplete description of the system; starting from this one studies the probable development. So, for instance in the theory of turbulence, of which I have already spoken, one can start with only the knowledge of the general distribution of the eddies. One just knows the spectrum of the motion, the intensity distribution at the different frequencies, without knowing the phase relations between these different amplitudes. In this case one can only ask what is the probable development of this liquid in the course of time.

Again at this point one gets into the difficulties at very small distances, and in a certain sense we meet what in quantum theory we call the ultraviolet catastrophe. Because physically what happens is this: If we have a turbulent motion, say in water, then we produce by external motions some big eddies. These big eddies develop into smaller eddies, the energy is dissipated from the bigger eddies into the smaller ones, so more and more smaller eddies are formed which get the energy, and finally the energy is dissipated into the infinitely many degrees of freedom which belong to the extremely small eddies. This process would go on to infinity if one would not introduce viscosity.

The concept of viscosity is in this case a nice way of getting around the fact that we get into the molecular region. Actually the energy is finally dissipated into thermal motion of the molecules, and this dissipation of energy into the motions of molecules can formally be replaced by including viscosity. But there may also be problems, which cannot be answered by the simple Navier-Stokes equation.

Generally, I believe that in the problems which you will study in this summer school you will notice these four characteristical features which I mentioned already in connection with point mechanics. You will frequently use two essential simplifications, one by means of the *symmetries* of the problem, the other one by the possibility of *linearization*, which in many cases give a first insight into the problem. Then you will find the singularities, the *unpre-*

dictability of nonlinear solutions, and the problem that you frequently can't continue solutions beyond a certain point in time where the singularity occurs. And finally, you will use *statistical methods* to get information not about a single system but about ensembles of systems, and thereby you will be able to predict the probability for certain events or the probable course of events.

Let me now come to one of the subjects in which I have been most interested — quantum theory. I would like to discuss briefly the question whether quantum theory is a linear or a nonlinear theory. There is no doubt that quantum mechanics in its conventional form is a linear theory. It is linear in that sense that in spite of the operator equations being nonlinear equations, these operator equations can be fulfilled by solving Schrödinger equations, that is, by looking for certain transformation matrices, and these Schrödinger equations are definitely linear equations. This fact in quantum theory has a very deep, an almost philosophical reason, and is not only connected with a certain approximation. In quantum theory we do not deal with facts but with possibilities — the square of the wave function describes the probability, and the superposition of wave functions, the possibility of adding two solutions to get a new solution, is absolutely essential for the whole foundation of quantum theory. Therefore it would be definitely wrong to say that this linear character of quantum theory is approximate in the same sense as the linearity of Maxwell's equations is approximate. The linearity of the quantum mechanical equations is essential for the understanding of quantum theory and for the interpretation of quantum theory as a statistical basis for calculating what happens to the atoms.

This fact already raises interesting mathematical problems on which I shall just touch. Let us, for instance, calculate solutions of the three-body problem, say, the helium atom, by means of quantum mechanics. We know that we get the complete system of all solutions by solving the linear Schrödinger equation in the coordinate space of the three bodies in the helium atom. So one could say that in quantum mechanics we have really solved the three-body problem which never has been solved in classical mechanics completely; and that seems somewhat strange, because in the limit $h \to 0$ the quantum mechanical solutions must — if prepared in a certain way — go over into the classical solutions. Therefore, it seems as if we could replace the solution of a nonlinear problem in classical theory by solving linear problems in quantum theory and then performing the limit $h \to 0$.

Let me discuss the connection between the two theories a little further. The limiting process would require that we start our solution with a wave packet in coordinate space. We would put the two electrons around the nucleus at some approximate positions and moving with an approximate velocity so that the uncertainty relations are fulfilled, and of course these wave packets can be made smaller and smaller when we let the quantity h tend to zero. But for any finite value of h we would have finite wave packets; it is quite clear that the wave packet after a long time will disperse. It will become bigger and less dense and finally it will be spread out over the whole system. So we should always compare with this wave packet not one single solution of the classical problem, but an ensemble of solutions, namely an ensemble belonging to the same uncertain initial conditions. It is just the nonlinear character of the classical equations which makes the deviations between two neighboring solutions become bigger and bigger when time increases. In the classical theory, when we start with an ensemble of solutions belonging to a wave-packet-like distribution of initial conditions, after some time this wave packet will also spread out over large parts of coordinate space. So one can see that if we first let h and the size of the wave packet tend to zero, and then go to larger times, then we do get a complete representation of the classical theory from quantum theory. If, however, we reverse the limiting processes, if we first go to infinite times and then let h tend to zero, then the situation is quite different, because then we always get after long times an infinitely big wave packet.

In spite of this relation between quantum theory and the nonlinear classical theory, quantum mechanics is definitely a linear theory. But is this still true when we come to the theory of elementary particles, that is, to a theory which does not start from given elementary particles but which tries to understand and to derive the elementary particles? We are interested in our Institute in an attempt which is called the nonlinear spinor theory; but I would like to emphasize at once that this term "nonlinear" does not necessarily mean nonlinear in the sense of this school, for the following reason. We start from an operator equation which looks more or less like Eq.(2), and therefore we can call it a nonlinear equation. But the operator equations in quantum theory always are nonlinear equations and the question is whether the solution of these operator equations corresponds to a series of linear equations or nonlinear equations. Generally we have learned that in quantum mechanics we can replace the nonlinear operator equation by a differential equation (Schrödinger equation) or a system of equations which are linear. In the same sense one could presume that also such a nonlinear equation like Eq. (2) can be replaced by a Schrödinger equation which in this case would be not a linear differential or integral equation but actually a linear functional equation. Alternatively, one could replace it by a system of infinitely many differential equations with infinitely many unknowns, and all these differential equations would be linear. This would be true if such a theory could be quantized along the ordinary rules of quantum theory. In this case we would have commutation relations between ψ (x) and $\bar{\psi}$ (x') which are essentially given by a delta-function in space at the time (t-t') = 0.

But this is still a controversial problem. From our present knowledge we can be rather certain that the commutator of such an equation does not look like such a delta-function. Such a conventional commutator has a strong singularity, and this strong singularity just allows one to reduce a nonlinear problem of the operators to the linear problem of the Schrödinger equation. In field theory, however, we can take it for granted that in the center of the commutator we have no delta-function, because if we had a delta-function an equation like Eq. (2) would have no meaning. Therefore the singularity of the 2-point function or the commutator at the origin is the real problem in any nonlinear field-operator equation like (2), and so we have to ask whether this problem can be treated in a similar way as in quantum mechanics with the help of delta-functions.

If one formulates the equations for the commutator or the 2-point function itself, then these equations are extremely complicated nonlinear integral equations. If it should be necessary to solve these equations, then definitely the problem would be a nonlinear problem, and then at the very basis of quantum theory we would again come to nonlinear mathematics. The trouble is that we don't know whether we really have to solve these equations. It may be that the following procedure will be sufficient: we just guess approximate solutions for such a commutator and then put the chosen solutions into approximation schemes of the Tamm-Dancoff type; then of course in every finite approximation the result will depend upon whether we have taken a good or poor approximation of the 2-point function. There are, however, examples which tend to show that in extremely high approximations the errors that we make in this way, by not having chosen the correct 2-point function, become less and less important, and it may be that even if we start with an incorrect commutator at the end we get the right results in infinitely high approximation. Such at least is the situation for simpler problems in quantum mechanics which have been studied by Stumpf and others. Therefore it may be that again the actual treatment of nonlinear equations may be replaced by the study of infinite processes concerning systems of linear differential equations with an arbitrary number of variables, and the solution of the nonlinear equation can be obtained by a limiting process from the solutions of linear equations. This situation resembles the other one which I have mentioned before, where by an infinite process from the linear 3-body problem of quantum mechanics one can approach the nonlinear 3-body problem in classical mechanics.

The present conclusion is that we do not know whether finally the fundamental problem in the quantum theory of elementary particles will be a nonlinear problem or a linear problem. But I cannot go into any details here.

Finally I may repeat what I have said in the introduction of my speech: that certainly the progress of physics will to a large extent depend on the progress of nonlinear mathematics, of methods to solve nonlinear equations. It may still be that every such problem is individual and requires individual methods. Still, as I told you, there are definitely some common features and therefore one can learn by comparing different nonlinear problems. I hope that this will be done very successfully in this summer school.

The Nonlinear Field Theories in Mechanics

C. Truesdell

Introduction

The purpose of the school has been outlined in the announcement by the Directors:

> In the past two decades increasing attention has been given to nonlinear problems - on the experimental side because of the availability of intense coherent and incoherent sources and reliable satellite data, and on the theoretical side because of the easy access to sophisticated computational results. The courses offered at the School are designed to present a uniform view of these developments.

I have been trying to see how my course can fit in. With one point I can agree fully. Since some of my contacts are with the Applied Mechanics Gang, I have indeed experienced damage from exposure to intensely incoherent sources. I cannot claim any knowledge of computation, since it has had no influence on the subject I shall present to you. This I can prove quantitatively by exhibiting this volume, *The Nonlinear Field Theories of Mechanics*, which contains, as I estimate, about 4,000,000 milligrams of theoretical mechanics with not one result requiring anything more than paper and pencil.

Experiment, indeed, has been and is important, but neither intense sources nor satellites have yet been used. Here is a specimen taken from a piece of standard apparatus for personal hygiene - a length of rubber tube. We may, with care and strength, turn this tube inside out. This second specimen was just like the first before it suffered eversion. The external forces applied to these configurations are the same, only gravity. Thus the specimen can be deformed severely, but smoothly, from one configuration of equilibrium into another corresponding to the same applied forces. The equilibrium state is not unique. Both these equilibria are stable in the sense that each specimen, if deformed a little - say, bent into a circle - snaps back to the state from which it started. Clearly we are outside the range of any principle of superposition. The strains are large. What was before the inside surface is now the outside, and vice versa. Also the length changes. With this precision instrument - a ruler given to my mother when she entered the first grade, an instrument which has done faithful service for three generations of household experiments - I can measure the change: 7 11/16" before, 7 13/16" afterward, a strain of about 1/64, enormously greater than can be produced in a steel bar before collapse. Yet I could turn the everted tube inside out again and get back the original configuration almost perfectly, and indeed repeat the experiment with the same specimen many times before the material would fatigue and tear.

I had hoped to perform the eversion before you, and hence I asked Professor James Bell to prepare this much larger specimen for me: a cylinder of sponge rubber two feet long, a foot in diameter, and an inch thick. I can indeed do the experiment, with a little patience. You see, however, that the everted shaped is a different one. Although subject to no force other than gravity, this everted cylinder looks like a metal tube crushed under great axial pressure: the wall is indented in several lobes. While a metal cylinder would experience these deformations plastically in the sense that the specimen would not spring back to its original shape even if brought near to it, our specimen only needs a little help along the right path, and it resumes its original form again. This "buckling by eversion" seems to be a new phenomenon, discovered last month by Professor Bell in the course of his attempt to provide a gross enough specimen for me to handle.

The same experiment on two similar specimens of different size and material has led to different results. In fact, the two cases are somewhat less different than they might seem. For the big sponge cylinder, an everted cylindrical shape may well be possible, but if so, it is certainly unstable. If I force the lobes flat, they spring right back. Perhaps there is a lobed equilibrium shape for the little tube, but if so, it is unstable, for if I push lobes in, they spring right out again. Since the researches of EULER on the stability of ships in the 1730's, we have grown to expect a "sequence" of equilibria, in which two stable equilibria always have an unstable one which is in some sense "between" them. In the tube experiment, presumably, there are three equilibria for each specimen, but the difference of dimensions and material causes two of them to exchange status.

These phenomena are typical of those the new continuum mechanics is designed to represent and correlate by precise, deductive theory. The eversion of cylinder into cylinder is fully explained by a solution due to RIVLIN. Although there is now a general theory of elastic stability, it has not been applied to many cases, and the phenomenon of buckling by eversion has not yet been studied in theory.

The experimental side of the new continuum mechanics remains simple and relatively inexpensive. The phenomena, mainly, are those of common, everyday experience with materials, phenomena which are easily seen and felt. For example, with the naked eye I can see a molecular structure in the specimen of sponge rubber. There seem to be roundish pieces of what I guess to be air, imbedded in a complex, twisted mess of rubber. Taking from the family laboratory another precision instrument - a pocket magnifier - I see even more confusion and complication. The obvious presence of these big molecules and this local ramification has no importance whatever for the area of physics I wish to discuss. For the phenomena, sponge rubber seems to behave just like a homogeneous, isotropic, elastic continuum. Of course, if I chop up this specimen into small chunks, I shall find that they have preferred directions; they will stretch much more easily in one direction than another. Thus a small piece, while it may behave like a homogeneous elastic continuum, will certainly not be isotropic. If I dissect further, I shall finally get pieces of pure air or pure rubber. Certainly the behavior of a sphere of air under load is entirely different from that of a sphere of sponge rubber one foot in diameter, but taking into account the well known physics of air will bring us further away from the phenomena of sponge rubber, not closer to them. It is as if in Professor Wheeler's chart of distances as a basis for space-time structure, we were to insist that the distance from Munich to Moscow really depends on which road you take. The statement is true, and persons who worry about such things are important, but they make road maps, not celestial atlases. Now I turn my instrument onto the specimen of apparently solid rubber and discern no molecules. When I know that the big, visible molecules make no difference in the behavior of sponge rubber, there is even less reason to devise powerful methods of determining the invisible molecules in the solid rubber, so long as gross phenomena are our interest. It is a pure illusion to think that learning all about tiny things first is the path to knowledge about big things. Continuum mechanics deals directly with the phenomena and represents them by mathematical quantities designed to reflect their gross properties.

Returning to the statement of purpose, "to present a uniform view" of nonlinear phenomena, I am not ashamed to admit my inability to do so. Professor Heisenberg in his opening address has admirably expressed the idea that all, or nearly all, of physics is nonlinear. Thus a "uniform view" amounts to a general, unified physics. I doubt if anyone today is anywhere near such a view. I have interpreted my task as that of explaining modern nonlinear continuum mechanics *as it is practised* by those who have created and developed it in the past few years. It does unify many branches of mechanics and thermodynamics that were previously studied as separate disciplines, but it is far from being a unified physics.

We may list the main areas of fundamental physical theory as follows:

1. Classical mechanics of "bodies"
2. Analytical dynamics
3. Classical continuum theories
4. Thermodynamics
5. Electromagnetism (including Optics)
6. Statistical mechanics
7. Relativity
8. Quantum mechanics

Modern continuum mechanics makes no use at all of the last, nor do I see such use in the immediate future; it has made very little use of the second. Statistical mechanics has remained for it a frequent source of suggestion and analogy, but no more. Continuum mechanics today largely replaces the mechanics of "bodies" by a unified theory which has grown out of classical hydrodynamics, elasticity, and thermodynamics. Bridges to electromagnetism and relativity have been built and are being strengthened, but I have interpreted my task here as that of showing you what continuum mechanics has done rather than trying to project its future. Thus in these eighteen lectures I shall try to outline the main features of the theory as I present it in a year's course to graduate students.

The prerequisites for continuum mechanics are:

I. The classical or "Newtonian" mechanics of bodies

II. Thermodynamics

While the former is standard, the foundations of thermodynamics have to be seen in a frame of ideas not yet customarily taught, so I shall begin with a brief review of the elementary theory of heat and temperature in a form fit for generalization to materials.

1. Thermodynamics of Homogeneous Processes

Introduction

Before taking up modern continuum mechanics, a student should be familiar with the classical mechanics of "bodies" and with a correspondingly elementary thermodynamics. While standard instruction suffices for the former, it does not for the latter. Therefore I shall begin this course with an outline of the thermodynamics of homogeneous processes in a form fit for later generalization to deformable continua. The approach is that which COLEMAN and NOLL created for thermo-elasticity.

Temperature and Heat

Experience shows that mechanical action does not always give rise to mechanical effects alone. Doing work upon a body may make it hotter, and heating a body may cause it to do work. This fact suggests that alongside the mechanical working or power P should be set a *heat working* Q, such that the total working is P + Q. If kinetic energy is the energy of motion, we recognize that not all work leads to motion and allow for storage of the remainder as *internal energy* E. Thus the *balance of energy* is asserted in the equation

$$\dot{K} + \dot{E} = P + Q, \tag{1.1}$$

where the dots denote time rates. K and P are definable in terms of mechanical quantities, but for the general scheme of thermodynamics their definitions need not be stated.

Much as the concept of place abstracts the idea of where a body is, the concept of temperature abstracts the idea of how hot it is.

The temperature θ is a real number indicated on a thermometer, just as distance is a real number indicated on a ruler. Experience suggests that no matter what thermometer is used, there is a temperature below which no body can be cooled. That is, the temperature is bounded below. If this greatest lower bound is assigned the temperature 0, the temperature is said to be absolute:

$$\theta > 0, \tag{1.2}$$

no matter what unit of temperature be selected. We shall always adopt (1.2) for convenience.

The third basic idea of thermodynamics expresses the fact that for any given body, the increase of energy not produced by mechanical work is limited. That is, there exists a least upper *bound* B for $\dot{K} + \dot{E} - P$:

$$\dot{K} + \dot{E} - P \leqq B. \tag{1.3}$$

Equivalently, the heat working **Q** is bounded: $Q \leqq B$.

Unlike the zero of temperature, this least upper bound B is not the same for all bodies. Rather, it is assignable in the same sense that force is assignable in mechanics. Different systems will have different bounds, and these bounds will depend on the circumstances in which the system finds itself.

In terms of the bound B, the *heat* H is defined by

$$H \equiv \int \frac{B}{\theta}\, dt. \tag{1.4}$$

Thus the heat is defined only to within an arbitrary constant. The heat is of importance for the theory only through $\dot{H}$, the *heating*. In terms of it, (1.3) assumes the form

$$Q \leqq \Theta \dot{H}. \tag{1.5}$$

A body stores heat, then, at least as fast as the heat working divided by the temperature. The hotter the body is, the greater must be the heat working in order to reach the maximum, which is determined by the heating.

Homogeneous Processes

The intended application of elementary thermodynamics is to blocks of material suffering no local deformation or variation of temperature. We describe the theory, therefore, as one of *homogeneous processes* in that all quantities occurring are *functions of time only*. Place plays no part. It is customary in the older studies to neglect motion and hence to set $K = 0$, but for later use we shall retain K, noting, however, that it occurs in the theory of homogeneous processes only in the combination $P - \dot{K}$. Since this quantity is the excess of the mechanical working over the kinetic energy rate, it may be called the *work storage*, W.

I think it unnecessary to supply motivation beyond that already given. For connection with other treatments, however, I remark that what I call *heat* H is usually called "entropy"; that (1.1) and (1.5), respectively, are among the various statements that physical writers call the "first and second laws of thermodynamics"; that, specifically, (1.5) is the *Clausius-Planck inequality*.

I shall now outline a mathematical theory of the thermodynamics of homogeneous processes, in the same way as I should present elementary mechanics to a beginning class, as a clean, self-contained mathematical theory, in which general theorems may be proved and specific problems may be solved.

The Thermodynamic State

Much as in mechanics the concept of place occupied at a given time is primitive, in thermodynamics we consider a *state* at a given time. The state is a set of k + 1 functions of time

$$\theta(t), \tau_1(t), \ldots, \tau_k(t),$$

or, briefly,

$$\theta(t), \tau(t).$$

The temperature $\theta(t)$ is a positive function: $\theta(t) > 0$. The k *parameters* $\tau_j(t)$ also are quantities given *a priori*. In the interpretation, τ_1 might be the volume of a body, and $\tau_2, \ldots, \tau_k$ might be the masses of k-1 constituents; but for the theory, no specific interpretation is necessary. We agree to consider only unconstrained states in the sense that the functions $\theta(t)$, $\tau(t)$ may be arbitrary. In particular, at any one time $\theta, \tau, \dot{\theta}$, and $\dot{\tau}$ may assume any real values whatever, provided only that $\theta > 0$.

Thermodynamic Processes

A *thermodynamic process* is a set of k + 5 functions

$$\theta(t),\ \tau(t),\ W(t),\ E(t),\ H(t),\ Q(t),$$

that satisfy the first axioms of thermodynamics:

$$\dot{E} = W + Q, \tag{1.1}$$

$$\theta > 0. \tag{1.2}$$

If all but the last of these functions are given, $Q(t)$ is uniquely determined by (1.1). Hence we may say equivalently that a thermodynamic process is a set of $k + 4$ functions

$$\theta(t),\ T(t),\ W(t),\ E(t),\ H(t)$$

such that $\theta > 0$.

If we eliminate Q from (1.5) by means of (1.1), we obtain the *reduced dissipation inequality:*

$$\dot{E} \leqq W + \theta\dot{H}. \tag{1.6}$$

The *free energy* Ψ is defined by

$$\Psi \equiv E - \theta H. \tag{1.7}$$

In terms of it, the reduced dissipation inequality becomes

$$\dot{\Psi} - W + H\dot{\theta} \leqq 0. \tag{1.8}$$

Histories

If $f(t)$ is a function of time, its restriction f^t to present and past values of t is called the *history* of f up to time t:

$$f^t(s) \equiv f(t - s),\ 0 \leqq s < \infty.$$

Thus, for example, θ^t, T^t is the history of the thermodynamic state up to time t.

A process in which

$$\left.\begin{array}{l} \dot{\theta} = 0 \\ Q = 0 \\ \dot{H} = 0 \\ \dot{E} = 0 \end{array}\right\} \text{ is called } \left\{\begin{array}{l} \text{isothermal.} \\ \text{adiabatic.} \\ \text{isentropic.} \\ \text{isoenergetic.} \end{array}\right.$$

A process in which equality holds in (1.8) at all time is called *reversible*. All other processes are called *irreversible*. In a reversible process, $Q = \theta\dot{H}$; therefore, in reversible processes "adiabatic" and "isentropic" have the same meaning. In an irreversible adiabatic process, the heat increases: $\dot{H} > 0$. In an irreversible isentropic process, the heat working is negative: $Q < 0$.

Thermodynamic Constitutive Equations

The *state history* is the set of k + 1 functions $\theta^t(s)$, $T_j^t(s)$. The principle of *thermodynamic determinism* asserts that the state history determines the work storage, the free energy, and the heat. That is,

$$\left.\begin{array}{l} W \\ \Psi \\ H \end{array}\right\} \text{ are functionals of } \theta^t \text{ and } T^t. \tag{1.9}$$

Such a set of three relations is called a *thermodynamic constitutive equation.*

Not all functionals are admissible, however. In order to define a *thermodynamic material,* we require that the constitutive equation be compatible with the reduced dissipation inequality (1.8) *for all thermodynamic processes.*

Example 1. The Classical Caloric Equations of State

The classical works in thermodynamics deal only, or almost only, with one extremely special class of constitutive equations. The work storage is thought of as being purely mechanical and also linear in the "velocities" $\dot{\tau}_a$

$$W = - \sum_{a=1}^{k} \pi_a(\Theta,\tau)\, \dot{\tau}_a, \tag{1.10}$$

where the coefficients π_a are called *thermodynamic pressures,* and the free energy and heat are given by *caloric equations of state:*

$$\begin{aligned} \Psi &= \Psi(\Theta,\tau), \\ H &= H(\Theta,\tau). \end{aligned} \tag{1.11}$$

Substitution of (1.10) and (1.11) into the reduced dissipation inequality (1.8) yields

$$(H + \partial_\Theta \Psi)\, \dot{\Theta} + \sum_{a=1}^{k} (\pi_a + \partial_{\tau_a} \Psi)\, \dot{\tau}_a \leqq 0. \tag{1.12}$$

The coefficients of $\dot{\Theta}$ and of $\dot{\tau}_a$ are functions of Θ and τ. Since Θ, τ, $\dot{\Theta}$, and $\dot{\tau}$ may be given arbitrary values, (1.12) holds if and only if each summand vanishes:

$$H = -\partial_\Theta \Psi,$$
$$\pi_a = -\partial_{T_a} \Psi. \tag{1.13}$$

We have proved, then, that the k + 2 functions π_a, T and H are not independent in a thermodynamic constitutive equation: The heat and the thermodynamic pressures are determined from the free-energy function $\Psi(\Theta,T)$ as a potential. Substitution back into (1.12) shows that equality holds always there. That is, *there are no irreversible processes*, and a process is adiabatic if and only if it is isentropic.

The contents of classical books on thermodynamics consist mainly in ringing the changes on (1.13).

Exercise 1.1

Assume that

$$W = -\sum_{a=1}^{k} \pi_a \dot{T}_a \tag{1.14}$$

without necessarily imposing the assumption that $\pi_a = \pi_a(\Theta,T)$, i.e., the π_a are certain functions of t. Define the *enthalpy* X and the *free enthalpy* Z as follows:

$$X = E + \sum_{a=1}^{k} \pi_a T_a,$$
$$Z = X - \Theta H, \tag{1.15}$$

so that

$$E + Z = \Psi + X. \tag{1.16}$$

Prove that

$$\dot{X} \leq \Theta\dot{H} + \sum_{a=1}^{k} \pi_a \dot{T}_a, \qquad \dot{Z} \leq -H\dot{\Theta} + \sum_{a=1}^{k} \pi_a \dot{T}_a, \tag{1.17}$$

and interpret these inequalities.

Exercise 1.2

Under the conditions of the theorem expressed by (1.13), assume that each of the functions $\Psi(\Theta,T)$, $H(\Theta,T)$, $\pi_a(\Theta,T)$ is invertible for any one of its arguments. Prove that then $E = E(H,T)$, $X = X(H,\pi)$, and $Z = Z(\Theta,\pi)$, and each of these three functions is also a thermodynamic potential in the sense that

$$\begin{aligned} \Theta &= \partial_H E, & \pi_a &= -\partial_{T_a} E, \\ \Theta &= \partial_H X, & T_a &= \partial_{\pi_a} X, \\ H &= -\partial_\Theta Z, & T_a &= \partial_{\pi_a} Z. \end{aligned} \tag{1.18}$$

The existence of the classical potentials, as we have just seen, is a consequence, not of any general thermodynamic ideas, but of the very special constitutive Eqs. (1.10) and (1.11) taken as the starting point. If we generalize these assumptions just a little, we no longer get the classical formulae. Namely, if we include the effect of "linear friction", the pressure π_a in (1.10) is no longer a function of Θ and T alone but is also a linear function of $\dot{T}$:

$$\pi_a = \pi_a^o(\Theta,T) + \sum_{b=1}^{k} \pi_{ab}(\Theta,T)\, \dot{T}_b, \tag{1.19}$$

while the caloric equations of state (1.11) are retained unchanged. Here $\pi_a^o(\Theta,T)$ is the "equilibrium pressure", namely, the value of π_a when $\dot{T} = 0$.

Exercise 1.3

Prove that under the assumption (1.19), the constitutive equation (1.10) defines a thermodynamic material if and only if

$$H = -\partial_\Theta \Psi,$$

$$\pi_a^o = -\partial_{T_a} \Psi, \tag{1.20}$$

$$\sum_{a,b=1}^{k} \pi_{ab}(\Theta,T)\, \dot{T}_a\, \dot{T}_b \leqq 0.$$

Collecting our results, we see that $\Psi(\Theta,T)$ is now a potential for H and the equilibrium pressure π_a^o but stands in no relation to the frictional pressures π_{ab}. The free-energy function, then, no longer determines all constitutive equations of the material, since the frictional pressure π_{ab} is independent of it. The frictional pressures, however, are not arbitrary. For each value of the thermodynamic state, the quadratic form of $\|\pi_{ab}\|$ must be non-positive definite. Conversely, if (1.20) holds, the assumed constitutive equations satisfy the reduced dissipation inequality. Therefore, equality need not hold in (1.20); indeed, generally it will not hold. Thus the constitutive equations (1.19), only slightly more general than the classical (1.10), allow irreversible processes and in fact generally lead to them. With (1.19), adiabatic processes are no longer generally isentropic, and conversely.

Looking back at (1.10), we see that π_a enters the constitutive equation only in the combination $\pi_a \dot{T}_a$. Hence the frictional pressures π_{ab} in (1.19) enter the constitutive equation only in the combination $\pi_{ab}\dot{T}_a\dot{T}_b$. Hence the quantity $\pi_{ab}-\pi_{ba}$ has no significance and cannot be determined by any argument within the present framework. Its value has no importance. There is a large literature concerning relations of the form

$$\pi_{ab} = \pi_{ba}. \tag{1.21}$$

Any meaningful remark on this subject necessarily rests on some prior ideas, not of purely thermodynamic origin, as to what π_{ab} is.

Classical thermodynamics need not be exhausted by the two cases just studied. According to (1.9), the thermodynamics of homogeneous processes is broad enough to allow memory effects of a very general nature. It is only in field theories, however, that such effects have received much attention.

2. Kinematics. Changes of Frame

Geometry

A *place*, denoted by $\underset{\sim}{x}$ or $\underset{\sim}{y}$, is a point in Euclidean 3-dimensional space. Generally the results we shall derive hold just as well in n-dimensional space, but in all cases where it makes a difference, n is taken as 3. The *time* t is a real variable, $-\infty<t<\infty$.

A *vector* $\underset{\sim}{v}, \underset{\sim}{w}, \ldots$ is an isometry of Euclidean space, that is, a distance-preserving transformation of space onto itself. The operation of transformation is indicated by a + sign:

$$\underset{\sim}{y} = \underset{\sim}{x} + \underset{\sim}{v}, \qquad \underset{\sim}{v} = \underset{\sim}{y} - \underset{\sim}{x},$$

where the second equation, defining the difference of two points, is merely an alternative for the first.

A *tensor* T is a linear transformation of the space of all vectors. The operation of transformation is written multiplicatively:

$$\underset{\sim}{w} = \underset{\sim}{T}\underset{\sim}{v}.$$

Components of vectors and tensors will be needed rarely. The notations v^k, v_k, T^{km}, $T^k{}_m$, $T_k{}^m$, T_{mk} are standard; for orthogonal co-ordinates it is sometimes helpful to introduce the physical components v<k>, T<km>, which are components with respect to an orthonormal frame tangent to the co-ordinate curves at the point in question. $\underset{\sim}{1}$ is the identity tensor. Its components of the above four types are g^{km}, $\delta^k{}_m$, $\delta_k{}^m$ and g_{mk}, respectively. The co-ordinates of $\underset{\sim}{x}$, in any system, are denoted in the lectures by x_1, x_2, x_3.

The scalar, tensor, and outer *products* of two vectors $\mathbf{v}$ and $\mathbf{w}$ are denoted by $\mathbf{v} \cdot \mathbf{w}$, $\mathbf{v} \otimes \mathbf{w}$, and $\mathbf{v} \wedge \mathbf{w}$. The composition or product of the tensors $\mathbf{S}$ and $\mathbf{T}$ is written $\mathbf{TS}$. The determinant, trace, transpose, and magnitude of $\mathbf{T}$ are written det $\mathbf{T}$, tr $\mathbf{T}$, $\mathbf{T}^T$, $|\mathbf{T}|$, where $|\mathbf{T}| \equiv \sqrt{\text{tr}\,\mathbf{T}\mathbf{T}^T}$.

A tensor $\mathbf{Q}$ is *orthogonal* if $\mathbf{Q}\mathbf{Q}^T = \mathbf{1}$. A tensor $\mathbf{T}$ is *symmetric* if $\mathbf{T} = \mathbf{T}^T$, *skew* if $\mathbf{T} = -\mathbf{T}^T$.

The *gradient* of a vector field $\mathbf{v}(\mathbf{x})$ is written as $\nabla\mathbf{v}$ and is defined by

$$(\nabla\mathbf{v})\mathbf{a} = \lim_{s\to 0} \frac{\mathbf{v}(\mathbf{x}+s\mathbf{a})-\mathbf{v}(\mathbf{x})}{s}$$

for all vectors $\mathbf{a}$. The covariant components of $(\nabla\mathbf{v})\mathbf{a}$ are $v_{k,m}a^m$, where the comma denotes the covariant derivative. The gradient of a tensor may similarly be defined by regarding the tensor as a vector in a space of higher dimension.

Primitive Elements of Thermomechanics

The primitive elements of thermomechanics are:

1. Bodies
2. Motions
3. Forces
4. Temperatures
5. Heats

These elements are governed by assumptions or *laws* which describe mechanics as a whole. The laws abstract the *common features* of all mechanical phenomena. The general principles are then illustrated by *constitutive equations*, which abstract the *differences* among bodies. Constitutive equations serve as models for different kinds of bodies. They define ideal materials, intended to represent aspects of the different behaviors of various materials in a physical world subject to simple and overriding laws.

Bodies, Configurations, Motions

A *body* $\mathcal{B}$ is a manifold of particles, denoted by X. In continuum mechanics the body manifold is assumed to be smooth, that is, a diffeomorph of a domain in Euclidean space. Thus, by assumption, the particles X can be set into one-to-one correspondence with triples of real numbers X_1, X_2, X_3, where the X_α run over a finite set of closed intervals.

The body $\mathcal{B}$ is assumed also to be a σ-finite *measure space* with non-negative measure $\mathcal{M}(\mathcal{P})$ defined over a σ-ring of subsets $\mathcal{P}$, which are called the *parts* of $\mathcal{B}$. Henceforth any subset of $\mathcal{B}$ to which we shall refer will be assumed to be measurable. The measure is called the *mass distribution* in $\mathcal{B}$. It is assigned, once and for all, to any body we shall consider.

Bodies are available to us only in their *configurations*, the regions they happen to occupy in Euclidean space at some time. These configurations are not to be confused with the bodies themselves. In analytical dynamics, the masses are discrete and hence stand in one-to-one correspondence with the numbers 1, 2,...,n. Nobody ever confuses the sixth particle with the number 6, or with the place the sixth particle happens to occupy at some time. The number 6 is merely a label attached to the particle, and any other would do just as well. Similarly, in continuum mechanics a body $\mathcal{B}$ may occupy infinitely many different regions, none of which is to be confused with the body itself.

The two fundamental properties of a continuum $\mathcal{B}$ are, then:

(1) each sufficiently small part of $\mathcal{B}$ can be mapped smoothly onto a cube in Euclidean space, and

(2) $\mathcal{B}$ is a measure space. These are the simplest, most fundamental things we can say in summary of our experience with actual bodies and with the older theories designed to represent physical bodies in mathematical terms.

We refrain from imputing any particular geometry to the body $\mathcal{B}$, because we have no basis for doing so, since a body is never encountered directly. While the above two assumptions imply some geometric structure in $\mathcal{B}$, it seems to me closer to physical experience and hence preferable to turn instead to the properties of $\mathcal{B}$ in its configurations, for it is the configurations that are available to us.

Since a body can be mapped smoothly onto a domain, it can be mapped smoothly onto any topological equivalent of that domain. A sequence of such mappings,

$$\underset{\sim}{x} = \underset{\sim}{\chi}(X,t), \tag{2.1}$$

is called a *motion* of $\mathcal{B}$. Here X is a particle, t is the time ($-\infty<t<\infty$), and $\underset{\sim}{x}$ is a place in Euclidean space. The value of $\underset{\sim}{\chi}$ is the place $\underset{\sim}{x}$ that the particle X comes to occupy at the time t. The notations $\mathcal{B}_{\underset{\sim}{\chi}}$ and $\mathcal{P}_{\underset{\sim}{\chi}}$ will indicate the configurations of $\mathcal{B}$ and $\mathcal{P}$ at the time t. We shall consider only motions that are smooth in the further sense that $\underset{\sim}{\chi}$ is differentiable with respect to t as many times as may be needed. The *velocity* $\dot{\underset{\sim}{x}}$, *acceleration* $\ddot{\underset{\sim}{x}}$ and n^{th} *acceleration* $\overset{(n)}{\underset{\sim}{x}}$ of the particle X at the time t are defined as usual:

$$\dot{\underset{\sim}{x}} \equiv \partial_t \underset{\sim}{\chi}(X,t),$$

$$\ddot{\underset{\sim}{x}} \equiv \partial_t^2 \underset{\sim}{\chi}(X,t), \tag{2.2}$$

$$\overset{(n)}{\underset{\sim}{x}} \equiv \partial_t^n \underset{\sim}{\chi}(X,t),$$

so that $\dot{\underset{\sim}{x}} \equiv \overset{(1)}{\underset{\sim}{x}}$, $\ddot{\underset{\sim}{x}} \equiv \overset{(2)}{\underset{\sim}{x}}$.

Any function defined over the body in an interval of time may be regarded, equivalently, as a *field* defined over the configurations of the body in that interval. Thus, e.g., $\dot{\underset{\sim}{x}} = \dot{\underset{\sim}{x}}(\underset{\sim}{x},t)$, $\ddot{\underset{\sim}{x}} = \ddot{\underset{\sim}{x}}(\underset{\sim}{x},t)$.

Mass Density

$\mathcal{B}$ is a continuum since it is a topological image of a domain. However, the mass distribution $\mathcal{M}(\cdot)$ is so far left arbitrary and might be discrete, or partially so. Of primary interest in continuum mechanics are masses which are absolutely continuous functions of volume. By volume I mean Borel measure[1] in the space of configurations, Euclidean space. To assume that $\mathcal{M}(\cdot)$ is absolutely continuous is to assume that in every configuration, a part $\mathcal{P}$ of sufficiently small volume has arbitrarily small mass. Thus, formally, concentrated masses are excluded, and analytical dynamics will not emerge directly as a special case of continuum mechanics.

By the Radon-Nikodym theorem, the mass of $\mathcal{P}$ may be expressed in terms of a *mass-density* $\rho_{\underset{\sim}{\chi}}$:

---

[1] The field of Lebesgue measurable sets contains certain null sets that are not Borel sets, but these seem not to be of interest in continuum mechanics, so Borel measure suffices.

$$\mathcal{M}(\mathcal{P}) = \int_{\mathcal{P}_{\chi}} \rho_{\chi} dv \tag{2.3}$$

where the integral is a Lebesgue integral. Clearly the function ρ_{χ} depends on χ and, as indicated, the integration is carried out over the configuration $\mathcal{P}_{\chi}$ of $\mathcal{P}$.

Reference Configuration

Often it is convenient to select one particular configuration and refer everything concerning the body to it. Let κ be such a configuration. Then the mapping

$$X = \kappa(X) \tag{2.4}$$

gives the place X occupied by the particle X in the configuration κ. Since this mapping is smooth, by assumption, the motion (2.1) may be written in the form

$$x = \chi(\kappa^{-1}(X),t) \equiv \chi_{\kappa}(X,t). \tag{2.5}$$

In the description furnished by this equation, the motion is a sequence of mappings of the *reference configuration* κ onto the actual configurations χ. Thus the motion is visualized as mapping of parts of space onto parts of space . A reference configuration is introduced so as to allow us to employ the apparatus of Euclidean geometry.

Deformation Gradient

The gradient of (2.5) is called the deformation gradient F:

$$F \equiv F_{\kappa}(X,t) \equiv \nabla\chi_{\kappa}(X,t). \tag{2.6}$$

It is the linear approximation to the mapping $\underset{\sim}{\chi}_{\underset{\sim}{\kappa}}$. More precisely, we should call it the gradient of the deformation from $\underset{\sim}{\kappa}$ to $\underset{\sim}{\chi}$, but when, as is usual, a single reference configuration $\underset{\sim}{\kappa}$ is laid down and kept fixed, no confusion should result from failure to remind ourselves that the very concept of a deformation gradient presumes use of a reference configuration.

Exercise 2.1

Set

$$J \equiv |\det F|, \tag{2.7}$$

and write ρ for $\rho_{\underset{\sim}{\chi}}$. From (2.3), derive *Euler's explicit equation for the density:*

$$\rho J = \rho_{\underset{\sim}{\kappa}}. \tag{2.8}$$

By differentiating (2.7), derive the *Euler identity:*

$$\dot{J} = J \operatorname{div} \dot{\underset{\sim}{x}}, \tag{2.9}$$

where the dot denotes the time derivative when $\underset{\sim}{X}$ is held constant. Hence derive *d'Alembert and Euler's differential equation for the density:*

$$\dot{\rho} + \rho \operatorname{div} \dot{\underset{\sim}{x}} = 0. \tag{2.10}$$

Conversely, starting from (2.10), derive (2.8).

Exercise 2.2

A motion is called *isochoric* if the volume $V(\mathcal{P})$ of each part $\mathcal{P}$ of the body remains constant in time. Show that any one of the following three equations is a necessary and sufficient condition for isochoric motion:

$$\operatorname{div} \dot{\underset{\sim}{x}} = 0, \quad \rho = \rho_{\underset{\sim}{\kappa}}, \quad J = 1. \tag{2.11}$$

Change of Reference Configuration

Let the same motion (2.1) be described in terms of two different reference configurations, $\underset{\sim}{\kappa}_1$ and $\underset{\sim}{\kappa}_2$:

$$\underset{\sim}{x} = \underset{\sim}{\chi}_{\underset{\sim}{\kappa}_1}(\underset{\sim}{X},t) = \underset{\sim}{\chi}_{\underset{\sim}{\kappa}_2}(\underset{\sim}{X},t). \tag{2.12}$$

The deformation gradients $\underset{\sim}{F}_1$ and $\underset{\sim}{F}_2$ at $\underset{\sim}{X},t$ are of course generally different. Let $\underset{\sim}{X}_1$ and $\underset{\sim}{X}_2$ denote the positions of X in $\underset{\sim}{\kappa}_1$ and $\underset{\sim}{\kappa}_2$:

$$\underset{\sim}{X}_1 = \underset{\sim}{\kappa}_1(X), \quad \underset{\sim}{X}_2 = \underset{\sim}{\kappa}_2(X). \tag{2.13}$$

Then

$$\underset{\sim}{X}_2 = \underset{\sim}{\kappa}_2(\underset{\sim}{\kappa}_1^{-1}(\underset{\sim}{X}_1)) \equiv \underset{\sim}{\lambda}(\underset{\sim}{X}_1). \tag{2.14}$$

The deformation from $\underset{\sim}{\kappa}_1$ to $\underset{\sim}{\chi}$ can be effected in two ways: either straight off by use of $\underset{\sim}{\chi}_{\underset{\sim}{\kappa}_1}$, or by using $\underset{\sim}{\lambda}$ to get to $\underset{\sim}{\kappa}_2$ and then using $\underset{\sim}{\chi}_{\underset{\sim}{\kappa}_2}$ to get to $\underset{\sim}{\chi}$. If a circle denotes the succession of mappings, then,

$$\underset{\sim}{\chi}_{\underset{\sim}{\kappa}_1} = \underset{\sim}{\chi}_{\underset{\sim}{\kappa}_2} \circ \underset{\sim}{\lambda}. \tag{2.15}$$

Since this relation holds among the three mappings, their linear approximations, the gradients, are related in the same way:

$$\underset{\sim}{F}_1 = \underset{\sim}{F}_2\underset{\sim}{P}, \tag{2.16}$$

where

$$\underset{\sim}{P} \equiv \nabla\underset{\sim}{\lambda}. \tag{2.17}$$

Of course, the relation (2.16) expresses the "chain rule" of differential calculus.

Current Configuration as Reference

To serve as a reference, a configuration need only be a diffeomorph of the body. Thus far, we have employed a reference configuration fixed in time, but we could just as well use a varying one. In this way a given motion may be described in terms of any other. The only variable reference configuration really useful in this way is the present one. If we take it as reference, we describe the past and future as they seem to an observer fixed to the particle X now at the place $\underset{\sim}{x}$.

To see how such a description is constructed, consider the configurations of $\mathcal{B}$ at the two times t and τ:

$$\begin{aligned} \underset{\sim}{\xi} &= \underset{\sim}{\chi}(X,\tau), \\ \underset{\sim}{x} &= \underset{\sim}{\chi}(X,t). \end{aligned} \tag{2.18}$$

That is, $\underset{\sim}{\xi}$ is the place occupied at time τ by the particle that at time t occupies $\underset{\sim}{x}$:

$$\underset{\sim}{\xi} = \underset{\sim}{\chi}(\underset{\sim}{\chi}^{-1}(\underset{\sim}{x},t),\tau), \qquad (2.19)$$
$$\equiv \underset{\sim}{\chi}_t(x,\tau),$$

say. The function $\underset{\sim}{\chi}_t$ just defined is called the *relative deformation function.*

When the motion is described by (2.19), we shall use a subscript t to denote quantities derived from $\underset{\sim}{\chi}_t$. Thus $\underset{\sim}{F}_t$, defined by

$$\underset{\sim}{F}_t \equiv \underset{\sim}{F}_t(\tau) \equiv \text{grad } \underset{\sim}{\chi}_t, \qquad (2.20)$$

is the *relative deformation gradient.*

By (2.16), at $\underset{\sim}{X}$,

$$\underset{\sim}{F}(\tau) = \underset{\sim}{F}_t(\tau)\ \underset{\sim}{F}(t). \qquad (2.21)$$

As the fixed reference configuration with respect to which $\underset{\sim}{F}(\tau)$ and $\underset{\sim}{F}(t)$ are taken we may select the configuration occupied by the body at time t'. Then (2.21) yields

$$\underset{\sim}{F}_{t'}(\tau) = \underset{\sim}{F}_t(\tau)\ \underset{\sim}{F}_{t'}(t). \qquad (2.22)$$

Stretch and Rotation

Since the motion $\underset{\sim}{\chi}_\kappa$ is continuous, $\underset{\sim}{F}$ is non-singular, so the polar decomposition theorem of CAUCHY enables us to write it in the two forms

$$\underset{\sim}{F} = \underset{\sim}{R}\underset{\sim}{U} = \underset{\sim}{V}\underset{\sim}{R} \tag{2.23}$$

where $\underset{\sim}{R}$ is an orthogonal tensor, while $\underset{\sim}{U}$ and $\underset{\sim}{V}$ are positive-definite symmetric tensors. $\underset{\sim}{R}$, $\underset{\sim}{U}$, and $\underset{\sim}{V}$ are unique. Cauchy's decomposition tells us that the deformation corresponding locally to $\underset{\sim}{F}$ may be obtained by first rotating a certain orthogonal triad and then effecting pure stretches along the resulting directions.

The *right* and *left Cauchy-Green tensors* $\underset{\sim}{C}$ and $\underset{\sim}{B}$ are defined as follows:

$$\begin{aligned} \underset{\sim}{C} &\equiv \underset{\sim}{U}^2 = \underset{\sim}{F}^T\underset{\sim}{F}, \\ \underset{\sim}{B} &\equiv \underset{\sim}{V}^2 = \underset{\sim}{F}\underset{\sim}{F}^T. \end{aligned} \tag{2.24}$$

While the fundamental decomposition (2.23) plays the major part in the proof of general theorems, calculation of $\underset{\sim}{U}$, $\underset{\sim}{V}$, and $\underset{\sim}{R}$ in special cases may be awkward, since irrational operations are usually required. $\underset{\sim}{C}$ and $\underset{\sim}{B}$, however, are calculated by mere multiplication of $\underset{\sim}{F}$ and $\underset{\sim}{F}^T$. E.g., if g_{km} and $g^{\alpha\beta}$ are the covariant and contravariant metric components in arbitrary selected co-ordinate systems in space and in the reference configuration, respectively, components of $\underset{\sim}{C}$ and $\underset{\sim}{B}$ are

$$\begin{aligned} C_{\alpha\beta} &= F^k_\alpha F^m_\beta gkm, \\ B^{km} &= F^k_\alpha F^m_\beta g^{\alpha\beta}, \end{aligned} \tag{2.25}$$

where

$$F^k_\alpha = x^k_{,\alpha} \equiv \partial_{X_\alpha} \chi_k(X_1, X_2, X_3, t).$$

If we begin with the relative deformation $\underset{\sim}{F}_t$, defined by (2.20), and apply to it the polar decomposition theorem, we obtain the *relative rotation* $\underset{\sim}{R}_t$, the *relative stretch tensors* $\underset{\sim}{U}_t$ and $\underset{\sim}{V}_t$ and the *relative Cauchy-Green tensors* $\underset{\sim}{C}_t$ and $\underset{\sim}{B}_t$:

$$\begin{aligned} \underset{\sim}{F}_t &= \underset{\sim}{R}_t\underset{\sim}{U}_t = \underset{\sim}{V}_t\underset{\sim}{R}_t, \\ \underset{\sim}{C}_t &= \underset{\sim}{U}_t^2, \\ \underset{\sim}{B}_t &= \underset{\sim}{V}_t^2. \end{aligned} \tag{2.26}$$

We recall the notation f^t introduced in the previous lecture for the history of f^t up to time t; i.e.,

$$f^t \equiv f^t(s) \equiv f(t-s), \qquad s \geqq 0. \tag{2.27}$$

Thus $\underset{\sim}{C}_t^t(s)$ is the history of the relative right Cauchy-Green tensor up to time t.

Stretching and Spin

For a tensor defined from the relative motion, for example $\underset{\sim}{F}_t$, we introduce the notation

$$\dot{\underset{\sim}{F}}_t(t) \equiv \partial_\tau \underset{\sim}{F}_t(\tau)|_{\tau=t} = -\partial_s \underset{\sim}{F}_t^t(s)|_{s=0}. \tag{2.28}$$

Set

$$\begin{aligned} \underset{\sim}{G} &= \dot{\underset{\sim}{F}}_t(t), \\ \underset{\sim}{D} &= \dot{\underset{\sim}{U}}_t(t) = \dot{\underset{\sim}{V}}_t(t), \\ \underset{\sim}{W} &= \dot{\underset{\sim}{R}}_t(t). \end{aligned} \tag{2.29}$$

$\mathbf{D}$, which is called the *stretching*, is the rate of change of the stretch of the configuration at time $t + \varepsilon$ with respect to that at time t, in the limit as $\varepsilon \to 0$. Likewise, $\mathbf{W}$, which is called the *spin*, is the ultimate rate of change of the rotation from the present configuration to one occupied just before or just afterward.

Exercise 2.3

Prove that $\mathbf{D}$ is symmetric, $\mathbf{W}$ is skew, and

$$\mathbf{G} = \text{grad}\, \dot{\mathbf{x}}, \tag{2.30}$$

where "grad" denotes the field gradient. Hence prove and interpret the *Cauchy-Stokes decomposition theorem:*

$$\mathbf{G} = \mathbf{D} + \mathbf{W}. \tag{2.31}$$

Changes of Frame

In classical mechanics we think of an observer as being a rigid body carrying a clock. Actually we do not need an observer as such, but only the concept of change of observer, or, as we shall say, change of frame. The ordered pair $\{x,t\}$, a place and a time, is called an *event*. The totality of events is *space-time*. A *change of frame* is a one-to-one mapping of space-time onto itself such that distances, time intervals, and the sense of time are preserved. We expect that every such transformation should be a time-dependent orthogonal transformation of space combined with a shift of the origin of time. This is so. The most general change of frame is given by

$$\begin{aligned} \mathbf{x}^* &= \mathbf{c}(t) + \mathbf{Q}(t)\,(\mathbf{x}-\mathbf{x}_0), \\ t^* &= t - a, \end{aligned} \tag{2.32}$$

where $\underset{\sim}{c}(t)$ is a time-dependent point, $\underset{\sim}{Q}(t)$ is a time-dependent orthogonal tensor, $\underset{\sim}{x}_o$ is a fixed point, and a is a constant. We commonly say that $\underset{\sim}{c}(t)$ represents a change of origin, since the fixed point $\underset{\sim}{x}_o$ is mapped into $\underset{\sim}{c}(t)$. $\underset{\sim}{Q}(t)$ represents a rotation and also, possibly, a reflection. Reflections are included since, although in most physics and engineering courses the student is taught to use a right-handed co-ordinate system, there is nothing in the divine order of nature to prevent two observers from orienting themselves oppositely. Of course, we may live to see the day when social democracy, in the interest of the greatest good for the greatest number, forbids comrades (citizens) in eastern (western) countries to think about right- (left-) handed systems, respectively.

A frame need not be defined and certainly must not be confused with a co-ordinate system. It is convenient, however, to describe (2.32) as a change from "the unstarred frame to the starred frame", since this wording promotes the interpretation in terms of two different observers.

A quantity is said to be *frame-indifferent* if it is invariant under all changes of frame (2.32). More precisely, the following requirements are laid down:

$$A^* = A \quad \text{for indifferent scalars}$$
$$\underset{\sim}{v}^* = \underset{\sim}{Q}\underset{\sim}{v} \quad \text{for indifferent vectors}$$
$$\underset{\sim}{S}^* = \underset{\sim}{Q}\underset{\sim}{S}\underset{\sim}{Q}^T \quad \text{for indifferent tensors (of second order)}$$

An indifferent scalar is a quantity which does not change its value. An indifferent vector is one which is the same "arrow" in the sense that if

$$\mathbf{v} = \mathbf{x} - \mathbf{y} \,, \quad \text{then} \quad \mathbf{v}^* = \mathbf{x}^* - \mathbf{y}^* . \tag{2.33}$$

By (2.32), then,

$$\mathbf{v}^* = \mathbf{Q}(\mathbf{x} - \mathbf{y}) = \mathbf{Q}\mathbf{v}, \tag{2.34}$$

as asserted. An indifferent tensor is one that transforms indifferent vectors into indifferent vectors. That is, if

$$\begin{gathered} \mathbf{v}^* = \mathbf{Q}\mathbf{v}, \quad \mathbf{w}^* = \mathbf{Q}\mathbf{w}, \quad \text{and if } \mathbf{v} = \mathbf{S}\mathbf{w}, \\ \text{then} \quad \mathbf{v}^* = \mathbf{S}^*\mathbf{w}^* . \end{gathered} \tag{2.35}$$

By substituting the first three equations into the last, we find that

$$\mathbf{Q}\mathbf{v} = \mathbf{S}^*\mathbf{Q}\mathbf{w} = \mathbf{Q}\mathbf{S}\mathbf{w}. \tag{2.36}$$

Since this relation is to hold for all $\mathbf{w}$, we infer the rule $\mathbf{S}^*\mathbf{Q} = \mathbf{Q}\mathbf{S}$, as stated above.

In mechanics we meet some quantities that are indifferent and some that are not. Sometimes we have a vector or tensor defined in one frame only. By using the above rules, we may extend the definitions to all frames in such a way as to obtain a frame-indifferent quantity. Such an extension is trivial. Usually, however, we are given a definition valid in all frames from the start. In that case, we have to find out what transformation law is obeyed and thus determine whether or not the quantity be frame-indifferent.

Exercise 2.4

Under a change of frame, the motion (2.1) of a body becomes

$$\mathbf{x}^* = \mathbf{c}(t) + \mathbf{Q}(t)[\boldsymbol{\chi}(X,t) - \mathbf{x}_0] = \boldsymbol{\chi}^*(X,t). \tag{2.37}$$

Prove *Euler's theorem:*

$$\dot{\mathbf{x}}^* - \mathbf{Q}\dot{\mathbf{x}} = \dot{\mathbf{c}} + \mathbf{A}(\mathbf{x}^* - \mathbf{c}), \tag{2.38}$$

and the *Euler-Coriolis theorem:*

$$\ddot{\mathbf{x}}^* - \mathbf{Q}\ddot{\mathbf{x}} = \ddot{\mathbf{c}} + 2\mathbf{A}(\dot{\mathbf{x}}^* - \dot{\mathbf{c}}) + (\dot{\mathbf{A}} - \mathbf{A}^2)(\mathbf{x}^* - \mathbf{c}), \tag{2.39}$$

where

$$\mathbf{A} = \dot{\mathbf{Q}}\mathbf{Q}^T = -\mathbf{A}^T. \tag{2.40}$$

Interpret $\mathbf{A}$.

In (2.37) we may refer both the motions $\boldsymbol{\chi}$ and $\boldsymbol{\chi}^*$ to the same reference configuration $\boldsymbol{\kappa}$ if we wish to. This amounts to replacing $\boldsymbol{\chi}$ and $\boldsymbol{\chi}^*$ by $\boldsymbol{\chi}_\kappa$ and $\boldsymbol{\chi}^*_\kappa$, respectively, since these functions have the same values as the former. By taking the gradient of the resulting formula we obtain

$$\mathbf{F}^* = \mathbf{Q}\mathbf{F}. \tag{2.41}$$

Exercise 2.5

Prove that

$$\mathbf{R}^* = \mathbf{Q}\mathbf{R}, \quad \mathbf{U}^* = \mathbf{U}, \quad \mathbf{V}^* = \mathbf{Q}\mathbf{V}\mathbf{Q}^T. \tag{2.42}$$

Prove and interpret the *Zaremba-Zorawski theorem:*

$$\mathbf{D}^* = \mathbf{Q}\mathbf{D}\mathbf{Q}^T, \quad \mathbf{W}^* = \mathbf{Q}\mathbf{W}\mathbf{Q}^T + \mathbf{A}. \tag{2.43}$$

3. Force and Work. Laws of Motion

Physical Principles of Mechanics

After bodies and motions, the most important primitive elements of mechanics are forces, torques, and energies. They are given to us *a priori*. They are related to one another by *laws* or general principles, common to all bodies. I shall use these laws in the following forms:

$$\dot{K} + \dot{E} = P + Q, \qquad (1.1)$$

$$\dot{\underset{\sim}{m}} = \underset{\sim}{f}, \qquad (3.1)$$

$$\dot{\underset{\sim}{M}}_{\underset{\sim}{x}_0} = \underset{\sim}{L}_{\underset{\sim}{x}_0}. \qquad (3.2)$$

The first, the principle of energy, we have already applied in the thermodynamics of homogeneous processes, explained in Lecture 1. The next two are called *Euler's laws of mechanics*. The former asserts that for any part $\mathcal{P}$ of any body, the total force $\underset{\sim}{f}$ applied to the body in the configuration $\mathcal{P}_{\underset{\sim}{\chi}}$ equals the rate of change of the total momentum $\underset{\sim}{m}$ in that configuration. The second asserts that for any part $\mathcal{P}$ of any body, the total torque $\underset{\sim}{L}_{\underset{\sim}{x}_0}$ with respect to a certain point $\underset{\sim}{x}_0$, applied to the body in the configuration $\mathcal{P}_{\underset{\sim}{\chi}}$, equals the rate of change of the total moment of momentum $\underset{\sim}{M}_{\underset{\sim}{x}_0}$ with respect to $\underset{\sim}{x}_0$ in that configuration. (Of course, Euler's laws generalize and include much earlier ones, due in varying forms and circumstances to mediaeval schoolmen, HUYGENS, NEWTON, JAMES BERNOULLI, and others. There is no justice in emphasizing our debt to any one of these predecessors of EULER to the exclusion of the rest.)

These laws are extremely general. They subsume all aspects of what is called "classical" mechanics. In continuum mechanics specific assumptions are made about the eight quantities entering these equations. In this course we shall be content with a statement that is special even for continuum mechanics today, although it is far more general than anything envisaged in the older books. I shall make this statement in parts, whose special character I shall emphasize by using the term *simple* to denote each one. I shall begin with Euler's laws and then turn afterward to the principle of energy. The reason, of course, is that K and P are mechanical quantities. The splitting of the two sides of (1.1) is done for convenience, not from principle, but the whole treatment would seem less familiar and hence less "physical" if I began, as is formally more natural, with the scalar equation.

Forces and Moments in Continuum Mechanics

While several kinds of forces and torques are considered in a more general mechanics, in these lectures we shall need only the classical ones. Forces act on the parts $\mathcal{P}$ of a body $\mathcal{B}$ in a configuration χ. We shall require two kinds of forces: *body force* $\mathbf{f}_b(\mathcal{P})$, which is an absolutely continuous function of the volume of $\mathcal{P}_\chi$, and *contact force* $\mathbf{f}_c(\mathcal{P})$, which is an absolutely continuous function of the surface area of the boundary $\partial\mathcal{P}_\chi$ of $\mathcal{P}_\chi$. The *resultant force* $\mathbf{f}(\mathcal{P})$ acting on $\mathcal{P}$ in χ is given by

$$\mathbf{f}(\mathcal{P}) = \mathbf{f}_b(\mathcal{P}) + \mathbf{f}_c(\mathcal{P}), \tag{3.3}$$

where

$$\mathbf{f}_b(\mathscr{P}) = \int_{\mathscr{P}_{\mathbf{\chi}}} \mathbf{b}\,dm = \int_{\mathscr{P}_{\mathbf{\chi}}} \rho \mathbf{b}\,dv,$$

$$\mathbf{f}_c(\mathscr{P}) = \int_{\partial \mathscr{P}_{\mathbf{\chi}}} \mathbf{t}\,ds. \tag{3.4}$$

The two densities, $\mathbf{b}$ and $\mathbf{t}$, are called the *specific body force* and the *traction*, respectively.

The torque $\mathbf{L}_{\mathbf{x}_o}(\mathscr{P})$ is taken to be the moment of the force:

$$\mathbf{L}_{\mathbf{x}_o} = \int_{\mathscr{P}_{\mathbf{\chi}}} (\mathbf{x}-\mathbf{x}_o) \wedge \mathbf{b}\,dm + \int_{\partial \mathscr{P}_{\mathbf{\chi}}} (\mathbf{x}-\mathbf{x}_o) \wedge \mathbf{t}\,ds. \tag{3.5}$$

A moment or *simple torque* is only a special case of *torque*, but more general torques are not needed in this course.

Exercise 3.1

Prove that

$$\mathbf{L}_{\mathbf{x}_o'}(\mathscr{P}) = \mathbf{L}_{\mathbf{x}_o}(\mathscr{P}) + (\mathbf{x}_o - \mathbf{x}_o') \wedge \mathbf{f}(\mathscr{P}). \tag{3.6}$$

We regard forces and moments of force as quantities given *a priori* and independent of the observer. Thus we assume that *forces and torques are indifferent*. That is, under a change of frame (2.32),

$$\mathbf{b}^* = \mathbf{Q}\mathbf{b} \qquad \text{and} \qquad \mathbf{t}^* = \mathbf{Q}\mathbf{t}. \tag{3.7}$$

Momentum and Moment of Momentum

The *momentum* $\underset{\sim}{m}(\mathcal{P})$ and the *moment of momentum* $\underset{\sim}{M}(\mathcal{P};\underset{\sim}{x}_o)$ of $\mathcal{P}$ in the configuration $\underset{\sim}{\chi}(\mathcal{P},t)$ are said to be *simple* if

$$\begin{aligned} \underset{\sim}{m}(\mathcal{P}) &= \int_{\mathcal{P}_{\underset{\sim}{\chi}}} \dot{\underset{\sim}{x}}\,dm, \\ \underset{\sim}{M}(\mathcal{P};\underset{\sim}{x}_o) &= \int_{\mathcal{P}_{\underset{\sim}{\chi}}} (\underset{\sim}{x}-\underset{\sim}{x}_o) \wedge \dot{\underset{\sim}{x}}\,dm. \end{aligned} \tag{3.8}$$

In this course we shall not need to consider any momenta more general than these.

Exercise 3.2

When the momenta are simple and Euler's first law (3.1) holds, then Euler's second law (3.2) holds for one $\underset{\sim}{x}_o$ if and only if it holds for all $\underset{\sim}{x}_o$.

It is possible to derive Euler's laws as a theorem from a single axiom proposed by NOLL: The rate of working is frame-indifferent for every part $\mathcal{P}$ of every body $\mathcal{B}$.

The Euler-Cauchy Stress Principle

As defined, the densities $\underset{\sim}{b}$ and $\underset{\sim}{t}$ in (3.4) may be extremely general:

$$\underset{\sim}{b} = \underset{\sim}{b}(\underset{\sim}{x},t,\mathcal{P},\mathcal{B}), \quad \underset{\sim}{t} = \underset{\sim}{t}(\underset{\sim}{x},t,\mathcal{P},\mathcal{B}). \tag{3.9}$$

We shall restrict attention to body force densities that are unaffected by the presence or absence of bodies in space:

$$\underset{\sim}{b} = \underset{\sim}{b}(\underset{\sim}{x},t), \tag{3.10}$$

whatever be $\mathcal{P}$ and $\mathcal{B}$. Such body forces are called *external*. The particular case when $\underset{\sim}{b}$ = constant pertains to *heavy bodies*. (A particular kind of more general body force called *mutual*, such as universal gravitation, is sometimes included in mechanics, but will not be needed in this course.) If $\underset{\sim}{b}$ is the gradient of a scalar field, the body force is said to be conservative. We shall restrict attention also to a particular kind of contact force, namely, one such that the traction $\underset{\sim}{t}$ at any given place and time has a common value for all parts $\mathcal{P}$ having a common tangent plane and lying upon the same side of it:

$$\underset{\sim}{t} = \underset{\sim}{t}(\underset{\sim}{x},t,\underset{\sim}{n}), \tag{3.11}$$

where $\underset{\sim}{n}$ is the outer normal to $\partial\mathcal{P}$ in the configuration $\underset{\sim}{\chi}$. Such tractions are called *simple*. The assumption embodied in (3.3), (3.4), and (3.11) is the *Euler-Cauchy stress principle*, which is the keystone of classical continuum mechanics. We shall not need to depart from it in this course.

The stress principle is put to use through *Cauchy's fundamental lemma:* There exists a tensor $\underset{\sim}{T}(\underset{\sim}{x},t)$, called the *stress tensor*, such that

$$\underset{\sim}{t}(\underset{\sim}{x},t,\underset{\sim}{n}) = \underset{\sim}{T}(\underset{\sim}{x},t)\underset{\sim}{n}. \tag{3.12}$$

That is, the traction $\underset{\sim}{t}$, which at the outset was allowed to depend arbitrarily on the normal $\underset{\sim}{n}$, is in fact a linear homogeneous function of it.

Exercise 3.3

Show that Cauchy's fundamental lemma follows from (3.11) and either $(3.1)_1$ or $(3.1)_2$.

Energy and Heat Work

We now turn back to the quantities K, E, P, and Q entering (1.1), the equation of energy balance.

The mechanical power P is said to be *simple* if it is the rate of working of the forces alone. By (3.4), then

$$P = \int_{\partial \mathcal{P}_{\underset{\sim}{\chi}}} \dot{\underset{\sim}{x}} \cdot \underset{\sim}{t}\, ds + \int_{\mathcal{P}_{\underset{\sim}{\chi}}} \dot{\underset{\sim}{x}} \cdot \underset{\sim}{b}\, dm. \tag{3.13}$$

The kinetic energy K is said to be *simple* if it is taken to have the classical form:

$$K = \frac{1}{2}\int_{\mathcal{P}_{\underset{\sim}{\chi}}} \dot{\underset{\sim}{x}}^2 dm. \tag{3.14}$$

We shall restrict our attention here to simple power and simple kinetic energy. We shall assume further that the internal energy is an absolutely continuous additive set function in the configurations $\mathcal{P}_{\underset{\sim}{\chi}}$ of $\mathcal{P}$:

$$E = \int_{\mathcal{P}_{\underset{\sim}{\chi}}} \varepsilon\, dm. \tag{3.15}$$

The density ε is called the *specific internal energy*. Finally, we shall assume that the heat working Q can be represented as the sum of two parts, due to contact working and internal working, respectively:

$$Q = \int_{\partial \mathcal{P}_{\chi}} q\,ds + \int_{\mathcal{P}_{\chi}} s\,dm. \tag{3.16}$$

The density s is the *heat supply;* the density q is the *heat influx.* We assume that these densities are frame-indifferent:

$$q^* = q, \qquad s^* = s. \tag{3.17}$$

Beyond this, they may be quite general:

$$q = q(\underset{\sim}{x}, t, \mathcal{P}, \mathcal{B}), \qquad s = s(\underset{\sim}{x}, t, \mathcal{P}, \mathcal{B}). \tag{3.18}$$

In this course, however, we shall be content with *external* heat supplies and with *simple* heat influxes:

$$s = s(\underset{\sim}{x}, t), \qquad q = q(\underset{\sim}{x}, t, \underset{\sim}{n}), \tag{3.19}$$

whatever be $\mathcal{P}$ and $\mathcal{B}$. According to a fundamental lemma due to FOURIER, $(3.19)_2$ implies the existence of a *heat efflux vector* $\underset{\sim}{h}(\underset{\sim}{x}, t)$.

Exercise 3.4

Prove *Fourier's lemma:* It follows from (1.1) that

$$q = \underset{\sim}{h} \cdot \underset{\sim}{n}\,, \quad \text{where} \quad \underset{\sim}{h} = \underset{\sim}{h}(\underset{\sim}{x}, t). \tag{3.20}$$

Equations of Balance

Collecting the results of substituting the expressions for our various assumptions, most of them labelled as "simple", into the fundamental principles (1.1), (3.1), and (3.2), we obtain the following final equations as basic for the special kind of continuum mechanics we shall consider in this course:

$$\left(\int_{\mathcal{P}}\left(\tfrac{1}{2}\dot{\mathbf{x}}^2+\varepsilon\right)dm\right)^{\bullet} = \int_{\partial\mathcal{P}_{\mathbf{\chi}}}\left(\dot{\mathbf{x}}\cdot\mathbf{T}\mathbf{n}+\mathbf{h}\cdot\mathbf{n}\right) + \int_{\mathcal{P}_{\mathbf{\chi}}}\left(\dot{\mathbf{x}}\cdot\mathbf{b}+s\right)dm, \tag{3.21}$$

$$\left(\int_{\mathcal{P}}\dot{\mathbf{x}}\,dm\right)^{\bullet} = \int_{\partial\mathcal{P}_{\mathbf{\chi}}}\mathbf{T}\mathbf{n}\,ds + \int_{\mathcal{P}_{\mathbf{\chi}}}\mathbf{b}\,dm, \tag{3.22}$$

$$\left(\int_{\mathcal{P}}(\mathbf{x}-\mathbf{x}_o)\wedge\dot{\mathbf{x}}\,dm\right)^{\bullet} = \int_{\partial\mathcal{P}_{\mathbf{\chi}}}(\mathbf{x}-\mathbf{x}_o)\wedge\mathbf{T}\mathbf{n}\,ds + \int_{\mathcal{P}_{\mathbf{\chi}}}(\mathbf{x}-\mathbf{x}_o)\wedge\mathbf{b}\,dm, \tag{3.23}$$

for every part $\mathcal{P}$ of every body $\mathcal{B}$, where the dots signify that the integration is to be performed over a fixed part of the body. The subscript $\mathbf{\chi}$ is omitted from $\mathcal{P}$ on the left-hand sides as an additional reminder that the time derivative is calculated for a fixed part $\mathcal{P}$ of the body, not for a fixed region of space.

These equations are both of the form called an *equation of balance* for a tensor field Ψ:

$$\left(\int_{\mathcal{P}_{\mathbf{\chi}}}\Psi\,dm\right)^{\bullet} = \int_{\partial\mathcal{P}_{\mathbf{\chi}}}\mathbf{E}\{\Psi\}\mathbf{n}\,ds + \int_{\mathcal{P}_{\mathbf{\chi}}}s\{\Psi\}\,dm. \tag{3.24}$$

$\mathbf{E}\{\Psi\}$, a tensor whose order is 1 greater than that of Ψ, is called an *efflux* of Ψ, while $s\{\Psi\}$ is called a *source* of Ψ. An equation of balance expresses the rate of growth of $\int_{\mathcal{P}}\Psi\,dm$ as the sum of two parts:

a rate of flow $-\underset{\sim}{E}\{\Psi\}$ inward through the boundary $\partial\mathcal{P}_{\underset{\sim}{X}}$ of the configuration $\mathcal{P}_{\underset{\sim}{X}}$ of $\mathcal{P}$ and a creation $s\{\Psi\}$ in the interior of that configuration. Equations of this form occur frequently in mathematical physics. Under the assumptions of smoothness made at the beginning of this course, an equation of balance is always equivalent to a differential equation.

Exercise 3.5

Prove that

$$\left(\int_{\mathcal{P}} \Psi dm\right)^{\bullet} = \int_{\mathcal{P}_{\underset{\sim}{X}}} \dot{\Psi} dm, \tag{3.25}$$

where $\dot{\Psi}$ is the time derivative of Ψ regarded as a function of $\underset{\sim}{X}$ and t. Show that the equation of balance (3.24) holds for every part $\mathcal{P}$ of every body $\mathcal{B}$ if and only if at each interior point of $\mathcal{B}$

$$\rho\dot{\Psi} = \operatorname{div} \underset{\sim}{E}\{\Psi\} + \rho s\{\Psi\}. \tag{3.26}$$

The general differential equation (3.26) is a consequence of the divergence theorem. Consequently (3.26) holds subject to specific assumptions about the region and about the fields. The assumptions about the region are satisfied here, because only interior points are considered; in fact, it would suffice to consider only those parts $\mathcal{P}$ whose configurations are spheres about the point in question. The fields, however, must be smooth. It suffices to assume that $\underset{\sim}{E}$ is continuously differentiable and that $\rho\dot{\Psi}$ and ρs are continuous in a sufficiently small sphere about the interior point considered. In general, (3.26) does not hold at points of $\partial\mathcal{B}_{\underset{\sim}{X}}$ or at interior points where the fields ρ, $\dot{\Psi}$, $\underset{\sim}{E}$, s fail to exist or to be sufficiently smooth.

Differential Equations of Continuum Mechanics

If we apply (3.26) to (3.22), we conclude at once that Euler's first law holds if and only if

$$\rho\ddot{\mathbf{x}} = \operatorname{div} \mathbf{T} + \rho\mathbf{b} \tag{3.27}$$

at every interior point of $\mathcal{B}_{\chi}$. This equation expresses *Cauchy's first law of continuum mechanics*. To apply (3.26) to (3.23), we let $\mathbf{M}$ stand for the tensor such that $(\mathbf{x}-\mathbf{x}_o) \wedge (\mathbf{T}\mathbf{n}) = \mathbf{M}\mathbf{n}$ for all $\mathbf{n}$ and conclude that Euler's second law holds if and only if

$$\begin{aligned}\rho(\mathbf{x}-\mathbf{x}_o) \wedge \ddot{\mathbf{x}} &= \operatorname{div} \mathbf{M} + \rho(\mathbf{x}-\mathbf{x}_o) \wedge \mathbf{b}, \\ &= \mathbf{T}^T - \mathbf{T} + (\mathbf{x}-\mathbf{x}_o) \wedge \operatorname{div} \mathbf{T} + \rho(\mathbf{x}-\mathbf{x}_o) \wedge \mathbf{b},\end{aligned} \tag{3.28}$$

where the second form follows by an easy identity. If (3.27) holds, then

$$\mathbf{T}^T = \mathbf{T}. \tag{3.29}$$

The stress tensor is symmetric, and conversely. This equation expresses *Cauchy's second law of continuum mechanics*.

Since this second law, in particular, has been questioned from time to time, I pause to emphasize its special character. We began with the general laws of EULER, but we applied them only subject to some special assumptions: (1) All torques are simple and (2) the traction is simple. (We assumed also that the body force is external, but that restriction is not important here.) Under these assumptions, *Euler's laws are equivalent to Cauchy's laws*, when $\mathbf{x}$, $\mathbf{b}$,

ρ_{κ}, and $\underset{\sim}{T}$ are sufficiently smooth. Cauchy's first law expresses locally the balance of linear momentum. Cauchy's second law, if the first is satisfied, expresses locally the balance of moment of momentum.

We turn now to the balance of energy. Our assumptions are sufficient to express the work storage W as an additive set function.

Exercise 3.6

Prove that for a material with simple kinetic energy and simple power (assumptions (3.13) and (3.14)), Cauchy's laws imply that

$$W \equiv P - \dot{K} = \int_{\mathcal{P}_{\underset{\sim}{\chi}}} w dv, \tag{3.30}$$

where w is the *stress working:*

$$w = \operatorname{tr}(\underset{\sim}{T}\underset{\sim}{D}). \tag{3.31}$$

By (3.30), (3.15), and (3.16), and Fourier's lemma, the balance of energy (1,1) assumes the form

$$\left(\int_{\mathcal{P}_{\underset{\sim}{\chi}}} \varepsilon dm\right)^{\cdot} = \int_{\partial\mathcal{P}_{\underset{\sim}{\chi}}} \underset{\sim}{h}\cdot\underset{\sim}{n} ds + \int_{\mathcal{P}_{\underset{\sim}{\chi}}} (w+\rho s) dv. \tag{3.32}$$

From (3.25) and (3.26) we may now read off the Fourier-Kirchhoff-Neumann equation of energy:

$$\rho\dot{\varepsilon} = w + \operatorname{div} \underset{\sim}{h} + \rho s. \tag{3.33}$$

Subject to sufficient assumptions of smoothness, this differential equation is equivalent to the principle of energy if the principles of momentum and moment of momentum are satisfied, for the case of simple forces, torques, kinetic energies, and heat influxes.

4. Constitutive Equations. Simple Materials

Nature of Constitutive Equations

We have derived, subject to a set of specific assumptions labelled "simple", equations expressing the local balance of energy, momentum, and moment of momentum:

$$\rho\dot{\varepsilon} = \mathrm{tr}(\underset{\sim}{T}\underset{\sim}{D}) + \mathrm{div}\ \underset{\sim}{h} + \rho s, \tag{3.33}$$

$$\rho\ddot{\underset{\sim}{x}} = \mathrm{div}\ \underset{\sim}{T} + \rho\underset{\sim}{b}, \tag{3.27}$$

$$\underset{\sim}{T}^{T} = \underset{\sim}{T}. \tag{3.29}$$

These express properties *common* to all bodies and motions we shall consider in this course. The number of quantities appearing in these equations far exceeds the number of equations, and indeed from the fundamental ideas they express we see easily that they cannot determine motions or forces. If we take any motion, any internal energy, any mass distribution, any vector $\underset{\sim}{h}$, and any symmetric tensor $\underset{\sim}{T}$, the equations (3.33), (3.27), and (3.29) are satisfied for some body force and some supply of heat.

The *differences* of bodies are expressed by *constitutive equations*. These equations, which define *ideal materials*, relate the variables entering the fundamental laws.

In this lecture I wish to explain constitutive equations for what might be called "pure" continuum mechanics. Heat and temperature and internal energy will be left out of account, and accordingly (3.33), the balance of energy, will not be used. The resulting theory, based on Cauchy's laws (3.27) and (3.29), turns out to include and describe accurately a good many important phenomena but of course is rather incomplete. My reason for limiting attention to it is to illustrate the concepts and mathematical methods first in a relatively simple context. Later, we shall see that a theory including heat and temperature may be constructed along the same lines by the addition of a single further basic idea.

Equivalent Processes

The motion $\underset{\sim}{\chi}(X,t)$ of a body $\mathcal{B}$ and the forces acting upon the corresponding configurations of $\mathcal{B}$ constitute a *dynamical process* if Cauchy's laws (3.27) and (3.29) are satisfied. If the body and its mass distribution are given, Cauchy's first law (3.27) determines a unique body force $\underset{\sim}{b}$ for any particular stress and motion. While in some specific case $\underset{\sim}{b}$ will be given, here we wish to consider rather the totality of all possible problems. There is then no reason to restrict $\underset{\sim}{b}$. Consequently, any pair of functions $\{\underset{\sim}{\chi},\underset{\sim}{T}\}$, where $\underset{\sim}{\chi}$ is a mapping of the body $\mathcal{B}$ onto configurations in space and where $\underset{\sim}{T}$ is any smooth field of symmetric tensors defined at each time t over the configuration $\mathcal{B}$, define a dynamical process.

Under a change of frame, $\underset{\sim}{\chi}$ is transformed into $\underset{\sim}{\chi}^*$ as given by (2.37). Our assumption (3.7) asserts that body force and contact force are indifferent. Since, of course, the normal $\underset{\sim}{n}$ is indifferent, Cauchy's lemma (3.12) implies that the stress tensor $\underset{\sim}{T}$ is indifferent, so that by a result in Lecture 2, $\underset{\sim}{T}^* = \underset{\sim}{Q}\underset{\sim}{T}\underset{\sim}{Q}^T$. Two dynamical processes $\{\underset{\sim}{\chi},\underset{\sim}{T}\}$ and $\{\underset{\sim}{\chi}^*,\underset{\sim}{T}^*\}$ related in this way are regarded as the same motion and associated contact forces as seen by two different observers. Thus, formally, two dynamical processes $\{\underset{\sim}{\chi},\underset{\sim}{T}\}$ and $\{\underset{\sim}{\chi}^*,\underset{\sim}{T}^*\}$ will be called *equivalent* if they are related as follows:

$$\underset{\sim}{\chi}^*(X,t^*) = \underset{\sim}{c}(t) + \underset{\sim}{Q}(t)\left[\underset{\sim}{\chi}(X,t) - \underset{\sim}{x}_o\right],$$

$$t^* = t - a, \tag{4.1}$$

$$T^*(X,t^*) = \underset{\sim}{Q}(t)\underset{\sim}{T}(X,t)\underset{\sim}{Q}(t)^T,$$

where the notations are those introduced in Lecture 2. We here regard the stress tensor as a function of the particle X and the time
t.

Constitutive Equations

As I have said several times, the general principles of mechanics
apply to all bodies and motions, and the diversity of materials in
nature is represented in the theory by *constitutive equations*. A constitutive equation in mechanics is a relation between forces and
motions. In popular terms, forces applied to a body "cause" it to
undergo a motion, and the motion "caused" differs according to the
nature of the body. In continuum mechanics the forces of interest
are contact forces, which are specified by the stress tensor $\underset{\sim}{T}$. Just
as different figures are defined in geometry as idealizations of certain important natural objects, in continuum mechanics *ideal materials* are defined by particular relations between the stress tensor
and the motion of the body. Some materials are important in themselves, but most of them are of more interest as members of a class
than in detail. Thus a general theory of constitutive equations is
needed. I shall now present certain aspects of NOLL's theory, which
was first published in 1958.

Axioms for Constitutive Equations

1. *Principle of Determinism.* The stress at the particle X in the body $\mathcal{B}$ at time t is determined by the history $\underset{\sim}{\chi}^t$ of the motion of $\mathcal{B}$ up to time t:

$$\underset{\sim}{T}(X,t) = \mathcal{F}\,(\underset{\sim}{\chi}^t;X,t). \qquad (4.2)$$

Here $\mathcal{F}$ is a functional in the most general sense of the term, namely, a rule of correspondence. The relation (4.2) asserts that the motion of the body up to and including the present time determines a unique symmetric stress tensor $\underset{\sim}{T}$ at each point of the body, and the manner in which it does so may depend upon X and t. The functional $\mathcal{F}$ is called the *constitutive functional*, and (4.2) is the *constitutive equation* of the ideal material defined by $\mathcal{F}$. Notice that the past, as much of it as need be, may affect the present stress, but in general past and future are not interchangeable. The common prejudice that mechanics concerns phenomena reversible in time is too naive to need refuting.

2. *Principle of Local Action.* In the principle of determinism, the motions of particles Z which lie far away from X are allowed to affect the stress at X. The notion of contact force makes it natural to exclude action at a distance as a material property. Accordingly, we assume as the second axiom of continuum mechanics that the motion of particles at a finite distance from X in some configuration may be disregarded in calculating the stress at X. (Of course, by the smoothness assumed for $\underset{\sim}{\chi}$, particles once a finite distance apart are always a finite distance apart.) Formally, if

$$\bar{\underset{\sim}{\chi}}^t(Z,s) = \underset{\sim}{\chi}^t(Z,s) \text{ when } s \geq 0 \text{ and } Z \varepsilon N(X), \qquad (4.3)$$

where N(X) is some neighborhood of X, then

$$\underset{\sim}{\mathcal{F}}(\bar{\underset{\sim}{\chi}}^t;Xt) = \underset{\sim}{\mathcal{F}}(\underset{\sim}{\chi}^t;X,t). \tag{4.4}$$

3. Principle of Material Frame-Indifference. We have said that we shall regard two equivalent dynamical processes as being really the same process, viewed by two different observers. We regard material properties as being likewise indifferent to the choice of observer. Since constitutive equations are designed to express idealized material properties, we require that they shall be frame-indifferent. That is, if (4.2) holds, *viz*

$$\underset{\sim}{T}(X,t) = \underset{\sim}{\mathcal{F}}(\chi^t;X,t), \tag{4.5}$$

then the constitutive functional $\underset{\sim}{\mathcal{F}}$ must be such that

$$\underset{\sim}{T}(X,t^*)^* = \underset{\sim}{\mathcal{F}}((\underset{\sim}{\chi}^{t*})^*;X,t^*), \tag{4.6}$$

where $\{\underset{\sim}{\chi}^*,\underset{\sim}{T}^*\}$ is any dynamical process equivalent to $\{\underset{\sim}{\chi},\underset{\sim}{T}\}$. Like the principle of local action, the principle of frame-indifference imposes a *restriction* on the functionals $\underset{\sim}{\mathcal{F}}$ to be admitted in constitutive equations. Namely, $\underset{\sim}{\mathcal{F}}$ must be such that the constitutive equation (3.20) is invariant under the transformations (3.19). Only those functionals $\underset{\sim}{\mathcal{F}}$ that satisfy the requirements of local action and frame-indifference are admissible as constitutive functionals.

Some steps may be taken to delimit the class of functionals satisfying these axioms, but in these lectures I shall treat only a special case, which is still general enough to include all the older theories of continua and most of the more recent ones. This special case is called the *simple material.*

Simple Materials

A motion $\underset{\sim}{\chi}$ is homogeneous with respect to the reference configuration $\underset{\sim}{\kappa}$ if

$$x = \underset{\sim}{\chi}_{\kappa}(\underset{\sim}{X},t) = \underset{\sim}{F}(t)\,(\underset{\sim}{X} - \underset{\sim}{X}_O) + \underset{\sim}{x}_O(t), \qquad (4.7)$$

where $\underset{\sim}{x}_O(t)$ is a place in $\underset{\sim}{\chi}$, possibly moving, $\underset{\sim}{X}_O$ is a fixed place in the reference configuration $\underset{\sim}{\kappa}$, and $\underset{\sim}{F}(t)$, which is the deformation gradient, is a nonsingular tensor that does not depend on $\underset{\sim}{X}$. A motion is homogeneous if and only if it carries every straight line at time O into a straight line at time t. A motion homogeneous with respect to one reference configuration generally fails to be so with respect to another.

Physically minded people almost always assume that everything there is to know about a material can be found out by performing experiments on homogeneous motions of that material from whatever state they happen to find it in. The materials in the special class that conforms to their prejudice are called *simple materials*. Formally, the material defined by (4.2) is called *simple* if there exists a reference configuration $\underset{\sim}{\kappa}$ such that

$$\underset{\sim}{T}(X,t) = \underset{\sim}{\mathcal{F}}\,(\underset{\sim}{\chi}^t;X,t) = \overset{\infty}{\underset{s=0}{\mathcal{G}}}_{\kappa}\,(\underset{\sim}{F}^t(\underset{\sim}{X},s);\underset{\sim}{X}). \qquad (4.8)$$

That is, the stress at the place $\underset{\sim}{x}$ occupied by the particle X at time t is determined by the totality of deformation gradients with respect to $\underset{\sim}{\kappa}$ experienced by that particle up to the present time. $\underset{\sim}{\mathcal{G}}_{\kappa}$ is called a *response functional* of the simple material. Ordinarily we shall write (4.8) in one of the simpler forms

$$\mathbf{T} = \mathop{\mathcal{G}_{\kappa}}_{s=0}^{\infty}(\mathbf{F}^t(s)) = \mathcal{G}_{\kappa}(\mathbf{F}^t) = \mathcal{G}(\mathbf{F}^t), \tag{4.9}$$

with exactly the same meaning.

The defining equation (4.9) asserts that at a given particle, $\mathcal{F}$ depends on the motion χ^t only through its deformation gradient $\mathbf{F}^t$ with respect to some fixed reference configuration. That is, all motions having the same gradient history at a given particle and time give rise to the same stress at that particle and time. Since, trivially, a homogeneous motion with any desired gradient history may be constructed, only homogeneous motions need be considered in determining any material properties that the original constitutive functional $\mathcal{F}$ may describe.

We remark first that it is unnecessary to mention a particular reference configuration in the above definition. In Lecture 2 we obtained the important formula

$$\mathbf{F}_1 = \mathbf{F}_2\mathbf{P} \tag{2.16}$$

connecting the gradients with respect to two reference configurations κ_1 and κ_2. By (4.9), then,

$$\mathbf{T} = \mathcal{G}_{\kappa_1}(\mathbf{F}_1^t) = \mathcal{G}_{\kappa_1}(\mathbf{F}_2^t\mathbf{P}). \tag{4.10}$$

Since $\mathbf{P}$ is the gradient of the transformation from κ_1 to κ_2, it is constant in time, and the right-hand side of (4.10) gives $\mathbf{T}$ as a functional of the history $\mathbf{F}_2^t$ of the deformation gradient with respect to κ_2. Thus if we write κ in (4.9) as κ_1, and if we set

$$\mathcal{G}_{\kappa_2}(\mathbf{F}^t) = \mathcal{G}_{\kappa_1}(\mathbf{F}^t\mathbf{P}), \tag{4.11}$$

we see that a relation of the form (4.9) holds again if we take $\underset{\sim}{\kappa}_2$ as reference. Therefore, we may speak of a simple material without mentioning any particular reference configuration, and usually we do not write the subscript $\underset{\sim}{\kappa}$ on $\underset{\sim}{\mathcal{G}}$. We must recall, however, that for a given simple material with constitutive functional $\underset{\sim}{\mathcal{F}}$ there are infinitely many different response functionals $\underset{\sim}{\mathcal{G}}_{\underset{\sim}{\kappa}}$, one for each choice of reference configuration $\underset{\sim}{\kappa}$.

The theory of simple materials includes all the common theories of continua studies in works on engineering, physics, applied mathematics, *etc. E.g.*, the *elastic material* is defined by the special case when the functional $\underset{\sim}{\mathcal{G}}$ reduces to a function $\underset{\sim}{\mathfrak{g}}$ of the present deformation gradient $\underset{\sim}{F}(X,t)$:

$$\underset{\sim}{T} = \underset{\sim}{\mathfrak{g}}(\underset{\sim}{F},\underset{\sim}{X}), \tag{4.12}$$

where $\underset{\sim}{\mathfrak{g}}$ is a function. The *linearly viscous material* is defined by the slightly more general case when $\underset{\sim}{\mathcal{G}}$ reduces to a function of $\underset{\sim}{F}(t)$ and $\dot{\underset{\sim}{F}}(t)$ which is linear in $\dot{\underset{\sim}{F}}$:

$$\underset{\sim}{T} = K(\underset{\sim}{F},\underset{\sim}{X})[\dot{\underset{\sim}{F}}] = L(\underset{\sim}{F},\underset{\sim}{X})[\underset{\sim}{G}], \tag{4.13}$$

where the second form follows from the first by the easy identity $\underset{\sim}{G} = \dot{\underset{\sim}{F}}\underset{\sim}{F}^{-1}$. The Boltzmann accumulative theory is obtained by supposing that $\underset{\sim}{\mathcal{G}}$ is expressible as an integral. It is customary to restrict these theories still further by assuming that $|\underset{\sim}{F} - \underset{\sim}{1}|$ is small or imposing requirements of material symmetry, or both, as we shall see later. Many but not all the recent nonlinear theories are included as special cases in the theory of simple materials. The simple material represents, in general, a material with long-range memory, so that stress relaxation, creep, and fatigue can occur.

Exercise 4.1 (Zaremba-Noll Theorem)

In the special case when (4.13) reduces to

$$\underset{\sim}{T} = K(\det \underset{\sim}{F},\underset{\sim}{X})[\underset{\sim}{G}], \tag{4.14}$$

prove that the principle of frame-indifference is satisfied if and only if in fact

$$\underset{\sim}{T} = -p\underset{\sim}{1} + \lambda(\operatorname{tr}\underset{\sim}{D})\underset{\sim}{1} + 2\mu\underset{\sim}{D}, \tag{4.15}$$

where p, λ, and μ are scalar functions of $\rho/\rho_{\underset{\sim}{\kappa}}$ and $\underset{\sim}{X}$.

If we were to assert that the theory of the simple material applies to the tube of sponge rubber on which I demonstrated a typical deformation in the first lecture, we could describe its mechanical nature as follows. To determine all the mechanical properties of this specimen, it would suffice to cut out a block and subject it to all possible time-dependent pure extensions along axes perpendicular to its faces. From the information so gained, we could predict the response of the tube to every other deformation; for example, we could, in principle, predict the whole sequence of stresses necessary to effect the eversion I demonstrated. Alternatively, we could start from the everted shape and subject it to all possible time-dependent stretches along and perpendicular to its axes. The stress we should measure would be different, in response to the same deformation history, from those we measured in the first place. Nevertheless, measurement of the totality of these stresses, too, would suffice to give us information sufficient, in principle, to determine the response of the material to any deformation history whatever.

In these remarks I have limited attention to deformation histories in which there is no rotation at all. Neglect of rotations is justified by our axioms, as we shall now see.

The definition of the single material satisfies trivially the principles of determinism and local action, but unless the functional $\underset{\sim}{\mathcal{G}}$ is suitably restricted, it does not generally satisfy the principle of frame-indifference. This fact is illustrated by Exercise 4.1, and now we shall consider it in general.

Reduction for Material Frame-Indifference

Under a change of frame, $\underset{\sim}{F}^{t} = \underset{\sim}{Q}\underset{\sim}{F}$, as we have proved in Lecture 2, let $\underset{\sim}{Q}^{t}(s)$ be the history up to time t of any orthogonal tensor function $\underset{\sim}{Q}(t)$. Then the principle of material frame-indifference states that the response functional $\underset{\sim}{\mathcal{G}}$ of a simple material must satisfy the equation

$$\underset{s=0}{\overset{\infty}{\underset{\sim}{\mathcal{G}}}} (\underset{\sim}{Q}^{t}(s)\underset{\sim}{F}^{t}(s)) = \underset{\sim}{Q}(t) \underset{s=0}{\overset{\infty}{\underset{\sim}{\mathcal{G}}}} (\underset{\sim}{F}^{t}(s))\underset{\sim}{Q}(t)^{T}, \qquad (4.16)$$

identically in the orthogonal tensor history $\underset{\sim}{Q}^{t}(s)$ and the nonsingular tensor history $\underset{\sim}{F}^{t}(s)$.

We can solve this equation. According to the polar decomposition theorem, $\underset{\sim}{F}^{t}(s) = \underset{\sim}{R}^{t}(s)\ \underset{\sim}{U}^{t}(s)$, so

$$\underset{s=0}{\overset{\infty}{\underset{\sim}{\mathcal{G}}}} (\underset{\sim}{F}^{t}(s)) = \underset{\sim}{Q}(t)^{T} \underset{s=0}{\overset{\infty}{\underset{\sim}{\mathcal{G}}}} (\underset{\sim}{Q}^{t}(s)\underset{\sim}{R}^{t}(s)\underset{\sim}{U}^{t}(s))\underset{\sim}{Q}(t). \qquad (4.17)$$

Since this equation must hold for all $\underset{\sim}{Q}^{t}$, $\underset{\sim}{R}^{t}$, $\underset{\sim}{U}^{t}$, it must hold in particular if $\underset{\sim}{Q}^{t} = (\underset{\sim}{R}^{t})^{T}$. Therefore

$$\mathcal{G}(\underset{\sim}{F}^t) = \underset{\sim}{R}(t)\ \mathcal{G}(\underset{\sim}{U}^t)\underset{\sim}{R}(t)^T. \tag{4.18}$$

Conversely, suppose $\mathcal{G}$ is of this form, and consider an arbitrary orthogonal tensor history $\underset{\sim}{Q}^t$. Since the polar decomposition of $\underset{\sim}{Q}^t\underset{\sim}{F}^t$ is $(\underset{\sim}{Q}^t\underset{\sim}{R}^t)\underset{\sim}{U}^t$,

$$\begin{aligned} \mathcal{G}(\underset{\sim}{Q}^t\underset{\sim}{F}^t) &= \underset{\sim}{Q}(t)\underset{\sim}{R}(t)\ \mathcal{G}(\underset{\sim}{U}^t)(\underset{\sim}{Q}(t)\underset{\sim}{R}(t))^T, \\ &= \underset{\sim}{Q}(t)\,\mathcal{G}(\underset{\sim}{F}^t)\underset{\sim}{Q}(t)^T, \end{aligned} \tag{4.19}$$

so that (4.16) is satisfied. Therefore, (4.18) gives the general solution of the functional equation (4.16). We have proved then, that the *constitutive equation of a simple material may be put into the form*

$$\underset{\sim}{T} = \underset{\sim}{R}\,\mathcal{G}(\underset{\sim}{U}^t)\underset{\sim}{R}^T, \tag{4.20}$$

and, conversely, that any functional $\mathcal{G}$ of positive-definite symmetric tensor histories, if its values are symmetric tensors, *serves to define a simple material through* (4.20). A constitutive equation of this kind, in which the functionals or functions occuring are not subject to any further restriction, is called a *reduced form*.

The result (4.20) shows us that while the stretch history $\underset{\sim}{U}^t$ of a simple material may influence its present stress in any way whatever, past rotations have no influence at all. The present rotation $\underset{\sim}{R}$ enters (4.20) explicitly. Thus, the reduced form enables us to dispense with considering rotation in determing the response of a motion.

The reduced form enables us also, in principle, to reduce the number of tests needed to determine the response function $\mathcal{G}$ by observation. Indeed, consider pure stretch histories: $R^t = 1$. If we know the stress T corresponding to an arbitrary homogeneous pure stretch history U^t, we have a relation of the form $T = \mathcal{G}(U^t)$. By (4.20) we then know T for all deformation histories. Thus, we may characterize simple materials in a more economical way: *A material is simple if and only if its response to all deformations is determined by its response to all homogeneous pure stretch histories.*

There are infinitely many other reduced forms for the constitutive equation of a simple material.

Exercise 4.2

Using (2.21), derive *Noll's reduced form* for the constitutive equation of a simple material:

$$\bar{T} = \mathfrak{K}(\bar{C}_t^t; C(t)), \tag{4.21}$$

where for any tensor K the tensor $\bar{K}$ is defined by

$$\bar{K} \equiv R(t)^T K R(t). \tag{4.22}$$

This last result, (4.21), shows that it is not possible to express the effect of the deformation history on the stress entirely by measuring deformation with respect to the present configuration. While the effect of all the *past* history, $0<s<\infty$, is accounted for in this way, a fixed reference configuration is required, in general, to allow for the effect of the deformation at the present instant, as shown by the appearance of $C(t)$ as a parameter in (4.21). The result itself is important in that it enables us to go as far as possible toward avoiding use of a fixed reference configuration. Roughly, it shows that memory effects can be accounted for entirely by use of the relative deformation, but finite-strain effects require use of some fixed reference configuration, any one we please.

Internal Constraints

So far, we have been assuming that the material is capable, if subjected to appropriate forces, of undergoing any smooth motion. If the class of possible motions is limited at interior points of $\mathcal{B}$, the material is said to be subject to an *internal constraint*. For internally constrained materials, the principle of determinism has to be modified, since certain motions cannot be produced at all; so the forces which we should expect to produce those motions in unconstrained materials cannot be determined by the motion of a constrained one. For example, a uniform pressure generally effects a change of volume. If a material is *incompressible*, such a pressure cannot be affected by its motion. That is, for incompressible materials

$$\underset{\sim}{T} = -p\underset{\sim}{1} + \underset{\sim}{\mathcal{G}}(\underset{\sim}{F}^t), \qquad (4.23)$$

where p is not determined by $\underset{\sim}{F}^t$, and where $\underset{\sim}{\mathcal{G}}$ need be defined only for $\underset{\sim}{F}^t$ such that $|\det \underset{\sim}{F}^t| = 1$. The reasoning I have just given is merely heuristic, but the result is correct and can be derived rigorously by a general method that time does not permit me to explain here, although the following exercise illustrates it.

Exercise 4.3

Prove that a stress tensor of the form $-p\underset{\sim}{1}$ does no work in any isochoric deformation, and that, conversely, every stress whose stress work is zero in every instantaneously isochoric deformation is of the form $-p\underset{\sim}{1}$.

Incompressible materials have proved to be of central importance in recent work. For example, even in the enormous deformations of rubber, illustrated in the first lecture, the change of volume is usually small, and results from the modern theory of incompressible elastic materials are rather well verified in experiments on rubbers with strains up to 200% over a broad range of circumstances.

5. The Isotropy Group

Isotropy Group

In the older literature a material is said to be "isotropic" if it is "unaffected by rotations". This means that if we first rotate a specimen of material and then do an experiment upon it, the outcome is the same as if the specimen had not been rotated. In other words, within the class of effects considered by the theory, rotations cannot be detected by any experiment. The response of the material with respect to the reference configuration $\underset{\sim}{\kappa}$ is the same as that with respect to any other obtained from it by rotation.

Noll's concept of *isotropy group* (1958) generalizes this old idea. Rather than lay down a statement of material symmetry, we consider the material defined by the response functional $\underset{\sim}{\mathcal{G}}_{\underset{\sim}{\kappa}}$ and ask what symmetries that material may have. Of course, those symmetries will depend on $\underset{\sim}{\kappa}$. In Lecture 4 we have derived an equation for the change of response functional under change of reference configuration:

$$\underset{\sim}{\mathcal{G}}_{\underset{\sim}{\kappa}_2}(\underset{\sim}{F}^t) = \underset{\sim}{\mathcal{G}}_{\underset{\sim}{\kappa}_1}(\underset{\sim}{F}^t\underset{\sim}{P}), \tag{4.11}$$

where $\underset{\sim}{P} = \nabla\underset{\sim}{\lambda}$, and $\underset{\sim}{\lambda}$ is the mapping from $\underset{\sim}{\kappa}_1$ to $\underset{\sim}{\kappa}_2$. This formula asserts that $\underset{\sim}{\mathcal{G}}_{\underset{\sim}{\kappa}_2}$ and $\underset{\sim}{\mathcal{G}}_{\underset{\sim}{\kappa}_1}$, if they are related in this way, describe the response of the *same material* to the deformation history $\underset{\sim}{F}^t$, in the one case when $\underset{\sim}{\kappa}_1$ is used as the reference configuration, in the other case when $\underset{\sim}{\kappa}_2$ is. We may consider (4.11) analogous to a transformation in analytic geometry, where the same figure is described by different equations in different co-ordinate systems.

We now ask, under what conditions does $\underset{\sim}{P}$ lead from a given $\underset{\sim}{\kappa}_1$ to a reference configuration $\underset{\sim}{\kappa}_2$ which is indistinguishable from it by experiments relating the stress to the deformation history $\underset{\sim}{F}^t$? Here I am using the term "experiment" in the ideal sense common in physics. Namely, a certain theory is assumed to describe some physical situation perfectly, and then in imagination one considers the various things that theory permits. By an "experiment" in the present instance I mean a measurement of the stress tensor throughout the body at time t after the body has been made to undergo a particular, known deformation with gradient history $\underset{\sim}{F}^t$ from the infinite past up to the present. First, simple experience in Eulerian hydrodynamics tells us that a change of density leads to different response, so we shall consider only those configurations with the same density. By (2.7) and (2.8), then, we restrict attention to unimodular tensors $\underset{\sim}{H}$: tensors such that det $\underset{\sim}{H} = \pm 1$. Suppose that $\underset{\sim}{H}$ is a unimodular tensor that satisfies the equation

$$\underset{\sim}{\mathcal{G}}_{\underset{\sim}{\kappa}_1}(\underset{\sim}{F}^t\underset{\sim}{H}) = \underset{\sim}{\mathcal{G}}_{\underset{\sim}{\kappa}_1}(\underset{\sim}{F}^t) \tag{5.1}$$

for every non-singular $\underset{\sim}{F}^t$. Note that $\underset{\sim}{\kappa}_1$ appears on both sides here. By (4.11), for all $\underset{\sim}{F}^t$,

$$\underset{\sim}{T} = \underset{\sim}{\mathcal{G}}_{\underset{\sim}{\kappa}_1}(\underset{\sim}{F}^t) = \underset{\sim}{\mathcal{G}}_{\underset{\sim}{\kappa}_2}(\underset{\sim}{F}^t), \tag{5.2}$$

where $\underset{\sim}{\kappa}_2$ is the reference configuration obtained from $\underset{\sim}{\kappa}_1$ by the transformation $\underset{\sim}{\lambda}$ such that $\underset{\sim}{H} = \nabla\underset{\sim}{\lambda}$. Thus $\underset{\sim}{H}$ is the gradient of a deformation that maps $\underset{\sim}{\kappa}_1$ onto another configuration $\underset{\sim}{\kappa}_2$ from which any deformation history $\underset{\sim}{F}^t$ leads to just the same stress as it does from $\underset{\sim}{\kappa}_1$. In other words, $\underset{\sim}{H}$ corresponds to a density-preserving transformation which cannot be detected by any experiment.

Suppose both $\underset{\sim}{H}_1$ and $\underset{\sim}{H}_2$ satisfy (5.1). Then by two applications of (5.1)

$$\underset{\sim}{\mathcal{G}}_{\underset{\sim}{\kappa}_1}(\underset{\sim}{F}^t\underset{\sim}{H}_1\underset{\sim}{H}_2) = \underset{\sim}{\mathcal{G}}_{\underset{\sim}{\kappa}_1}(\underset{\sim}{F}^t\underset{\sim}{H}_1) = \underset{\sim}{\mathcal{G}}_{\underset{\sim}{\kappa}_1}(\underset{\sim}{F}^t). \tag{5.3}$$

Hence $\underset{\sim}{H}_1\underset{\sim}{H}_2$ is a solution. If $\underset{\sim}{H}$ is a particular unimodular tensor, $\underset{\sim}{F}^t\underset{\sim}{H}^{-1}$ is a non-singular deformation history for every non-singular $\underset{\sim}{F}^t$, and any non-singular deformation history may be written in the form $\underset{\sim}{F}^t\underset{\sim}{H}^{-1}$. Replacing $\underset{\sim}{F}^t$ by $\underset{\sim}{F}^t\underset{\sim}{H}^{-1}$ in (5.1) yields

$$\underset{\sim}{\mathcal{G}}_{\underset{\sim}{\kappa}_1}(\underset{\sim}{F}^t\underset{\sim}{H}^{-1}\underset{\sim}{H}) = \underset{\sim}{\mathcal{G}}_{\underset{\sim}{\kappa}_1}(\underset{\sim}{F}^t) = \underset{\sim}{\mathcal{G}}_{\underset{\sim}{\kappa}_1}(\underset{\sim}{F}^t\underset{\sim}{H}^{-1}), \tag{5.4}$$

for every non-singular history $\underset{\sim}{F}^t$. That is, if $\underset{\sim}{H}$ is a solution, so is $\underset{\sim}{H}^{-1}$. Hence the totality of all solutions $\underset{\sim}{H}$ forms a group, the *isotropy group* $\mathcal{g}_{\underset{\sim}{\kappa}}$ of the material at the particle whose place in the reference configuration $\underset{\sim}{\kappa}$ is X. From its definition, $\mathcal{g}_{\underset{\sim}{\kappa}}$ is a subgroup of the unimodular group u:

$$\mathcal{g}_{\underset{\sim}{\kappa}} \subset u. \tag{5.5}$$

The isotropy group is the collection of all static density preserving deformations from $\underset{\sim}{\kappa}$ at $\underset{\sim}{X}$ that cannot be detected by experiment. Alternatively we may describe the isotropy group as the group of material symmetries. We have proved that every material has a non-empty isotropy group for every $\underset{\sim}{\kappa}$ and $\underset{\sim}{X}$.

Orthogonal Part of the Isotropy Group

Suppose an orthogonal tensor $\underset{\sim}{Q} \in \underset{\sim}{\mathcal{g}_\kappa}$. Since $\underset{\sim}{Q}^{-1} = \underset{\sim}{Q}^T$ and since $\underset{\sim}{\mathcal{g}_\kappa}$ is a group, $\underset{\sim}{Q}^T \in \underset{\sim}{\mathcal{g}_\kappa}$. Therefore, if $\underset{\sim}{Q} \in \underset{\sim}{\mathcal{g}_\kappa}$, we may set $\underset{\sim}{H} = \underset{\sim}{Q}^T$ in (5.1). Moreover, since $\underset{\sim}{Q}$ is a given orthogonal tensor, any non-singular deformation history may be written as $\underset{\sim}{Q}\underset{\sim}{F}^t$ for an appropriate choice of $\underset{\sim}{F}^t$. Thus we may, if we like, replace $\underset{\sim}{F}^t$ by $\underset{\sim}{Q}\underset{\sim}{F}^t$ in (5.1). The resulting equation is

$$\underset{\sim}{\mathcal{G}_\kappa}(\underset{\sim}{Q}\underset{\sim}{F}^t) = \underset{\sim}{\mathcal{G}_\kappa}(\underset{\sim}{Q}\underset{\sim}{F}^t\underset{\sim}{Q}^T), \qquad (5.6)$$

which must be satisfied for all non-singular $\underset{\sim}{F}^t$ if $\underset{\sim}{Q} \in \underset{\sim}{\mathcal{g}_\kappa}$. Recall (4.16), which expresses the principle of material frame-indifference:

$$\underset{s=0}{\overset{\infty}{\underset{\sim}{\mathcal{G}_\kappa}}}(\underset{\sim}{Q}^t(s)\underset{\sim}{F}^t(s)) = \underset{\sim}{Q}(t) \underset{s=0}{\overset{\infty}{\underset{\sim}{\mathcal{G}_\kappa}}}(\underset{\sim}{F}^t(s))\underset{\sim}{Q}(t)^T, \qquad (4.16)$$

for all non-singular histories $\underset{\sim}{F}^t$ and all orthogonal histories $\underset{\sim}{Q}^t$. In particular, (4.16) must hold when we choose for $\underset{\sim}{Q}^t$ the history that has remained always at a particular value $\underset{\sim}{Q}$. Thus every $\underset{\sim}{\mathcal{G}_\kappa}$ satisfies the identity

$$\underset{\sim}{\mathcal{G}_\kappa}(\underset{\sim}{Q}\underset{\sim}{F}^t) = \underset{\sim}{Q}\, \underset{\sim}{\mathcal{G}_\kappa}(\underset{\sim}{F}^t)\underset{\sim}{Q}^T. \qquad (5.7)$$

If for a particular $\underset{\sim}{\mathcal{G}_\kappa}$ a particular $\underset{\sim}{Q} \in \underset{\sim}{\mathcal{g}_\kappa}$ we may use *both* (5.6) and (5.7) and obtain

$$\underset{\sim}{\mathcal{G}_\kappa}(\underset{\sim}{Q}\underset{\sim}{F}^t\underset{\sim}{Q}^T) = \underset{\sim}{Q}\, \underset{\sim}{\mathcal{G}_\kappa}(\underset{\sim}{F}^t)\underset{\sim}{Q}^T, \qquad (5.8)$$

for this particular $\underset{\sim}{Q}$ and for all non-singular histories $\underset{\sim}{F}^t$.

Exercise 5.1

Prove, conversely, that if $\underset{\sim}{Q}$ satisfies (5.8), then $\underset{\sim}{Q} \in \underset{\sim}{g_\kappa}$.

Therefore, (5.8) is the equation that determines all the orthogonal members of $\underset{\sim}{g_\kappa}$.

Clearly the central inversion, $-\underset{\sim}{1}$, satisfies (5.8). Hence $-\underset{\sim}{1} \in \underset{\sim}{g_\kappa}$ always. The central inversion cannot be visualized as a deformation of the material but corresponds to a change from a right-handed to a left-handed frame of reference, or *vice-versa*. We have shown, then, that no experiment can distinguish the difference between a right-handed and a left-handed frame of reference. (In a more general theory, involving *e.g.* heat conduction or electromagnetism, an isotropy group may be defined, but $-\underset{\sim}{1}$ generally will fail to be a member of it.)

Since $\underset{\sim}{1}$ and $-\underset{\sim}{1}$ form a group by themselves, the group $\{\underset{\sim}{1}, -\underset{\sim}{1}\}$ is the smallest possible isotropy group:

$$\{\underset{\sim}{1}, -\underset{\sim}{1}\} \subset \underset{\sim}{g_\kappa} \subset u. \tag{5.9}$$

Exercise 5.2

Construct special functionals having $\{\underset{\sim}{1}, -\underset{\sim}{1}\}$ and u as their isotropy groups. (Hint: refer to the classical theories of elastic crystals and perfect fluids.)

Change of Reference Configuration

The isotropy group $\underset{\sim}{g_\kappa}$ depends upon the choice of the reference configuration $\underset{\sim}{\kappa}$. This fact reflects the common experience that material symmetry may be changed by deformation. Suppose, for example, that we take a block having the directions of symmetry normal to its faces and then bend and twist it severely. Its new directions of symmetry will clearly be quite different. However, they can be determined from the old ones, as we shall now see.

First, we have the relation (4.11) connecting the response functionals $\mathfrak{G}_{\kappa_1}$ and $\mathfrak{G}_{\kappa_2}$ for the two reference configurations κ_1 and κ_2:

$$\mathfrak{G}_{\kappa_2}(\mathbf{F}^t) = \mathfrak{G}_{\kappa_1}(\mathbf{F}^t\mathbf{P}), \tag{4.11}$$

where $\mathbf{P} = \nabla\lambda$ and λ maps κ_1 onto κ_2. Equivalently, since $\mathbf{P}$ is non-singular,

$$\mathfrak{G}_{\kappa_2}(\mathbf{F}^t\mathbf{P}^{-1}) = \mathfrak{G}_{\kappa_1}(\mathbf{F}^t). \tag{5.10}$$

Second, $\mathbf{H} \in \mathcal{g}_{\kappa_1}$ if and only if

$$\mathfrak{G}_{\kappa_1}(\mathbf{F}^t\mathbf{H}) = \mathfrak{G}_{\kappa_1}(\mathbf{F}^t). \tag{5.1}$$

By (5.10) this means

$$\mathfrak{G}_{\kappa_2}(\mathbf{F}^t\mathbf{H}\mathbf{P}^{-1}) = \mathfrak{G}_{\kappa_2}(\mathbf{F}^t\mathbf{P}^{-1}). \tag{5.11}$$

Equivalently,

$$\mathfrak{G}_{\kappa_2}(\mathbf{F}^t\mathbf{P}\mathbf{H}\mathbf{P}^{-1}) = \mathfrak{G}_{\kappa_2}(\mathbf{F}^t). \tag{5.12}$$

By (5.1), we have proved that if $\underset{\sim}{H} \in \mathcal{g}_{\underset{\sim}{\kappa}_1}$, then $\underset{\sim}{P}\underset{\sim}{H}\underset{\sim}{P}^{-1} \in \mathcal{g}_{\underset{\sim}{\kappa}_2}$. Since the labels 1 and 2 are attached arbitrarily, and since $\nabla\underset{\sim}{\lambda}^{-1} = \underset{\sim}{P}^{-1}$, the converse follows. Thus we obtain *NOLL's rule:*

$$\underset{\sim}{H} \in \mathcal{g}_{\underset{\sim}{\kappa}_1} \rightleftarrows \underset{\sim}{P}\underset{\sim}{H}\underset{\sim}{P}^{-1} \in \mathcal{g}_{\underset{\sim}{\kappa}_2}, \tag{5.13}$$

or, symbolically

$$\mathcal{g}_{\underset{\sim}{\kappa}_2} = \underset{\sim}{P}\, \mathcal{g}_{\underset{\sim}{\kappa}_1} \underset{\sim}{P}^{-1}, \tag{5.14}$$

where $\underset{\sim}{P} = \nabla\underset{\sim}{\lambda}$ and λ maps $\underset{\sim}{\kappa}_1$ onto $\underset{\sim}{\kappa}_2$. Noll's rule shows that if $\mathcal{g}_{\underset{\sim}{\kappa}}$ is known for one configuration $\underset{\sim}{\kappa}$, it is known for every other and may be calculated explicitly.

Exercise 5.3

Prove that dilatations and inversions leave the isotropy group invariant. Prove that if $\{\underset{\sim}{1},-\underset{\sim}{1}\}$ or u is the isotropy group for one configuration, it is the isotropy group for every configuration, and interpret these two facts.

6. Solids, Isotropic Materials, Fluids, Fluid Crystals

Main Properties of the Isotropy Group

The isotropy group g_κ, as we have seen, is the group of all static deformations from a given reference configuration κ that cannot be detected by any experiment on the material whose response functional is $\underset{\sim}{\mathcal{G}}_\kappa$. Every material has an isotropy group g_κ in every configuration, and

$$\{1, -1\} \subset g_\kappa \subset u. \tag{5.9}$$

If $\underset{\sim}{P} = \nabla \underset{\sim}{\lambda}$ and $\underset{\sim}{\lambda}$ maps κ_1 onto κ_2, then

$$g_{\kappa_2} = \underset{\sim}{P} \, g_{\kappa_1} \underset{\sim}{P}^{-1}. \tag{5.14}$$

With these simple facts in hand, we may construct an invariant classification of materials.

Isotropic Materials

The concept of isotropy is one of Cauchy's finest creations. In the present framework, it corresponds precisely to the following definition: *A material is isotropic if there exists a reference configuration κ such that*

$$g_\kappa \subset \mathcal{O}, \tag{6.1}$$

where $\mathcal{O}$ is the orthogonal group. Such a reference configuration is called an *undistorted state* of the isotropic material. According to the definition, no orthogonal deformation from an undistorted state is discernible by measurement of stress. By Noll's rule (5.14), any orthogonal transformation carries one undistorted state into another.

A particularly simple reduced form for the constitutive equation of an isotropic material holds if an undistorted state is taken as reference. By Noll's result (4.21), for any simple material

$$
\begin{aligned}
\mathbf{T} &= \mathfrak{G}(\mathbf{F}^t) = \mathbf{R}\mathfrak{K}(\mathbf{R}^T\mathbf{C}_t^t\mathbf{R};\mathbf{C})\mathbf{R}^T, \\
&= \mathbf{R}\mathfrak{K}(\mathbf{R}^T(\mathbf{F}_t^t)^T\mathbf{F}_t^t\mathbf{R};\mathbf{F}^T\mathbf{F})\mathbf{R}^T.
\end{aligned}
\tag{6.2}
$$

But for an isotropic material in an undistorted state, (5.1), instead of being an equation to solve for certain $\mathbf{Q}$, becomes an identity satisfied by all $\mathbf{Q}$:

$$\mathfrak{G}(\mathbf{F}^t) = \mathbf{Q}^T\mathfrak{G}(\mathbf{Q}\mathbf{F}^t\mathbf{Q}^T)\mathbf{Q}. \tag{6.3}$$

Hence (6.2) yields

$$\mathbf{T} = \mathbf{Q}^T\mathbf{R}\mathfrak{K}(\mathbf{R}^T\mathbf{Q}(\mathbf{F}_t^t)^T\mathbf{F}_t^t\mathbf{Q}^T\mathbf{R};\mathbf{Q}\mathbf{F}^T\mathbf{F}\mathbf{Q}^T)\mathbf{R}^T\mathbf{Q}, \tag{6.4}$$

for all $\mathbf{Q}$. In particular, we can choose $\mathbf{Q} = \mathbf{R}$ and obtain

$$\mathbf{T} = \mathfrak{K}(\mathbf{C}_t^t;\mathbf{B}) \tag{6.5}$$

since $\mathbf{R}\mathbf{F}^T\mathbf{F}\mathbf{R}^T = \mathbf{B}$, by (2.23) and (2.24). This is not all, since for an isotropic material in an undistorted state (5.8) also becomes an identity in $\mathbf{Q}$.

Exercise 6.1

Prove that (5.8) implies the identity

$$\mathfrak{K}(Q C_t^t Q^T; Q B Q^T) = Q \mathfrak{K}(C_t^t; B) Q^T. \tag{6.6}$$

(Notice that here $Q = Q(t)$, a function of the present time only, not a history.) Prove that, conversely, if $\mathfrak{K}$ satisfies (6.6), then (6.5) defines an isotropic material referred to an undistorted state.

These results constitute *Noll's theorem on isotropic materials*. This theorem connects the older, only partly formalized ideas concerning isotropy with precise mathematical concepts. A functional satisfying (6.6) is called *isotropic*. According to (6.5), not only past, but also present rotation has no effect in determining the stress, providing we use C_t^t to measure the deformation history and B to measure the present deformation. As the concept of isotropy suggests, rotation can be altogether eliminated.

These results require that we use an undistorted state as reference. If we use some other reference configuration, as of course we may, we shall not have the simple form (6.5) and the identity (6.6) for the constitutive equation. This simple and concrete remark explains the common vague claim that a deformation causes an isotropic material to "lose" its isotropy. Isotropy of a material, like the symmetry of a figure with respect to an axis, is an intrinsic property, which can neither be gained nor be lost. An unfortunate choice of reference, like an unfortunate choice of co-ordinate system, may obscure this property, but it does not change it.

While (6.1) embodies a natural concept of isotropy, it seems more general than in fact it is. According to a theorem of group theory, the orthogonal group is maximal in the unimodular group. That is, if $\mathcal{g}$ is a group such that $\mathcal{O} \subset \mathcal{g} \subset u$ then

$$\text{either} \quad \mathcal{g} = \mathcal{O} \quad \text{or} \quad \mathcal{g} = u. \tag{6.7}$$

Thus the isotropy group of an isotropic material in an undistorted state is either the orthogonal group or the unimodular group.

Solids

A material is thought of as "solid" if it has some "preferred configuration", a change of shape from which changes some of its apparent properties. A change of shape is non-orthogonal deformation. Thus a solid has some configuration from which any non-orthogonal deformation is detectable by some experiment measuring the stress arising in response to subsequent deformation histories. Formally, then, *a simple material is a solid if there exists a reference configuration* $\underset{\sim}{\kappa}$ *such that*

$$\mathcal{g}_{\underset{\sim}{\kappa}} \subset \mathcal{O}. \tag{6.8}$$

Such a $\underset{\sim}{\kappa}$ is called an *undistorted state*. According to this definition, no non-orthogonal transformation belongs to the isotropy group $\mathcal{g}_{\underset{\sim}{\kappa}}$ when $\underset{\sim}{\kappa}$ is an undistorted state.

The material for which $\mathcal{g}_{\underset{\sim}{\kappa}} = \{\underset{\sim}{1}, -\underset{\sim}{1}\}$ is a solid. Indeed, it is called *triclinic*, and it furnishes an example of a *crystalline solid* in the classical sense. All the classical crystallographic groups, provided they be extended so as to include $-\underset{\sim}{1}$, correspond to solids. So also do the groups defining "transversely isotropic" and "orthotropic" materials, and many others.

For solids, no particularly simple form of the constitutive equation has been found.

For an isotropic solid, by hypothesis, there exist configurations $\underset{\sim}{\kappa}$ and $\bar{\underset{\sim}{\kappa}}$ such that

$$\mathcal{g}_{\underset{\sim}{\kappa}} \supset \mathcal{O} \quad , \quad \mathcal{g}_{\bar{\underset{\sim}{\kappa}}} \subset \mathcal{O}. \qquad (6.9)$$

(Here $\underset{\sim}{\kappa}$ is an undistorted state in the sense of "isotropic material", while $\bar{\underset{\sim}{\kappa}}$ is an undistorted state of "solid".) By (6.7), either $\mathcal{g}_{\underset{\sim}{\kappa}} = \mathcal{O}$ or $\mathcal{g}_{\underset{\sim}{\kappa}} = u$. If $\mathcal{g}_{\underset{\sim}{\kappa}} = u$, then by (6.11) $\mathcal{g}_{\bar{\underset{\sim}{\kappa}}} = u$, contradicting (6.9). Hence $\mathcal{g}_{\underset{\sim}{\kappa}} = \mathcal{O}$. In particular, $\underset{\sim}{\kappa}$ is also an undistorted state of the solid.

Exercise 6.2

Let $\underset{\sim}{\kappa}$ and $\underset{\sim}{\kappa}^*$ be two undistorted states of a solid, and let $\underset{\sim}{P} = \nabla\underset{\sim}{\lambda}$, where $\underset{\sim}{\lambda} : \underset{\sim}{\kappa} \to \underset{\sim}{\kappa}^*$. If the polar decomposition of $\underset{\sim}{P}$ is $\underset{\sim}{P} = \hat{\underset{\sim}{R}}\hat{\underset{\sim}{U}}$, prove that

$$\mathcal{g}_{\underset{\sim}{\kappa}^*} = \hat{\underset{\sim}{R}}\, \mathcal{g}_{\underset{\sim}{\kappa}} \hat{\underset{\sim}{R}}^{-1}. \qquad (6.10)$$

That is, $\mathcal{g}_{\underset{\sim}{\kappa}^*}$ is an orthogonal conjugate of $\mathcal{g}_{\underset{\sim}{\kappa}}$. Hence prove that $\mathcal{g}_{\bar{\underset{\sim}{\kappa}}} = \mathcal{O}$.

These results show that for an isotropic solid, both definitions of "undistorted state" yield the same implication, namely, *a material is an isotropic solid if and only if there exists a configuration* $\underset{\sim}{\kappa}$, *called an undistorted state, in which*

$$\mathcal{g}_{\kappa} = \mathcal{O};$$ (6.11)

any orthogonal deformation from an undistorted state carries the solid into another undistorted state, and all undistorted states are obtained by orthogonal deformations from any given one.

Fluids

There are various physical motions concerned with fluids. One is that a fluid is a substance which can flow. "Flow" is itself a vague term. One meaning of "flow" is simply deformation under stress, which does not distinguish a fluid from a solid. Another is that steady velocity results from constant stress, which seems to be special and inapplicable except to particular cases. Another is the inability to support shear stress when in equilibrium. Formally, within the theory of simple materials, such a definition would yield

$$\mathbf{T} = -p(\rho)\mathbf{1} + \mathfrak{F}(\mathbf{F}^t),$$ (6.12)

where $\mathfrak{F}(\mathbf{1}) = \mathbf{0}$. Since the material so defined may have any isotropy group whatever, including one of those already considered to define a solid, this definition does not lend itself to a criterion in terms of material symmetry. Finally, a fluid is regarded as a material having no preferred configurations. In terms of isotropy groups, $\mathcal{g}_{\kappa_1} = \mathcal{g}_{\kappa_2}$ for all κ_1 and κ_2.

Accordingly, we adopt the following definition: *A fluid is a non-solid material with no preferred configurations.* By Noll's rule (5.14), the isotropy group $\mathcal{g}$ of a fluid satisfies

$$\mathcal{g} = \mathbf{H}\,\mathcal{g}\,\mathbf{H}^{-1},$$ (6.13)

for all unimodular $\underset{\sim}{H}$. We have seen already that $\mathscr{g} = \{\underset{\sim}{1}, -\underset{\sim}{1}\}$ and $\mathscr{g} = u$ satisfy this equation. A theorem in group theory asserts that u is a "simple group", which means that there are no solutions of (6.13) beyond the two trivial ones just specified. The former solution corresponds to a solid. Hence the only group compatible with the definition is $\mathscr{g} = u$: *The isotropy group of a fluid, in every configuration, is the unimodular group.* As a corollary, *every fluid is isotropic,* and every configuration of a fluid is undistorted.

Since fluids are isotropic, we can apply (6.15). Since for a fluid $\underset{\sim}{T}$ cannot be changed by a static deformation from one configuration to another with the same density, the dependence upon $\underset{\sim}{B}(t)$ must reduce to dependence on det $\underset{\sim}{B}(t)$, or, what is the same thing, dependence on ρ:

$$\underset{\sim}{T} = \underset{\sim}{\mathfrak{K}}(\underset{\sim}{C}_t^t; \rho). \tag{6.14}$$

Thus the stress in a fluid, at a given density, is determined by the history of the relative deformation alone. Of course, $\underset{\sim}{\mathfrak{K}}$ must satisfy (6.6). For the particular case of the rest history, $\underset{\sim}{C}_t^t = \underset{\sim}{1}$, so that (6.6) yields

$$\underset{\sim}{T} = \underset{\sim}{\mathfrak{K}}(\underset{\sim}{1}; \rho) = \underset{\sim}{Q}\underset{\sim}{\mathfrak{K}}(\underset{\sim}{1}; \rho)\underset{\sim}{Q}^T = \underset{\sim}{Q}\underset{\sim}{T}\underset{\sim}{Q}^T. \tag{6.15}$$

That is, the stress $\underset{\sim}{T}$ in a fluid at rest commutes with every orthogonal tensor. The only tensor satisfying this requirement is $\underset{\sim}{T} = -p(\rho)\underset{\sim}{1}$. Therefore the stress in a fluid at rest is a hydrostatic pressure which depends on the density alone, and (6.14) may be put into the alternative form

$$\mathbf{T} = -p(\rho)\mathbf{1} + \mathfrak{F}(\mathbf{K}_t^t;\rho), \tag{6.16}$$

where

$$\mathbf{K}_t^t(s) \equiv \mathbf{C}_t^t(s) - \mathbf{1},$$
$$\mathfrak{F}(\mathbf{0};\rho) = \mathbf{0}. \tag{6.17}$$

This result constitutes *Noll's fundamental theorem* on fluids. It asserts, among other things, that a fluid is a substance which can flow in the sense expressed by (6.12).

A fluid may react to its entire deformation history, yet such reaction cannot be different in different configurations with the same density. A fluid reconciles these two seemingly contradictory qualities - ability to remember all its past and inability to regard one configuration as different from another - by reacting to the past only in so far as it differs from the everchanging present. This statement is the content of Noll's theorem.

Fluid Crystals

To exhaust the possible types of simple materials, we define any non-solid as being a *fluid crystal*. For a fluid crystal, then, there exists no reference configuration κ such that $g_\kappa \subset \mathcal{O}$. Of course, every isotropy group has orthogonal elements. For a fluid crystal, the isotropy group with respect to every configuration has some non-orthogonal elements. That is, from every configuration of a fluid crystal there is *some* undetectable change of shape. In this regard a fluid crystal is like a fluid, for which all changes of shape without change of density are undetectable. Since it is impossible that

$$\underset{\sim}{\mathcal{G}_{\kappa}} \supset \mathcal{O}$$

unless the fluid crystal is in fact a fluid, for an isotropic fluid crystal there are also some rotations that are detectable. Clearly *a fluid crystal is a fluid if and only if it is isotropic.* An anisotropic fluid crystal is a substance that can never be brought into any configuration such as to render its isotropy group comparable with the orthogonal group: For every $\underset{\sim}{\kappa}$,

$$\underset{\sim}{\mathcal{G}_{\kappa}} \not\supset \mathcal{O}\,, \quad \text{and} \quad \underset{\sim}{\mathcal{G}_{\kappa}} \not\subset \mathcal{O}. \tag{6.18}$$

7. Motions with Constant Stretch History

Introduction

After following the general developments just presented, the student may well ask, can anything be done with a theory so complicated? In fact, the historical order is the reverse. A number of concrete, explicit results were obtained in special ways, and the general theory was then created so as to unify them. This lecture and the next three will concern an area of theory which grew up around some striking phenomena occurring in polymer solutions. In the 1940's it was noticed that certain fluids swell on extrusion from a pipe, and that when agitated by a rotor they tend to climb it. These effects, seen with the naked eye, are certainly non-classical in the sense that they are not predicted by the Navier-Stokes equations. In fact, the climbing effect was not new, as witnessed by the fact that no-one stirs paint with an eggbeater. In *The Non-Linear Field Theories* you may see a few pictures showing typical "normal-stress effects" of "nonlinear viscosity", and better ones are to be found in the book by LODGE. A theory, which has proved very successful, was developed in stages. First, RIVLIN obtained solutions for all the classical viscometric problems in a simple and special nonlinear theory. Then he found they could be extended to much more general fluids. Finally COLEMAN and NOLL calculated the solutions in the full generality of the simple fluid. In these lectures I shall sketch the results in reverse order, beginning with a mathematical treatment which explains, in general terms, why such solutions are possible.

Definition

The simple material may have a long memory. Deformations undergone an arbitrarily long time ago may continue to affect and alter the present stress. With such complicated material response possible, the solution of particular problems may be extremely difficult, indeed unfeasible. There are particular motions, however, in which memory effects are given little chance to manifest themselves, because there is little to remember.

Consider, for example, the constitutive equation of a simple fluid in the form (6.5):

$$\mathbf{T} = \boldsymbol{\mathfrak{K}}(\mathbf{C}_t^t;\rho). \tag{6.5}$$

In the particular case when ρ = const. and $\mathbf{C}_t^t(s)$ is the same function of s for all t, the stress becomes constant in time for a given particle. The particle may have experienced deformation for all past time, but as it looks backward it sees the entire sequence of past deformations relative to its present configuration remain unchanged.

More generally, in view of the principle of frame-indifference, essentially the same simplification will result if there exists an orthogonal tensor $\mathbf{Q}(t)$ such that

$$\mathbf{C}_t^t(s) = \mathbf{Q}(t)\mathbf{C}_o^o(s)\mathbf{Q}(t)^T. \tag{7.1}$$

Here, of course, $\mathbf{C}_o^o$ denotes $\mathbf{C}_t^t$ when t = 0. Such motions were introduced and called *substantially stagnant* by COLEMAN. In them, an observer situated upon the moving particle may orient himself in such a way as to see behind him always the same deformation history relative to the present configuration. The proper numbers of $\mathbf{C}_t^t(s)$ are the same as those of $\underset{\sim}{C}_o^o(s)$, although the principal axes of the one tensor may rotate arbitrarily with respect to those of the other. Thus substantially stagnant motions may be defined alternatively as those having *constant stretch history*.

Noll's Theorem

All such motions are characterized by a *fundamental theorem* of *NOLL:* A motion has constant stretch history if and only if there exists an orthogonal tensor $\underset{\sim}{Q}(t)$, a scalar κ, and a constant tensor $\underset{\sim}{N}_o$ such that

$$\begin{aligned} \underset{\sim}{F}_o(\tau) &= \underset{\sim}{Q}(\tau)e^{\tau\kappa\underset{\sim}{N}_o}, \\ \underset{\sim}{Q}(0) &= \underset{\sim}{1}, \quad |\underset{\sim}{N}_o| = 1. \end{aligned} \tag{7.2}$$

To prove Noll's theorem, we begin by converting the hypothesis (7.1) into a difference equation.

Exercise 7.1

Set

$$\begin{aligned} \underset{\sim}{H}(s) &\equiv \underset{\sim}{C}_o(-s) = \underset{\sim}{Q}(t)^T\underset{\sim}{C}_t(t-s)\underset{\sim}{Q}(t), \\ \underset{\sim}{E}(t) &\equiv \underset{\sim}{Q}(t)^T\underset{\sim}{F}_o(t). \end{aligned} \tag{7.3}$$

Using the fact that $\underset{\sim}{F}_t(\tau) \equiv \underset{\sim}{F}_o(\tau)\underset{\sim}{F}_o(t)^{-1}$, show that

$$\underset{\sim}{H}(s-t) = \underset{\sim}{E}(t)^T\underset{\sim}{H}(s)\underset{\sim}{E}(t). \tag{7.4}$$

To obtain a necessary condition for a solution $\underset{\sim}{H}(s)$, we differentiate (7.4) with respect to t and put t = 0, obtaining the first-order linear differential equation

$$-\dot{\underset{\sim}{H}}(s) = \underset{\sim}{M}^T \underset{\sim}{H}(s) + \underset{\sim}{H}(s)\underset{\sim}{M}, \tag{7.5}$$

where $\underset{\sim}{M} \equiv \dot{\underset{\sim}{E}}(0)$ and where the dot denotes differentiation with respect to s. The unique solution such that $\underset{\sim}{H}(0) = \underset{\sim}{1}$ is easily verified to be

$$\underset{\sim}{H}(s) = e^{-s\underset{\sim}{M}^T} e^{-s\underset{\sim}{M}}. \tag{7.6}$$

Since histories are defined only when s ≧ 0, this result has been derived only for that range. Nevertheless, the difference equation (7.4) serves to define $\underset{\sim}{H}(s)$ for negative s as well and shows that $\underset{\sim}{H}(s)$ is analytic. By the principle of analytic continuation, since (7.6) is analytic, it is the unique solution for all s when $\underset{\sim}{E}(t)$ is given. If we substitute (7.6) back into (7.4), we obtain

$$\underset{\sim}{E}(t)e^{t\underset{\sim}{M}}[\underset{\sim}{E}(t)e^{t\underset{\sim}{M}}]^T = \underset{\sim}{1}. \tag{7.7}$$

Hence $\underset{\sim}{E}(t)e^{t\underset{\sim}{M}}$ is an orthogonal tensor, say $\bar{\underset{\sim}{Q}}(t)$. By (7.2), then,

$$\underset{\sim}{F}_O(t) = \underset{\sim}{Q}(t)\bar{\underset{\sim}{Q}}(t)e^{-t\underset{\sim}{M}}. \tag{7.8}$$

The form asserted by Noll's theorem holds trivially if $\underset{\sim}{M} = \underset{\sim}{0}$; if $\underset{\sim}{M} \neq \underset{\sim}{0}$, it follows if we set

$$\kappa \equiv |\underset{\sim}{M}|, \quad \underset{\sim}{N}_O \equiv \frac{1}{\kappa}\underset{\sim}{M}. \tag{7.9}$$

Conversely, if (7.2) holds, an easy calculation shows that the motion is one of constant stretch history.

Exercise 7.2

Prove that in a motion with constant stretch history

$$\mathbf{F}_t(\tau) = \mathbf{Q}(\tau)\mathbf{Q}(t)^T e^{(\tau-t)\mathbf{G}}, \tag{7.10}$$

where

$$\mathbf{G} = \kappa\mathbf{N} \equiv \kappa\mathbf{Q}(t)\mathbf{N}_o\mathbf{Q}(t)^T, \quad |\mathbf{N}| = 1; \tag{7.11}$$

also

$$\mathbf{C}_t^t(s) = e^{-s\mathbf{G}^T} e^{-s\mathbf{G}}. \tag{7.12}$$

The *Rivlin-Ericksen tensors* $\mathbf{A}_n$ are defined as follows:

$$\mathbf{A}_n \equiv \overset{(n)}{\mathbf{C}_t}(t), \tag{7.13}$$

where (n) stands for n dots, and where the notation (2.2) is used. Prove that these tensors are frame-indifferent, and that in a substantially stagnant motion

$$\begin{aligned}
\mathbf{A}_1 &= \mathbf{G}^T + \mathbf{G} = \kappa(\mathbf{N} + \mathbf{N}^T), \\
\mathbf{A}_2 &= \mathbf{G}^T\mathbf{A}_1 + \mathbf{A}_1\mathbf{G} = \kappa^2(2\mathbf{N}^T\mathbf{N} + \mathbf{N}^2 + (\mathbf{N}^T)^2), \\
\mathbf{A}_3 &= \mathbf{G}^T\mathbf{A}_2 + \mathbf{A}_2\mathbf{G}, \\
&\vdots \\
\mathbf{A}_k &= \mathbf{G}^T\mathbf{A}_{k-1} + \mathbf{A}_{k-1}\mathbf{G}.
\end{aligned} \tag{7.14}$$

A motion with constant stretch history is isochoric if and only if

$$\operatorname{tr} \underset{\sim}{N}_0 = 0. \tag{7.15}$$

Determination of the Deformation History from the First Three Rivlin-Ericksen Tensors

With the aid of these results we see easily the extremely special nature of motions with constant stretch history, which is expressed by WANG's *corollary: The relative deformation history* $\underset{\sim}{C}_t^t(s)$ *of a motion with constant stretch history is determined uniquely by its first three Rivlin-Ericksen tensors*. That is, if three tensors $\underset{\sim}{A}_1(t)$, $\underset{\sim}{A}_2(t)$, and $\underset{\sim}{A}_3(t)$ are given, they can be the first three Rivlin-Ericksen tensors corresponding to at most one constant relative stretch history $\underset{\sim}{C}_t^t$.

The proof rests upon a simple lemma. Let $\underset{\sim}{S}$ be a symmetric tensor and $\underset{\sim}{W}$ a skew tensor in 3-dimensional space. Without loss of generality we can take the matrices of these tensors as being

$$[\underset{\sim}{S}] = \begin{Vmatrix} a & 0 & 0 \\ 0 & b & 0 \\ 0 & 0 & c \end{Vmatrix}, \qquad [\underset{\sim}{W}] = \begin{Vmatrix} 0 & x & y \\ -x & 0 & z \\ -y & -z & 0 \end{Vmatrix}. \tag{7.16}$$

Then

$$[\underset{\sim}{S}\underset{\sim}{W} - \underset{\sim}{W}\underset{\sim}{S}] = \begin{Vmatrix} 0 & (a-b)x & (a-c)y \\ (a-b)x & 0 & (b-c)z \\ (a-c)y & (b-c)z & 0 \end{Vmatrix}. \tag{7.17}$$

Hence $\underset{\sim}{S}$ and $\underset{\sim}{W}$ commute if and only if

$$(a-b)x = 0, \quad (a-c)y = 0, \quad (b-c)z = 0.$$

Consequently, if $\underset{\sim}{S}$ has distinct proper numbers, it commutes with no skew tensor other than $\underset{\sim}{O}$. If $a = b \neq c$, $\underset{\sim}{S}$ commutes with $\underset{\sim}{W}$ if and only if $y = z = 0$. If $a = b = c$, $\underset{\sim}{S}$ commutes with all $\underset{\sim}{W}$.

Wang's corollary may now be proved in stages. Assume first that the proper numbers of $\underset{\sim}{A}_1$ are distinct. If two constant relative stretch histories $\underset{\sim}{c}_t^t$ can correspond to $\underset{\sim}{A}_1$ and $\underset{\sim}{A}_2$, then there exist tensors $\underset{\sim}{G}$ and $\bar{\underset{\sim}{G}}$ such that, by $(7.14)_{3,6}$,

$$\underset{\sim}{G} + \underset{\sim}{G}^T = \bar{\underset{\sim}{G}} + \bar{\underset{\sim}{G}}^T,$$

$$\underset{\sim}{G}^T\underset{\sim}{A}_1 + \underset{\sim}{A}_1\underset{\sim}{G} = \bar{\underset{\sim}{G}}^T\underset{\sim}{A}_1 + \underset{\sim}{A}_1\bar{\underset{\sim}{G}}. \tag{7.18}$$

The first of these equations asserts that $\underset{\sim}{G} - \bar{\underset{\sim}{G}}$ is skew; the second, that $\underset{\sim}{G} - \bar{\underset{\sim}{G}}$ commutes with $\underset{\sim}{A}_1$. By the lemma, $\underset{\sim}{G} - \bar{\underset{\sim}{G}} = \underset{\sim}{O}$.

Exercise 7.3

Complete the proof of Wang's corollary.

First, suppose that $\underset{\sim}{A}_1$ has two and only two distinct proper numbers. Then relative to a suitable orthonormal basis

$$[\underset{\sim}{A}_1] = \left\|\begin{array}{ccc} a & 0 & 0 \\ 0 & a & 0 \\ 0 & 0 & b \end{array}\right\|, \qquad a \neq b. \tag{7.19}$$

Case 1. If, relative to the same basis,

$$[\underset{\sim}{A}_2] = \begin{Vmatrix} u & 0 & 0 \\ 0 & u & 0 \\ 0 & 0 & v \end{Vmatrix}, \tag{7.20}$$

then by (7.14) and (7.19) it follows that

$$[\underset{\sim}{G}] = \begin{Vmatrix} \frac{1}{2}a & x & 0 \\ -x & \frac{1}{2}a & 0 \\ 0 & 0 & \frac{1}{2}b \end{Vmatrix}, \quad u = a^2, \ v = b^2. \tag{7.21}$$

Hence by (7.11)

$$\underset{\sim}{C}_t^t(s) = e^{-s\underset{\sim}{A}_1}. \tag{7.22}$$

Case 2. If (7.20) does not hold, $\underset{\sim}{G} = \bar{\underset{\sim}{G}}$ in (7.18).

Finally, if the three proper numbers of $\underset{\sim}{A}_1$ are the same, (7.22) holds.

Accordingly, then, three given tensors $\underset{\sim}{A}_1(t)$, $\underset{\sim}{A}_2(t)$, and $\underset{\sim}{A}_3(t)$ can be the Rivlin-Ericksen tensors corresponding to *at most one* $\underset{\sim}{C}_t^t(s)$ belonging to a motion with constant stretch history. In general, three tensors taken arbitrarily will fail to belong to *any* motion with constant stretch history.

While Noll's theorem clearly is independent of the dimension of the space, Wang's corollary rests heavily on use of the dimension 3.

Equivalence of Simple Materials with Rivlin-Ericksen Materials in Motions with Constant Stretch History

In view of Wang's corollary, any information that can be determined in a motion with constant stretch history from $\tilde{C}_t^t(s)$ can be determined also from $\tilde{A}_1(t)$, $\tilde{A}_2(t)$, $\tilde{A}_3(t)$. Therefore any *functional* of $\tilde{C}_t^t(s)$ equals, in these motions, a *function* of $\tilde{A}_1(t)$, $\tilde{A}_2(t)$, $\tilde{A}_3(t)$. Consequently the general constitutive equation (4.21)

$$\bar{\tilde{T}} = \tilde{\mathfrak{K}}(\bar{\tilde{C}}_t^t; \tilde{C}), \qquad (4.21)$$

where $\bar{\tilde{K}} = \tilde{R}^T\tilde{K}\tilde{R}$, may be replaced in motions with constant stretch history by

$$\bar{\tilde{T}} = \tilde{f}(\bar{\tilde{A}}_1, \bar{\tilde{A}}_2, \bar{\tilde{A}}_3, \tilde{C}), \qquad (7.23)$$

where $\tilde{f}$ is a function. A material which satisfies (7.26) for all motions is called a *Rivlin-Ericksen material of complexity 3*. According to (7.26), then, *in motions with constant stretch history, the class of simple materials cannot be distinguished from the far more special class of Rivlin-Ericksen materials of complexity 3*. For an isotropic material, (7.23) becomes

$$\tilde{T}(t) = \tilde{f}(\tilde{A}_1(t), \tilde{A}_2(t), \tilde{A}_3(t), \tilde{B}(t)), \qquad (7.24)$$

and for a fluid

$$\tilde{T}(t) = -p\tilde{1} + \tilde{f}(\tilde{A}_1(t), \tilde{A}_2(t), \tilde{A}_3(t), \rho), \qquad (7.25)$$

where the functions $\tilde{f}$, in the two cases, are isotropic in the sense that for all symmetric $\tilde{A}_1$, $\tilde{A}_2$, $\tilde{A}_3$, $\tilde{B}$, and all orthogonal Q,

$$\underset{\sim}{f}(\underset{\sim}{Q}\underset{\sim}{A}_1\underset{\sim}{Q}^T, \underset{\sim}{Q}\underset{\sim}{A}_2\underset{\sim}{Q}^T, \underset{\sim}{Q}\underset{\sim}{A}_3\underset{\sim}{Q}^T, \underset{\sim}{Q}\underset{\sim}{B}\underset{\sim}{Q}^T, \text{ or } \rho),$$
$$= \underset{\sim}{Q}\underset{\sim}{f}(\underset{\sim}{A}_1, \underset{\sim}{A}_2, \underset{\sim}{A}_3, \underset{\sim}{B} \text{ or } \rho)\underset{\sim}{Q}^T. \quad (7.26)$$

These results, which express *Wang's theorem*, may be interpreted in two ways. On the one hand, they enable us to solve easily special problems concerned with motions of constant stretch history. However complicated may be in general the response of a material, in these particular motions we need consider only a simple special constitutive equation. On the other hand, they show that observation of this class of flows is insufficient to tell us much about a material, since most of the complexities of material response are prevented from manifesting themselves.

Classification of Motions with Constant Stretch History

Returning to Noll's theorem (7.2), we see at once that it suggests an invariant classification of all motions with constant stretch history into three mutually exclusive types:

1. $\underset{\sim}{N}_o^2 = \underset{\sim}{0}$. These motions are called *viscometric flows*.
2. $\underset{\sim}{N}_o^3 = \underset{\sim}{0}$ but $\underset{\sim}{N}_o^2 \neq \underset{\sim}{0}$.
3. $\underset{\sim}{N}_o$ is not nilpotent.

In types 1 and 2, since tr $\underset{\sim}{N}_o = \underset{\sim}{0}$, the motion is isochoric.

In the next lecture we shall consider viscometric flows in greater detail.

8. The Stress System in Viscometric Flows of Incompressible Fluids

Recapitulation

From the theorems of NOLL and WANG, stated and proved in the last lecture, it follows that in a viscometric flow of an incompressible fluid,

$$\underset{\sim}{T} = -p\underset{\sim}{1} + \underset{\sim}{\mathfrak{g}}(\underset{\sim}{A}_1, \underset{\sim}{A}_2), \tag{8.1}$$

where

$$\underset{\sim}{A}_1 = \kappa(\underset{\sim}{N} + \underset{\sim}{N}^T),$$

$$\underset{\sim}{A}_2 = 2\kappa^2 \underset{\sim}{N}^T \underset{\sim}{N}, \tag{8.2}$$

$$\underset{\sim}{A}_n = \underset{\sim}{0}, \text{ if } n \geqq 3,$$

and

$$\underset{\sim}{N}^2 = \underset{\sim}{0}, \quad \text{tr } \underset{\sim}{N} = 0, \quad |\underset{\sim}{N}| = 1, \tag{8.3}$$

and

$$\underset{\sim}{\mathfrak{g}}(\underset{\sim}{Q}\underset{\sim}{A}_1\underset{\sim}{Q}^T, \underset{\sim}{Q}\underset{\sim}{A}_2\underset{\sim}{Q}^T) = \underset{\sim}{Q}\, \underset{\sim}{\mathfrak{g}}(\underset{\sim}{A}_1, \underset{\sim}{A}_2)\underset{\sim}{Q}^T, \tag{8.4}$$

for all symmetric tensors $\underset{\sim}{A}_1$ and $\underset{\sim}{A}_2$ and for all orthogonal tensors $\underset{\sim}{Q}$.

Following the method of COLEMAN and NOLL, we shall now determine the most general stress system compatible with the equations just stated.

Functional Equations for the Response Function

First of all, (8.1) and (8.2) show that

$$\mathbf{T} = -p\mathbf{1} + \mathbf{f}(\kappa,\mathbf{N}). \tag{8.5}$$

Second, if we replace κ by $\pm\kappa$ and $\mathbf{N}$ by $\pm\mathbf{Q}\mathbf{N}\mathbf{Q}^T$, by (8.2) $\mathbf{A}_1$ is replaced by $\mathbf{Q}\mathbf{A}_1\mathbf{Q}^T$, and $\mathbf{A}_2$ is replaced by $\mathbf{Q}\mathbf{A}_2\mathbf{Q}^T$. Therefore, (8.4) is equivalent to

$$\mathbf{f}(\pm\kappa,\pm\mathbf{Q}\mathbf{N}\mathbf{Q}^T) = \mathbf{Q}\mathbf{f}(\kappa,\mathbf{N})\mathbf{Q}^T, \tag{8.6}$$

where $\mathbf{f}$ is the function occurring in (8.5) and where the $\pm$ signs are associated. In view of (8.1) we may describe as follows the invariance asserted by the functional equation (8.6): To replace κ by $\pm\kappa$ and $\mathbf{N}$ by $\pm\mathbf{Q}\mathbf{N}\mathbf{Q}^T$, where $\mathbf{Q}$ is any orthogonal tensor, results in replacing $\mathbf{T}$ by $\mathbf{Q}\mathbf{T}\mathbf{Q}^T$.

In view of (8.3), we may choose an orthonormal basis such that

$$[\mathbf{N}] = \begin{Vmatrix} 0 & 0 & 0 \\ 1 & 0 & 0 \\ 0 & 0 & 0 \end{Vmatrix}. \tag{8.7}$$

By (7.11), the basis with respect to which $[\mathbf{N}]$ has this special form is generally a rotating one, and it need not be the natural basis of any co-ordinate system.

Illustration by Means of Shearing Flow

While the results we shall deduce follow from the algebraic formulae just given, it is easier to visualize them in terms of a special case. To this end we consider a *shearing flow*, given in a suitable rectangular Cartesian system by the velocity components

$$\dot{x}_1 = 0, \quad \dot{x}_2 = v(x_1), \quad \dot{x}_3 = 0. \tag{8.8}$$

Exercise 8.1

Show from (2.19) and (2.20) that for the flow (8.8)

$$\underset{\sim}{F}_t(\tau) = \underset{\sim}{1} + (\tau-t)\kappa\underset{\sim}{N} = e^{(\tau-t)\kappa\underset{\sim}{N}}, \tag{8.9}$$

where $\underset{\sim}{N}$ has the form (8.7) with respect to the co-ordinate basis and where

$$\kappa = v(x_1). \tag{8.10}$$

In this example of a viscometric flow, the particles move in straight lines at uniform speed, and κ is twice the only non-vanishing principal stretching. It is customary to refer to κ as the *shearing*, not only in the special case (8.10) but also for any viscometric flow.

Consequences of Invariance Under Reflections

We shall now determine the most general stress system compatible with the constitutive equation (8.1) in a viscometric flow, characterized by (8.2). We shall motivate the results by describing them in terms of the special case (8.7).

First, a reflection in the plane normal of the flow should be expected to leave the whole stress system invariant. To see if it does, take $\mathbf{Q}$ such that

$$[\mathbf{Q}] = \begin{Vmatrix} 1 & 0 & 0 \\ 0 & 1 & 0 \\ 0 & 0 & -1 \end{Vmatrix} \tag{8.11}$$

and leave κ unchanged. From (8.7), $\mathbf{QNQ}^T = \mathbf{N}$. The assertion embodied in (8.6) is that this same transformation carries $\mathbf{T}$ into $\mathbf{QTQ}^T$. But by (8.11)

$$[\mathbf{QTQ}^T] = \begin{Vmatrix} T\langle 11\rangle & T\langle 12\rangle & -T\langle 13\rangle \\ \cdot & T\langle 22\rangle & -T\langle 23\rangle \\ \cdot & \cdot & T\langle 33\rangle \end{Vmatrix}, \tag{8.12}$$

where the $\mathbf{T}\langle ij\rangle$ are the components of $\mathbf{T}$ with respect to the basis in which $\mathbf{N}$ and $\mathbf{Q}$ have the forms (8.7) and (8.11). Thus in order that $\mathbf{QTQ}^T = \mathbf{T}$, it is necessary that

$$T\langle 13\rangle = 0, \quad T\langle 23\rangle = 0. \tag{8.13}$$

By (8.5), the remaining components of $\mathbf{T} + p\mathbf{1}$ with respect to this basis are functions of κ only:

$$\begin{aligned} T\langle 12\rangle &= \tau(\kappa), \\ T\langle 11\rangle - T\langle 33\rangle &= \sigma_1(\kappa), \\ T\langle 22\rangle - T\langle 33\rangle &= \sigma_2(\kappa). \end{aligned} \tag{8.14}$$

If we prefer an invariant form, without mention of a basis, it may be written down by combining (8.13), (8.14), and (8.7):

$$\mathbf{T} = -p\mathbf{1} + \tau(\kappa)(\mathbf{N}+\mathbf{N}^T) + \sigma_1(\kappa)\mathbf{N}^T\mathbf{N} + \sigma_2(\kappa)\mathbf{N}\mathbf{N}^T, \tag{8.15}$$

where p is not determined by the constitutive equation. (In general, $p = -T\langle 33\rangle \neq -\frac{1}{3}\operatorname{tr}\mathbf{T}$.) That this form is general follows from (8.5).

Now we consider a reflection in the direction of flow. We expect that such a reflection, which amounts to replacing κ by $-\kappa$, should reverse the shear stress $T\langle 12\rangle$ but leave all normal tractions $T\langle 11\rangle$, $T\langle 22\rangle$, $T\langle 33\rangle$ unchanged. Accordingly, we take $\mathbf{Q}$ such that

$$[\mathbf{Q}] = \begin{Vmatrix} 1 & 0 & 0 \\ 0 & -1 & 0 \\ 0 & 0 & 1 \end{Vmatrix}. \tag{8.16}$$

By (8.7),

$$-\mathbf{Q}\mathbf{N}\mathbf{Q}^T = \mathbf{N}. \tag{8.17}$$

Under this transformation, κ goes into $-\kappa$ and $\mathbf{T}$ goes into $\mathbf{Q}\mathbf{T}\mathbf{Q}^T$. By (8.16),

$$[\mathbf{Q}\mathbf{T}\mathbf{Q}^T] = \begin{Vmatrix} T\langle 11\rangle & -T\langle 12\rangle & 0 \\ \cdot & T\langle 22\rangle & 0 \\ \cdot & \cdot & T\langle 33\rangle \end{Vmatrix}. \tag{8.18}$$

We have proved, then, that change of the sign of κ changes $T\langle 12\rangle$ into $-T\langle 12\rangle$ but leaves the remaining stresses unaffected. By (8.14), then,

$$\begin{aligned} \tau(-\kappa) &= -\tau(\kappa), \\ \sigma_1(-\kappa) &= \sigma_1(\kappa), \\ \sigma_2(-\kappa) &= \sigma_2(\kappa). \end{aligned} \tag{8.19}$$

That is, τ is an even function of the shearing, while σ_1 and σ_2 are odd functions. From (8.19) it follows that

$$\tau(0) = 0. \tag{8.20}$$

If, as is customary, p is normalized so that $\underset{\sim}{f}(0,\underset{\sim}{N}) = \underset{\sim}{0}$, then also

$$\begin{aligned} \sigma_1(0) &= 0, \\ \sigma_2(0) &= 0. \end{aligned} \tag{8.21}$$

Exercise 8.2

Prove that if (8.15) and (8.19) hold, then (8.4) and (8.5) are satisfied.

The Viscometric Functions. Normal-Stress Effects

What has been shown, then, is that the stress system in any incompressible simple fluid in a viscometric flow is given by (8.15), with the functions τ, σ_1 and σ_2 restricted by (8.19) and (8.21).

The functions τ, σ_1, and σ_2 are the *viscometric functions* of the simple fluid whose constitutive equation reduces to (8.15) in a viscometric flow. Obviously, infinitely many different fluids share the same set of three viscometric functions.

In the particular case when the velocity field is given by (8.8), the basis with respect to which $\underset{\sim}{N}$ has the form (8.7) is the natural basis of the fixed Cartesian co-ordinate system, and the component relations (8.14) may be interpreted in terms of Cartesian co-ordinates. In the further special case when

$$v(x_1) = \kappa x_1, \quad \kappa = \text{const.}, \tag{8.22}$$

the motion (8.8) is called *simple shearing.*

Exercise 8.3

Prove that in any given homogeneous simple material, a simple shearing may be produced by the application of suitable surface tractions alone.

In particular, by specializing this result, we conclude that no body force is needed in order to produce a simple shearing in any given homogeneous simple fluid. In this case, the stress system (8.14) may be interpreted immediately. $\tau(\kappa)$ gives the shear stress that must be supplied on a plane x_1 = const. in order to produce the flow. The normal traction on this same plane is $T\langle 11\rangle$, and that on the flow plane is $T\langle 33\rangle$. In view of the term $-p\underset{\sim}{1}$ in (8.1), either of these, but not both, may be given any value, at pleasure. If we choose to leave the planes x_3 = const. free, then $T\langle 33\rangle = 0$, and (8.14) yields

$$\begin{aligned} T\langle 11\rangle &= \sigma_1(\kappa) \\ T\langle 22\rangle &= \sigma_2(\kappa). \end{aligned} \tag{8.23}$$

Thus a fixed normal traction on the plane x_1 = const., determined by κ and by the nature of the fluid, must be supplied, and likewise a normal traction in the plane x_2 = const., which are normal to the flow. The necessity for these normal tractions is an example of what are called normal-stress effects. In particular, the result (8.23) shows that shear stress alone is insufficient to produce simple shearing. In addition, suitable and generally unequal normal tractions, determined by the nature of the fluid, are necessary in order for the flow to occur.

Position of the Classical Theory

According to the classical or *Navier-Stokes theory* of fluids,

$$\begin{aligned} &\tau(\kappa) \equiv \mu\kappa\,, \quad \text{where } \mu = \text{const.}, \\ &\sigma_1(\kappa) \equiv \sigma_2(\kappa) \equiv 0. \end{aligned} \tag{8.24}$$

If we assume that τ and σ_1 and σ_2 have three continuous derivatives at $\kappa = 0$, then, by (8.19) and (8.21),

$$\begin{aligned} \tau(\kappa) &= \mu\kappa + \mu'\kappa^3 + O(\kappa^5), \\ \sigma_1(\kappa) &= s_1\kappa^2 + O(\kappa^4), \\ \sigma_2(\kappa) &= s_2\kappa^2 + O(\kappa^4), \end{aligned} \tag{8.25}$$

where μ, μ', s_1 and s_2 are constants.

Thus the effects of second order in κ are normal-stress effects, while departure from the classical proportionality of shear stress to shearing is an effect of third order in κ. Roughly speaking, departures from the classical behavior (8.24) may be expected to be observed for σ_1 and σ_2 at lower shearings κ than for τ. Still more roughly, normal-stress effects can be expected to manifest themselves within the range in which the response of the shear stress remains classical.

In the example used to interpret the results so far, $v(x_1)$ in (8.8) has been assumed linear. This velocity field is identical with the one that follows in the same circumstances according to the Navier-Stokes theory, but the stress system required in order to produce it has been shown to be different. Nevertheless, the flow is a possible one, subject to boundary tractions alone.

When, on the other hand, the velocity profile $v(x_1)$ in (8.8) fails to be linear, or when, as is the case in general, the basis with respect to which $\underset{\sim}{N}$ has the form (8.7) is a rotating one, we have no reason to expect a viscometric flow to be possible unless suitable non-conservative body force be supplied. In the next lecture we shall consider some particular cases in which the dynamical equations can indeed be satisfied when $\underset{\sim}{b} = \underset{\sim}{0}$. As will be seen, the assignment of speeds to the streamlines will generally be entirely different from that required by the Navier-Stokes theory.

9. Dynamical Conditions in Viscometric Flows

Recapitulation

The stress corresponding to any viscometric flow of an incompressible simple fluid is given by (8.15):

$$\underset{\sim}{S} \equiv \underset{\sim}{T} + p\underset{\sim}{1} = \tau(\kappa)(\underset{\sim}{N}+\underset{\sim}{N}^T) + \sigma_1(\kappa)\underset{\sim}{N}^T\underset{\sim}{N} + \sigma_2(\kappa)\,\underset{\sim}{N}\underset{\sim}{N}^T. \tag{8.15}$$

The viscometric functions $\tau(\kappa)$, $\sigma_1(\kappa)$, $\sigma_2(\kappa)$ are determined uniquely by the constitutive equation and hence are the same for all viscometric flows of any given fluid. κ and $\underset{\sim}{N}$ generally are functions of both place and time. κ is a scalar, which may be called the *shearing*, and $\underset{\sim}{N}$ is a tensor such that $|\underset{\sim}{N}| = 1$, $\underset{\sim}{N}^2 = \underset{\sim}{0}$ (and hence tr $\underset{\sim}{N} = 0$). The orthonormal basis with respect to which $\underset{\sim}{N}$ has the special component matrix (8.7) may vary with place and time and need not be the natural basis of any co-ordinate system. The scalar p, which equals $-T\langle 33\rangle$ in the special basis, is not determined by the deformation history. In general, (8.15) will fail to satisfy Cauchy's law (3.27) unless a suitable body force be supplied.

Dynamic Compatibility

Of greatest interest are such particular viscometric flows as may be effected by applying suitable boundary tractions alone. One such has already been exhibited, the rectilinear shearing, defined by (8.8):

$$\dot{x}_1 = 0, \quad \dot{x}_2 = v(x_1), \quad \dot{x}_3 = 0. \tag{8.8}$$

in the special case when $v(x_1) = \kappa x_1$, κ = const. This case is not typical in that one and the same velocity field is possible in all fluids. Generally the velocity field that meets given dynamic requirements depends upon the viscometric functions, as we shall see.

Shearing Flow

We shall now find the most general shearing flow (8.8) that can be effected by boundary tractions and conservative body force in a simple fluid whose viscometric functions are $\tau(\kappa)$, $\sigma_1(\kappa)$, and $\sigma_2(\kappa)$. For shearing flow, the basis with respect to which $\underset{\sim}{N}$ has the special form (8.7) is the natural basis of the co-ordinate system used, and $\underset{\sim}{N}$ = const. As has been shown already,

$$\kappa = v(x_1). \tag{8.10}$$

We assume that ρ = const. We shall consider only conservative body force, so that

$$\underset{\sim}{b} = - \operatorname{grad} \upsilon(\underset{\sim}{x},t). \tag{9.1}$$

If we set

$$\phi \equiv \frac{p}{\rho} + \upsilon, \tag{9.2}$$

then Cauchy's first law (3.27) may be put into the form

$$\operatorname{div} \underset{\sim}{S} - \rho \operatorname{grad} \phi = \rho \ddot{\underset{\sim}{x}}. \tag{9.3}$$

The scalar ϕ is not determined by the motion, but $\underset{\sim}{S}$ and $\ddot{\underset{\sim}{x}}$ are. In fact, by (8.8), $\ddot{\underset{\sim}{x}} = \underset{\sim}{0}$, and by (8.15), $\underset{\sim}{S}$ is a function of x_1 only.

Hence (9.3) is equivalent to the following three differential equations:

$$\partial_{x_1} S\langle 11\rangle - \rho\partial_{x_1}\phi = 0,$$

$$\partial_{x_1} S\langle 12\rangle - \rho\partial_{x_2}\phi = 0, \tag{9.4}$$

$$\rho\partial_{x_3}\phi = 0.$$

Exercise 9.1

Prove that the general solution of (9.4) is given by

$$\rho\phi = p + \rho\upsilon = -ax_2 + k(x_1) + h(t),$$

$$T\langle 12\rangle = S\langle 12\rangle = -ax_1 + c, \tag{9.5}$$

$$S\langle 11\rangle = k(x_1) + b,$$

where the functions $k(x_1)$ and $h(t)$ are arbitrary, and where a, b, and c are arbitrary constants.

The entire stress system may be calculated as follows. First, by $(9.5)_5$ and $(9.5)_2$,

$$T\langle 11\rangle = S\langle 11\rangle - p = S\langle 11\rangle - (p+\rho\upsilon) + \rho\upsilon,$$

$$= ax_2 + b + \rho\upsilon - h(t). \tag{9.6}$$

Since $\partial_{x_2}[T\langle 11\rangle - \rho\upsilon] = a$, the constant a is the gradient of $T\langle 11\rangle - \rho\upsilon$ in the direction of flow. Thus we may interpret the assignable quantity, a, as the specific driving force in the direction of flow. Second, by this result and (8.14),

$$T\langle 22\rangle = (T\langle 22\rangle - T\langle 11\rangle) + T\langle 11\rangle,$$

$$= \sigma_2(\kappa) - \sigma_1(\kappa) + ax_2 + b + \rho\upsilon - h(t),$$

$$T\langle 33\rangle = (T\langle 33\rangle - T\langle 11\rangle) + T\langle 11\rangle, \tag{9.7}$$

$$= -\sigma_1(\kappa) + ax_2 + b + \rho\upsilon - h(t).$$

With $v(x_1)$ arbitrary and κ given by (8.10), the normal stresses are delivered by these formulae. Combining $(9.5)_4$ and $(8.14)_1$ yields a differential equation to determine $v(x_1)$:

$$\tau(\kappa) = \tau(v'(x_1)) = -ax_1 + c. \tag{9.8}$$

The arbitrary constants a and c are to be assigned, and then $v(x_1)$ is determined by integrating (9.8).

If we take a = 0 and assume that τ is invertible, (9.8) requires that κ = const., and we recover the results already obtained for simple shearing. The foregoing analysis shows simple shearing to be the only rectilinear shearing flow (8.8) that can be produced with zero specific driving force, provided the body force be conservative.

Channel Flow

We shall now seek a solution that represents the flow of a mass of material adhering to stationary infinite plates $x_1 = \pm d$. Thus we require $v(x_1)$ to be such that

$$v(d) = v(-d) = 0. \qquad (9.9)$$

We assume that the shear viscosity function $\tau(\kappa)$ is invertible with inverse, say, ζ. Any inverse is necessarily odd. Then (9.8) yields

$$\kappa = v'(x_1) = \zeta(-ax_1+c). \qquad (9.10)$$

Exercise 9.2

Prove that, since ζ is an odd function, (9.10) is compatible with (9.9) for arbitrary a if and only if $c = 0$.

Consequently

$$v(x_1) = \int_{x_1}^{d} \zeta(a\xi)d\xi. \qquad (9.11)$$

The velocity profile $v(x_1)$ is thus determined uniquely by the shear-viscosity function $\tau(\kappa)$. In contrast to the case of simple shearing, however, the profile is generally not at all the same as that predicted by the Navier-Stokes equations. Indeed, if $(8.24)_1$ holds, then $\rho(\xi) = (1/\mu)\xi$, and (9.11) yields

$$v(x_1) = \frac{1}{\mu}\int_{x_1}^{d} a\xi d\xi = \frac{a}{2\mu}[d^2-x_1^2], \qquad (9.12)$$

the classical parabolic form. Conversely, if (9.12) holds, $\zeta(a\xi)$ is a linear function of a, and the classical linear formula (8.24) for shear viscosity results.

The discharge D, the volume of fluid passing through unit depth of channel in unit time, is given by

$$D \equiv \int_{-d}^{d} v(x)dx = 2\int_0^d dx \int_x^d \zeta(a\xi)d\xi,$$
$$= \frac{2}{a^2}\int_0^{ad} \xi\zeta(\xi)d\xi. \tag{9.13}$$

Conversely, if the discharge D(a,d) is known as a function of a and d, (9.13) yields

$$\tau^{-1}(ad) = \zeta(ad) = \frac{1}{2ad^2}\partial_a[a^2 D(a,d)]. \tag{9.14}$$

Thus D(a,d) determines the shear-viscosity function uniquely. In particular, the classical formula

$$D = \frac{2ad^3}{3\mu} \tag{9.15}$$

holds if and only if the shear-viscosity function is linear.

Velocity profile, discharge, and shear-viscosity function determine one another and are unaffected by the normal-stress functions σ_1 and σ_2. Thus if (9.15) holds, there is no reason at all to expect the remaining classical formulae $(8.24)_{3,4}$ to hold. Therefore, the classical viscometric tests, which refer to shear viscosity alone, do not tell much about the fluid being tested. If in a particular case a classical formula such as (9.15) emerges, this fact not only fails to show that the fluid tested obeys the Navier-Stokes equations but even is insufficient to establish the Navier-Stokes theory of viscometry. Additional measurements are necessary. In the present case, by (9.6), the normal tractions on the fixed plates do not differ from those predicted by the classical theory. By (9.7), however, those on the flow planes (x_3 = const.) and those on planes normal to the flow (x_2 = const.) may be entirely different.

Since these normal tractions are difficult to interpret, we turn to a different class of flows, in which normal-stress effects are more striking.

Helical Flows

Consider the velocity field whose contravariant components in a cylindrical polar co-ordinate system r, θ, z are given by

$$\dot{r} = 0, \quad \dot{\theta} = \omega(r), \quad \dot{z} = u(r). \tag{9.16}$$

Each particle remains upon a fixed cylinder r = const., on which it describes a helix, whose pitch is the same for all the particles on any one cylinder. We set

$$f(r) \equiv \omega'(r), \quad h(r) \equiv u'(r). \tag{9.17}$$

Exercise 9.3

Prove that a helical flow is a flow with constant stretch history and that

$$\kappa = \sqrt{r^2 f(r)^2 + h(r)^2}. \tag{9.18}$$

Let $\underset{\sim}{e}_i(\underset{\sim}{x})$, $i = 1, 2, 3$, be an orthonormal basis tangent to the co-ordinate curves at $\underset{\sim}{x}$, and let

$$\underset{\sim}{i}_1 = \underset{\sim}{e}_1, \quad \underset{\sim}{i}_2 = \alpha \underset{\sim}{e}_2 + \beta \underset{\sim}{e}_3, \quad \underset{\sim}{i}_3 = -\beta \underset{\sim}{e}_2 + \alpha \underset{\sim}{e}_3, \tag{9.19}$$

where

$$\alpha = \frac{r}{\kappa}f(r), \quad \beta = \frac{1}{\kappa}h(r), \quad \alpha^2+\beta^2 = 1. \tag{9.20}$$

Prove that $\underset{\sim}{N}_o$ has the form (8.7) with respect to this basis.

From the results of the foregoing problem, the formulae (8.14) are valid for the components of $\underset{\sim}{T}$ relative to the basis $\underset{\sim}{i}_1, \underset{\sim}{i}_2, \underset{\sim}{i}_3$:

$$\begin{aligned} &T\langle 12\rangle = \tau(\kappa), \quad T\langle 13\rangle = 0, \quad T\langle 23\rangle = 0, \\ &T\langle 11\rangle - T\langle 33\rangle = \sigma_1(\kappa), \\ &T\langle 22\rangle - T\langle 33\rangle = \sigma_2(\kappa). \end{aligned} \tag{9.21}$$

The physical components of $\underset{\sim}{T}$ in cylindrical co-ordinates are its components with respect to the orthonormal basis $\underset{\sim}{e}_1$. Denoting these components by $T\langle rr\rangle$, $T\langle r\theta\rangle$, *etc.*, we find that

$$\begin{aligned} T\langle r\theta\rangle &= \underset{\sim}{e}_1\cdot\underset{\sim}{T}\underset{\sim}{e}_2 = \underset{\sim}{i}_1\cdot(\alpha\underset{\sim}{T}\underset{\sim}{i}_2-\beta\underset{\sim}{T}\underset{\sim}{i}_3), \\ &= \alpha T\langle 12\rangle - \beta T\langle 13\rangle, \\ T\langle\theta z\rangle &= \underset{\sim}{e}_2\cdot\underset{\sim}{T}\underset{\sim}{e}_3 = (\alpha\underset{\sim}{i}_2-\beta\underset{\sim}{i}_3)\cdot(\beta\underset{\sim}{T}\underset{\sim}{i}_2+\alpha\underset{\sim}{T}\underset{\sim}{i}_3), \\ &= \alpha\beta(T\langle 22\rangle-T\langle 33\rangle) + (\alpha^2-\beta^2)T\langle 23\rangle, \end{aligned} \tag{9.22}$$

etc. From these results and (9.21) we find the stress system in terms of the viscometric functions:

$$T\langle r\theta\rangle = \alpha\tau(\kappa),$$

$$T\langle rz\rangle = \beta\tau(\kappa),$$

$$T\langle\theta z\rangle = \alpha\beta\sigma_2(\kappa), \qquad (9.23)$$

$$T\langle rr\rangle - T\langle zz\rangle = \sigma_1(\kappa) - \beta^2\sigma_2(\kappa),$$

$$T\langle\theta\theta\rangle - T\langle zz\rangle = (\alpha^2-\beta^2)\sigma_2(\kappa).$$

It remains now to see whether the functions f(r) and h(r) can be chosen in such a way as to make these stresses compatible with Cauchy's first law of motion.

Exercise 9.4

Since $\underset{\sim}{S}$ is a function of r only, prove that for a helical flow of an incompressible fluid Cauchy's first law assumes the form

$$\partial_r S\langle rr\rangle + \frac{1}{r}(S\langle rr\rangle - S\langle\theta\theta\rangle) - \rho\partial_r\phi = -\rho r\omega^2,$$

$$r\partial_r S\langle r\theta\rangle + 2S\langle r\theta\rangle - \rho\partial_\theta\phi = 0, \qquad (9.24)$$

$$\partial_r S\langle rz\rangle + \frac{1}{r}S\langle rz\rangle - \rho\partial_z\phi = 0.$$

Hence

$$\partial_r(r^2 T\langle r\theta\rangle) = -rd,$$

$$\partial_r(rT\langle rz\rangle) = -ra, \qquad (9.25)$$

$$T\langle rr\rangle = \rho\upsilon + k(r,t) + az + d\theta, \tag{9.25}$$

$$\partial_r k(r,t) + \frac{1}{r}(T\langle rr\rangle - T\langle\theta\theta\rangle) = -\rho r\omega^2,$$

where a and d are arbitrary constants and where k(r,t) is an arbitrary function.

Integration of the first two of these equations yields

$$T\langle r\theta\rangle = \frac{c}{r^2} - \frac{d}{2},$$

$$T\langle rz\rangle = \frac{b}{r} - \frac{ra}{2}, \tag{9.26}$$

where b and c are arbitrary constants. These equations are compatible with $(9.23)_{1,2}$ if and only if

$$\tau(\kappa) = \gamma, \quad \kappa = \zeta(\gamma), \tag{9.27}$$

where

$$\gamma = \gamma(r) = \sqrt{\left(\frac{c}{r^2} - \frac{d}{2}\right)^2 + \left(\frac{b}{r} - \frac{ra}{2}\right)^2}. \tag{9.28}$$

By $(9.23)_{1,2}$, then,

$$\alpha = \frac{T\langle r\theta\rangle}{\tau(\kappa)} = \frac{1}{\gamma}\left(\frac{c}{r^2} - \frac{d}{2}\right),$$

$$\beta = \frac{T\langle rz\rangle}{\tau(\kappa)} = \frac{1}{\gamma}\left(\frac{b}{r} - \frac{ra}{2}\right). \tag{9.29}$$

Finally, from (9.20) and (9.17),

$$\omega'(r) = f = \frac{\kappa\alpha}{r} = \frac{\zeta(\gamma)}{\gamma r}\left(\frac{c}{r^2} - \frac{d}{2}\right),$$

$$\qquad (9.30)$$

$$u'(r) = h = \kappa\beta = \frac{\zeta(\gamma)}{\gamma}\left(\frac{b}{r} - \frac{ra}{2}\right).$$

When the four constants a, b, c, d are fixed, γ becomes a known function of r by (9.28). Hence the two functions $\omega(r)$ and $u(r)$ occurring in the velocity field (9.16) are determined to within six arbitrary constants by the inverse $\zeta(\gamma)$ of the shear-viscosity function $\tau(\kappa)$. Conversely, if $\omega(r)$ and $u(r)$ satisfy (9.30), the helical flow may be produced by the aid of suitable boundary tractions in the fluid whose shear-viscosity function is $\tau(\kappa)$.

Exercise 9.5

Prove from (9.25) and (9.27) that

$$T\langle rr\rangle - T\langle zz\rangle = \hat{\sigma}_1(\gamma) - \beta^2\hat{\sigma}_2(\gamma),$$

$$T\langle\theta\theta\rangle - T\langle zz\rangle = (\alpha^2-\beta^2)\hat{\sigma}_2(\gamma), \qquad (9.31)$$

$$T\langle rr\rangle = \rho\upsilon + \int\{\frac{1}{r}[\alpha^2\hat{\sigma}_2(\gamma)-\hat{\sigma}_1(\gamma)] - \rho r\omega(r)^2\}dr + az + d\theta + g(t),$$

where

$$\hat{\sigma}_\Gamma(\gamma) \equiv \sigma_\Gamma(\zeta(\gamma)), \quad \Gamma = 1,2. \qquad (9.32)$$

We shall now interpret these results in two major special cases.

Flow Between Rotating Cylinders

In $(9.30)_2$, set $a = b = 0$; then $\beta = 0$, and we may take $u(r) \equiv 0$ in (9.16). The fluid particles move in concentric circles with angular speeds $\omega(r)$ given by $(9.30)_1$. In order that radial stress $T\langle rr\rangle$ be single-valued, it is necessary by $(9.31)_3$ that $d = 0$. By (9.28),

$$\gamma(r) = \frac{c}{r^2}. \tag{9.33}$$

The one remaining arbitrary constant c is easily interpreted, since the torque M per unit height applied to the cylinder r = const. is given by $M = (2\pi r)r\, T\langle r\theta\rangle$, which by $(9.26)_1$ is $2\pi c$. That is, $c = M/(2\pi)$. This torque M is to be so adjusted that the cylinders $r = R_1$ and $r = R_2$ move with prescribed angular speeds Ω_1 and Ω_2:

$$\omega(R_1) = \Omega_1, \quad \omega(R_2) = \Omega_2. \tag{9.34}$$

By $(9.30)_1$,

$$\Omega_2 - \Omega_1 = \int_{R_1}^{R_2} \frac{1}{r}\, \frac{M}{2\pi r^2}\, dr. \tag{9.35}$$

A flow of this kind is often called "Couette flow". Such a flow is approximated in a common type of viscometer in which the torque applied to one cylinder is measured as a function of the difference of angular velocities. For any simple fluid, the corresponding relation is given by (9.35). If (9.35) is inverted, the inverse ζ of the shear-viscosity function τ is determined as a function of M for given $\Omega_2 - \Omega_1$.

While according to the classical theory the surfaces z = const. sustain an almost uniform pressure when $\upsilon = 0$ and $\rho\omega^2$ is negligibly small, we see from (9.31) that in general fluid the normal traction $T\langle zz\rangle$ is a function of r. If this traction is not supplied, as for example on the top surface of the fluid in a Couette viscometer, the free surface will tend to rise or fall, according to the signs of the normal-stress functions.

Exercise 9.6

Let the flow be thought of as terminated by a free surface z = const. on which the atmospheric pressure is a constant p_o. Balance of total force requires that

$$2\pi\int_{R_1}^{R_2} T\langle zz\rangle r\,dr = -p_o\pi(R_2^2-R_1^2). \qquad (9.36)$$

Use this relation to evaluate $g(t)$ in $(9.31)_3$. Denoting the excess of p_o over the normal pressure $-T\langle zz\rangle$ by N, show that

$$\partial_r N = -\rho r\omega(r)^2 + \frac{1}{r}\left[\hat{\sigma}_2\left(\frac{M}{2\pi r^2}\right) - \hat{\sigma}_1\left(\frac{M}{2\pi r^2}\right)\right] + \frac{M}{\pi r^3}\hat{\sigma}_1'\,\frac{M}{2\pi r^2}\,. \qquad (9.37)$$

<u>Flow in a Circular Pipe</u>

For a second special case of helical flow, we now consider a flow straight down a cylindrical pipe of infinite length, and we assume that the fluid adheres to the wall, $r = R$. In (9.30) we take $c = d = 0$, so that $\alpha = 0$, and to keep the velocity finite at $r = 0$ we take also $b = 0$. Then by (9.28)

$$\gamma(r) = \frac{ra}{2}, \tag{9.38}$$

and a is the specific driving force. From (9.30) we obtain the velocity profile:

$$u(r) = \int_r^R \zeta\left(\frac{1}{2}a\xi\right) d\xi, \tag{9.39}$$

whence it is plain that the classical parabolic form holds if and only if ζ is linear. The discharge D is given by

$$\begin{aligned} D(a,R) &= 2\pi\int_0^R rdr\int_r^R \zeta\left(\frac{1}{2}a\xi\right) d\xi, \\ &= \pi\int_0^R r^2\zeta\left(\frac{1}{2}ar\right) dr. \end{aligned} \tag{9.40}$$

This, the famous "Hagen-Poiseuille formula" or "law of the fourth power", is valid if and only if the shear-viscosity function $\tau(\kappa)$ is linear. If it is found to be valid, we have no assurance that even the Navier-Stokes theory of viscometry is justified, since the nature of the normal-stress functions has no effect on the discharge and hence cannot be determined from measurements of it.

The presence of a radial tension $T\langle rr\rangle$ which is generally different from the other normal tension suggests that a column of fluid emerging after flowing through a long pipe will tend to swell or shrink in diameter.

Exercise 9.7

Evaluate g(t) in (9.31) by supposing that the total axial force at the exit cross section is that exerted by a uniform atmospheric pressure p_0. Let P be the excess of that pressure over the radial pressure -T<rr> at the exit cross section. Prove that

$$P = az + \frac{1}{R^2}\int_0^R r[\hat{\sigma}_1(\tfrac{1}{2}ar) - \hat{\sigma}_2(\tfrac{1}{2}ar)]dr. \qquad (9.41)$$

Hence a sufficient condition that the fluid shall swell upon emergence is

$$2\hat{\sigma}_2(\gamma) - \hat{\sigma}_1(\gamma) > 0. \qquad (9.42)$$

Other Viscometric Flows

There are other interesting special cases of helical flow. Also two other kinds of viscometric flow have been studied: torsional flow, given in cylindrical co-ordinates by the velocity field,

$$\dot{r} = 0, \quad \dot{\theta} = \omega(z), \quad \dot{z} = 0, \qquad (4.43)$$

and *cone-and-plate flow*, given in spherical polar co-ordinates by the velocity field

$$\dot{r} = 0, \quad \dot{\theta} = 0, \quad \dot{\phi} = \omega(\theta). \qquad (9.44)$$

Since these flows are viscometric, the stress system is easily expressed in terms of the viscometric functions. Neither, however, can satisfy the dynamical equations exactly unless non-conservative body forces be supplied. To make them agree roughly with Cauchy's first law when $\mathbf{b} = \mathbf{0}$, it is necessary to suppose the accelerations negligible, and for cone-and-plate flow it is further necessary to suppose θ limited to a very small interval about $\theta = 0$.

In any viscometric flow, the tensor $\mathbf{S}$ is completely determined by κ, $\mathbf{N}$, and the three viscometric functions τ, σ_1, and σ_2. Thus all phenomena in the entire class of viscometric flows are simply related to one another. The only problem comes in adjusting κ and $\mathbf{N}$ in such a way as to make the flow dynamically possible, subject to boundary tractions and conservative body force. In this lecture we have considered in detail the two special classes of viscometric flows for which this adjustment is known to be possible exactly, and I have mentioned the other two in which it is known that an approximate solution is possible. In each case, $\mathbf{N}$ and κ were shown to depend upon the nature of the function $\tau(\kappa)$. To find all viscometric flows that are compatible for all simple fluids is an unsolved problem.

10. Impossibility of Rectilinear Flow in Pipes

Problem

In Lecture 8 the most general form of the stress possible in a simple fluid undergoing viscometric flow was determined, and in Lecture 9 certain classes of viscometric flows were shown to be dynamically possible without bringing to bear non-conservative body force. In these classes, the streamlines are the same as for a Navier-Stokes fluid in the same circumstances, but the distribution of speeds upon them is different, being determined by the shear-viscosity function $\tau(\kappa)$ of the fluid.

Up to the present, very few fixed or simply moving boundaries are known to correspond to flows for which the Navier-Stokes equations can be solved exactly. Nearly all of these give rise to viscometric flows. Those for which the analysis is elementary are exhausted by the cases analysed, at least in outline, in the last lecture. The next easiest class is defined by flow in an infinitely long tube of cross section $\mathcal{A}$, a simply-connected region of the plane. The customary procedure begins by assuming the motion to be an accelerationless linear flow in the direction of a unit vector $\underset{\sim}{k}$ normal to the cross section $\mathcal{A}$:

$$\dot{\underset{\sim}{x}} = v(\underset{\sim}{p})\underset{\sim}{k}, \quad \text{where} \quad v(\underset{\sim}{p}) = \underset{\sim}{0} \quad \text{if} \quad \underset{\sim}{p} \in \partial\mathcal{A}, \tag{10.1}$$

$\underset{\sim}{p}$ being the position vector of a point in the plane. The particles are thus assumed to move at constant speed down the lines parallel to the generators of the pipe wall. The assumption (10.1) is easily shown to be compatible with the Navier-Stokes equations, a partial differential equation for the function $v(\underset{\sim}{p})$ is derived, and it is proved to have a unique solution corresponding to the boundary condition $(10.1)_2$. Thus a unique rectilinear solution exists. Whether it is the only solution of the problem is a far more difficult matter, today unsettled.

We shall now approach the problem in the same spirit for general incompressible fluids. We shall show that in general, no rectilinear flow exists. For proof it suffices to refer to a particular case treated by ERICKSEN, who discovered this remarkable fact. A single counter-example, of course, disproves a general assertion. However, use of the apparatus set up in the previous two lectures makes it easier to see just why no rectilinear flow can be expected in general and to characterize the special cases when rectilinear flow is possible.

Explicit Constitutive Equation

In the cross section $\mathscr{A}$, let the curves θ = const. be the orthogonal trajectories of the isovels, v = const., so that a three-dimensional co-ordinate system is given by

$$x_1 = v(\underset{\sim}{p}), \quad x_2 = z, \quad x_3 = \theta(\underset{\sim}{p}), \tag{10.2}$$

where z is the distance from some fixed cross section and

$$\nabla\theta \cdot \nabla v = 0. \tag{10.3}$$

(Since only functions of place $\underset{\sim}{x}$ are used here, the longer symbol "grad" for the gradient operator need not be used.) The covariant and contravariant components g_{km} and g^{km} of the unit tensor are

$$g^{11} = g_{11}^{-1} = (\nabla v)^2, \quad g^{22} = g_{22} = 1, \quad g^{33} = g_{33}^{-1} = (\nabla\theta)^2. \qquad (10.4)$$

Exercise 10.1

Prove that the flow is viscometric, that

$$\kappa = |\nabla v|, \qquad (10.5)$$

and that $\underset{\sim}{N}$ has the form (8.7) with respect to an orthonormal basis tangent to the co-ordinate curves.

Because of this result, we may apply (8.14) and obtain

$$T\langle vz\rangle = \tau(\kappa), \quad T\langle z\theta\rangle = 0, \quad T\langle v\theta\rangle = 0,$$

$$T\langle vv\rangle - T\langle\theta\theta\rangle = \sigma_1(\kappa), \qquad (10.6)$$

$$T\langle zz\rangle - T\langle\theta\theta\rangle = \sigma_2(\kappa).$$

Let $\underset{\sim}{i}$ and $\underset{\sim}{j}$ be unit vectors in the directions of ∇v and $\nabla\theta$, so that in particular

$$\nabla v = \underset{\sim}{i}\kappa, \qquad (10.7)$$

and

$$\underset{\sim}{T} = T\langle vz\rangle(\underset{\sim}{i}\otimes\underset{\sim}{k}+\underset{\sim}{k}\otimes\underset{\sim}{i}) + T\langle vv\rangle(\underset{\sim}{i}\otimes\underset{\sim}{i}) + T\langle zz\rangle(\underset{\sim}{k}\otimes\underset{\sim}{k})$$
$$+ T\langle\theta\theta\rangle(\underset{\sim}{j}\otimes\underset{\sim}{j}). \qquad (10.8)$$

Replacing $\underset{\sim}{j}\otimes\underset{\sim}{j}$ in the last term by $\underset{\sim}{1} - \underset{\sim}{i}\otimes\underset{\sim}{i} - \underset{\sim}{k}\otimes\underset{\sim}{k}$ and using (10.6), we find that

$$\underset{\sim}{T} = (\tau\underset{\sim}{i}+\sigma_2\underset{\sim}{k})\otimes\underset{\sim}{k} + \underset{\sim}{k}\otimes\tau\underset{\sim}{i} + \kappa\underset{\sim}{i}\otimes\frac{\sigma_1}{\kappa}\underset{\sim}{i} + T\langle\theta\theta\rangle\underset{\sim}{1}, \tag{10.9}$$

where the argument κ of τ, σ_1, and σ_2 is not written. The last term is not determined by the flow.

Dynamical Equation

Exercise 10.2

Prove that

$$\kappa\nabla\kappa = \kappa\nabla(\kappa\underset{\sim}{i})\underset{\sim}{i}. \tag{10.10}$$

Using this fact and the identity

$$\operatorname{div}(\underset{\sim}{u}\otimes\underset{\sim}{v}) = (\nabla\underset{\sim}{u})\underset{\sim}{v} + \underset{\sim}{u}\operatorname{div}\underset{\sim}{v}, \tag{10.11}$$

show that

$$\operatorname{div}\underset{\sim}{T} = \underset{\sim}{k}\operatorname{div}(\tilde{\mu}\nabla v) + \nabla v\operatorname{div}\left(\frac{\sigma_1}{\kappa^2}\nabla v\right) + \frac{\sigma_1}{\kappa}\nabla\kappa + \nabla T\langle\theta\theta\rangle, \tag{10.12}$$

where

$$\tilde{\mu}(\kappa) \equiv \frac{\tau(\kappa)}{\kappa}. \tag{10.13}$$

If we set

$$h = T\langle\theta\theta\rangle + \int \frac{\sigma_1(\kappa)}{\kappa} d\kappa - \rho\upsilon, \tag{10.14}$$

then

$$\nabla h = \nabla T\langle\theta\theta\rangle + \frac{\sigma_1(\kappa)}{\kappa}\nabla\kappa + \rho\underset{\sim}{b}, \tag{10.15}$$

where $\underset{\sim}{b} = -\operatorname{grad}\upsilon$. Accordingly, Cauchy's first law (3.27) assumes the form

$$\underset{\sim}{k}\ \operatorname{div}(\tilde{\mu}\nabla v) + \nabla v \operatorname{div} \frac{\sigma_1}{\kappa^2}\nabla v + \nabla h = 0. \tag{10.16}$$

From this result, then, ∇h is independent of z. Therefore h is of the form

$$h = za + g(\underset{\sim}{p}), \tag{10.17}$$

where a = const., and (10.16) splits into the following two equations:

$$\operatorname{div}(\tilde{\mu}(\kappa)\nabla v) = -a, \tag{10.18}$$

where a is a constant, and

$$\nabla v \operatorname{div} \frac{\sigma_1(\kappa)}{\kappa^2}\nabla v + \nabla g = 0. \tag{10.19}$$

The second equation states that g is constant along the isovels v = const. That is, $g(\underset{\sim}{p}) = f(v(\underset{\sim}{p}))$, and (10.19) becomes

$$\operatorname{div} \frac{\sigma_1(\kappa)}{\kappa^2}\nabla v = -f'(v). \tag{10.20}$$

Compatibility

For any given fluid, the two functions $\tilde{\mu}(\kappa)$ and $\sigma_1(\kappa)$ are given. Accordingly, we have derived *two* nonlinear partial differential equations, (10.18) and (10.20), to be satisfied by the two functions $v(\underset{\sim}{p})$ and $f(v)$. The function $v(\underset{\sim}{p})$ is to be found; we may choose $f(v)$ at will, if we can.

In some particular cases, these two equations are compatible.For example, if the fluid is such that

$$\sigma_1(\kappa) = c\kappa^2\tilde{\mu}(\kappa), \tag{10.21}$$

then by the choice $f'(v) = ca$, (10.20) becomes identical with (10.18). This case includes the Navier-Stokes theory, for which $c = 0$, and, more generally, any theory for which $\sigma_1(\kappa) = 0$. In the Navier-Stokes theory, (10.18) becomes $\mu\nabla v = -a$, where μ is the viscosity and a is the specific driving force, both assigned. The equation has a unique solution satisfying the boundary condition $(10.1)_2$.

Exercise 10.3

Prove that if (10.18) and (10.20) are compatible, then

$$T\langle zz\rangle - \rho\upsilon = za + \sigma_2(\kappa) - \int\frac{\sigma_1(\kappa)}{\kappa}d\kappa + f(v). \tag{10.22}$$

Hence

$$\partial_z[T\langle zz\rangle - \rho\upsilon] = a. \tag{10.23}$$

Thus a is the specific driving force.

More generally, we expect, though it has not been proved, that (10.18) by itself, with assigned $\tilde{\mu}(\kappa)$, is again sufficient to determine a unique v(p) satisfying the boundary condition. If this is so, then such a v will generally fail to satisfy (10.20). Again there are exceptions. If the curves v = const. are concentric circles or parallel straight lines, then $\kappa = |\nabla v| = g(v)$ and (10.20) is always satisfied. ERICKSEN has shown that if $\tilde{\mu}$ is analytic and if (10.21) does not hold, there are always solutions of (10.18) for which (10.20) fails to hold, and he has conjectured that in fact (10.18) and (10.20) are compatible if and only if either (10.21) holds or $|\nabla v| = f(v)$.

In summary, *for a general fluid rectilinear flow in a tube is possible only for exceptional cross sections; for general cross sections, only for exceptional fluids.*

ERICKSEN conjectures that in the general case, a non-rectilinear solution will exist. A departure from a classical streamline pattern is generally described as a "secondary flow". The secondary flow in this case will be a component of velocity normal to the generators of the pipe, as a result of which the fluid moves along spiraliform streamlines.

We may see in advance that calculation of such a flow will be intricate. Indeed, if we assume that the shear-viscosity function $\tau(\kappa)$ and the normal-stress function $\sigma_1(\kappa)$ may be expanded in series in κ, *e.g.* (8.25), then (10.21) is always satisfied to the second order. The effect of incompatibility, then, must be of at least third order in some parameter whose smallness keeps the shearings small. It is natural to seek such a parameter in the specific driving force $\partial_z[T\langle zz\rangle - \rho\upsilon]$. Unfortunately no general method of solving the problem is now known. A procedure of approximation has been found, but I cannot include it in this brief course. According to the result of this calculation, the effect is of the fourth order in the specific driving force, and the pattern of secondary flow is the same for all fluids. Details have been worked out for an elliptical tube, and the effect has been observed in the laboratory.

11. Elastic Materials

Statics of Simple Materials

If a material is at rest and has been so for all time, the deformation history is a constant for each particle:

$$\underset{\sim}{F}^t(\underset{\sim}{X};s) = \underset{\sim}{F}(\underset{\sim}{X}). \tag{11.1}$$

In this class of histories, therefore, anything determined by $\underset{\sim}{F}^t$ is determined by $\underset{\sim}{F}$. In particular, the constitutive equation (4.9) of a simple material reduces to

$$\underset{\sim}{T} = \underset{\sim}{\mathfrak{g}}_{\underset{\sim}{\kappa}}(\underset{\sim}{F},\underset{\sim}{X}), \tag{11.2}$$

where $\underset{\sim}{\mathfrak{g}}_{\underset{\sim}{\kappa}}$ is a tensor-valued function, the *response function* of the material with respect to $\underset{\sim}{\kappa}$. A material whose constitutive equation is of the form (11.2) for *all* deformation histories, not merely rest histories, is said to be *elastic*. In such a material, the stress at each particle is uniquely determined by the present deformation from a fixed reference configuration.

As a special case of (4.11), we see that

$$\underset{\sim}{\mathfrak{g}}_{\underset{\sim}{\kappa}_2}(\underset{\sim}{F}) = \underset{\sim}{\mathfrak{g}}_{\underset{\sim}{\kappa}_1}(\underset{\sim}{F}\underset{\sim}{P}), \tag{11.3}$$

where $\underset{\sim}{P}$ is the gradient of the deformation that carries $\underset{\sim}{\kappa}_1$ into $\underset{\sim}{\kappa}_2$. Hence the definition of an elastic material which seems to employ a particular $\underset{\sim}{\kappa}$ is invariant under change of reference configuration.

A similar definition holds for incompressible elastic materials:

138

$$\underset{\sim}{T} = -\, p\underset{\sim}{1} + \underset{\sim}{\mathfrak{g}}(\underset{\sim}{F}), \quad |\det \underset{\sim}{F}| = 1, \tag{11.4}$$

where, as usual henceforth, the argument $\underset{\sim}{X}$ and the subscript $\underset{\sim}{\kappa}$ are not written.

What has been shown is, *the class of simple materials is indistinguishable from the class of elastic materials by static experiments*. In other words, the theory of static elasticity is at the same time the *statics of simple materials*.

Reduction for Frame-Indifference

The functional Eq. (4.16) expressing frame-indifference reduces for elastic materials to

$$\underset{\sim}{\mathfrak{g}}(\underset{\sim}{Q}\underset{\sim}{F}) = \underset{\sim}{Q}\, \underset{\sim}{\mathfrak{g}}(\underset{\sim}{F})\underset{\sim}{Q}^{T}, \tag{11.5}$$

which holds identically in the non-singular tensor $\underset{\sim}{F}$ and the orthogonal tensor $\underset{\sim}{Q}$. The solution of this equation may be read off from (4.20):

$$\underset{\sim}{T} = \underset{\sim}{R}\, \underset{\sim}{\mathfrak{g}}(\underset{\sim}{U})\underset{\sim}{R}^{T}. \tag{11.6}$$

Many other reduced forms are possible.

Elastic Fluids

An elastic material may be a fluid, a crystal, or a solid. If it is isotropic but not solid, of course it is a fluid. From (6.7) we see at once that the constitutive equation of an elastic fluid is

$$\tilde{T} = -p(\rho)\tilde{1}, \tag{11.7}$$

while for an incompressible elastic fluid

$$\tilde{T} = -p\tilde{1}, \tag{11.8}$$

where p is arbitrary.

Natural States

A stress-free configuration is called a *natural state*. If such a configuration is used as reference,

$$\tilde{\mathcal{G}}(\tilde{1}) = \tilde{0}. \tag{11.9}$$

In the applications to solids it is customary, though often not necessary, to assume that the elastic material does have a natural state. In fluid mechanics it is customary to impose conditions such as to insure that $p(\rho) > 0$ if $\rho > 0$. If $\rho_{\kappa} \neq 0$ in some $\tilde{\kappa}$, the result (2.3) shows that $\rho \neq 0$ in all configurations obtainable by smooth deformations. Hence such fluids have no natural states. In these lectures we shall not assume that a material has a natural state without saying so expressly.

Isotropic Elastic Materials

By (6.5) an isotropic elastic material has a constitutive equation of the form

$$\tilde{T} = \tilde{f}(\tilde{B}), \tag{11.10}$$

where by (6.6)

$$\mathbf{f}(\mathbf{Q}\mathbf{B}\mathbf{Q}^T) = \mathbf{Q}\mathbf{f}(\mathbf{B})\mathbf{Q}^T, \tag{11.11}$$

identically in the positive-definite symmetric tensor $\mathbf{B}$ and the orthogonal tensor $\mathbf{Q}$.

Exercise 11.1

Without recourse to the general theorems given earlier, prove (11.10) and (11.11) directly from the definitions of "elastic material" and "isotropic material". Prove, conversely, that if (11.10) and (11.11) hold, the material is an isotropic elastic material referred to an undistorted state.

The functional Eq.(11.11) is solved, in general and explicitly, by the *Rivlin-Ericksen representation theorem,*[1] yielding *Finger's theorem:*

$$\mathbf{T} = \aleph_0\mathbf{1} + \aleph_1\mathbf{B} + \aleph_2\mathbf{B}^2, \tag{11.12}$$

where the response coefficients $\aleph_\Gamma$ are scalar invariants of $\mathbf{B}$ and hence have representations of the form

$$\aleph_\Gamma = \aleph_\Gamma(I, II, III), \quad \Gamma = 1,2,3, \tag{11.13}$$

where I, II and III are the principal invariants of $\mathbf{B}$:

[1]The Rivlin-Ericksen theorem, which refers to tensor functions in general, is not to be confused with special cases concerning tensor polynomials or polynomial functions, which were obtained in old researches on invariant theory and were rediscovered by REINER and others. A proof of the Rivlin-Ericksen theorem is outlined in § 12 of *The Non-Linear Field Theories of Mechanics*.

$$\mathrm{I} \equiv \operatorname{tr} \underset{\sim}{\mathrm{B}}, \quad \mathrm{II} \equiv \tfrac{1}{2}[(\operatorname{tr} \underset{\sim}{\mathrm{B}})^2 - \operatorname{tr} \underset{\sim}{\mathrm{B}}^2], \quad \mathrm{III} \equiv \det \underset{\sim}{\mathrm{B}}. \tag{11.14}$$

In view of the Hamilton-Cayley equation, Finger's theorem may be put into the alternative form

$$\underset{\sim}{\mathrm{T}} = \beth_0 \underset{\sim}{1} + \beth_1 \underset{\sim}{\mathrm{B}} + \beth_{-1} \underset{\sim}{\mathrm{B}}^{-1}, \tag{11.15}$$

where the response coefficients $\beth_\Gamma$ are represented by functions of I, II, and III.

Likewise, for an incompressible isotropic elastic material we have *Rivlin's theorem:*

$$\underset{\sim}{\mathrm{T}} = -p\underset{\sim}{1} + \beth_1 \underset{\sim}{\mathrm{B}} + \beth_{-1} \underset{\sim}{\mathrm{B}}^{-1}, \tag{11.16}$$

where

$$\beth_\Gamma = \beth_\Gamma(\mathrm{I}, \mathrm{II}), \quad \Gamma = \pm 1. \tag{11.17}$$

These explicit formulae played a great part in the early work on explicit solutions in finite elasticity, 1948 - 1952. Before illustrating their use, we remark that from them it can be demonstrated that *the stress on any undistorted state is hydrostatic*, and that *a conformal deformation always carries one undistorted state into another*. (A converse of the latter statement can be proved for solids: all undistorted states of any given solid may be obtained by conformal deformations from any one undistorted state; however, the converse of the former statement is false for solids.)

Exercise 11.2

Develop the clasical theory of infinitesimal elastic deformation from a natural state by introducing suitable approximations into the equations given above.

The particular homogeneous deformation called *simple shear* serves as illustration of the type of tractions required. Such a deformation is given by (8.8) at any one fixed instant; $\kappa t = K$, say. Thus

$$x_1 = X_1,$$

$$x_2 = X_2 + KX_1, \tag{11.18}$$

$$x_3 = X_3.$$

The constant κ is called the *amount* of shear.

Since

$$[\underset{\sim}{F}] = \begin{Vmatrix} 1 & 0 & 0 \\ \kappa & 1 & 0 \\ 0 & 0 & 1 \end{Vmatrix}, \tag{11.19}$$

it follows that

$$[\underset{\sim}{B}] = [\underset{\sim}{F}\underset{\sim}{F}^T] = \begin{Vmatrix} 1 & \kappa & 0 \\ \kappa & 1+\kappa^2 & 0 \\ 0 & 0 & 1 \end{Vmatrix},$$

$$[\underset{\sim}{B}^{-1}] = \begin{Vmatrix} 1+\kappa^2 & -\kappa & 0 \\ -\kappa & 1 & 0 \\ 0 & 0 & 1 \end{Vmatrix}, \tag{11.20}$$

$$I = \text{tr}\, \underset{\sim}{B} = 3 + \kappa^2 = II = \text{tr}\, \underset{\sim}{B}^{-1}, \quad III = 1.$$

In the theory of infinitesimal elastic deformation from a natural state, a simple shear may be effected by applying a proportional shear stress in the same direction. In an isotropic elastic body in finite strain such a simple correlation of effects is no longer possible. As we shall see, shear stress alone can never produce simple shear.

We shall consider only homogeneous isotropic bodies referred to a homogeneous configuration.

Exercise 11.3

Prove that any static homogeneous deformation may be effected in any homogeneous simple material by application of suitable surface tractions alone.

Unconstrained Bodies

To find the stress required to effect the simple shear (11.18) in an unconstrained body, we substitute (11.20) into (11.16). If we set

$$\hat{\beth}_\Gamma \equiv \hat{\beth}_\Gamma(\kappa^2) \equiv \hat{\beth}_\Gamma(3+\kappa^2, 3+\kappa^2, 1), \tag{11.21}$$

the result is

$$\underset{\sim}{T} = (\hat{\beth}_0 + \hat{\beth}_1 + \hat{\beth}_{-1})[\underset{\sim}{1}] + \kappa(\hat{\beth}_1 - \hat{\beth}_{-1}) \begin{Vmatrix} 0 & 1 & 0 \\ 1 & 0 & 0 \\ 0 & 0 & 0 \end{Vmatrix}$$

$$+ \kappa^2 \beth_1 \begin{Vmatrix} 0 & 0 & 0 \\ 0 & 1 & 0 \\ 0 & 0 & 0 \end{Vmatrix} + \kappa^2 \beth_{-1} \begin{Vmatrix} 1 & 0 & 0 \\ 0 & 0 & 0 \\ 0 & 0 & 0 \end{Vmatrix} . \qquad (11.22)$$

In particular, the relation between shear stress $T\langle 12\rangle$ and amount of shear κ is

$$T\langle 12\rangle = \kappa\hat{\mu}(\kappa^2), \qquad (11.23)$$

where

$$\hat{\mu}(\kappa^2) \equiv \hat{\beth}_1(\kappa^2) - \hat{\beth}_{-1}(\kappa^2). \qquad (11.24)$$

Thus, as expected, $T\langle 12\rangle$ is an odd function of κ, and any such function is compatible with the theory based on (11.16). The function $\hat{\mu}(\kappa^2)$ is the generalized *shear modulus* of the material in the particular undistorted state used as reference. The classical shear modulus $\mu \equiv \hat{\mu}(0)$. If the response coefficients $\beth(I,II,III)$ are differentiable at the argument $(1,1,1)$,

$$\hat{\mu}(\kappa^2) = \mu + O(\kappa^2). \qquad (11.25)$$

Therefore any departure from the classical proportionality of shear stress to the amount of shear produced is an effect of at least third order in the latter.

For an elastic fluid, $\hat{\mu}(\kappa^2) = 0$ for all κ. While it is not known whether any non-fluids can satisfy this condition, in the interpretation of results we shall exclude any such material.

Returning to (11.22), we see that in order for shear stress to suffice for maintenance of simple shear, it would be necessary that $\beth_1 = \beth_{-1} = 0$, so $\hat{\mu}(\kappa^2) = 0$ by (11.24). Thus the shear stress also would vanish. That is, the response predicted by the infinitesimal theory cannot hold exactly for *any* isotropic elastic material with non-vanishing shear modulus. In general, normal tractions on the planes of shear X_3 = const., the shearing planes X_1 = const., and the normal planes X_2 = const. must be supplied in order for the shear to be effected.

From (11.22) these tractions are easily written down:

$$\frac{T\langle 33\rangle}{\kappa^2} \equiv \tau(\kappa^2) = \frac{\hat{\beth}_0 + \hat{\beth}_1 + \hat{\beth}_{-1}}{\kappa^2},$$

$$\frac{T\langle 11\rangle - T\langle 33\rangle}{\kappa^2} = \hat{\beth}_{-1}(\kappa^2), \tag{11.26}$$

$$\frac{T\langle 22\rangle - T\langle 33\rangle}{\kappa^2} = \hat{\beth}_1(\kappa^2).$$

It is impossible that all three normal tractions be equal, or even that $T\langle 11\rangle = T\langle 22\rangle$, for again $\hat{\mu}(\kappa^2) = 0$ for all shears, a case we have agreed to set aside. Thus of the normal tractions required to effect the simple shear, at most two can be equal. Indeed, from (11.26) and (11.23) we find the following *universal relation:*

$$T\langle 22\rangle - T\langle 11\rangle = \kappa T\langle 12\rangle, \tag{11.27}$$

a relation connecting three of the components of $\underset{\sim}{T}$, *no matter what* be the response coefficients $\beth_\Gamma$. A relation of this kind serves as a check on the theory as a whole. If (11.27) is not satisfied in a particular case, then no choice of the response coefficients in (11.16) can force the case in question into conformity with the theory of isotropic elastic materials.

Exercise 11.4

Let N and T be the normal and tangential tractions on the deformed configuration of a plane X_2 = const. Prove that

$$
\begin{aligned}
(1+\kappa^2)T &= \kappa(T\langle 22\rangle - T\langle 11\rangle) + (1-\kappa^2)T\langle 12\rangle, \\
(1+\kappa^2)N &= T\langle 22\rangle - 2\kappa T\langle 12\rangle + \kappa^2 T\langle 11\rangle.
\end{aligned}
\tag{11.28}
$$

Hence the universal relation (11.27) may be expressed in the form

$$\kappa T\langle 12\rangle = (1+\kappa^2)(T\langle 11\rangle - N). \tag{11.29}$$

If κ is small, $N \approx T\langle 11\rangle$. However, if $N = T\langle 11\rangle$, again we conclude that $T\langle 12\rangle = 0$. Thus, rigorously, it is impossible for the two normal tractions $T\langle 11\rangle$ and N to be equal. More generally, if the response coefficients $\beth_\Gamma$ are differentiable at the reference configuration,

$$\hat{\beth}_\Gamma(\kappa^2) = \beth_\Gamma(0) + O(\kappa^2), \tag{11.30}$$

and by (11.26) the inequality of the normal tractions is an effect, generally, of *second order* in the amount of shear. The presence of these inequalities is called the *Poynting effect*. Using a very special theory of elasticity, POYNTING noticed formulae such as (11.26). He inferred that if shear stress without these normal tractions is applied to a block, the faces will draw together or spread apart by an amount proportional to κ^2.

The mean tension is given by

$$\tfrac{1}{3}\,\mathrm{tr}\,\mathbf{T} = [\tau+\tfrac{1}{3}(\hat{\beth}_1+\hat{\beth}_{-1})]\kappa^2. \tag{11.31}$$

The presence of such a tension is called the *Kelvin effect*. If the undistorted reference configuration is a natural state, by (11.22)

$$\hat{\beth}_0(0) + \hat{\beth}_1(0) + \hat{\beth}_{-1}(0) = 0. \tag{11.32}$$

If also (11.30) holds, then $\frac{1}{3}\,\mathrm{tr}\,\mathbf{T} = O(\kappa^2)$. KELVIN obtained results of this kind from a very special theory. He inferred that if, conversely, the hydrostatic tension corresponding to (11.31) is not supplied, a sheared isotropic body will tend to contract or expand, according to the sign of (11.31), in proportion to the square of the amount of shear. Both these effects are easy to demonstrate on a block of sponge rubber.

Incompressible Bodies

Since simple shear is isochoric, it may be effected also in incompressible bodies, although the stress system required will be of a somewhat different kind.

Exercise 11.5

Find the tractions necessary to effect simple shear in an incompressible elastic body. Discuss the Poynting effect so as to illustrate the different responses of compressible and incompressible bodies in the same deformation.

Remarks

While simple shear is the most illuminating homogeneous static deformation, other cases are important as well, notably simple extension and uniform dilatation. More important than any one case is the fact that homogeneous static deformations are possible in *all* homogeneous elastic bodies, subject to surface tractions alone. Deformation fields which can be assigned in advance, without solving any differential equations, allow an experimental program for determining the nature of the stress relation. Such a program would be difficult if confined to deformation fields not known in advance to be possible, for the experimenter would have first to conjecture a form for the function he is measuring and solve the resulting problem to find the tractions such as to produce it, then apply those tractions, and check the result. With homogeneous strains it is much easier, since any homogeneous strain corresponds to *some* tractions, and these need only be measured directly. Empirical tables of $\beth_1$ and $\beth_{-1}$ for certain rubbers have been compiled in this way.

12. Non-Homogeneous Universal Solutions for Incompressible Elastic Bodies

Universal Solutions

As remarked at the end of the last lecture, static homogeneous deformations from homogeneous configurations are universal solutions in the sense that they can be effected in all homogeneous simple bodies, whatever be their functionals $\mathfrak{G}$. It is natural to ask if further static universal solutions exist. According to a theorem of ERICKSEN, for isotropic bodies subject to no internal constraints, no such solutions exist. That is, in general a static deformation field $\mathbf{F}$ which can be produced by surface tractions alone in one simple body will generally require specially adjusted body forces if it is to be produced in some other.

In incompressible isotropic materials, the situation is different. In 1949-52 RIVLIN and others discovered, by methods of trial, a number of universal solutions for cases of great interest and application. It was this work that opened the great revival of finite elasticity in the past fifteen years.

List of the Universal Solutions

Once a particular solution is known, it is a simple matter to verify it. Below are listed four families of solutions, each depending on several constants A, B, C, *etc*. In this list, capital letters denote co-ordinates in the undistorted reference configuration: X, Y, Z are rectangular Cartesian; R, Θ, Z, cylindrical polar; R, Θ, Φ, spherical polar. Small letters denote co-ordinates in the deformed configuration: x, y, z; r, θ, z; and r, θ, ϕ, with standard meanings.

In each case the list gives the mapping $x = \chi(X)$ in terms of components with respect to the co-ordinate systems indicated.

Family 1: Bending, stretching, and shearing of a rectangular block.

$$r = \sqrt{2AX}, \quad \theta = BY, \quad z = \frac{Z}{AB} - BCY, \quad AB \neq 0. \qquad (12.1)$$

Family 2: Straightening, stretching, and shearing of a sector of a circular-cylindrical tube.

$$x = \frac{1}{2}AB^2R^2, \quad y = \frac{\Theta}{AB}, \quad z = \frac{Z}{B} + \frac{C\Theta}{AB}, \quad AB \neq 0. \qquad (12.2)$$

Family 3: Inflation or eversion, bending, torsion, extension, and shear of a sector of a circular-cylindrical tube.

$$r = \sqrt{AR^2+B}, \quad \theta = c\Theta + DZ, \quad z = E\Theta + FZ, \quad A(CF-DE) = 1. \qquad (12.3)$$

Family 4: Inflation or eversion of a sector of a spherical shell.

$$r = (\pm R^3+A)^{\frac{1}{3}}, \quad \theta = \pm\Theta, \quad \phi = \Phi. \qquad (12.4)$$

These families are important because the stress systems needed to produce them illustrate the interaction of different kinds of deformations. In the infinitesimal theory, the stress corresponding to two displacements together is the sum of the stress required to produce each separately. In finite deformation, of course, superposition fails. These families of universal solutions allow us to understand just how it fails in certain cases, as we see below.

Inflation or Eversion, Torsion, and Extension of a Cylinder

Family 3 is certainly the most interesting. As an illustration of method, we shall work out the details for the most important special case included, namely, that when there is no angular shear (C = 1) and no shear of the generators of the cylinders (E = 0):

$$r = \sqrt{AR^2+B}, \quad \theta = \Theta+DZ, \quad z = FZ, \quad AF = 1. \tag{12.5}$$

First we calculate the contravariant components of B, recalling that $B^{km} = x^k{}_{,\alpha}\, x^m{}_{,\beta}\, g^{\alpha\beta}$:

$$\|B^{km}\| = \begin{Vmatrix} \frac{A^2R^2}{r^2} & 0 & 0 \\ 0 & \frac{1}{R^2}+D^2 & DF \\ 0 & DF & F^2 \end{Vmatrix} . \tag{12.6}$$

To calculate $(B^{-1})_{km}$ we may invert the matrix (12.6), or we may use the formula $(B^{-1})_{km} = X^\alpha{}_{,k}\, X^\beta{}_{,m}\, g_{\alpha\beta}$.

$$\|(B^{-1})_{km}\| = \begin{Vmatrix} \frac{r^2}{A^2R^2} & 0 & 0 \\ 0 & R^2 & -ADR^2 \\ 0 & -ADR^2 & A^2(1+D^2R^2) \end{Vmatrix}, \tag{12.7}$$

$$I = \operatorname{tr} \underset{\sim}{B} = g_{km}B^{km} = \frac{A^2R^2}{r^2} + r^2\left(\frac{1}{R^2} + D^2\right) + F^2,$$

$$II = \operatorname{tr} \underset{\sim}{B}^{-1} = g^{km}(B^{-1})_{km} = \frac{r^2}{A^2R^2} + \frac{R^2}{r^2} + A^2(1+D^2R^2).$$

The components of $\underset{\sim}{T} + p\underset{\sim}{1}$ follow by putting (12.6) and (12.7) into (11.16).

For the moment, not needing the full results, we remark only on three facts concerning them:

$$\underset{\sim}{T} + p\underset{\sim}{1} = \underset{\sim}{f}(r), \qquad\qquad (12.8)$$

$$T\langle r\theta\rangle = T\langle rz\rangle = 0.$$

We proceed to show that in any case when (12.8) holds, it is possible to choose p in such a way as to render the stress system $\underset{\sim}{T}$ compatible with Cauchy's first law (3.27) when $\underset{\sim}{b} = \underset{\sim}{0}$. This is so because in cylindrical co-ordinates that law then assumes the form

$$\partial_r T\langle rr\rangle + \frac{1}{r}(T\langle rr\rangle - T\langle\theta\theta\rangle) = 0,$$

$$\partial_\theta p = 0, \qquad\qquad (12.9)$$

$$\partial_z p = 0.$$

Hence $p = p(r)$, and

$$T\langle rr\rangle = -\int \frac{T\langle rr\rangle - T\langle\theta\theta\rangle}{r} dr. \qquad\qquad (12.10)$$

Conversely, if (12.10) is satisfied and $p = p(r)$, the stress system is equilibrated, subject to boundary tractions alone. An alternative form of (12.10) is

$$T\langle\theta\theta\rangle = (rT\langle rr\rangle)', \qquad\qquad (12.11)$$

since $\underset{\sim}{T}$ has now been shown to be a function of r only.

Exercise 12.1

Prove that the resultant normal traction N on the part of the plane end z = const. bounded by the circles $r = r_1$ and $r = r_2$ is

$$N = \pi r^2 T_{\langle rr\rangle}\Big|_{r_1}^{r_2} + \pi\int_{r_1}^{r_2}(2T_{\langle zz\rangle}-T_{\langle rr\rangle}-T_{\langle\theta\theta\rangle})r\,dr. \tag{12.12}$$

Choice of the constant of integration in (12.10) allows us to render any one cylinder, say $r = r_1$ or $r = r_2$, free of traction, but generally not more than one. By (12.12), it is not generally possible to produce the deformation without supplying definite normal tensions on the planes z = const., and these tensions generally have a non-zero resultant N.

The full solution of the problem is now obtained by substituting (12.6) and (12.7) into (11.16). The shear stress is given directly:

$$T_{\langle\theta z\rangle} = DFR^2\left(\frac{r}{R^2}\beth_1-\frac{A^2}{r}\beth_{-1}\right). \tag{12.13}$$

Explicit formulae result also for, say, $T_{\langle rr\rangle} - T_{\langle\theta\theta\rangle}$ and for $T_{\langle zz\rangle} - T_{\langle rr\rangle}$. To calculate $T_{\langle rr\rangle}$, we substitute the former in the right-hand side of (12.10).

The results will be illustrated now in major special cases.

Torsion and Tension of a Solid Cylinder

Set B = 0, $r_1 = R_1 = 0$ in (12.5), and choose the constant of integration in (12.10) so that $T_{\langle rr\rangle} = 0$ when $R = R_2$. The deformation then represents a twist D/F superimposed upon a longitudinal stretch F in a solid cylinder with free mantle. The resultant torque T and resultant normal tension N on the plane ends are given by (12.13) and (12.12):

$$\frac{T}{D} = \frac{1}{D}\int_0^{r_2} rT\langle z\theta\rangle \cdot 2\pi r dr,$$

$$= \frac{2\pi}{F}\int_0^{R_2} (\beth_1 - \frac{1}{F}\beth_{-1})R^3 dR. \tag{12.14}$$

$$N = 2\pi(F - \frac{1}{F^2})\int_0^{R_2} (\beth_1 - \frac{1}{F}\beth_{-1})R dR - \frac{\pi D^2}{F^2}\int_0^{R_2} (\beth_1 - \frac{2}{F}\beth_{-1})R^3 dR.$$

Since the arguments of $\beth_1$ and $\beth_{-1}$ are I and II as given by (12.7), T and N are both sums of two functions of F and D which are not functions of either variable alone. True, the first integral in the expression for N vanishes if the extension vanishes (F = 1), and the second one vanishes if the twist vanishes (D = 0), but each is a function of both F and D in general. Thus it is impossible to separate the normal tension into a part "due" to the twist and another "due" to the stretch.

Three major problems are solved by (12.15). In *Coulomb's problem* we seek the relation between torque and twist for small twist. Thus we hold F fixed and let D approach 0. If we set

$$\tilde{\beth}_\Gamma(F) \equiv \beth_\Gamma(F^2 + \frac{2}{F}, \frac{1}{F^2} + 2F), \tag{12.15}$$

then (12.15) yields

$$\tau(F) \equiv \lim_{D\to 0} \frac{T}{D/F} = \tfrac{1}{2}\pi R_2^4(\tilde{\beth}_1 - \tfrac{1}{F}\tilde{\beth}_{-1}),$$
$$N_0(F) \equiv \lim_{D\to 0} N = \pi R_2^2(F-\tfrac{1}{F^2})(\tilde{\beth}_1 - \tfrac{1}{F}\tilde{\beth}_{-1}). \tag{12.16}$$

N_0 is the resultant tension needed to produce the stretch F when there is no twist, while τ is the torsional modulus at the extension corresponding to N_0. Both are determined as rather simple functions of F when the response coefficients $\beth_\Gamma$ are known. One important result may be obtained without knowing $\beth_\Gamma$, namely,

$$\frac{R_2^2 N_0(F)}{\tau(F)} = 2(F-\frac{1}{F^2}). \tag{12.17}$$

This elegant *universal relation* of RIVLIN may be regarded as solving *Coulomb's problem* for a circular cylinder of incompressible isotropic elastic material. If $N_0(F)$ is determined in any way, *e.g.* empirically, $\tau(F)$ may be calculated. The results of a tensile test enable us to predict the results of a torsion test. It is theorems of this kind, connecting one phenomenon with another, that continuum mechanics prizes above everything else.

In *Poynting's problem* we seek the elongation that results from torsion of a cylinder with free ends. To obtain the solution from (12.15), hold D fixed, set N = 0, and solve for F. It is not known whether a solution exists in general, but for small twist it does.

Exercise 12.2

Prove that

$$\lim_{D\to 0} \frac{F-1}{(D/F)^2} = \frac{1}{12}R_2^2 \frac{\mu - \tilde{\beth}_{-1}(1)}{\mu} . \tag{12.18}$$

This result, also, is of central importance for the theory. First, it shows that the effect of small torsion is to produce extension ultimately proportional to the square of the twist. Second, there have been many attempts, using special and spurious arguments, to calculate the magnitude of the Poynting effect in terms of the concepts of the infinitesimal theory. In that theory, for isotropic incompressible materials, there is but a single elastic modulus, namely, μ. The exact and general result (12.18) shows that any such attempt is certain to fail, for *two* moduli, μ and $\beth_{-1}(1)$, are required. Thus it is impossible to describe the Poynting effect properly without information from *outside* the infinitesimal theory. Third, (12.18) predicts that torsion of a solid incompressible isotropic elastic cylinder results in

$$\begin{matrix}\text{elongation}\\ \text{contraction}\end{matrix} \quad \text{if} \ \frac{\tilde{\beth}_{-1}(1)}{\mu} \begin{matrix}<\\ >\end{matrix} 1. \qquad (12.19)$$

Experiments on homogeneous strains of rubber sheets give values of $\beth_{-1}(I,II)$ which are negative for all values of I and II. We expect, then, that cylinders of these same rubbers will *always elongate* in torsion, and they do, as POYNTING observed in 1913.

In some books for engineers may be found an argument, deriving from YOUNG and MAXWELL, which shows that if a wire is idealized as being a bundle of perfectly slippery inextensible rods, it will always shorten when twisted. Neither YOUNG nor MAXWELL claimed this model to be a good one, but the textbooks, always drawn to a bad "intuitive" theory even if contradicted by experiment, continue to present it. If a rod is idealized as a bundle of wires, those wires are not perfectly smooth; they exert both normal and tangential tractions upon one another, and the only known way to calculate the effects of such unknown tractions is to forget about the wires and solve the problem by real elasticity theory. The result, for incompressible materials is (12.18). It shows that there is no *a priori* reason to expect one sign or the other, but it enables us to *correlate* two classes of experiments. The data taken *empirically* or homogeneous strains enable us to *predict* lengthening, not only in quality but in amount. Again, it is results of this type that continuum mechanics is designed to deliver.

Finally, we consider *pure torsion* by setting F = 1. Comparison of (12.13) with (11.24) now yields

$$T_{\langle\theta z\rangle} = Dr\hat{\mu}(D^2r^2), \tag{12.20}$$

where $\hat{\mu}(\kappa^2)$ is the generalized shear modulus. This result confirms the common intuitive claim that torsion is "equivalent" or "analogous" to shear, since the shear stress in torsion is determined by the shear modulus alone. Similarly, the torsional modulus determines the resultant torque by $(12.14)_2$. While "intuition" has been confirmed, its triumph stops at this point. In an unconstrained material the shear modulus is well defined because simple shear is a universal solution. In view of Ericksen's theorem, stated at the beginning of this lecture, torsion is not a universal solution for unconstrained materials. *A fortiori*, response in torsion cannot be determined by any quantity defined by a universal solution. In particular, it *cannot* be determined by the shear modulus.

Eversion

A natural problem for study in elasticity is furnished by eversion. If an infinitely long hollow cylinder is turned inside out, it will assume some everted shape, presumably again a circular cylinder, subject to vanishing surface tractions on its mantles. This phenomenon was demonstrated in the first lecture. The problem, then, for a tube of inner and outer radii R_1 and R_2, where $R_1 < R_2$, is to determine the possible radii r_1 and r_2 on which the traction vanishes. The trivial solution is $r_1 = R_1$, $r_2 = R_2$. If a second solution exists in which $r_1 > r_2$, it corresponds to eversion. To solve this problem we return to (12.5) and set D = 0, leaving A and B to be adjusted. We choose the constant of integration in (12.11) so that T<rr> = 0 when $R = R_1$. The condition that T<rr> = 0 when $R = R_2$ is

$$0 = \int_{R_1}^{R_2} \left[\frac{R^2}{(R^2+B/A)^2} - \frac{1}{R^2}\right]\left[\beth_1 - \frac{1}{A^2}\beth_{-1}\right] R\,dR. \tag{12.21}$$

A second equation is obtained by setting N = 0 in the appropriate special case of (12.12).

Exercise 12.3

Prove that the condition N = 0 takes the form

$$0 = \int_{R_1}^{R_2} \left[\left(\frac{A^2R^2}{AR^2+B} - \frac{2}{A^2} + \frac{AR^2+B}{R^2}\right)\beth_1 + \left(\frac{AR^2+B}{A^2R^2} - 2A^2 + \frac{R^2}{AR^2+B}\right)\beth_{-1}\right] R\,dR. \tag{12.22}$$

A and B are to be found by solving (12.21) and (12.22). If a solution exists in which A < 0, it represents eversion. The corresponding stretch is given by F = 1/A. In special cases, such a solution is known to exist.

Exercise 12.4

In the special case when $\beth_1$ is a positive constant and $\beth_{-1} = 0$, calculate F explicitly.

As yet, no explanation is known for the second phenomenon demonstrated in the first lecture and there called "buckling by eversion".

Remarks on Method

Most of the solutions listed were first found by a method of trial. Once one has an idea that a certain deformation should be a universal solution, one has only to try it out. ERICKSEN set up and pursued a systematic attack on the problem of finding *all* universal solutions. He published his analysis of the problem, a difficult piece of work, in 1954. In the course of it, he discovered Family 2. His results show that beyond homogeneous strain and the four families listed, there are only two possibilities, highly overdetermined, for further identical solutions. I did not read his paper so carefully as I ought have, and in my Dallas lectures of 1960 I claimed that he had proved further universal deformation impossible. Since then, I have tried to correct this misrepresentation, but in any case, until quite recently it was commonly conjectured that only proof was lacking to fill the gaps, and that no more universal solutions could be found. Long unproved conjectures, however, often turn out to be false, and last spring two students at Brown, M. KLINGBEIL and M. SYNGH, working with SHIELD and PIPKIN, respectively, found a new universal solution:

$$r = AR, \quad \theta = B \log R + C\Theta, \quad z = Z/(A^2C). \qquad (12.23)$$

An easy calculation shows that the principal stretches for this deformation are constant, but $\underset{\sim}{F}$ is not homogeneous, since the principal axes of $\underset{\sim}{U}$ and the axis of $\underset{\sim}{R}$ are functions of r.

13. Hyperelastic Materials

The Piola-Kirchhoff Stress Tensor

For problems concerning elastic non-fluids a description from start to finish in terms of a reference configuration is often useful. To this end we write the resultant contact force on $\mathcal{P}_\chi$ as an integral of a traction vector $\underset{\sim}{t}_\kappa$ over the boundary $\partial \mathcal{P}_\kappa$ of $\mathcal{P}$ in the reference configuration $\underset{\sim}{\kappa}$:

$$\underset{\sim}{f}_c(\mathcal{P}) = \int_{\partial \mathcal{P}_\chi} \underset{\sim}{t} ds(\underset{\sim}{x}) = \int_{\partial \mathcal{P}_\kappa} \underset{\sim}{t}_\kappa ds(\underset{\sim}{X}), \qquad (13.1)$$

where $\underset{\sim}{x} = \underset{\sim}{\chi}(\underset{\sim}{X})$ as usual. Thus $\underset{\sim}{t}_\kappa$ is parallel to $\underset{\sim}{t}$, but its magnitude is adjusted according to the local change of area:

$$\underset{\sim}{t}_\kappa = \frac{ds(\underset{\sim}{x})}{ds(\underset{\sim}{X})} \underset{\sim}{t} . \qquad (13.2)$$

Because of Cauchy's fundamental lemma (3.12) there exists a tensor $\underset{\sim}{T}_\kappa(\underset{\sim}{X},t)$ such that

$$\underset{\sim}{t}_\kappa = \underset{\sim}{T}_\kappa \underset{\sim}{n}_\kappa = \frac{ds(\underset{\sim}{x})}{ds(\underset{\sim}{X})} \underset{\sim}{T} \underset{\sim}{n}, \qquad (13.3)$$

where $\underset{\sim}{n}_\kappa$ is the outer unit normal at $\underset{\sim}{X}$ in $\underset{\sim}{\kappa}$. But from integral calculus

$$\underset{\sim}{n} = J(\underset{\sim}{F}^{-1})^T \underset{\sim}{n}_\kappa \frac{ds(\underset{\sim}{X})}{ds(\underset{\sim}{x})} . \qquad (13.4)$$

Hence

$$\underset{\sim}{T}_\kappa = J\underset{\sim}{T}(\underset{\sim}{F}^{-1})^T, \qquad \underset{\sim}{T} = J^{-1}\underset{\sim}{T}_\kappa \underset{\sim}{F}^T. \qquad (13.5)$$

The tensor $\underset{\sim}{T}_{\underset{\sim}{\kappa}}$ is the *Piola-Kirchhoff stress tensor*.

It is important to notice that while $\underset{\sim}{T}$ and $\underset{\sim}{T}_{\underset{\sim}{\kappa}}$ serve to determine the traction upon every part of the body, neither determines the other unless the deformation $\underset{\sim}{F}$ from $\underset{\sim}{\kappa}$ to the present configuration $\underset{\sim}{\chi}$ be specified. The Cauchy stress $\underset{\sim}{T}$ delivers the present tractions alone and completely; the Piola-Kirchhoff stress $\underset{\sim}{T}_{\underset{\sim}{\kappa}}$ delivers those same tractions only if $\underset{\sim}{F}$ is specified.

Exercise 13.1

Prove that Cauchy's laws (3.27) and (3.29) assume the following forms in terms of $\underset{\sim}{T}_{\underset{\sim}{\kappa}}$:

$$\text{Div}\, \underset{\sim}{T}_{\underset{\sim}{\kappa}} + \rho_{\underset{\sim}{\kappa}}\underset{\sim}{b} = \rho_{\underset{\sim}{\kappa}}\ddot{\underset{\sim}{x}}, \qquad \underset{\sim}{T}_{\underset{\sim}{\kappa}}\underset{\sim}{F}^T = \underset{\sim}{F}\underset{\sim}{T}_{\underset{\sim}{\kappa}}^T, \tag{13.6}$$

where the capital D on "Div" serves to remind us that $\underset{\sim}{T}_{\underset{\sim}{\kappa}}$ is regarded as a function of $\underset{\sim}{X}$ here.

The differential equation (13.6) is to be satisfied in the interior of the reference configuration $\mathcal{B}_{\underset{\sim}{\kappa}}$; the constitutive equation of the material is assumed to be such that (13.6) is satisfied identically.

Exercise 13.2

Prove that the work storage W may be expressed in the following form, alternative to (3.30):

$$W = \int_{\mathcal{B}_{\underset{\sim}{\kappa}}} w_{\underset{\sim}{\kappa}}\, dV_{\underset{\sim}{\kappa}}, \quad \text{where} \quad w_{\underset{\sim}{\kappa}} = \text{tr}(\underset{\sim}{T}_{\underset{\sim}{\kappa}}\dot{\underset{\sim}{F}}^T). \tag{13.7}$$

In terms of the Piola-Kirchhoff tensor, the constitutive Eq.(11.2) of elasticity assumes the form

$$\mathbf{T}_{\kappa} = \mathfrak{h}_{\kappa}(\mathbf{F},\mathbf{X}) = \mathfrak{h}(\mathbf{F}), \tag{13.8}$$

say, while the functional Eq.(11.5) expressing frame-indifference becomes

$$\mathfrak{h}(\mathbf{Q}\mathbf{F}) = \mathbf{Q}\,\mathfrak{h}(\mathbf{F}). \tag{13.9}$$

<u>Definition of a Hyperelastic Material</u>

An elastic material is said to be *hyperelastic* if there exists a scalar function $\sigma(\mathbf{F})$ such that in all dynamic processes

$$w_{\kappa} = \rho_{\kappa}\dot{\sigma}, \tag{13.10}$$

or, equivalently,

$$w = \rho\dot{\sigma}. \tag{13.11}$$

In such a material, the work storage per unit mass is the time-derivative of a potential, $\sigma(\mathbf{F})$, which is called the *stored-energy function*.

For a hyperelastic material, then,

$$\operatorname{tr}(\mathbf{T}_{\kappa}\dot{\mathbf{F}}^{T}) = \rho_{\kappa}\operatorname{tr}[\partial_{\mathbf{F}}\sigma(\mathbf{F})\dot{\mathbf{F}}^{T}]. \tag{13.12}$$

That is,

$$0 = \mathrm{tr}[(\mathbf{T}_\kappa - \rho_\kappa \partial_{\mathbf{F}}\sigma(\mathbf{F}))\mathbf{F}^T(\dot{\mathbf{F}}\mathbf{F}^{-1})^T],$$
$$= \mathrm{tr}[(\mathbf{T}_\kappa - \rho_\kappa \partial_{\mathbf{F}}\sigma(\mathbf{F}))\mathbf{F}^T\mathbf{G}^T], \tag{13.13}$$

since $\mathbf{G} = \dot{\mathbf{F}}\mathbf{F}^{-1}$. This equation must hold for all tensors $\mathbf{G}$. Since $\mathbf{G}$ and $\mathbf{F}$ may be selected independently in a dynamic process, and since $\mathrm{tr}(\mathbf{A}\mathbf{B}^T)$ defines an inner product in the nine-dimensional space of tensors regarded as vectors, (13.13) asserts that

$$[\mathbf{T}_\kappa - \rho_\kappa \partial_{\mathbf{F}}\sigma(\mathbf{F})]\mathbf{F}^T \tag{13.14}$$

is orthogonal to every $\mathbf{G}$. Hence the former tensor is $\mathbf{0}$. That is,

$$\mathbf{T}_\kappa = \rho_\kappa \partial_{\mathbf{F}}\sigma(\mathbf{F}), \tag{13.15}$$

or, equivalently,

$$\mathbf{T} = \rho\partial_{\mathbf{F}}\sigma(\mathbf{F})\mathbf{F}^T. \tag{13.16}$$

This form of the stress relation is due to KIRCHHOFF.

Kirchhoff's result does not by itself yield a symmetric stress.

Exercise 13.3

Prove that a scalar function $\sigma(\mathbf{F})$ is frame-indifferent if and only if

$$\sigma(\mathbf{F}) = \sigma(\mathbf{U}). \tag{13.17}$$

Hence prove *Noll's theorem on hyperelastic materials:*

$$\text{Frame indifference of } \mathfrak{h} \Longleftrightarrow \text{Frame indifference of } \sigma \Longleftrightarrow \underset{\sim}{T} = \underset{\sim}{T}^T. \tag{13.18}$$

Hint: Frame indifference of $\underset{\sim}{\mathfrak{h}}$ yields

$$\sigma(\underset{\sim}{Q}\underset{\sim}{F}) - \sigma(\underset{\sim}{F}) = \sigma(\underset{\sim}{Q}) - \sigma(\underset{\sim}{1}), \tag{13.19}$$

for all orthogonal $\underset{\sim}{Q}$. Use the compactness of the orthogonal group to show that $\sigma(\underset{\sim}{Q}) = \sigma(\underset{\sim}{1})$.

Physicists usually seem to prefer to elude direct use of the principle of moment of momentum, preferring to derive it from something else. Noll's theorem affords them two ways of doing so if they restrict attention to hyperelastic materials.

The Two Isotropy Groups

Let $\mathcal{g}$ be the isotropy group of a hyperelastic material with respect to a particular reference configuration. If $\underset{\sim}{H} \in \mathcal{g}$, then by (5.1) and (13.16)

$$\begin{aligned} \partial_{\underset{\sim}{F}}\sigma(\underset{\sim}{F})\underset{\sim}{F}^T &= [\partial_{\underset{\sim}{F}}\sigma(\underset{\sim}{F}\underset{\sim}{H})](\underset{\sim}{F}\underset{\sim}{H})^T, \\ &= [\partial_{\underset{\sim}{F}}\sigma(\underset{\sim}{F}\underset{\sim}{H})]\underset{\sim}{H}^T\underset{\sim}{F}^T, \end{aligned} \tag{13.20}$$

for arbitrary $\underset{\sim}{F}$. Cancelling $\underset{\sim}{F}^T$ and integrating, we find that

$$\sigma(\underset{\sim}{F}) = \sigma(\underset{\sim}{F}\underset{\sim}{H}) + \psi(\underset{\sim}{H}), \tag{13.21}$$

for all $\underset{\sim}{F}$. Setting $\underset{\sim}{F} = \underset{\sim}{1}$ yields $\psi(\underset{\sim}{H}) = \sigma(\underset{\sim}{1}) - \sigma(\underset{\sim}{H})$. Hence (13.21) becomes

$$\sigma(\underset{\sim}{F}) = \sigma(\underset{\sim}{F}\underset{\sim}{H}) + \sigma(\underset{\sim}{1}) - \sigma(\underset{\sim}{H}). \tag{13.22}$$

Conversely, differentiation of this equation yields (13.20). We have shown that $\underset{\sim}{H} \in \mathcal{g}$ if and only if (13.22) is satisfied for all non-singular $\underset{\sim}{F}$.

We recall that $\mathcal{g}$ is the group of all deformations that cannot be detected by experiments on the stress. We now define $\mathcal{g}_\sigma$ as the group of all unimodular transformations that cannot be detected by experiments on the stored-energy σ. Formally, $\underset{\sim}{H} \in \mathcal{g}_\sigma$ if and only if

$$\sigma(\underset{\sim}{F}) = \sigma(\underset{\sim}{F}\underset{\sim}{H}) \tag{13.23}$$

for all non-singular $\underset{\sim}{F}$.

If (13.23) is satisfied, we may put $\underset{\sim}{F} = \underset{\sim}{1}$ and obtain $\sigma(\underset{\sim}{1}) = \sigma(\underset{\sim}{H})$. Hence (13.22) is satisfied also. That is,

$$\mathcal{g}_\sigma \subset \mathcal{g} \cdot \tag{13.24}$$

Furthermore, by (13.13) $\sigma(\underset{\sim}{Q}) = \sigma(\underset{\sim}{1})$. Therefore if $\underset{\sim}{H}$ is an orthogonal tensor $\underset{\sim}{Q}$, (13.22) reduces to (13.23). That is, the orthogonal subsets of $\tilde{\mathcal{g}}$ and $\mathcal{g}_\sigma$ are identical:

$$\mathcal{g} \cap \mathcal{O} = \mathcal{g}_\sigma \cap \mathcal{O}. \tag{13.25}$$

Now these results hold for every reference configuration. If the material is a solid, there exists some reference configuration for which $\mathcal{G} \subset \mathcal{O}$. By (13.24), $\mathcal{G}_\sigma \subset \mathcal{O}$. Hence $\mathcal{G} \cap \mathcal{O} = \mathcal{G}$ and $\mathcal{G}_\sigma \cap \mathcal{O} = \mathcal{G}_\sigma$. By (13.25), $\mathcal{G} = \mathcal{G}_\sigma$. If the material is a fluid, $\sigma(\underset{\sim}{F}) = f(\det \underset{\sim}{F})$, so (13.22) is satisfied for all $\underset{\sim}{H}$, and again it follows that $\mathcal{G} = \mathcal{G}_\sigma$. We have proved that *in a hyperelastic solid or fluid, the two isotropy groups* $\mathcal{G}$ *and* $\mathcal{G}_\sigma$ *are the same.* In a fluid crystal, generally, $\mathcal{G} \neq \mathcal{G}_\sigma$. That is, there are some deformations which cannot be detected by measurements on the stress but can be detected by measurements of the stored-energy function. This property does not characterize fluid crystals, however, since in some of them, too, $\mathcal{G} = \mathcal{G}_\sigma$.

Minimum of the Stored-Energy Function

If $\underset{\sim}{H} \in \mathcal{G}$, (13.22) is satisfied in particular when $\underset{\sim}{F} = \underset{\sim}{H}^{-1}$. Therefore

$$\sigma(\underset{\sim}{H}^{-1}) - \sigma(\underset{\sim}{1}) = \sigma(\underset{\sim}{1}) - \sigma(\underset{\sim}{H}). \tag{13.26}$$

If $\underset{\sim}{H}$ is orthogonal, then $\sigma(\underset{\sim}{H}) = \sigma(\underset{\sim}{1})$, and this equation reduces to $0 = 0$. If $\sigma(\underset{\sim}{H}) > \sigma(\underset{\sim}{1})$, then (13.26) shows that $\sigma(\underset{\sim}{H}^{-1}) < \sigma(\underset{\sim}{1})$, and if $\sigma(\underset{\sim}{H}) < \sigma(\underset{\sim}{1})$, then $\sigma(\underset{\sim}{H}^{-1}) > \sigma(\underset{\sim}{1})$. Suppose now the stored-energy function has a minimum in the sense that $\sigma(\underset{\sim}{F}) > \sigma(\underset{\sim}{1})$ when $\underset{\sim}{F}$ is not orthogonal. Then there can be no non-orthogonal $\underset{\sim}{H}$ satisfying (13.26). At a minimum, the gradient of σ vanishes, and hence so does $\underset{\sim}{T}$, by (13.16).

We have proved the following *fundamental theorem on hyperelastic solids:* If the stored-energy function has a minimum in a certain configuration, the material is a solid, and that configuration is an undistorted natural state.

We have shown that a strict minimum of σ implies the existence of a natural state, which is of course unique to within a rotation.

The converse, however, does not hold. Without some further condition, a natural state need not correspond to minimum energy and hence need not be unique.

Need for an *A Priori* Inequality

In the classical infinitesimal theory of elasticity, $\sigma(\underset{\sim}{F})$ is expressed by a quadratic form in $\underset{\sim}{U} - \underset{\sim}{1}$. There are various reasons for imposing the *a priori* restriction that this quadratic form be positive-definite. One of these reasons is that without this inequality, the classic theorems of uniqueness and stability fail. In the finite theory, neither uniqueness nor stability in the classic senses is desirable. The example of eversion of a tube shows that for a physically appropriate theory, the solution to the traction boundary-value problem should not always be unique. The example of the equivalence of clockwise and counterclockwise rotations of a hollow cylinder shows that the solution to the place boundary-value problem should not be unique. Since one of the main aims of finite elasticity theory is to explain instability, there should be no theorem of universal stability. On the other hand, the results of the infinitesimal are just for the intended range. These remarks show that for a good finite theory, an *a priori* inequality of *some* kind is necessary, and that it should be one that in infinitesimal strains renders the stored-energy function positive-definite. These requirements are far from sufficient to determine a finite inequality. There is now a fair body of important work on the problem. In these lectures I cannot do more than mention the most important proposal so far, an inequality developed in 1959 by COLEMAN and NOLL.

Coleman and Noll's Inequality

COLEMAN & NOLL proposed the inequality

$$\sigma(\underset{\sim}{S}\underset{\sim}{F}) - \sigma(\underset{\sim}{F}) - \mathrm{tr}[(\underset{\sim}{S}-\underset{\sim}{1})\underset{\sim}{F}(\partial_{\underset{\sim}{F}}\sigma(\underset{\sim}{F}))^T] > 0, \tag{13.27}$$

for all positive-definite symmetric tensors $\underset{\sim}{S}$ other than $\underset{\sim}{1}$; we call it the *C-N condition*. It asserts a type of restricted convexity of the function $\sigma(\underset{\sim}{F})$.

Exercise 13.4

Prove that for infinitesimal deformation from a natural state, Coleman and Noll's inequality holds if and only if the work done in any non-orthogonal deformation is positive.

If the material has a natural state, we may take it as reference configuration and conclude from (13.27) that

$$\sigma(\underset{\sim}{S}) > \sigma(\underset{\sim}{1}), \tag{13.28}$$

if $\underset{\sim}{S}$ is positive-definite, symmetric, and not equal to $\underset{\sim}{1}$. By (13.17), we have shown that

$$\text{C-N} \rightarrow \text{in a natural state, the stored-energy function has a strict minimum.} \tag{13.29}$$

This is a result complementary to the fundamental theorem on solids. Together, they show that

$$\text{C-N} \rightarrow \text{only solids can have natural states.} \tag{13.30}$$

We must recall, however, that while the fundamental theorem is perfectly general, the C-N condition is a supplementary *a priori* inequality, the merits of which are still a subject of discussion.

Exercise 13.5

The *principal stretches* v_i are defined as the proper numbers of $\underset{\sim}{U}$. Prove that in an isotropic hyperelastic material the stored-energy function equals a function of the principal stretches:

$$\sigma(\underset{\sim}{F}) = \hat{\sigma}(v_1, v_2, v_3). \tag{13.31}$$

Prove that

$$\text{C-N} \rightarrow \sigma \text{ is convex}, \tag{13.32}$$

but the converse is false. For this last, consider the example

$$\rho_{\underset{\sim}{\kappa}} \hat{\sigma}(v_1, v_2, v_3) = \kappa[\tfrac{1}{2}(v_1^2 + v_2^2 + v_3^2) - (v_1 + v_2 + v_3)]. \tag{13.33}$$

14. Work Theorems in Hyperelasticity

Nonsense About Perpetual Motion

In most presentations of the infinitesimal theory of elasticity, the stored-energy function plays a great part. Commonly, some sort of argument about work done or perpetual motion avoided is brought in to "prove" that a stored-energy function exists. As illustrated in Lectures 11 and 12, for most aspects of elasticity theory, including special problems, it makes no difference whether or not an energy function exists. In giving the stored-energy function only a subordinate place, I have wished to exert my small influence against the diffusion of the muddy verbiage that spouts forth as soon as thermodynamics is mentioned. The formal structure of the infinitesimal theory has been well known for a long time and stands up uninjured by the bad reasoning and confusion of principle various authors, especially paedagogic ones, feel themselves compelled to use in "explaining" it. The finite theory is another matter. Here, incorrect thermodynamic arguments can lead to incorrect results.

In the first place, perpetual motion has nothing to do with the matter. According to the infinitesimal theory for isotropic materials, with which critics of every school are perfectly satisfied, and in which no thermodynamic argument is needed to infer the existence of a stored-energy function, perpetual motion occurs, as it does in any theory taking no account of friction. Therefore, a person who does not like perpetual motion in a theory must reject elasticity *in toto*, with or without a stored-energy function. Second, the possibility of extracting work indefinitely, sometimes called a perpetual motion of the second kind, also has nothing to do with the presence or absence of a stored-energy function, as we shall see below.

In the general theory of elasticity, there are various means of motivating the assumption that a stored-energy function exists, or even of proving its existence within a more general framework of ideas. In each case, naturally, a *proved theorem*, not just a lot of physical talk, forms the basis.

Virtual Work of the Traction on the Boundary

Consider a one-parameter family of deformations:

$$x = \chi_{\kappa}(X,\alpha), \qquad \alpha_1 \leqq \alpha \leqq \alpha_2. \tag{14.1}$$

Since α need not be the time, such a family is called a virtual motion. We shall write

$$\dot{\chi} \equiv \partial_{\alpha}\chi_{\kappa}(X,\alpha) \tag{14.2}$$

for the *virtual velocity*. If α is in fact the time, $\dot{\chi} = \dot{x}$, the velocity. In an elastic body $\mathcal{B}$ a traction $t_{\kappa}(X,\alpha)$ on $\partial\mathcal{B}_{\kappa}$ is determined from (14.1) by Cauchy's lemma (13.3) and the constitutive equation (13.8). The *virtual work* done by this traction in the virtual motion is, by definition,

$$W_{12} \equiv \int_{\alpha_1}^{\alpha_2}\left(\int_{\partial\mathcal{B}_{\kappa}} \dot{\chi}\cdot t_{\kappa}\, ds\right)d\alpha. \tag{14.3}$$

Before going a step further we must post a blazing placard: Even if α is the time and the elastic body is made to undergo (14.1) as a real motion,

W_{12} is generally NOT the actual work done

by the forces effecting the motion. Why not? Because (14.1) generally gives rise to an acceleration $\ddot{\mathbf{x}}$ and also by (13.8) to a stress $\mathbf{T}$ in the elastic body, and by Cauchy's first law (3.27) a particular and uniquely determined body force $\mathbf{b}$ will be required in order to make the motion occur. In the unlikely case that the right $\mathbf{b}$ could be produced in the laboratory, it, too, will do work. In any honest attempt to discuss perpetual motion of the second kind, *all the work done* must be taken into account. Therefore, as stated above, *no argument about the virtual work of the traction alone can determine whether or not a perpetual motion of the second kind occurs.*

Homogeneous Processes in Homogeneous Bodies

If the deformation gradient $\mathbf{F}$ as calculated from (14.1) is independent of $\mathbf{X}$, we regard the virtual motion from α_1 to α_2 as a curve $\mathbf{F}(\alpha)$ from the point $\mathbf{F}(\alpha_1)$ to the point $\mathbf{F}(\alpha_2)$ in the space of non-singular tensors $\mathbf{F}$. If $\mathbf{F}(\alpha_1) = \mathbf{F}(\alpha_2)$, the virtual motion is said to be *closed*. If the elastic body is homogeneous, then $\mathbf{T}_{\kappa}$ is independent of $\mathbf{X}$ also: $\mathbf{T}_{\kappa} = \mathbf{T}_{\kappa}(\alpha)$. Since Div $\mathbf{T}_{\kappa} = \mathbf{0}$, (14.3) yields

$$W_{12} = \int_{\alpha_1}^{\alpha_2} \int_{\mathcal{B}_{\kappa}} \operatorname{Div}(\dot{\mathbf{\chi}}\mathbf{T}_{\kappa}) dv \, d\alpha,$$

$$= \int_{\alpha_1}^{\alpha_2} \int_{\mathcal{B}_{\kappa}} \operatorname{tr}(\mathbf{T}_{\kappa}^{T}\dot{\mathbf{F}}) dv \, d\alpha,$$

$$= \mathcal{V}(\mathcal{B}_{\underset{\sim}{\kappa}}) \int_{\alpha_1}^{\alpha_2} \operatorname{tr}(\underset{\sim}{T}_{\kappa}^{T} \dot{\underset{\sim}{F}}) d\alpha,$$

$$= \mathcal{V}(\mathcal{B}_{\underset{\sim}{\kappa}}) \int_{F(\alpha_1)}^{F(\alpha_2)} \operatorname{tr}(\underset{\sim}{T}_{\kappa}^{T} d\underset{\sim}{F}), \tag{14.4}$$

where $(13.3)_1$ has been used, where $\mathcal{V}(\mathcal{B}_{\underset{\sim}{\kappa}})$ is the volume of $\mathcal{B}_{\underset{\sim}{\kappa}}$, and where the last integral is a line integral along the path from $\underset{\sim}{F}(\alpha_1)$ to $\underset{\sim}{F}(\alpha_2)$. In what follows, we shall assume the reference configuration so chosen that $\det \underset{\sim}{F} > 0$.

The First Work Theorem

The *first work theorem* may be read off from (14.4) by using well known facts about line integrals: *For a homogeneous elastic body, the following three assertions are equivalent:*

1) The virtual work of the traction on the boundary is non-negative in any closed motion.

2) The virtual work of the traction on the boundary is the same in any two homogeneous virtual motions with the *same* initial and final deformation gradients.

3) The material is hyperelastic, and in a homogeneous virtual motion from $\underset{\sim}{F}(\alpha_1)$ to $\underset{\sim}{F}(\alpha_2)$

$$W_{12} = \mathcal{M}(\mathcal{B})[\sigma(\underset{\sim}{F}(\alpha_2)) - \sigma(\underset{\sim}{F}(\alpha_1))], \tag{14.5}$$

where $\mathcal{M}(\mathcal{B})$ is the mass of $\mathcal{B}$ and σ is the stored-energy function.

We must notice that in a hyperelastic body in an actual motion, the volume integral of $\rho_{\underset{\sim}{\kappa}}\sigma$ does *not* generally give the total work done. W_{12} may be calculated from $(14.4)_1$, which no longer reduces to $(14.4)_2$.

There is sure to be someone who rises and says, "I don't accept your claim that W_{12} is not the actual work done. In a homogeneous process in a homogeneous body, $\underset{\sim}{T}$ is homogeneous, so div $\underset{\sim}{T} = \underset{\sim}{0}$, so the body force $\underset{\sim}{b}$ equals the acceleration $\ddot{\underset{\sim}{x}}$, and if the process is slow enough, accelerations can be neglected, so there is no work done beyond W_{12}." Such people cannot be answered. It is best to bow to their wisdoms and let them depart. When they are gone, we can remind each other of a simple theorem of kinematics.

Exercise 14.1

Prove that a motion is accelerationless if and only if

$$\underset{\sim}{F}(t) = \underset{\sim}{F}_0(\underset{\sim}{1} + t\underset{\sim}{F}_1), \tag{14.6}$$

where $\underset{\sim}{F}_0$ and $\underset{\sim}{F}_1$ are constant tensors.

As a result of this theorem, the only accelerationless paths in the space of non-singular tensors $\underset{\sim}{F}$ are straight lines: $\underset{\sim}{F}(\alpha) = \underset{\sim}{F}_0(\underset{\sim}{1} + \alpha\underset{\sim}{F}_1)$. Thus if the accelerations are "negligible", in order to apply statement 1 of the theorem we must find a "nearly" closed straight line. To apply statement 2, we must find two different straight line segments that "nearly" connect the same two endpoints, in defiance of Euclid. If we consider two paths which are both "nearly" straight from $\underset{\sim}{F}(\alpha_1)$ to $\underset{\sim}{F}(\alpha_2)$, W_{12} will have "nearly" the same value for each by mere continuity, without any theorem. People who can handle all these "nearlys" have no need to call the theory of finite elastic strain to their aid. They can prove that 2 = 1 without it.

The Second Work Theorem

According to Caprioli's work theorem, *if an elastic body has a homogeneous configuration* $\underset{\sim}{\kappa}$ *such that the virtual work done in every homogeneous motion from* $\underset{\sim}{\kappa}$ *is non-negative, then the material is hyperelastic, its stored-energy function has a weak minimum at* $\underset{\sim}{\kappa}$, *and* $\underset{\sim}{\kappa}$ *is a natural state.* To prove this theorem, let the deformation gradient $\underset{\sim}{F}$ be taken with respect to $\underset{\sim}{\kappa}$, so that $\underset{\sim}{F} = \underset{\sim}{1}$ there, and let $\mathscr{C}$ be any closed homogeneous virtual motion. If the point $\underset{\sim}{F} = \underset{\sim}{1}$ does not lie on $\mathscr{C}$, connect it by a path $\mathscr{P}$ to some point $\underset{\sim}{F}_0$ which does lie on $\mathscr{C}$. Then $\mathscr{P} + \mathscr{C} - \mathscr{P}$ is a closed virtual motion from $\underset{\sim}{\kappa}$ back to $\underset{\sim}{\kappa}$. By hypothesis, for this motion $W_{12} \geqq 0$. But W_{12} for this motion is W_{12} for $\mathscr{C}$. Thus alternative 1 in the first work theorem holds. By alternative 3, the material is hyperelastic, and by (14.5) the hypothesis of Caprioli's theorem takes the form

$$\sigma(\underset{\sim}{F}) \geqq \sigma(\underset{\sim}{1}). \tag{14.7}$$

Hence $\partial_{\underset{\sim}{F}}\sigma(\underset{\sim}{F}) = \underset{\sim}{0}$ when $\underset{\sim}{F} = \underset{\sim}{1}$. Therefore $\underset{\sim}{\kappa}$ is a natural state.

The second work theorem gives an energetic criterion sufficient that there be a natural state as well as a stored-energy function. The inequality (14.7) is weaker than (13.25): The minimum has not been proved to be strict, nor can it be, as is shown by the result of the following exercise.

Exercise 14.2

Prove that the stored-energy function

$$\sigma(\underset{\sim}{F}) = \kappa(\frac{1}{J} - 1)^2 \tag{14.8}$$

does not satisfy (13.25) but does satisfy (14.7) and defines an elastic fluid with a natural state.

The Third Work Theorem

Let W_{12} be the virtual work corresponding to a homogeneous virtual motion $\underset{\sim}{F}(\alpha)$ from $\underset{\sim}{F}(\alpha_1)$ to $\underset{\sim}{F}(\alpha_2)$ when, disregarding the constitutive equation, we hold $\underset{\sim}{T}_{\underset{\sim}{\kappa}}$ at a *fixed* value $\underset{\sim}{T}^*_{\underset{\sim}{\kappa}}$. By (14.4)

$$W^*_{12} = \mathcal{V}(\mathcal{B}_{\underset{\sim}{\kappa}})\operatorname{tr}\{\underset{\sim}{T}^{*T}_{\underset{\sim}{\kappa}}[\underset{\sim}{F}(\alpha_2)-\underset{\sim}{F}(\alpha_1)]\}. \tag{14.9}$$

If $\underset{\sim}{T}^*_{\underset{\sim}{\kappa}} = \underset{\sim}{\mathfrak{h}}(\underset{\sim}{F}(\alpha_1))$, its initial value, W^*_{12} is the virtual work corresponding to *dead load*. According to Coleman's work theorem, *the following two statements are equivalent for an elastic material deformed from a homogeneous configuration:*

1) In any motion whose termini differ by a non-identical pure stretch,

$$W_{12} > W^*_{12} . \tag{14.10}$$

2) The material is hyperelastic, and its stored-energy function satisfies the C-N inequality (13.27).

To prove Coleman's theorem, we notice first that for a closed motion, $W^*_{12} = 0$ by (14.9), while by (14.10) $W_{12} \geqq W^*_{12}$. Hence in any closed motion $W_{12} \geqq 0$. By the first work theorem, the material is hyperelastic. By (14.5), (14.9), and (13.15), the hypothesis (14.10) assumed the form

$$\sigma(\bar{\mathbf{F}}) - \sigma(\mathbf{F}) > \frac{1}{\rho_{\kappa}} \operatorname{tr}\{(\bar{\mathbf{F}}-\mathbf{F})\partial_{\mathbf{F}}\sigma(\mathbf{F})^T\} \tag{14.11}$$

if $\bar{\mathbf{F}} = \mathbf{S}\mathbf{F}$, where $\mathbf{S}$ is a positive-definite symmetric tensor other than $\mathbf{1}$. This is Coleman and Noll's inequality (13.27).

Coleman's theorem serves to give a static motivation for assuming that an elastic material is hyperelastic and satisfies the C-N inequality.

We must not join the "physical" writers on elasticity in confusing the work theorems with thermodynamic arguments. While energy is concerned, heat and temperature are not. It is impossible to prove thermodynamic theorems without first introducing thermodynamics; that is, more quantities and assumptions are needed before any link between existence of a stored-energy function and the "second law of thermodynamics" can be made. A thermodynamic basis for hyperelasticity, along with results of a much more general nature, will be given in Lecture 16. It is time, therefore, to set forth the general theory of heat and temperature in deformable media.

15. Thermomechanics. Equipresence

Fading Memory

The examples we have developed and analysed concern motions with constant stretch history and the still more restricted case of statics. One could well ask why I went to the trouble of constructing a theory of materials with memory if the only cases amenable to treatment are those in which memory effects are given no chance to manifest themselves. In order to analyse real memory effects, we must recognize that the theory so far does not take account of a further property of real materials, namely, that their memory of deformation fades with time. A theory which truly requires that the entire past history of a specimen be known in order to determine its reaction to further deformation has little predictive value, since we are never in a position to know what has happened in the long past to any body we encounter. We reduce a specimen to some sort of standard state for a few years, days, or even seconds and then presume that what happened to it before that time does not matter. In a word, we assume that the effect of long past experiences is small in comparison with that of events in the recent past and at the present. The classic experiment is that of ROBERVALE, who imprisoned air in a bottle for fifteen years and found that it suffered no loss of spring. This air, it would seem, had no memory at all for the compression suffered. The large tube of sponge rubber I showed you in the first lecture was not quite cylindrical, because it had been squashed in the packing box for a week or two. At the time of the opening lecture, it still remembered that it had been in the box. Now you see it again, still everted. It has been out of the box for fifteen days. Upon release from eversion, you see it snap back to a perfect cylindrical form. Thus the material has now forgotten entirely the effect of having been deformed so as to get into the box, but it has not lost its elasticity at all.

Continuum mechanics must take account of the property of fading memory and incorporate it within the mathematical structure. There are different precise concepts of fading memory, similar, though not equivalent. The first persons to achieve a theory were COLEMAN & NOLL in 1960. They proved a general theorem on stress relaxation and found a method for approximating a general constitutive functional by certain functions in the limit of slow motions, and their theory has led to many fine results and has been put on an axiomatic basis recently. In these lectures I shall not be able to develop it, but I shall present instead a simpler substitute for one important part of it.

While it is possible to discuss fading memory in strictly mechanical situations, we know that effects of heat and temperature are often essential in deformations occurring over long periods of time, so I shall turn first to construction of the thermomechanics of simple materials.

Thermodynamics and Continuum Mechanics Reviewed

In the thermodynamics of homogeneous processes we considered the state θ, T to be a function of time only. The general principle of balance of energy was laid down, the temperature was restricted from the start by the inequality

$$\theta > 0, \tag{1.2}$$

and the constitutive equations were assumed to be of the form

$$\left.\begin{array}{l} W \\ \Psi \\ H \end{array}\right\} \text{ are functionals of } \theta^t, T^t, \tag{1.9}$$

subject to the restrictions implied by the dissipation principle:

$$\theta \dot{H} \geqq Q. \tag{1.5}$$

In continuum mechanics we laid down as general not only the balance of energy, but also the balances of momentum and moment of momentum, and the principle of determinism was taken as

$$\underset{\sim}{T} = \underset{\sim}{\mathfrak{F}} (\underset{\sim}{\chi}^t), \tag{4.2}$$

subject to the restriction of local action and material frame-indifference.

General Principles of the Thermomechanics of Continua

The general principles of the thermomechanics of continua are obtained by combining the two special theories just summarized. The field equations have been obtained already:

$$\rho \dot{\varepsilon} = \operatorname{tr}(\underset{\sim}{T}\underset{\sim}{D}) + \operatorname{div} \underset{\sim}{h} + \rho s, \tag{3.33}$$

$$\rho \ddot{\underset{\sim}{x}} = \operatorname{div} \underset{\sim}{T} + \rho \underset{\sim}{b}, \tag{3.27}$$

$$\underset{\sim}{T} = \underset{\sim}{T}^T. \tag{3.29}$$

We retain the axiom

$$\theta > 0, \tag{1.2}$$

and we suppose that the heat , H, is an additive set function:

$$H = \int_{\mathcal{P}} \eta dm, \tag{15.1}$$

where the density η is called the *specific entropy* or the *heat density*. The free-energy density ψ is defined by

$$\psi \equiv \varepsilon - \eta\theta. \tag{15.2}$$

We have proved already that

$$W = \int_{\mathcal{P}} w dv, \tag{3.30}$$

where

$$w = \operatorname{tr}(\underset{\sim}{T}\underset{\sim}{D}). \tag{3.31}$$

The *internal dissipation* δ is defined by

$$\delta \equiv w - \rho(\dot{\varepsilon}-\theta\dot{\eta}). \tag{15.3}$$

Thus δ is the stress power not compensated by growth of the internal energy in excess of the heating. By (3.33)

$$\begin{aligned} \delta &= \rho\theta\dot{\eta} - \operatorname{div} h - \rho s, \\ &= w - \rho(\dot{\psi}+\eta\dot{\theta}). \end{aligned} \tag{15.4}$$

Exercise 15.1

Show that in a Navier-Stokes fluid (8.24), if the pressure $p(\rho)$ and the specific entropy η are obtained from the specific free energy by formulae analogous to (1.13), *viz*

$$p(\rho) = \Pi(\upsilon) = -\partial_\upsilon\psi, \qquad \eta = -\partial_\theta\psi, \tag{15.5}$$

where $\psi = \psi(\upsilon,\theta)$ and υ is the specific volume $1/\rho$, then

$$\delta = \lambda(\operatorname{tr} \underset{\sim}{D})^2 + 2\mu \operatorname{tr} \underset{\sim}{D}^2. \tag{15.6}$$

Show that the *Duhem-Stokes inequalities:*

$$\mu \geqq 0, \qquad 3\lambda + 2\mu \geqq 0, \tag{15.7}$$

are necessary and sufficient that $\delta \geqq 0$ for all $\underset{\sim}{D}$.

Thermodynamic Process

Homogeneous thermodynamic processes were defined in Lecture 1, dynamic processes in Lecture 3. A *thermodynamic* process combines these two concepts. Consider a set of functions consisting of, first, a motion $\underset{\sim}{x} = \underset{\sim}{\chi}(\underset{\sim}{X},t)$ of a body $\mathcal{B}$, and, second, frame-indifferent fields

$$\eta(\underset{\sim}{x},t), \quad \theta(\underset{\sim}{x},t), \quad \underset{\sim}{T}(\underset{\sim}{x},t), \quad \underset{\sim}{h}(\underset{\sim}{x},t), \quad \psi(\underset{\sim}{x},t),$$

defined over the configuration $\mathcal{B}_{\underset{\sim}{\chi}}$ of $\mathcal{B}$ at time t. These fields are said to constitute a *thermodynamic process* if they satisfy the principles of balance of momentum, moment of momentum, and energy, for every part of $\mathcal{B}$. By proper choice of body force $\underset{\sim}{b}$ and supply of heat working, the first and third principles, expressed in the forms (3.27) and (3.33), can always be satisfied, and the only restriction is imposed by Cauchy's second law, (3.29), *viz*, the stress tensor $\underset{\sim}{T}$ is symmetric: $\underset{\sim}{T} = \underset{\sim}{T}^T$.

Constitutive Equations. Equipresence

The first step in obtaining combined constitutive equations is to replace the homogeneous histories θ^t, τ^t by field histories. We shall use the same symbol θ^t to denote the history of the temperature field $\theta(\underset{\sim}{x},t)$. For the parameters τ_k we choose, naturally, the three components of $\underset{\sim}{\chi}^t$. Thus we have the independent variables θ^t, $\underset{\sim}{\chi}^t$.

By (3.30) we see that a constitutive equation for $\underset{\sim}{T}$ determines a constitutive equation for $\underset{\sim}{W}$. Thus there remain only the following dependent variables for constitutive equations: $\underset{\sim}{T}$, $\underset{\sim}{h}$, ψ, η. The *principle of determinism* is now taken as the statement that the histories θ^t, $\underset{\sim}{\chi}^t$ determine $\underset{\sim}{T}$, $\underset{\sim}{h}$, ψ, η:

$$\left.\begin{matrix} \underset{\sim}{T} \\ \underset{\sim}{h} \\ \psi \\ \eta \end{matrix}\right\} \text{ at } \underset{\sim}{X},\ t \text{ are functionals of } \theta^t,\ \underset{\sim}{\chi}^t \text{ in } \mathcal{B}\ . \qquad (15.8)$$

A relation of this kind is a *thermomechanical constitutive equation*. Such an equation satisfies the guiding principle for formulation of constitutive equations called the *principle of equipresence:* The independent variables in all constitutive equations of a theory shall be the same. In this case those variables are $\underset{\sim}{\chi}^t$ and θ^t.

The functionals entering a thermodynamic constitutive equation (15.8) are subject to requirements generalizing and including those laid down in the two simpler theories discussed earlier: (1) The principle of local action, (2) the principle of material frame-indifference, and (3) the principle of dissipation. These requirements are to be satisfied *identically* in every thermodynamic process. The first two principles are easy to state and apply, but the third, the principle of dissipation, requires some thought.

Exercise 15.2

Show that a constitutive equation of the form

$$\underset{\sim}{h} = \underset{\sim}{f}(\underset{\sim}{F}, \text{grad } \phi), \tag{15.9}$$

where ϕ is a frame-indifferent scalar, satisfies the principle of material frame-indifference if and only if it is equivalent to

$$\underset{\sim}{h} = \underset{\sim}{R}\underset{\sim}{f}(\underset{\sim}{U}, \underset{\sim}{R}^T \text{grad } \phi). \tag{15.10}$$

The Clausius-Duhem Inequality

The classical dissipation principle in Planck's form,

$$\theta\dot{H} \geqq Q, \tag{1.5}$$

was generalized to fields by DUHEM. An idea somewhat more general than his is suggested by the form of Q adopted in continuum mechanics:

$$Q = \int_{\partial\mathcal{P}_{\underset{\sim}{\chi}}} q\,ds + \int_{\mathcal{P}_{\underset{\sim}{\chi}}} s\,dm. \tag{3.16}$$

We regard the heat working per unit temperature, in whatever form or place it occurs, as contributing to the lower bound for the heating:

$$\dot{H} \geq \int_{\partial \mathcal{P}_{\underset{\sim}{\chi}}} \frac{q}{\theta}\, ds + \int_{\mathcal{P}_{\underset{\sim}{\chi}}} \frac{s}{\theta}\, dm. \tag{15.11}$$

This inequality, called the *Clausius-Duhem inequality* or *dissipation principle*, is the postulate of irreversibility used in continuum thermodynamics.

By (15.1), (3.20), and (3.33), a local equivalent for the Clausius-Duhem inequality is

$$\rho\theta\dot{\eta} \geqq \theta \operatorname{div}\left(\frac{\underset{\sim}{h}}{\theta}\right) + \rho s. \tag{15.12}$$

If θ is uniform, (15.11) reduces to (1.5). More generally, the classical heat-conduction inequality asserts that heat never flows against a temperature gradient:

$$\frac{\underset{\sim}{h}\cdot\underset{\sim}{g}}{\theta} \geqq 0, \tag{15.13}$$

where

$$\underset{\sim}{g} \equiv \operatorname{grad} \theta. \tag{15.14}$$

We shall not adopt (15.13), for in fact heat may flow against a temperature gradient, just as water may flow uphill. Sufficient local supply of energy may certainly turn its flow toward any desired direction, and since such supplies are allowed here, (15.13) cannot generally hold.

The Reduced Dissipation Inequality

By (15.4) we see at once that the Clausius-Duhem inequality is equivalent to

$$\begin{aligned} \frac{\underset{\sim}{h}\cdot\underset{\sim}{g}}{\theta} &\geqq -\delta \\ &= \rho(\dot{\psi}+\eta\dot{\theta}) - w. \end{aligned} \tag{15.15}$$

These forms are called *reduced dissipation inequalities*, since s has been eliminated by use of the balance of energy.

Simple Thermodynamic Materials

Let Φ stand for the collection ($\underset{\sim}{T}$, $\underset{\sim}{h}$, ψ, η). A thermodynamic material is *simple* if (15.8) reduces to

$$\Phi = \mathcal{F}\,(\underset{\sim}{F}^t, \theta^T, \underset{\sim}{g}^t), \tag{15.16}$$

where, as before, $\underset{\sim}{F} = \nabla\underset{\sim}{\chi}$, $\underset{\sim}{g} = \text{grad}\ \theta$. Clearly the simple materials of Lecture 3 are included as a special case, and clearly the principle of local action is satisfied.

Exercise 15.3

For the four constitutive equations represented schematically by (15.16), find reduced forms such as to satisfy the principle of material frame-indifference.

16. Thermodynamics of Simple Materials with Fading Memory

Quasi-Elastic Response. Fading Memory

Any history $f^t(s)$ for $s \geq 0$ may be regarded as consisting in its *past* history f^t_+, which is defined as $f^t(s)$ for $s > 0$, and its present value $f^t(0) = f(t)$. Therefore we may write the constitutive equation (15.16) of a simple material in the form

$$\Phi = \mathscr{F}\,(\underset{\sim}{F},\theta,\underset{\sim}{g};\underset{\sim}{F}^t_+,\theta^t_+,\underset{\sim}{g}^t_+), \tag{16.1}$$

where $\mathscr{F}$ is a functional of its last three arguments and a function of the first three. At any given instant t, the past histories at a particular particle have taken place, and according to (16.1) Φ depends upon the present values $\underset{\sim}{F}$, θ, $\underset{\sim}{g}$. That is,

$$\Phi = f(\underset{\sim}{F},\theta,\underset{\sim}{g},t) \tag{16.2}$$

where f is a function. Now for a given functional $\mathscr{F}$ in (16.1), *i.e.*, for given material, the function f in (16.2) will depend upon what the past history is in (16.1). Thus f does not express a material property alone, but rather the effect of a particular past history upon a certain material. The quantity Φ for two different particles in a homogeneous body will generally be given by two different f's in (16.2) though by the same $\mathscr{F}$ in (16.1), because the past history differs from one particle to another.

If the four functions symbolized by f in (16.2) are continuously differentiable with respect to all four arguments, the response of the material to the particular past deformation-temperature history is said to be *quasi-elastic*. The term arises from the fact that the dependence of $\underset{\sim}{T}$ on $\underset{\sim}{F}$ is just like that of an elastic material (11.2) except that the response is smoothly time-dependent. Here, as we shall see shortly, it is the *smoothness* of the four functions denoted collectively by f that makes all the difference.

If f is continuously differentiable with respect to t, a small change in the past history effects an approximately linear change in Φ, and likewise for changes in the present values $\underset{\sim}{F}$, θ, and $\underset{\sim}{g}$. This smoothness in itself is an assertion of fading memory in one sense. In a difficult analysis COLEMAN has shown that if $\mathfrak{F}$ satisfies a generalized principle of fading memory in Coleman & Noll's sense, the response of the material is quasi-elastic. This theorem is far too difficult for me to prove here. As was shown by WANG & BOWEN, quasi-elastic response by itself, for whatever cause, suffices to yield a formal theory similar to the thermodynamics of homogeneous processes.

Before showing how this is so, we remark that the Navier-Stokes fluid does not exhibit quasi-elastic response. For it, if we hold $\underset{\sim}{F}$, θ, and $\underset{\sim}{g}$ fixed but change t, we do not get even a small change in Φ itself, since $\underset{\sim}{T}$ is determined by $\dot{\underset{\sim}{F}}\underset{\sim}{F}^{-1}$. Likewise, more general Rivlin-Ericksen materials do not exhibit quasi-elastic response. Thus quasi-elastic response is not a general property of materials but rather a distinguishing attribute of an important special class.

Dissipation and Quasi-Elastic Response

We now write (16.2) explicitly:

$$\underset{\sim}{T} = \underset{\sim}{\mathfrak{G}}(\underset{\sim}{F},\theta,\underset{\sim}{g},t)$$

$$\underset{\sim}{h} = \underset{\sim}{\mathfrak{h}}(\underset{\sim}{F},\theta,\underset{\sim}{g},t),$$

$$\psi = \mathfrak{p}\,(\underset{\sim}{F},\theta,\underset{\sim}{g},t),$$
$$\eta = \mathfrak{h}(\underset{\sim}{F},\theta,\underset{\sim}{g},t), \tag{16.3}$$

and substitute these formulae into the reduced dissipation inequality (15.15), obtaining

$$-(\partial_{\underset{\sim}{g}}\mathfrak{p})\dot{\underset{\sim}{g}} - \rho(\partial_\theta\mathfrak{p} + \eta)\dot{\theta} + \operatorname{tr}\{[\underset{\sim}{F}^{-1}\underset{\sim}{T} - \rho(\partial_{\underset{\sim}{F}}\mathfrak{p})^T]\dot{\underset{\sim}{F}}\}$$
$$+ \frac{1}{\theta}\,\underset{\sim}{h}\cdot\underset{\sim}{g} - \rho\partial_t\,\mathfrak{p} \geqq 0. \tag{16.4}$$

With any values we please for $\underset{\sim}{F}$, θ, and $\underset{\sim}{g}$, we can find a process such that at a given place and time, $\dot{\underset{\sim}{g}} = \underset{\sim}{0}$, $\dot{\theta} = 0$, $\dot{\underset{\sim}{F}} = \underset{\sim}{0}$. Hence (16.4) yields the necessary condition

$$\frac{1}{\theta}\,\underset{\sim}{h}\cdot\underset{\sim}{g} - \rho\partial_t\,\mathfrak{p} \geqq 0. \tag{16.5}$$

The left-hand side is a function of $\underset{\sim}{F}$, θ, $\underset{\sim}{g}$, and t alone, by (16.3). Giving $\underset{\sim}{F}$, θ, $\underset{\sim}{g}$ any values we please, we can find a process in which $\dot{\underset{\sim}{F}} = 0$, $\dot{\theta} = 0$, and $\dot{\underset{\sim}{g}}$ has any magnitude and direction we please. Hence the first term in (16.4) may be given a value such as to violate (16.4) unless $\partial_{\underset{\sim}{g}}\,\mathfrak{p} = \underset{\sim}{0}$. That is, the free energy does not depend on the temperature gradient:

$$\psi = \mathfrak{p}\,(\underset{\sim}{F},\theta,t). \tag{16.6}$$

Similar reasoning shows that the next two terms in (16.4) vanish:

$$\eta = -\partial_\theta\mathfrak{p} \tag{16.7}$$

and

$$\mathrm{tr}\{[\underset{\sim}{F}^{-1}\underset{\sim}{T}-\rho(\partial_{\underset{\sim}{F}}\mathfrak{p})^T]\dot{\underset{\sim}{F}}\} = 0. \tag{16.8}$$

This last equation is essentially the same as (13.13) and may be solved in the same way, by an appeal to the frame-indifference of $\mathfrak{p}$, with the result that

$$\underset{\sim}{T} = \rho\underset{\sim}{F}(\partial_{\underset{\sim}{F}}\mathfrak{p})^T = \rho\partial_{\underset{\sim}{F}}\mathfrak{p}\underset{\sim}{F}^T. \tag{16.9}$$

These results state that the free-energy function $\mathfrak{p}(\underset{\sim}{F}, \theta, t)$ is a *thermodynamic potential* for the specific entropy η and the stress $\underset{\sim}{T}$. That is, the effects of memory upon the free energy determine all effects of memory on η and $\underset{\sim}{T}$. Moreover, comparison with (13.16) shows that $\mathfrak{p}$ plays the role of a stored-energy function in hyperelasticity.

We have not finished with the consequences of the reduced dissipation inequality. If we substitute (16.7), (16.9), and (16.6) into (15.4), we obtain

$$\delta = -\rho\partial_t\mathfrak{p}. \tag{16.10}$$

That is, the internal dissipation is the rate of decrease of the free energy per unit volume at constant deformation and temperature. Since we interpret the time dependence of quasi-elastic response as an overall effect of material memory, the result (16.10) may be regarded as stating, loosely, that δ is the rate of decrease of free energy per unit volume due to effects of memory alone. Naturally, we do not expect the material to gain free energy as a result of remembering the past; that is, we expect this decrease to be non-positive, and shortly we shall prove that it is so, indeed.

Namely, we look back at our starting assumption (16.3) and observe that by (16.9) and (16.6), the right-hand side is independent of $\underset{\sim}{g}$. Hence the left-hand side is an upper bound for $-\delta$ for every $\underset{\sim}{g}$, and in particular when $\underset{\sim}{g} = \underset{\sim}{0}$. Hence

$$\delta \geqq 0. \tag{16.11}$$

That is, the internal dissipation is non-negative.

Exercise 16.1

Prove that, conversely, if

1. $\mathfrak{p}\,(\underset{\sim}{F}, \theta, t)$ is a frame-indifferent scalar function,
2. $\partial_t\, \mathfrak{p} \leqq 0$,
3. η and $\underset{\sim}{T}$ are defined by (16.7) and (16.9),
4. $\underset{\sim}{h}$ satisfies $(15.15)_1$ with δ defined by (16.20),

the equations (16.3) of quasi-elastic response satisfy the reduced dissipation inequality $(15.15)_2$ in every thermodynamic process.

The foregoing results constitute the *grand thermodynamic theorem* of COLEMAN, as generalized by WANG & BOWEN. That theorem states, roughly, that any material with quasi-elastic response always consistent with the Clausius-Duhem dissipation inequality obeys essentially the equations of the classical, special thermodynamics of homogeneous processes, based on caloric equations of state, provided that these equations relate the densities of the state variables and be time-dependent. In addition, the internal dissipation is non-negative, and the classical heat-conduction inequality is replaced by one giving a lower bound for $\underset{\sim}{h}\cdot\underset{\sim}{g}$ in terms of the internal dissipation. If the product $\theta\delta$ is large enough, indicating that there is sufficiently great internal dissipation at sufficiently high temperature, the heat efflux vector $\underset{\sim}{h}$ may point even in the opposite direction from $\underset{\sim}{g}$, indicating that heat may be drawn straight from a cold part to a hot part, the opposite of the behavior we have been led to expect by experience with materials in which $\delta = 0$ always, by definition.

Applications

To develop alternative forms and applications of Coleman's theorem would require a course of lectures. I remark here on only two. First, under suitable hypotheses of inversion, the functions $\varepsilon(\underset{\sim}{F},\eta,t)$ may be used as a potential, just as in the thermodynamics of homogeneous processes. Second, the fact that $\delta \geqq 0$ makes it possible to prove, rigorously and without further assumptions, explicit theorems on minimum free energy, maximum entropy, and minimum internal energy, and a general theorem on cyclic processes, corresponding to the rather vague assertions of this kind in thermodynamics books.

A Final Remark on the Stored-Energy Function in Elasticity

It is commonly believed that thermodynamics proves every elastic material to be hyperelastic. In order to apply thermodynamic reasoning, we must introduce heat and temperature, which are never mentioned in elasticity. Thus to get results from thermodynamics we must first accept more assumptions than are necessary to obtain the theory of elasticity. *A fortiori,* thermodynamics cannot prove *anything* about elasticity itself, though it can and does deliver results about thermo-elasticity formulated within a thermodynamic framework:

$$\begin{aligned}
\underset{\sim}{T} &= \underset{\sim}{\mathfrak{G}}(\underset{\sim}{F},\theta),\\
\psi &= \psi(\underset{\sim}{F},\theta),\\
\eta &= \eta(\underset{\sim}{F},\theta),\\
\underset{\sim}{h} &= \underset{\sim}{\mathfrak{h}}(\underset{\sim}{F},\theta,\underset{\sim}{g}).
\end{aligned} \tag{16.12}$$

Clearly, in the terms used before in these lectures, the thermo-elastic material so defined is *not even an elastic material*, in general, since $\underset{\sim}{T}$ depends on θ as well as $\underset{\sim}{F}$. The theorems on quasi-elastic response may be applied. They yield the following results: The constitutive equations of the thermo-elastic material (16.12) are consistent with the reduced dissipation inequality for all thermodynamic processes if and only if they reduce to

$$\begin{aligned}
&\psi = \psi(\underset{\sim}{F},\theta),\\
&\underset{\sim}{T}_{\kappa} = \rho_{\kappa}\partial_{\underset{\sim}{F}}\psi = \rho_{\kappa}\partial_{\underset{\sim}{F}}\varepsilon,\\
&\eta = -\partial_{\theta}\psi, \quad \theta = \partial_{\eta}\varepsilon,\\
&\underset{\sim}{h}\cdot\underset{\sim}{g} \geqq 0,
\end{aligned} \tag{16.13}$$

where

$$\varepsilon(\underset{\sim}{F},\eta) \equiv \psi(\underset{\sim}{F}, \mathfrak{t}(\underset{\sim}{F},\eta)) + \eta\, \mathfrak{t}(\underset{\sim}{F},\eta)$$

and $\mathfrak{t}(\underset{\sim}{F}, \eta)$ is the function obtained by solving (16.13) for θ. In particular, the thermo-elastic material is a hyperelastic material for two special cases:

1. In isothermal deformations, with $\sigma(\underset{\sim}{F}) = \psi(\underset{\sim}{F}, \theta)$ and θ = const.

2. In isentropic deformation, with $\sigma(\underset{\sim}{F}) = \varepsilon(\underset{\sim}{F}, \eta)$ and η = const.

The very fact that the potential relations $(16.13)_{2,3}$ hold in *all* deformations of the thermo-elastic material shows that in general, when θ or η varies from point to point, the thermo-elastic material is *not elastic* and hence, *a fortiori*, cannot have a stored-energy function.

Exercise 16.2

Find analogues of (16.13) when η is given as a function of $\underset{\sim}{F}$ and ε. Show that a thermo-elastic material is not elastic in iso-energetic deformations unless its temperature function $\mathfrak{t}(\underset{\sim}{F}, \eta)$ is in fact independent of $\underset{\sim}{F}$, in which case it is hyperelastic.

17. Wave Motions. Compatibility

The Nature of Wave Motions

One way to gain insight into the behavior of complicated materials is to determine the conditions under which a disturbance may move through a body. No specific boundary-value problem is solved, since only necessary, and hence descriptive, conditions are obtained. Such conditions, however, may suggest ways in which more complicated and general motions occur.

A wave is idealized as a surface of discontinuity. The main tool for the study of such surfaces is Hadamard's lemma.

Hadamard's Lemma

Let $\mathcal{S}$ be a part of the boundary of a region, which we shall denote by $\mathcal{R}_+$, and let $\underset{\sim}{x}$ be a point on $\mathcal{S}$. The field $\psi(\underset{\sim}{y})$ is said to be *smooth* in $\mathcal{R}_+$ if it is continuously differentiable in $\mathcal{R}_+$, if for every point $\underset{\sim}{x}$ on $\mathcal{S}$ the fields of $\psi(\underset{\sim}{y})$ and $\partial_{\underset{\sim}{y}}\psi(\underset{\sim}{y})$ approach limits $\psi^+(\underset{\sim}{x})$ and $\partial_{\underset{\sim}{x}}\psi^+(\underset{\sim}{x})$ as $\underset{\sim}{y} \to \underset{\sim}{x}$, and if $\psi^+(\underset{\sim}{x})$ is differentiable on any path $\mathcal{P}$ lying on $\mathcal{S}$. *Hadamard's lemma* asserts that for a smooth field $\psi(\underset{\sim}{x})$, the theorem of the total differential holds for the limit functions ψ^+ and $\overset{+}{\partial}_{\underset{\sim}{x}}\psi$. That is, if the path $\mathcal{P}$ is described by the parametric equation $\underset{\sim}{x} = \underset{\sim}{x}(\ell)$, then

$$\psi^{+\prime}(\ell) = (\overset{+}{\partial}_{\underset{\sim}{x}}\psi)\cdot\underset{\sim}{x}'(\ell). \qquad (17.1)$$

The field ψ may be a scalar, vector, or tensor. The result (17.1) is written in a form appropriate for a scalar ψ; the notation needs to be modified in other cases.

Exercise 17.1

For the case when ψ is a scalar, prove Hadamard's lemma.

Singular Surfaces

Let the orientable surface $\mathscr{S}$ be a part of the common boundary separating two regions $\mathscr{R}_+$ and $\mathscr{R}_-$, in each of which ψ is smooth. In this case, at a point $\underset{\sim}{x}$ on $\mathscr{S}$ the limits ψ^+ and ψ^-, and likewise $\partial^+_{\underset{\sim}{x}}\psi$ and $\partial^-_{\underset{\sim}{x}}\psi$, exist but need not be equal. The *jumps* of ψ and $\partial_{\underset{\sim}{x}}\psi$ are defined as the differences in these values:

$$[\psi](\underset{\sim}{x}) \equiv [\psi] \equiv \psi^+ - \psi^-, \; [\partial_{\underset{\sim}{x}}\psi] \equiv \partial^+_{\underset{\sim}{x}}\psi - \partial^-_{\underset{\sim}{x}}\psi. \tag{17.2}$$

If one or both of these jumps is not zero, $\mathscr{S}$ is said to be *singular* with respect to ψ.

A singular surface, then, is not merely one where some property of continuity or differentiability fails, since ψ is required to be smooth on each side. The jumps possible across a singular surface are strictly limited in kind. Since $[\psi](\underset{\sim}{x})$ is a differentiable function of $\underset{\sim}{x}$ on $\mathscr{S}$, applying Hadamard's lemma to ψ^+ and ψ^- and subtracting yields

$$[\psi]'(\underset{\sim}{x}) = [\partial_{\underset{\sim}{x}}\psi]\cdot\underset{\sim}{x}'(\ell). \tag{17.3}$$

This is *Hadamard's fundamental condition of compatibility*, which relates the jumps possible in ψ and $\partial_{\underset{\sim}{x}}\psi$.

An important corollary follows when ψ is continuous: $[\psi] = 0$. Then (17.3) yields

$$[\partial_{\underset{\sim}{x}}\psi]\cdot\underset{\sim}{x}'(\ell) = 0 \qquad (17.4)$$

for all paths on $\mathscr{S}$. Since $\underset{\sim}{x}'(\ell)$ may be any vector tangent to $\mathscr{S}$, (17.4) requires that there exist a quantity $a(\underset{\sim}{x})$, defined if $\underset{\sim}{x} \in \mathscr{S}$, such that

$$[\partial_{\underset{\sim}{x}}\psi] = a\underset{\sim}{n} \qquad (17.5)$$

where $\underset{\sim}{n}$ is a vector normal to $\mathscr{S}$. This result expresses *Maxwell's theorem:* The jump of the gradient of a continuous field is normal to the singular surface. It is convenient to adopt a convention of sign; *e.g.*, let $\underset{\sim}{n}$ be the unit normal pointing from $\mathscr{R}_-$ into $\mathscr{R}_+$. Then a, which is called the *amplitude* of the discontinuity, is uniquely determined. The form (17.5) is appropriate to a scalar field ψ. If ψ is a vector field, the amplitude a becomes a vector, $\underset{\sim}{a}$, and the right-hand side of (17.5) should be written as $\underset{\sim}{a} \otimes \underset{\sim}{n}$.

When $\underset{\sim}{a}$ is a vector, the singularity is called *longitudinal* if $\underset{\sim}{a} \parallel \underset{\sim}{n}$, *transverse* if $\underset{\sim}{a} \perp \underset{\sim}{n}$. In general, a singularity is neither.

Singular Surfaces for a Motion

We consider the motion with respect to a reference configuration $\underset{\sim}{\kappa}$:

$$\underset{\sim}{x} = \underset{\sim}{\chi}_{\underset{\sim}{\kappa}}(\underset{\sim}{X},t), \qquad (2.5)$$

and its derivatives $\dot{\underset{\sim}{x}}$, $\ddot{\underset{\sim}{x}}$, $\underset{\sim}{F}$, $\nabla\underset{\sim}{F}$, *etc.* The surface $\mathscr{S}$ in the reference configuration $\underset{\sim}{\kappa}$ is said to be a *singular surface of nth order,* if it is singular with respect to some n^{th} derivative of $\underset{\sim}{\chi}$, but all derivatives of lower order exist and are continuous in a region containing $\mathscr{S}$ in its interior. The surface $\mathscr{S}$ is allowed to move in the reference configuration. Moreover, we consider only singular surfaces that persist throughout an interval of time. Thus they may be regarded as surfaces in a 4-dimensional space whose points are pairs $(\underset{\sim}{x}, \underset{\sim}{t})$, and Hadamard's condition (17.3) may be applied in that space.

Singular surfaces of orders 0 and 1 are called *strong;* shock waves, vortex sheets, tears, and welds are included. Singular surfaces of order 2 or greater are called *weak*. The mathematical theory of strong singular surfaces is rather intricate and will not be considered in these lectures.

Compatibility Conditions for Second-Order Singular Surfaces

Application of Maxwell's theorem (17.5) to second-order singular surfaces in $\underset{\sim}{\kappa}$ yields a formally simple relation in the reference configuration, with $\underset{\sim}{n}$ being the 4-dimensional unit normal. Separation into spatial and temporal parts, followed by transformation to the configuration $\underset{\sim}{\chi}$, yields

$$
\begin{aligned}
[\nabla\underset{\sim}{F}] &= \underset{\sim}{a}\otimes(\underset{\sim}{F}^T\underset{\sim}{n}) \otimes (\underset{\sim}{F}^T\underset{\sim}{n}), &\qquad [x^k_{;\alpha\beta}] &= a^k x^p_{,\alpha} x^q_{,\beta} n_p n_q, \\
[\dot{\underset{\sim}{F}}] &= -u\underset{\sim}{a}\otimes(\underset{\sim}{F}^T\underset{\sim}{n}), &\qquad [\dot{x}^k_{,\alpha}] &= -ua^k x^p_{,\alpha} n_p, \qquad (17.6)\\
[\ddot{\underset{\sim}{x}}] &= u^2\underset{\sim}{a}, &\qquad [\ddot{x}^k] &= u^2 a^k,
\end{aligned}
$$

where now $\underset{\sim}{n}$ is the unit normal to the present configuration of $\mathscr{S}$, $\underset{\sim}{a}$ is a vector called the amplitude, and u is a scalar called the local speed of propagation.

The first of these equations, called *Hadamard's geometrical condition of compatibility,* reflects the assumption that the discontinuity is spread out over a surface at the instant in question. The second two, called *Hadamard's kinematical conditions of compatibility,* reflect the assumption that the singular surface instantaneously persists. If a point $\underset{\sim}{x}$ on the surface is moving with velocity $\underset{\sim}{v}$, and if the velocity of the particle instantaneously at $\underset{\sim}{x}$ is $\dot{\underset{\sim}{x}}$, then

$$U = (\underset{\sim}{v}-\dot{\underset{\sim}{x}})\cdot\underset{\sim}{n}. \tag{17.7}$$

That is, U at a place and time is the normal speed of advance of the singular surface relative to the particle instantaneously situated at $\underset{\sim}{x}$. If $U \neq 0$, the singular surface *propagates* through the material and hence is called a *wave*. If $U = 0$ over an interval of time, the singular surface divides two portions of material. From $(17.6)_3$ it is clear that every second-order wave carries a non-zero jump of the acceleration. For this reason, such waves are called *acceleration waves*.

Exercise 17.2

Filling in the argument outlined above, prove (17.6). Prove also that the jump in the velocity gradient $\underset{\sim}{G}$ satisfies

$$[\underset{\sim}{G}] = -U\underset{\sim}{a} \otimes \underset{\sim}{n}, \tag{17.8}$$

and hence derive *Weingarten's theorem:*

$$[W] = -U\underset{\sim}{a} \wedge \underset{\sim}{n}, \qquad [\operatorname{div} \dot{\underset{\sim}{x}}] = -U\underset{\sim}{a} \cdot \underset{\sim}{n}. \tag{17.9}$$

By use of Weingarten's theorem, interpret the normal and tangential components of the amplitude $\underset{\sim}{a}$.

<u>Equation of Balance at a Singular Surface</u>

In Lecture 3 we considered a general equation of balance:

$$\left(\int_{\mathcal{P}_{\chi}} \psi \; dm\right)^{\cdot} = \int_{\partial\mathcal{P}_{\chi}} \tilde{E} \; n \; ds + \int_{\mathcal{P}_{\chi}} s \; dm. \tag{3.24}$$

This equation may be applied also in the case when $\mathcal{P}$ contains or is divided by a weak singular surface $\mathcal{S}$. If $\mathcal{S}$ is a singular surface also with respect to ψ,as defined above, then the jump of ψ is subjected to the following requirement, called *Kotchine's theorem:*

$$U\rho\,[\psi] + [E]\cdot n = 0, \tag{17.10}$$

and conversely, if (17.10) holds at each point of $\mathcal{S}$ and (3.26) holds at interior points of a region containing $\mathcal{S}$ in its interior, then (3.24) holds in that region.

Exercise 17.3

Under the assumption that s is bounded in a region containing $\mathcal{S}$, and bearing in mind the restrictions on ψ already stated in the definition of "singular surface", prove *Kotchine's theorem.*

By applying Kotchine's theorem to the equations of balance of energy and linear momentum, derive the conditions of FOURIER and POISSON, respectively:

$$[h]\cdot n = 0, \qquad [T]n = 0, \tag{17.11}$$

on the assumption that the singular surface is of second (or higher) order and that b, ε, and s are continuous across it. According to the conditions of FOURIER and POISSON, the normal heat flux and the traction vector are continuous across a weak singular surface.

Acceleration Waves in Elasticity

To consider acceleration waves in elasticity, we write the constitutive equation in the form

$$\underset{\sim}{T}_{\kappa} = \underset{\sim}{\mathfrak{h}}(\underset{\sim}{F}, \underset{\sim}{X}) \tag{13.8}$$

and assume that $\underset{\sim}{\mathfrak{h}}$ is continuously differentiable with respect to each of its arguments. At a weak singular surface Poisson's condition $(17.11)_2$ is then satisfied. On each side of the singular surface, the deformation satisfied the differential equation obtained by substituting (13.8) into (13.6), *viz:*

$$A^{\alpha\beta}_{pm} x^m_{,\alpha;\beta} + q_p + \rho_{\kappa} b_p = \rho_{\kappa} \ddot{x}_p, \qquad (17.12)_1$$

where

$$A^{\alpha\beta}_{pm} \equiv \partial_{x^m_{,\beta}} \mathfrak{h}^{\alpha}_p , \qquad q_p \equiv \partial_{X^\alpha} \mathfrak{h}^{\alpha}_p . \qquad (17.12)_{2,3}$$

From our hypothesis regarding $\underset{\sim}{\mathfrak{h}}$ it follows that $\underset{\sim}{A}$ and $\underset{\sim}{q}$ as just defined are continuous across $\mathscr{S}$. We shall assume that $\underset{\sim}{b}$ also is continuous. At a point on $\mathscr{S}$, we may approach from the + side and obtain then in the limit

$$A^{\alpha\beta}_{pm} (x^m_{,\alpha;\beta})^+ + q_p + \rho_{\kappa} b_p = \rho_{\kappa} \ddot{x}^+_p, \tag{17.13}$$

while approach from the - side yields the limit relation

$$A^{\alpha\beta}_{pm}(x^m_{,\alpha;\beta})^- + q_p + \rho_{\underset{\sim}{\kappa}} b_p = \rho_{\underset{\sim}{\kappa}} \ddot{x}^-_p. \qquad (17.14)$$

Taking the difference yields

$$A^{\alpha\beta}_{pm}[x^m_{,\alpha;\beta}] = \rho_{\underset{\sim}{\kappa}} [\ddot{x}_p]. \qquad (17.15)$$

In this result we substitute Hadamard's conditions $(17.6)_{2,6}$. If we set

$$Q_{pm}(\underset{\sim}{n}) \equiv \frac{\rho}{\rho_{\underset{\sim}{\kappa}}} A^{\alpha\beta}_{pm} x^r_{,\alpha} x^s_{,\beta} n_r n_s, \qquad (17.16)$$

then the result of the substitution is

$$(\underset{\sim}{Q}(\underset{\sim}{n}) - \rho U^2 \underset{\sim}{1})\underset{\sim}{a} = \underset{\sim}{0}. \qquad (17.17)$$

$\underset{\sim}{Q}(\underset{\sim}{n})$ is called the *acoustic tensor* in the direction $\underset{\sim}{n}$ for the elastic material at $\underset{\sim}{X}$ when subjected to the deformation $\underset{\sim}{F}$. According to (17.17), which is called the *propagation condition*, any amplitude $\underset{\sim}{a}$ of a second-order singular surface with normal $\underset{\sim}{n}$ must be a right proper vector of $\underset{\sim}{Q}(\underset{\sim}{n})$, and its speed of propagation U is such that ρU^2 is the corresponding proper number. The directions corresponding to the proper numbers $\underset{\sim}{a}$ are called the *acoustic axes* for waves traveling in the direction $\underset{\sim}{n}$ at $\underset{\sim}{x}$, t. In the generality here maintained, little can be said about the number and nature of the acoustic axes and the corresponding speeds of propagation. In the special case when $\underset{\sim}{Q}(\underset{\sim}{n})$ is symmetric,

$$\underset{\sim}{Q}(\underset{\sim}{n}) = \underset{\sim}{Q}(\underset{\sim}{n})^T, \qquad (17.18)$$

application of a standard theorem tells us that at least one orthogonal triple of real acoustic axes for the direction $\underset{\sim}{n}$ exists, and that the corresponding squared speeds U^2 are real. It remains possible, however, that the speeds U may be purely imaginary, so that the corresponding singular surfaces do not exist. The foregoing remarks, summarizing the implications of the propagation condition (17.17), constitute the *Fresnel-Hadamard theorem*.

Exercise 17.4

By linearizing the differential Eq. (17.12) about a homogeneous configuration, prove that the speeds and amplitudes of free plane sinusoidal oscillations about a homogeneous configuration satisfy the propagation condition (17.17).

Exercise 17.5

Prove that the acoustic tensor $\underset{\sim}{Q}(\underset{\sim}{n})$ for the deformation gradient $\underset{\sim}{F}$ is symmetric for *every* $\underset{\sim}{n}$ if and only if the operator $\underset{\sim}{A}$ is self-adjoint at the argument $\underset{\sim}{F}$:

$$A_{pm}^{\alpha\beta} = A_{mp}^{\beta\alpha} . \tag{17.19}$$

Prove that (17.19) holds for all $\underset{\sim}{F}$ if and only if the material is hyperelastic.

Exercise 17.6

Let $\underset{\sim}{n}$ be a unit proper vector of $\underset{\sim}{B}$. Prove that in an isotropic elastic material, whether or not it be hyperelastic, $\underset{\sim}{Q}(\underset{\sim}{n}) = \underset{\sim}{Q}(\underset{\sim}{n})^T$ for this $\underset{\sim}{n}$. Hence prove that any such wave is either longitudinal or transverse.

Weak Singular Surfaces in General

The Fresnel-Hadamard theorem applies not only to acceleration waves but to all weak surfaces of any order.

Exercise 17.7

Prove that the geometrical and kinematical conditions of compatibility for a third-order singular surface are

$$[x^m_{,\alpha;\beta\gamma}] = a^m x^r_{,\alpha} n_r x^s_{,\beta} n_s x^u_{,\gamma} n_u, \qquad (17.20)$$

$$[\dddot{x}_p] = - U^3 a_p.$$

Hence show that the amplitudes and speeds of third-order waves satisfy the propagation condition (17.17).

Since the propagation condition (17.17) expresses properties common to so many kinds of disturbances, it is customary to call the theory based upon it *acoustics* and to refer, loosely, to any wave described by it as a *sound wave*. The speeds U are often called *speeds of sound*.

After this preliminary study of waves according to the theory of elasticity, we are ready to consider waves in dissipative media.

18. Wave Propagation in Dissipative Materials

The "Smoothing" Effect of Dissipation

There is a widespread belief that dissipation "smoothes out" discontinuities. The source of this belief seems to lie in a fact about linearly viscous fluids. Indeed, consider the Navier-Stokes formula (8.24), *viz*

$$\underset{\sim}{T} = -p(\rho)\underset{\sim}{1} + \lambda(\mathrm{tr}\,\underset{\sim}{D})\underset{\sim}{1} + 2\mu\underset{\sim}{D}, \tag{8.24}$$

and assume that the fluid is non-trivially viscous: $\mu > 0$. By (17.8) and (17.9), the geometrical and kinematical conditions at an acceleration wave with normal $\underset{\sim}{n}$ imply that

$$[\mathrm{tr}\,\underset{\sim}{D}] = -U\underset{\sim}{a}\cdot\underset{\sim}{n}, \quad [2\underset{\sim}{D}] = -U(\underset{\sim}{a}\otimes\underset{\sim}{n}+\underset{\sim}{n}\otimes\underset{\sim}{a}), \tag{18.1}$$

where $\underset{\sim}{a}$ is the amplitude and U is the local speed of propagation. Hence

$$\begin{aligned} [\underset{\sim}{T}] &= -U\{\lambda(\underset{\sim}{a}\cdot\underset{\sim}{n})\underset{\sim}{1} + \mu(\underset{\sim}{a}\otimes\underset{\sim}{n}+\underset{\sim}{n}\otimes\underset{\sim}{a})\}, \\ \underset{\sim}{0} = [\underset{\sim}{T}]\underset{\sim}{n} &= -U\{(\lambda+\mu)(\underset{\sim}{a}\cdot\underset{\sim}{n})\underset{\sim}{n} + \mu\underset{\sim}{a}\}, \end{aligned} \tag{18.2}$$

where the vanishing of this last expression is a consequence of Poisson's condition $(17.11)_2$. If $U = 0$, there is no wave. If $U \neq 0$, taking the scalar products of (18.2) by $\underset{\sim}{n}$ and $\underset{\sim}{a}$ implies that

$$\begin{aligned} (\lambda+2\mu)(\underset{\sim}{a}\cdot\underset{\sim}{n}) &= 0, \\ (\lambda+\mu)(\underset{\sim}{a}\cdot\underset{\sim}{n})^2 + \mu a^2 &= 0. \end{aligned} \tag{18.3}$$

Since $\mu > 0$, it follows from the Duhem-Stokes inequalities (15.7) that $\lambda + 2\mu > 0$, and hence (18.3) requires that $\underset{\sim}{a} = \underset{\sim}{0}$. That is, the Navier-Stokes constitutive equation is incompatible with the existence of acceleration waves.

True though this fact is, it does not justify any sweeping conclusions about the effects of "dissipative mechanisms" on surfaces of discontinuity. For example, the partial differential equation governing the infinitesimal displacement u of a vibrating string subject to linear viscosity is

$$\sigma \partial_t^2 u = T \partial_x^2 u - K \partial_t u, \qquad (18.4)$$

where σ is the line density, T is the tension, and K is a positive constant. Certainly a "dissipative mechanism" is represented here, yet (18.4) admits discontinuous solutions of all orders, and all have the classical speed of propagation given by $U^2 = T/\sigma$, no matter what be the value of K.

Though in both these examples the dissipative mechanism is called "linear friction", clearly such different results must correspond to ideas that are quite different from one another. We have seen in Lecture 16 that the Navier-Stokes fluid does not show quasi-elastic response. Eq. (18.4) does not fall into any of the types of constitutive equations we have studied.

However, as we shall see now quasi-elastic response generally compatible with the propagation of weak waves.

Waves and Quasi-Elastic Response

The theory of quasi-elastic response from the previous lecture and the theory of wave propagation in elastic materials, presented in Lecture 17, may be combined. The results are due in principle to COLEMAN & GURTIN, but the simple approach I shall follow here was first outlined by WANG & BOWEN.

First, although in all the analysis in Lectures 15 and 16 a single particle was considered, we must now take account of the variation of material properties from one particle to another. Even in a homogeneous material with fading memory, the quasi-elastic response function giving the specific free energy will generally vary from one particle to another because the deformation-temperature history will not generally be homogeneous. Thus (16.6) and (16.9) are replaced by

$$\psi = \mathfrak{p}\,(\underset{\sim}{F}, \theta, t, \underset{\sim}{X}) \tag{18.5}$$

and

$$\underset{\sim}{T}_{\kappa} = \rho_{\kappa} \partial_{\underset{\sim}{F}} \mathfrak{p}. \tag{18.6}$$

In the definition of quasi-elastic response, *both* ψ and $\underset{\sim}{T}$ were required to be given by continuously differentiable functions of $\underset{\sim}{F}$, θ, and t. By the result (18.6), also $\partial_{\underset{\sim}{F}}\mathfrak{p}$ is shown to be continuously differentiable in these same variables. We shall assume also that the quasi-elastic response varies smoothly from one particle to the next in the sense that $\mathfrak{p}$ is continuously differentiable with respect to $\underset{\sim}{X}$.

We recall Cauchy's first law of motion assumes in the form (13.6), *viz*

$$\operatorname{Div} \underset{\sim}{T}_{\kappa} + \rho_{\kappa} \underset{\sim}{b} = \rho_{\kappa} \ddot{\underset{\sim}{x}}. \tag{13.6}$$

By (18.5) and (18.6), the tensor $\underset{\sim}{T}_{\kappa}$ now depends on θ and t as well as $\underset{\sim}{F}$ and $\underset{\sim}{X}$. To spare indices, we shall use a schematic direct notation. Setting

$$\begin{aligned} \tilde{A} &\equiv \rho_{\kappa} \partial^2_{\underset{\sim}{F}} \mathfrak{p}, \\ \underset{\sim}{\tilde{q}} &\equiv \partial_{\underset{\sim}{X}} \cdot (\rho_{\kappa} \partial_{\underset{\sim}{F}} \mathfrak{p}), \end{aligned} \tag{18.7}$$

then by substitution in (13.6) we obtain

$$\operatorname{tr}(\tilde{A}[\nabla \underset{\sim}{F}]) + \rho_{\kappa} \operatorname{tr}((\partial_{\theta} \partial_{\underset{\sim}{F}} \mathfrak{p}) \nabla \theta) + \underset{\sim}{\tilde{q}} + \rho_{\kappa} \underset{\sim}{b} = \rho_{\kappa} \ddot{\underset{\sim}{x}}, \tag{18.8}$$

differing in form from the corresponding Eq. (17.12) for elasticity only by the term involving $\nabla\theta$. In content, however, we must take account also of the fact that the first and third terms on the left-hand side depend also on θ. We are to calculate the jump of this equation at an acceleration wave. As I mentioned above, even in a homogeneous material with quasi-elastic response we shall generally have $\underset{\sim}{\tilde{q}} \neq \underset{\sim}{0}$. However, by $(18.7)_2$ and our assumptions, $\underset{\sim}{\tilde{q}}$ is a continuous function of $\underset{\sim}{F}$, θ, and $\underset{\sim}{X}$. By the definition of an acceleration wave, $\underset{\sim}{F}$ is continuous across it. We shall add the assumption that θ is also continuous. That is, the wave is not a thermal shock. Then

$$[\underset{\sim}{\tilde{q}}] = \underset{\sim}{0}, \tag{18.9}$$

even though $\underset{\sim}{\tilde{q}} \neq 0$. So as to get a simple result, we assume further that the temperature gradient suffers no discontinuity at the wave:

$$[\nabla\theta] = \underset{\sim}{0}. \tag{18.10}$$

Such a wave is called *homothermal*, because it is most easily visualized as travelling through a region of uniform temperature. For a homothermal acceleration wave, taking the jump of (18.8) yields a propagation condition of just the same form as (17.17), *viz*

$$(\tilde{\underset{\sim}{Q}}(\underset{\sim}{n}) - \rho\tilde{U}^2\underset{\sim}{1})\underset{\sim}{a} = \underset{\sim}{0}, \tag{18.11}$$

where the tildes indicate that the elasticity is $\tilde{A}$, calculated from $\mathfrak{p}(\underset{\sim}{F}, \theta, t, \underset{\sim}{X})$ by $(18.7)_1$. Hence, corresponding to the result for elasticity in Lecture 17, $\tilde{\underset{\sim}{Q}}(\underset{\sim}{n})$ is symmetric:

$$\tilde{\underset{\sim}{Q}}(\underset{\sim}{n}) = \tilde{\underset{\sim}{Q}}(\underset{\sim}{n})^T. \tag{18.12}$$

The results so far may be summarized as follows: the propagation condition for homothermal waves in a material with quasi-elastic response is that of a hyperelastic material with time-dependent stored-energy function. The corresponding acoustical tensor is calculated from the isothermal elasticity $\tilde{A}$.

Thus the dissipative mechanism embodied in quasi-elastic response subject to the Clausius-Duhem dissipation inequality lead to constitutive equations in no way incompatible with wave motion. Quite the reverse, the laws of wave propagation are at each place and time the same as that of some hyperelastic material. In the interpretation for materials with fading memory, we may say that the past history of temperature and deformation as well as their present values determine the acoustic tensor.

As mentioned in the preceding lecture, we can just as well use the specific internal-energy function $\mathfrak{e}$ ($\underset{\sim}{F}$, η, t, $\underset{\sim}{X}$) as a thermodynamic potential. Let quantities calculated from this potential be denoted by a caret: $\hat{A}$, $\hat{\underset{\sim}{Q}}(\underset{\sim}{n})$, *etc.* Parallel reasoning, based on the assumptions that not only η but also $\nabla\eta$ is continuous,

$$[\nabla\eta] = \underset{\sim}{0}, \tag{18.13}$$

leads to a result of the same form as (18.11) with carets replacing tildes:

$$(\hat{\underset{\sim}{Q}}(\underset{\sim}{n}) - \rho\hat{U}^2\underset{\sim}{1})\hat{\underset{\sim}{a}} = \underset{\sim}{0}. \tag{18.14}$$

Waves satisfying (18.13) are called *homentropic*, and (18.14) is the propagation condition for them.

In general, waves will not be both homentropic and homothermal. The results (18.11) and (18.14) refer to *different* waves, and conceivably there are still other kinds in which neither assumption is valid. Before determining circumstances in which the results apply, we shall find a relation between the two acoustic tensors $\hat{\underset{\sim}{Q}}$ and $\tilde{\underset{\sim}{Q}}$.

Relation Between the Homothermal and Homentropic Acoustic Tensors

By the results in Lecture 16, the Piola-Kirchhoff stress is given not only by (18.6) but also by its counterpart when the internal energy function $\mathfrak{e}$ is used as a potential:

$$\underset{\sim}{T}_{\kappa} = \rho_{\kappa}\partial_{\underset{\sim}{F}}\mathfrak{p}(\underset{\sim}{F},\theta,t) = \rho_{\kappa}\partial_{\underset{\sim}{F}}\mathfrak{e}(\underset{\sim}{F},\eta,t). \tag{18.15}$$

These equations are valid in all motions of the material with quasi-elastic response subject to the Clausius-Duhem inequality. We are assuming further that the relation (16.7), *viz* $\eta = -\partial_\theta \mathfrak{p}(\underset{\sim}{F},\theta,t)$, may be solved for θ:

$$\theta = \mathfrak{t}(\underset{\sim}{F},\eta,t), \tag{18.16}$$

and that $\mathfrak{t}$ is continuously differentiable.

Exercise 18.1

Calculate $\hat{A} - \tilde{A}$, and hence prove *Duhem's theorem:*

$$\hat{\underset{\sim}{Q}} = \tilde{\underset{\sim}{Q}} + \underset{\sim}{p} \otimes \underset{\sim}{p}, \tag{18.17}$$

where

$$\underset{\sim}{p} \equiv \sqrt{\frac{\rho\theta}{\kappa_{(\underset{\sim}{F})}}}\,\mathrm{tr}\{(\partial_\theta \partial_{\underset{\sim}{F}} \mathfrak{p})\underset{\sim}{F}^T \underset{\sim}{n}\},$$

$$\kappa_{(\underset{\sim}{F})} \equiv \theta\partial_\theta \mathfrak{h}, \qquad \eta = \mathfrak{h}(\underset{\sim}{F},\theta,t). \tag{18.18}$$

The quadratic form of $\underset{\sim}{p} \otimes \underset{\sim}{p}$ is non-negative definite. Hence if $\tilde{\underset{\sim}{Q}}$ is positive definite, so is $\hat{\underset{\sim}{Q}}$. This result, too, is due to DUHEM. Since $\tilde{\underset{\sim}{Q}}(\underset{\sim}{n})$ is symmetric, it has real principal axes; therefore, if the proper numbers of $\tilde{\underset{\sim}{Q}}(\underset{\sim}{n})$ are non-negative, then $\tilde{\underset{\sim}{Q}}(\underset{\sim}{n})$ is non-negative definite. By (18.17), so is $\hat{\underset{\sim}{Q}}(\underset{\sim}{n})$. Thus follows a major corollary of Duhem's theorem: *If for a given direction of propagation the homothermal wave speeds are real, so are the homentropic ones.* The special case of this theorem in gas dynamics is familiar, since the two squared speeds are $\partial_\rho \tilde{p}(\rho,\theta)$ and $\partial_\rho \hat{p}(\rho,\eta)$, and the ratio of these quantities is the positive constant denoted usually by γ.

Waves in Non-Conductors

To obtain $\underset{\sim}{Q}$ as the acoustic tensor, we assumed that both η and $\nabla\eta$ are continuous across the wave, and we called such waves homentropic. In some materials, as we shall see now, all acceleration waves are necessarily homentropic. We shall lay down the conditions

$$[\eta] = 0, \qquad [\theta] = 0, \tag{18.19}$$

as part of the definition of a second-order singular surface. That is, we assume that the singularity, already assumed weak in regard to deformation, is not a shock with respect to entropy or temperature. For materials with quasi-elastic response, the former condition follows from the latter because of the results (16.6) and (16.7) of Coleman's theorem; for materials having caloric equations of state in the ordinary sense, it follows directly.

We consider *non-conductors of heat*, namely, those materials for which the constitutive equation $(16.3)_2$ takes the form

$$\underset{\sim}{h} \equiv \underset{\sim}{0}. \tag{18.20}$$

(In fact, for the argument I shall give it suffices that

$$[\operatorname{div} \underset{\sim}{h}] = 0, \tag{18.21}$$

but this condition does not seem to correspond to any natural model except under the stronger assumption in (18.20). By (18.19) and Maxwell's theorem (17.5), there exists a scalar B such that

$$[\operatorname{grad} \eta] = B\underset{\sim}{n}, \tag{18.22}$$

and by Hadamard's kinematical condition

$$[\dot{\eta}] = -UB. \tag{18.23}$$

By (15.4) and the result (16.10) of Coleman's theorem, in quasi-elastic response

$$-\rho\partial_t \mathfrak{p} = \rho\theta\dot{\eta} - \operatorname{div} \underset{\sim}{h} - \rho s. \tag{18.24}$$

Moreover, the left-hand side is a function of ρ, θ, $\underset{\sim}{F}$, and t alone, by (16.6). Taking the jump of (18.24) yields by (18.21) and (18.23)

$$0 = \rho\theta[\dot{\eta}] = -\rho\theta UB. \tag{18.25}$$

If the discontinuity is indeed a wave, $U \neq 0$, so $B = 0$, so that $[\text{grad } \eta] = 0$ by (18.22). Thus we have the following theorem: *Every acceleration wave in a non-conductor with quasi-elastic response is homentropic*. This result, due in stages to DUHEM and to COLEMAN & GURTIN, includes and generalizes a classic result of gas dynamics: When heat conduction is neglected, sound waves are isentropic, and their squared speed is $\partial_\rho \hat{p}(\rho,\eta)$.

Waves in Definite Conductors

To obtain $\tilde{\underset{\sim}{Q}}$ as the acoustic tensor, we assumed that both θ and $\nabla\theta$ are continuous across the wave, and we called such waves homothermal. In some materials, as we shall see now, all acceleration waves are necessarily homothermal. Naturally these materials must exclude the non-conductors, just considered.

We consider the constitutive equation $(16.3)_2$ for the heat flux.

Defining the *conductivity* by

$$\underset{\sim}{K} \equiv \partial_{\underset{\sim}{g}} \underset{\sim}{\mathfrak{h}}(\underset{\sim}{F},\theta,\underset{\sim}{g},t) \qquad (18.26)$$

we notice that in Fourier's theory of heat conduction $\underset{\sim}{h} = \mathfrak{h}(\underset{\sim}{F},\theta,\underset{\sim}{g},t) = \underset{\sim}{K}\underset{\sim}{g}$, so that $\underset{\sim}{K}$ reduces then to what is ordinarily called the "thermal conductivity". If the quadratic form of $\underset{\sim}{K}$ is positive-definite,

$$\underset{\sim}{m}\cdot\underset{\sim}{K}\underset{\sim}{m} > 0 \qquad (18.27)$$

for every non-zero vector $\underset{\sim}{m}$, we shall say that the material is a *definite conductor of heat*. (In the special case of Fourier's theory, a definite conductor is a material satisfying the classical heat-conduction inequality, but for more general materials no such interpretation of (18.27) suggests itself.) From Fourier's condition $(17.11)_2$,

$$(\underset{\sim}{\mathfrak{h}}(\underset{\sim}{F},\theta,\underset{\sim}{g}+[\underset{\sim}{g}],t) - \underset{\sim}{\mathfrak{h}}(\underset{\sim}{F},\theta,\underset{\sim}{g},t))\cdot\underset{\sim}{n} = 0. \qquad (18.28)$$

At an acceleration wave, again, (18.19) holds, so by Maxwell's theorem there exists a scalar B such that

$$[\underset{\sim}{g}] = B\underset{\sim}{n}. \qquad (18.29)$$

Therefore (18.28) may be written as

$$(\underset{\sim}{\mathfrak{h}}(\underset{\sim}{F},\theta,\underset{\sim}{g}+B\underset{\sim}{n},T) - \underset{\sim}{\mathfrak{h}}(\underset{\sim}{F},\theta,\underset{\sim}{g},t))\cdot\underset{\sim}{n} = 0. \qquad (18.30)$$

But if $B \neq 0$ and $\underset{\sim}{\mathfrak{h}}$ is differentiable,

$$\underset{\sim}{n}\cdot d_B \underset{\sim}{\mathfrak{h}}(\underset{\sim}{F},\theta,\underset{\sim}{g}+B\underset{\sim}{n},t) = \underset{\sim}{n}\cdot\partial_{\underset{\sim}{g}} \underset{\sim}{\mathfrak{h}}(\underset{\sim}{F},\theta,\underset{\sim}{g}+B\underset{\sim}{n},t)\underset{\sim}{n}. \qquad (18.31)$$

By (18.30), the left-hand side vanishes. For a definite conductor, the right-hand side is positive. Hence $B = 0$. By (18.29), $[\underset{\sim}{g}] = \underset{\sim}{0}$. We have shown, then, that *every acceleration wave in a definite conductor of heat with quasi-elastic response is homothermal.* This theorem derives ultimately from FOURIER; in the generality given here, it is due to COLEMAN & GURTIN.

Closure

Taken together, these two theorems show how important is heat conduction in determining the nature of wave motion. In general, of course, there is little reason to expect a material with quasi-elastic response to be either a non-conductor or a definite conductor, and hence we cannot justly regard either of the two foregoing theorems as indicating the kind of waves that do occur according to the general theory. The acoustic tensors $\hat{\underset{\sim}{Q}}$ and $\tilde{\underset{\sim}{Q}}$ have been shown to pertain to homentropic and homothermal waves, but acceleration waves in a material with quasi-elastic response cannot generally be expected to be of these kinds.

Introduction to Nonequilibrium Statistical Physics

I. Prigogine

1. Introduction

Nonequilibrium statistical mechanics is a very new subject. Important pioneering work is due to YVON, BORN, GREEN, KIRKWOOD and BOGOLIUBOV (for Ref. see [1]). However, this work was based on the free use of probabilistic assumptions together with mechanical concepts as was also the classic work by BOLTZMANN, MAXWELL and GIBBS.

The success of Boltzmann's kinetic theory of gases is well known [2]. However, to understand the conditions of validity of the kinetic theory of gases and more generally of statistical mechanics, as well as to extend the applications beyond the classical examples, one has clearly to use a more rigourous and general starting point. This now becomes possible thanks to the progress realised mainly in the last ten years. Here we shall not go into the history of these developments. References can be found in Ref. 1 as well as in the excellent review paper by CHESTER [3] and in the monograph by RICE and GRAY [4].

We have now a general evolution equation for the distribution function valid both for classical and quantum systems which can be used as a convenient starting point for the investigation of the dynamical behaviour of an N-body system. Chap. 3 is devoted to a brief derivation of this equation. We have tried to emphasize physical ideas: The main point is that we are concerned with a new classification of dynamical effects through the effect they have on *correlations*. We therefore have to distinguish between vacuum of correlations to vacuum of correlations transitions, and creation (or destruction) of correlations starting from the vacuum of correlations. Moreover, this equation separates dynamical effects from memory effects.

It is in terms of these concepts that our evolution equation is formulated. If in this equation we neglect memory effects and retain only the lowest order contributions we come back to well known classes of transport equations (i.e., FOKKER-PLANCK, BOLTZMANN, ...) for which a general theory is really not essential. We study very briefly two examples of this type in Chap. 2. But of course the main interest is to go beyond these classical situations. The still relatively simple case corresponds to "Generalized Boltzmann" type of situations, in which concepts as duration of a collision and relaxation time still retain their meaning. We still expect a kind of kinetic equation. But we now are interested in higher order effects (either in the concentration or in the coupling constant). Here already many unexpected features appear. The higher order contributions are not even of the same structure as the lowest order contributions, since processes already considered in the lower order terms have to be *subtracted*. This appeared for the first time clearly in the beautiful work of CHOH and UHLENBECK on classical three-body systems [5]. Recent investigations, due mainly to RÉSIBOIS [6] have even shown that there is no simple connection between these higher order contributions and S-matrix theory. Independently of this subtraction problem, higher order terms lead to a bewildering variety of contributions. It is only very recently that we begin to be able to give, at least in simple cases, some interpretation to all of them. In this interpretation a close relation appears, most unexpectedly, with the renormalisation techniques of field theory. However, the connection is essentially with damping problems and in general with problems involving unstable particles. This is not surprising as in all these problems as well as in nonequilibrium statistical mechanics we deal with an *initial value* problem.

In Chap. 5 we discuss a few of such problems. We give a minimum of mathematical details which can be found in the original literature, but we emphasize the physical ideas as well as the range of problems (which include the problems of line width, resonances, ...) which may be studied from the kinetic point of view we have adopted.

In Chap. 6 we go over to problems which imply a complete departure from the Boltzmann formulation. This is the case when the Hamiltonian cannot be written as the sum of an unperturbed term plus a perturbation but is given essentially by a single contribution. In these circumstances there are no longer two separate time scales (collision time and relaxation time) and a simple perturbational approach is surely impossible. I believe it very significant that in such cases (e.g. the case of spins interacting through a Heisenberg-Hamiltonian) RÉSIBOIS and DE LEENER [6] have been able to deduce from this general theory a kinetic equation of a new non-Boltzmann type.

In gravitational plasmas we have also a typically non-Boltzmann situation related to the infinite range of the forces (which leads to an infinite duration of the "collisions"). One cannot be sure that in such a case initial correlations would decay and that the second law of thermodynamics remains valid. This problem is also briefly discussed in Chap. 6.

It is precisely to the problem of irreversibility and entropy that we devote the last two chapters. Here we go into more detail than in the other chapters of these notes because this remains a question of much controversy. The problem of irreversibility was realized in a most remarkable way about 50 years ago by SMOLUCHOWSKI [7]. However, his analysis was based on stochastic theory and it is therefore not out of place to try to reformulate the problems on a strictly mechanical basis.

What we show in Chap. 7 is that our general evolution equation applies *both* to systems which evolve towards equilibrium as to systems which go away from it. The difference between these two cases depends essentially on the correlations at the *initial time*.

For mechanically closed systems and *for times much shorter than Poincaré's recurrence time* it is then possible to understand in detail the dominance of initial conditions which drive the system towards equilibrium (at least for simple laws of interactions).

The final chapter is devoted to the discussion of the microscopic meaning of entropy. At least in a simple example we can now analyse the meaning of entropy and show that its identification with "disorder" is valid only for weakly coupled or dilute systems. This then permits us to show exactly why Gibbs' definition of canonical entropy is valid only at equilibrium and what types of terms have to be eliminated from Gibbs' definition to extend it to the nonequilibrium range.

One may hope that as a result of the development of nonequilibrium statistical mechanics two main results will be reached: (a) a better integration of thermal physics in physics as a whole; (b) a better understanding of the limits and possibilities of thermodynamic methods.

From the latter point of view it is clear that phenomenological methods of a thermodynamic character may be developed quite beyond what has been done in classical physics. This is probably even essential to understand notions as organization, dissipation, structure, and so on, and for this reason we have devoted Chap. 2 of these lectures to such methods. On the other hand thermodynamics is probably severely limited by the type of interactions, at present, as already mentioned, only systems interacting through short-range forces (or fields leading to at least effective short-range forces) lead to the usual statistical formulation of the second law.

There are of course many omissions in these lectures. No study of transport processes is made, as well as no mention of the autocorrelation formula (Kubo's formula).

Also at some points these lectures take the form of a research report: For example at present some results are proved only to low orders in a perturbation parameter, while others are given to all orders.

I regret especially the omission from our discussion of Lorentz invariance*. It is indeed clear, thanks mainly to Dirac's work [8], that the problem of time evolution (described by an Hamiltonian) and the transformation properties with respect to Lorentz transformations are in fact closely related. Therefore, as has been shown recently explicitly by BALESCU, *et al.* [9], it is possible to apply the techniques described in these lectures to the problem of explicit Lorentz transformation of dynamical quantities in many body systems.

I believe this direction of research deserves special attention. It is indeed clear that a satisfactory formulation of the problem

*A short introduction to this problem is given in the lectures by M. BAUS on "Statistical Mechanics of a Classical Relativistic Plasma", published in this volume.

of time and irreversibility has to take into account *both* the many body aspect (inherent either in a phenomenological way through thermodynamics or through statistical physics) as well as the Minkowski structure of the space-time of the observer.

References

[1] I. PRIGOGINE: Non-Equilibrium Statistical Mechanics; Wiley-Interscience, New York-London, 1962.

[2] S. CHAPMAN and T.G. COWLING: Mathematical Theory of Non-Uniform Gases; University Press, Cambridge, 1939.

[3] G.V. CHESTER: Rep. Progr. Physics 26, 411 (1963).

[4] St. RICE and P. GRAY: Statistical Mechanics of Simple Liquids; Wiley, New York, 1965.

[5] S.T. CHOH and G.E. UHLENBECK: The Kinetic Theory of Phenomena in Dense Gases; Univ. of Michigan, 1958.

[6] P. RÉSIBOIS: Physica 31, 645 (1965).
P. RÉSIBOIS and DE LEENER: Phys. Rev. 152, 305 (1966)

[7] M. VON SMOLUCHOWSKI: Physik. 13, 1069 (1912); 14, 261 (1913).

[8] P.A.M. DIRAC: Can. J. Math. 2, 129 (1950); 3, 1 (1951).
See also D.G. CURRIE, T.F. JORDAN and E.C.G. SUDARSHAN: Rev. Mod. Phys. 35, 350 (1963).

[9] R. BALESCU and T. KOTERA: Physica 33, 558 (1967); R. BALESCU, T. KOTERA and E. PIÑA: Physica 33, 587 (1967); R. BALESCU: to appear Physica 35, (1967); M. BAUS: to appear.

2. Phenomenological Approach - Thermodynamics of Irreversible Processes

In classical thermodynamics one can define for an isolated system an entropy S which always increases with time, reaching a maximum in the equilibrium state. Similarly for non-isolated systems with given volume and temperature we may define the Helmholtz free energy F:

$$F = E - TS$$

where E is the internal energy and T the absolute temperature. F decreases with time until equilibrium is reached. It is clear that the position of the minimum is determined by the energy when the temperature is sufficiently low. (Note that a solid has a lower entropy than a liquid).

Now, the question is: Can one generalize such statements to non-equilibrium conditions? For example, does there still exist a quantity describing the steady state of a system whose surface is maintained at a given temperature varying along the boundary? This can be done and the result is summarized in the theorem of minimum entropy production (see Ref. 2). Let us introduce the entropy production P per unit time:

$$P = \frac{d_i S}{dt} = \int dV \sigma \geq 0. \quad ^{*} \qquad (2.1)$$

Here σ is the entropy production per unit time and volume. The theorem of minimum entropy production states that under well defined

---

* In (2.1) $d_i S$ is the entropy change within the system due to irreversible processes during the time interval dt (see Ref. 2).

assumptions P reaches its minimum in the steady state. These assumptions are that the system is in some sense near equilibrium, i.e., that S is a function of the same arguments (E, V, molar fraction,...) as in equilibrium (see Ref. 2). Then σ is a bilinear form in the generalized forces X_i and fluxes J_i of the irreversible process:

$$\sigma = \sum_i J_i X_i . \tag{2.2}$$

Moreover one has to make the specific assumptions:

1) linear phenomenological laws:

$$J_i = \sum_j L_{ij} X_j \; ; \tag{2.3}$$

2) validity of Onsager's reciprocity relations:

$$L_{ij} = L_{ji} \; ; \tag{2.4}$$

3) treat the phenomenological coefficients L_{ij} as constants. The boundary conditions are assumed time independent.

Let us apply this, for example, to the case of heat conduction:

$$\underset{\sim}{J} \equiv \underset{\sim}{W} = -\lambda \frac{\partial T}{\partial \underset{\sim}{x}} , \qquad \underset{\sim}{X} = \frac{\partial}{\partial \underset{\sim}{x}} (T^{-1}) . \tag{2.5}$$

We have:

$$P = \frac{d_i S}{dt} = \int dV \frac{\lambda}{T^2} \left(\frac{\partial T}{\partial \underset{\sim}{x}}\right)^2 . \tag{2.6}$$

If the temperature in our system deviates from some average T_o, i.e., $T = T_o + \delta T$, then we have, neglecting higher orders:

$$P = \frac{\lambda}{T_o^2} \int dV \left(\frac{\partial T}{\partial \underset{\sim}{x}}\right)^2 \tag{2.7}$$

and P reaches its minimum for:

$$\lambda(\underset{\sim}{\nabla}^2 T) = 0 \tag{2.8}$$

which is the steady state as obtained from Fourier's law. As a matter of fact the equation being linear there is no great interest in the method here. The interesting situations are the *nonlinear* ones:

$$\lambda = \lambda(T), \tag{2.9}$$

i.e., a temperature dependent heat conductivity λ. Moreover what happens when the assumptions (2.3), (2.4) are no longer valid? In these situations, theorem (2.1) fails but some of its aspects remain valid. Indeed, let us write from (2.2):

$$\begin{aligned} d\sigma &= d_X\sigma + d_J\sigma \\ d_X\sigma &= \sum_i J_i dX_i \\ d_J\sigma &= \sum_i X_i dJ_i \end{aligned} \tag{2.10}$$

and similarly: $$dP = d_X P + d_J P. \tag{2.11}$$

Then one can prove for time independent boundary conditions the inequality:

$$d_X P = \int dV \sum_i J_i dX_i \leq 0 \tag{2.12}$$

without reference to any phenomenological relation between J_i and X_i. We can sketch the proof for heat conduction as follows. The equation of energy conservation reads:

$$\rho \frac{\partial e}{\partial t} = - \frac{\partial}{\partial x_j} W_j \quad \text{(sum over j)} \tag{2.13}$$

ρ being the density and e the energy per unit mass. Multiplying (2.13) by $(\frac{\partial T^{-1}}{\partial t})$ we obtain for the left-hand side:

$$\Psi = \rho (\frac{\partial T^{-1}}{\partial t}) \frac{\partial e}{\partial t} = -\rho \frac{c_V}{T^2} (\frac{\partial T}{\partial t})^2 \tag{2.14}$$

and for the right-hand side:

$$\Psi = \frac{\partial}{\partial x_j} (-W_j \frac{\partial T^{-1}}{\partial t}) + W_j \frac{\partial}{\partial t} (\frac{\partial T^{-1}}{\partial x_j}). \tag{2.15}$$

If the specific heat c_V is positive, Ψ is negative from Eq. (2.14), and Eq. (2.15) yields after integration over the volume for time independent boundary conditions:

$$\int dV \Psi = \int dV \, W_j \frac{\partial}{\partial t} (\frac{\partial T^{-1}}{\partial x_j}) \;, \tag{2.16}$$

which is the proof of (2.12) for heat conduction. More generally one can include mechanical convection and flow terms and assert that there exists a quantity $d\Phi$ of the form of (2.12) for which we have:

$$d\Phi = \int dV \sum_i J_i' dX_i' \leq 0 \tag{2.17}$$

where X_i' and J_i' are forces and fluxes including mechanical terms (see Ref. 1). Time dependent boundary conditions can also be included (see Ref. 1). We may therefore say that (2.17) yields a *"universal"* evolution criterion. However, this should not be confused with a universal potential. Such a quantity does not always exist because $d\Phi$ is not necessarily a total differential. The non-existence of a universal potential is related to the fact that not all macroscopic systems can forget their initial conditions.

Consider again the example of heat conduction (2.5) but in the nonlinear domain (2.9). We have from (2.16):

$$\int dV\lambda(T)T^2\sum_j \frac{\partial T^{-1}}{\partial x_j}\frac{\partial}{\partial t}\left(\frac{\partial T^{-1}}{\partial x_j}\right) \leq 0 . \tag{2.18}$$

Using the same procedure as in (2.7), let us call $T_o(x)$ the solution of the time independent Fourier equation:

$$\frac{\partial}{\partial x_j}\lambda(T_o)\frac{\partial T_o}{\partial x_j} = 0 . \tag{2.19}$$

Replacing λT^2 by $\lambda_o T_o^2$ in (2.18), we obtain:

$$\frac{1}{2}\frac{\partial}{\partial t}\int dV\lambda(T_o)T_o^2\sum_j\left(\frac{\partial T^{-1}}{\partial x_j}\right)^2 \leq 0 . \tag{2.20}$$

Now we can introduce the "local potential" $\Phi(T,T_o)$ for heat conduction:

$$\Phi(T,T_o) = \frac{1}{2}\int dV\lambda_o T_o^2\left(\frac{\partial T^{-1}}{\partial \underset{\sim}{x}}\right)^2 . \tag{2.21}$$

The essential point is that $\Phi(T,T_o)$ is a functional of both T and T_o. In the problem here T_o is time independent. We may notice that:

1) $\Phi(T,T_o)$ decreases in time until it reaches its minimum value $\Phi(T_o,T_o)$ and

2) $\Phi(T_o,T_o) = \frac{1}{2}\frac{d_iS}{dt}$

The local potential appears, therefore, as a generalization of the usual thermodynamic entropy production (2.1). If we minimize with respect to T (at constant T_o) we obtain from Eq. (2.21):

$$\left(\frac{\delta\Phi}{\delta T}\right)_{T_o} = 0 \qquad \frac{\partial}{\partial x_j}\lambda_o T_o^2\left(\frac{\partial T^{-1}}{\partial x_j}\right) = 0. \tag{2.22}$$

If *after* minimization we use the subsidiary condition

$$T = T_o \tag{2.23}$$

we obtain

$$\frac{\partial W_j}{\partial x_j} = 0 \tag{2.24}$$

which is identical to Eq. (2.19).

The interpretation of T and T_o can be given in terms of a trivial generalization of the classical Einstein theory of fluctuations; T appears in this way as the "fluctuating" temperature and T_o as the average temperature.

Let us indeed consider Einstein's formula which gives the probability of a fluctuation of an isolated system around its equilibrium state for a given change in entropy(ΔS), namely:

$$\text{Prob} \sim e^{\Delta S/k} \quad \text{(with } k = \text{Boltzmann constant)} \tag{2.25}$$

For a fluctuation around an equilibrium state (see Fig.2.1) we can thus compute the entropy of the fluctuating distribution and from (2.25) we obtain the probability. For our example of heat conductivity this would yield:

$$\text{Prob} \sim \exp\left[-\frac{1}{2k}\int dV\,\frac{c_V}{T_o^2}\,(\delta T)^2\right]. \tag{2.26}$$

This formula permits us to calculate the probability of the temperature distribution $T(\underset{\sim}{x},t) = T_o + \delta T(\underset{\sim}{x},t)$ knowing the equilibrium temperature T_o. For the fluctuation around a steady nonequilibrium state $T_o(\underset{\sim}{x})$ (see Fig. 2.2), one can no longer use Einstein's formula because the entropy is not maximum in a nonequilibrium state. Still we may use Eq. (2.26) in the slightly generalized form:

$$\text{Prob} \sim \exp\left[-\frac{1}{2k}\int dV\,\frac{c_V^o}{T_o(\underset{\sim}{x})}\,(\delta T)^2\right] \tag{2.27}$$

in which the fluctuation is related to the *local* steady state value $T_o(\underset{\sim}{x})$. It can be shown (see Ref. 1) that Eq. (2.27) can also be written:

$$\text{Prob} \sim \exp\left[\frac{1}{k}\int dt\Delta\Phi\right] \quad \text{with} \quad \Delta\Phi = \Phi(T_o,T_o)-\Phi(T,T_o)\ , \tag{2.28}$$

where the time integration goes from some initial time to the time at which the fluctuation has disappeared.

From (2.28) one can compute, knowing the local potential (2.21), the probability of a temperature distribution $T(\underset{\sim}{x},t)$ for a given steady state temperature distribution $T_o(\underset{\sim}{x})$. This can be extended to time dependent phenomena, i.e., $T_o(\underset{\sim}{x},t)$ (see Ref.1).

An interesting feature of Φ for systems where both dissipative and flow processes are involved is that it often takes the form:

$$\Phi = (\text{dissipative processes}) - (\text{convection processes}) \qquad (2.28)$$

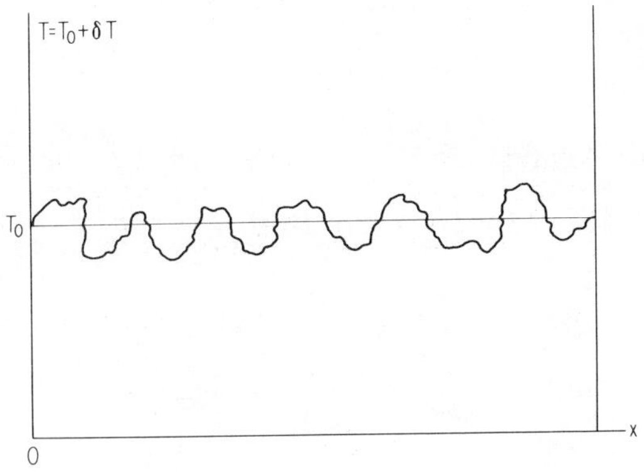

Fig. 2.1.
Fluctuations around an equilibrium state

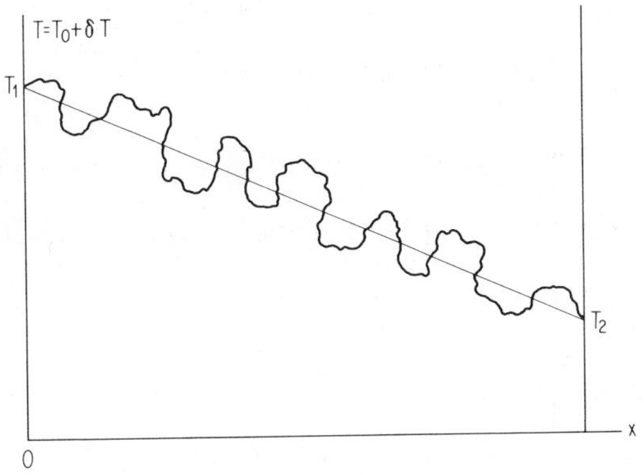

Fig. 2.2.
Fluctuations around a steady nonequilibrium state

This formula has a structure which is reminiscent of the structure of classical free energy ($F = E - TS$). This analogy gives rise to interesting similarities between thermodynamic phase transitions and hydrodynamic stability problems. To make this analogy clearer let us consider the well known Bénard problem (onset of thermal instability in a horizontal fluid layer heated from below). Let us plot the corresponding local potential as a function of the imposed adverse gradient (see Fig. 2.3) for the liquid at rest as well as for the liquid in motion (more precisely the first possible state of thermal convection).

For the fluid at rest (curve 1, Fig. 2.3) the only quantity which enters into Φ is dissipation due to thermal conduction. This increases of course with increasing temperature gradient (or increasing characteristic Rayleigh number R).

For the fluid in motion we have viscous dissipation as well. Therefore Φ starts at a positive value even for R = 0 (see curve 2, Fig. 2.3). It increases more slowly with R, however, because as seen in formula (2.28), convective processes as a rule decrease the value of Φ. As a result the two curves intersect at a critical value R_c of the Rayleigh number.

The analogy with the liquid-solid transition as described by the crossing of the free energy line of the liquid with that of the solid is striking (see Fig. 2.4).

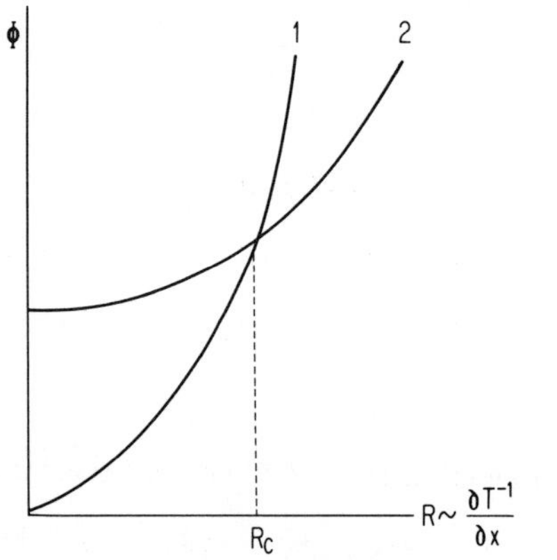

Fig. 2.3.

Local potential versus Rayleigh number (proportional to the adverse temperature gradient)
1:for the system without motion
2:for the system with motion

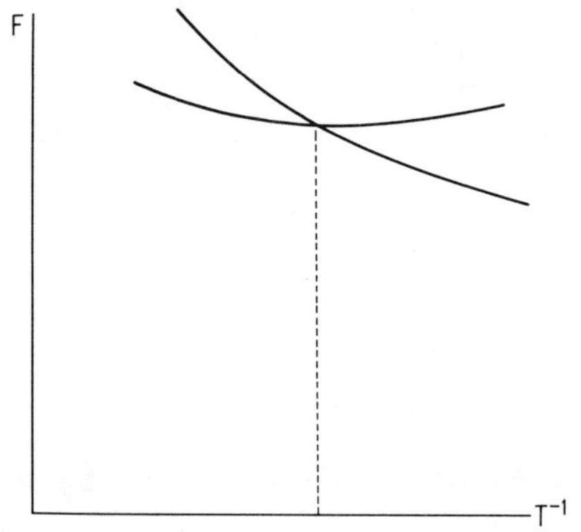

Fig. 2.4.

Schematic representation of the free energy versus $\frac{1}{T}$

A decrease of temperature leads to a transition from the less-ordered phase to the ordered phase. Similarly in the hydrodynamical case, the increase of the *constraint* (which is here the Rayleigh number) leads from a situation in which the whole of the energy is in thermal motion to a much more organized state in which part of it is in the form of macroscopic motion which is, of course, a highly cooperative phenomenon from the molecular point of view. We have here a very interesting example in which we see how the energy received by the system is transformed into organization. An example of chemical autocatalytic reactions in which instability may lead to a nonhomogeneous situation has been studied recently [6]. This is a case of a symmetry-breaking instability which leads to a highly organized steady state which is maintained through a delicate balance of chemical reactions and diffusion.

The method we have used may also be extended to problems of kinetic theory. The local potential then becomes a functional $F(f,f_o)$ of both the average distribution function f_o and of a fluctuating distribution function f. Again we may then determine the most probable distribution function f^* through the variational condition:

$$\left(\frac{\delta F}{\delta f}\right)_{f_o} = 0 \qquad (2.29)$$

and use the self-consistency condition:

$$f^* = f_o. \qquad (2.30)$$

In this way we identify the most probable distribution with the average one. Through Eq. (2.29) we apply a variational technique to fluctuations in phase space.

Very interesting applications of this method have been presented by NICOLIS [1],[3] to stability problems involving dissipation processes (as for example the run-away problem in plasmas). The importance of this technique originates in the fact that kinetic equations can be derived from ordinary "true variational" techniques only in exceptional cases (i.e., when the deviation from local equilibrium is very small).

It is interesting that thermodynamics of irreversible processes leads us to a new type of variational approach valid for a large class of dissipative nonlinear problems (for a comparison with other variational methods see especially SCHECHTER [1]).

From the point of view of computation, variational techniques lead generally to equations similar to the equations which are obtained from Galerkin's method for direct approximation of the original differential equation. However, a variational method is often a very powerful tool, as emphasized for example by VON NEUMANN [4]. From the physical point of view it leads to a very synthetic and compact formulation of the problem, and from the mathematical point of view it leads to a much more precise formulation of the convergence of self-consistent or successive approximation schemes (for the convergence problem involved in the local potential method see KRUSKAL [1], GLANSDORFF [5]).

References *

[1] For the recent development in Nonlinear Thermodynamics of Irreversible Processes consult: Non-Equilibrium Thermodynamics, Variational Techniques and Stability, ed. by R.J. DONNELLY, R. HERMAN and I. PRIGOGINE, The University of Chicago Press, Chicago, London (1966). In the papers collected in this volume the references to the original publications may be found.

[2] The most complete and lucid exposition of linear thermodynamics of irreversible processes is due to: S.R. DE GROOT and P. MAZUR: Non-Equilibrium Thermodynamics, Amsterdam, North Holland Publishing Co. (1962). See also for a short introduction: I. PRIGOGINE: Introduction to Thermodynamics of Irreversible Processes, (2nd ed.), New York, Wiley (1961).

[3] G. NICOLIS: Local Potential Methods in the Study of Kinetic Equations, Adv. Chem. Phys. 13, 299 (1967). See also Ref. 1.

[4] J. VON NEUMANN: Coll. Works, Pergamon Press, London (1961-1963).

[5] P. GLANSDORFF: Physica 32, 1745 (1966).

[6] I. PRIGOGINE and G. NICOLIS: J. Chem. Phys. 46, 3542 (1967); I. PRIGOGINE and R. LEFEVER: to appear J. Chem. Phys.; R. LEFEVER, G. NICOLIS and I. PRIGOGINE: J. Chem. Phys. 47, 1045 (1967); I. PRIGOGINE: Communication at the International Conference "Physical Théorique et Biologie", Institut de la Vie, Versailles, France.

---

*In the monograph, The Variational Methods in Engineering (McGraw-Hill, 1967) which recently appeared, S.R. SCHECHTER has given an excellent survey of the applications of the local potential method. A short introduction to the problem of dissipative structures is given in I. PRIGOGINE, Introduction to Thermodynamics of Irreversible Processes, (3rd. ed.), Wiley, New York (1967). PRIGOGINE and GLANSDORFF are preparing a detailed monograph devoted to these problems.

3. Statistical Mechanics - General Method

a) Introduction

Before presenting the general theory of irreversible processes based on the Liouville equation, let us briefly make some comments on the two approaches which have been introduced by the founders of statistical mechanics, MAXWELL and BOLTZMANN, to justify the use of this method.

We first have the ergodic theory. We know today, thanks to the works mainly of BIRKHOFF, VON NEUMANN and others, what the conditions of ergodicity are (see KHINCHIN [1]). We have to express that the whole surface of constant energy in phase space is available (metric indecomposibility). What is lacking is the proof that the systems studied in statistical mechanics satisfy these conditions. However, some progress seems to have been realized recently (see Ref. 2). But even so it is very unlikely that ergodic theory has anything to do with the irreversibility we generally observe. Indeed the time necessary to "cover" the ergodic surface is very large, presumably of the order of the Poincaré recurrence time. This is an immense time span (i.e., $10^{10^{10}}$ years) for systems involving about 10^{23} degrees of freedom. On the contrary the relaxation time as observed in dilute gases is of the order of the time between two collisions, i.e., of the order of 10^{-8} sec. This difference in time scale is reflected in the fact that in ergodic theory one studies the approach to equilibrium of the whole system of N ($\simeq 10^{23}$) particles, whereas in the Boltzmann-like kinetic approach one studies the approach to equilibrium of the one-particle velocity distribution function (d.f.). One can say that all physical quantities (which depend on 1, 2, 3, ... particles) reach equilibrium in a time which has nothing to do with the time scale of ergodic theory (10^{23} particles). On the other hand Boltzmann's kinetic proof of the approach to equilibrium is restricted to dilute systems and in fact it proves too much. Indeed systems which satisfy the Boltzmann kinetic equation can only approach equilibrium and never go away from it.

This can, however, not always be true. As an example let us consider some recent computer calculations due to BELLEMANS and ORBAN [3] on "two-dimensional" hard spheres (hard discs). They start with the discs on lattice sites with an isotropic velocity distribution and calculate the well known Boltzmann $\mathcal{H}$-quantity:

$$\mathcal{H} = \int f(\underset{\sim}{v}) \log f(\underset{\sim}{v}) \, d\underset{\sim}{v}.$$

The results are represented on Fig. 3.1

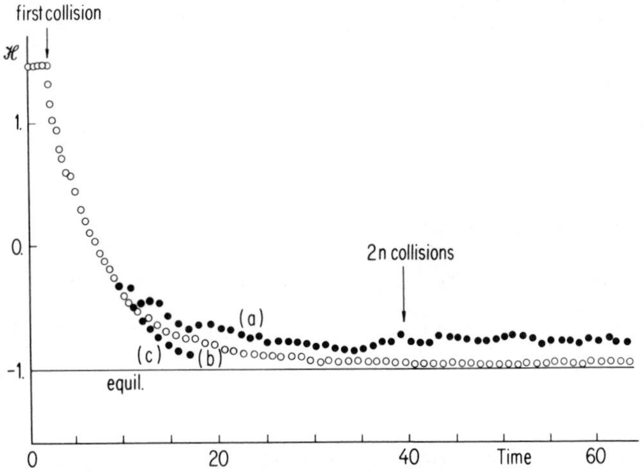

Fig. 3.1.

Evolution of $\mathcal{H}$ with time

(a) Total number of discs (n) = 100

(b) " " " " " = 484

(c) " " " " " = 1225

As in earlier calculations by ALDER and WAINWRIGHT [4] we have a fluctuating approach to equilibrium. Now if after 50 or 100 collisions we invert the velocities, we obtain a new ensemble. We may then again follow the evolution of the corresponding $\mathcal{H}$-quantity in time. The results are represented on Fig. 3.2.

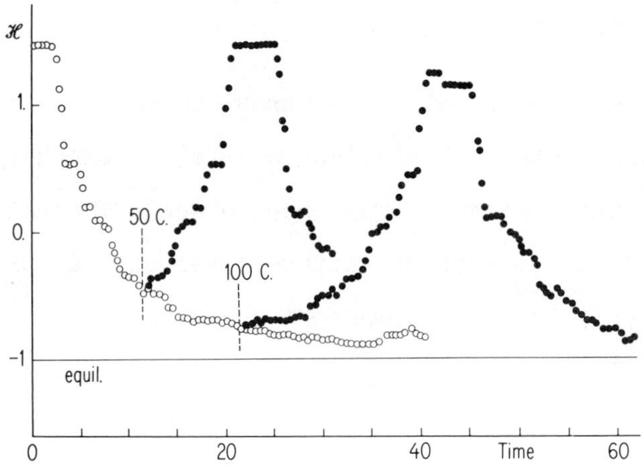

Fig. 3.2.

Evolution of $\mathcal{H}$ with time for a system of 100 discs when velocities are inverted after

(a) 50 collisions

(b) 100 collisions

We see that indeed the entropy (that is, $-\mathcal{H}$) first decreases after the velocity inversion. The system deviates from equilibrium on a period of 50 to 60 collisions (this would correspond in a dilute gas to about 10^{-6} sec).

The computer calculations by BELLEMANS and ORBAN provide a nice illustration of Loschmidt's reversibility paradox [5] (see the excellent discussion by CHANDRASEKHAR [6])*. We shall not go into the historical development of this question. But what is clear is that a consistent description of the time evolution of many body systems must include situations in which a system goes near to equilibrium as well as situations in which it goes away from

* We shall come back to the discussion of Loschmidt's paradox in Chap. 7 of these lectures.

equilibrium. We shall see that our general evolution equation, derived later, satisfies this condition (see Chap. 7).

b) Liouville Equation

We will consider now the general mechanical theory of irreversibility. We will develop the method for a classical system. The extensions to relativistic and quantum systems can be found in the references 7, 8, 9, 14. We start from the Hamilton equations:

$$\dot{q}_i = \frac{\partial H}{\partial p_i}, \qquad \dot{p}_i = -\frac{\partial H}{\partial q_i} \qquad (i = 1,\dots 3N), \tag{3.1}$$

H being the Hamiltonian of the system of N particles described by the set of canonical variables $\{q_i, p_i\}$ spanning the corresponding phase space of the system. The system as a whole can then be described by a normalized density in phase space $\rho(\{q_i\},\{p_i\},t)$ obeying the Liouville equation (equation of continuity in phase space):

$$\frac{\partial \rho}{\partial t} = \left[H, \rho\right], \tag{3.2}$$

where $[\ ,\ ]$ is the Poisson bracket:

$$\left[H, \rho\right] = \sum_i \left(\frac{\partial H}{\partial q_i}\frac{\partial \rho}{\partial p_i} - \frac{\partial H}{\partial p_i}\frac{\partial \rho}{\partial q_i}\right). \tag{3.3}$$

We rewrite Eq. (3.2) as:

$$i\frac{\partial \rho}{\partial t} = L\rho; \qquad L = i\left[H,\ \right], \tag{3.4}$$

introducing a "Liouville operator" which is Hermitian in the space of all integrable functions which vanish at the boundaries of phase space, as can be shown easily by partial integration (see Ref. 7).

The integration of (3.4) is a problem equivalent to the integration of the system (3.1), and Eq. (3.4) is only a compact way of writing the evolution of the system. In Eq. (3.4), linear in phase space, all nonlinear features of Hamilton's equation are absorbed in the linear differential operator L. We will consider now simple examples.

c) Free Particles

Let us first consider the case of a free particle:

$$H = \frac{\underset{\sim}{p}^2}{2m} \qquad L = -i\,\frac{\underset{\sim}{p}}{m}\cdot\frac{\partial}{\partial \underset{\sim}{x}} \;. \tag{3.5}$$

The Liouville equation reads:

$$\frac{\partial \rho}{\partial t} + \frac{\underset{\sim}{p}}{m}\cdot\frac{\partial \rho}{\partial \underset{\sim}{x}} = 0. \tag{3.6}$$

The eigenfunctions of L are given by:

$$L\phi_{\underset{\sim}{k}} \equiv -i\,\frac{\underset{\sim}{p}}{m}\cdot\frac{\partial}{\partial \underset{\sim}{x}}\,\phi_{\underset{\sim}{k}} = \lambda_{\underset{\sim}{k}}\phi_{\underset{\sim}{k}}, \tag{3.7}$$

and thus:

$$\begin{aligned} \phi_{\underset{\sim}{k}}(\underset{\sim}{p}) &= \Omega^{-1/2}\, e^{i\underset{\sim}{k}\cdot\underset{\sim}{x}}\, f(\underset{\sim}{p}) \\ \lambda_{\underset{\sim}{k}} &= \underset{\sim}{k}\cdot\underset{\sim}{v} \qquad m\underset{\sim}{v} = \underset{\sim}{p}, \end{aligned} \tag{3.8}$$

where $f(\underset{\sim}{p})$ is a normalized function and where Ω is the volume in which the system is enclosed. Assuming for example a cubic box with periodic boundary conditions, we have:

$$\underset{\sim}{k} = \frac{2\pi}{\Omega^{1/3}} \underset{\sim}{n} , \tag{3.9}$$

with $\underline{n}$ a vector with integer components. The solution of the Liouville equation can now be written as:

$$\rho(\underset{\sim}{p},\underset{\sim}{x},t) = \Omega^{-\frac{1}{2}} \sum_{\underset{\sim}{k}} e^{-i\lambda_{\underset{\sim}{k}} t} \phi_{\underset{\sim}{k}} = \Omega^{-1} \sum_{\underset{\sim}{k}} e^{i\underset{\sim}{k}\cdot(\underset{\sim}{x}-\underset{\sim}{v}t)} \rho_{\underset{\sim}{k}}(\underset{\sim}{p}) . \tag{3.10}$$

This equation expresses that ρ is an arbitrary function of $\underline{p}$ and the free motion $\underset{\sim}{x}-\underset{\sim}{v}t$. This can be generalized to a system of free particles by introducing adequate summations in (3.5) and (3.10) but remains of no interest, because without interactions the system cannot reach equilibrium. However, this shows how we will proceed in the interacting case, namely, just as in (3.10) we will develop ρ in eigenfunctions of the unperturbed Liouville operator. Therefore one can say that we use the method developed in quantum mechanics on an equation (3.4) of the Schrödinger type but in phase space instead of in configuration space. As second example we will consider the problem of potential scattering.

d) Potential Scattering

Consider the following situation (see Fig.3.3).

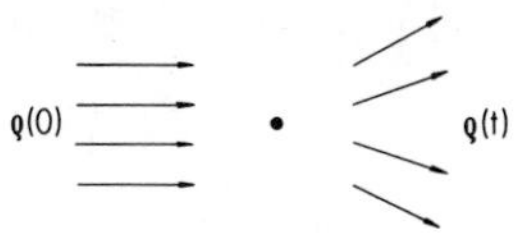

Fig. 3.3.

Schematical Representation of a Scattering Experiment

An incident beam of independent particles interacts with a center of force at $\underset{\sim}{x} = 0$. The distribution of the beam (= the ensemble) will be $\rho(\underset{\sim}{x},\underset{\sim}{p};t)$ and the Hamiltonian is, assuming central forces,

$$H = \frac{\underset{\sim}{p}^2}{2m} + \lambda V(|\underset{\sim}{x}|). \tag{3.11}$$

The Liouville operator can be split according to its dependence on the coupling constant λ just as in (3.11):

$$L = L_o + \lambda \delta L$$
$$\text{where} \qquad L_o = -i\,\underset{\sim}{v}\cdot\frac{\partial}{\partial \underset{\sim}{x}} \qquad \delta L = +i\,\frac{\partial V}{\partial \underset{\sim}{x}}\,\frac{\partial}{\partial \underset{\sim}{p}}. \tag{3.12}$$

We now expand ρ in eigenfunctions of L_o, namely (3.10), and introduce this expansion into the Liouville Eq. (3.4), obtaining:

$$i\,\frac{\partial}{\partial t}\,\rho_{\underset{\sim}{k}}(\underset{\sim}{p};t) = \sum_{\underset{\sim}{k}'} e^{i\underset{\sim}{k}\cdot\underset{\sim}{v}t}\langle \underset{\sim}{k}|\delta L|\underset{\sim}{k}'\rangle e^{-i\underset{\sim}{k}'\cdot\underset{\sim}{v}t}\rho_{\underset{\sim}{k}'}(\underset{\sim}{p};t). \tag{3.13}$$

The matrix element is defined as in quantum mechanics:

$$\langle \underset{\sim}{k}|\delta L|\underset{\sim}{k}'\rangle = \Omega^{-1}\int d\underset{\sim}{x}\, e^{-i\underset{\sim}{k}\cdot\underset{\sim}{x}}\delta L\; e^{+i\underset{\sim}{k}'\cdot\underset{\sim}{x}} \tag{3.14}$$

$$= -\frac{8\pi^3}{\Omega}\sum_{\underset{\sim}{\ell}} V_{\underset{\sim}{\ell}}\;\underset{\sim}{\ell}\cdot\frac{\partial}{\partial \underset{\sim}{p}}\;\delta_{\underset{\sim}{k}',\underset{\sim}{k}-\underset{\sim}{\ell}}$$

where $V_{\underset{\sim}{k}}$ is the Fourier component of the potential V:

$$V(|\underset{\sim}{x}|) = \frac{8\pi^3}{\Omega}\sum_{\underset{\sim}{\ell}} V_{\underset{\sim}{\ell}}\; e^{i\underset{\sim}{\ell}\cdot\underset{\sim}{x}} \tag{3.15}$$

and the delta function is a Kronecker delta function. Eq. (3.13) is the analogue of Dirac's equation of variation of constants:

$$i\,\dot{c}_n = \sum_{n'} e^{iE_n t/\hbar} \langle n|V|n'\rangle e^{-iE_{n'} t/\hbar}\, c_{n'} \tag{3.16}$$

However, Eq. (3.16) and (3.13) are written in different spaces: c_n is only a function of time and $\langle n|V|n'\rangle$ is a c-number, whereas $\rho_{\underset{\sim}{k}}$ is still a function of $\underset{\sim}{p}$ and $\langle \underset{\sim}{k}|\delta L|\underset{\sim}{k}'\rangle$ is an operator in $\underset{\sim}{p}$ space. The relation between (3.16) and (3.13) is that whereas in (3.16) the matrix element $\langle n|V|n'\rangle$ induces transitions between states characterized by different quantum numbers n, n', the operator matrix element $\langle \underset{\sim}{k}|\delta L|\underset{\sim}{k}'\rangle$ induces transitions between states with different wave vectors $\underset{\sim}{k}$, $\underset{\sim}{k}'$, i.e., different states of *correlation* of the particle with the scattering center. This example of potential scattering, which in fact corresponds to the elementary evolution mechanism in more complicated situations, already introduces what will be an overall characteristic, namely, that this way of describing the evolution of the system corresponds to a *dynamics of correlations*. The evolution Eq. (3.13) reads with the help of (3.14):

$$i\,\frac{\partial}{\partial t}\,\rho_{\underset{\sim}{k}}(\underset{\sim}{p};t) = \lambda\,\frac{8\pi^3}{\Omega} \sum_{\underset{\sim}{\ell}} e^{i\underset{\sim}{k}\cdot\underset{\sim}{v}t} V_{\underset{\sim}{\ell}}\, \underset{\sim}{\ell}\cdot\frac{\partial}{\partial \underset{\sim}{p}}\, e^{-i(\underset{\sim}{k}+\underset{\sim}{\ell})\cdot\underset{\sim}{v}t} \rho_{\underset{\sim}{k}+\underset{\sim}{\ell}}(\underset{\sim}{p};t), \tag{3.17}$$

where we used the central character of the potential, i.e., $V_{\underset{\sim}{\ell}} = V_{-\underset{\sim}{\ell}}$. We can assume the forces to be small ($\lambda \ll 1$) and solve (3.17) by iteration. Restricting to the lowest orders in λ we have:

$$\rho_{\underset{\sim}{k}}(t) = \rho_{\underset{\sim}{k}}(0) + \left(\frac{\lambda}{i}\right)\left(\frac{8\pi^3}{\Omega}\right) \sum_{\underset{\sim}{\ell}} \int_0^t dt_1 e^{i\underset{\sim}{k}\cdot\underset{\sim}{v}t_1} V_{\underset{\sim}{\ell}}\, \underset{\sim}{\ell}\cdot\frac{\partial}{\partial \underset{\sim}{p}}\, e^{-i(\underset{\sim}{k}+\underset{\sim}{\ell})\cdot\underset{\sim}{v}t_1} \rho_{\underset{\sim}{k}+\underset{\sim}{\ell}}(0)$$

$$+ \left(\frac{\lambda}{i}\right)^2\left(\frac{8\pi^3}{\Omega}\right)^2 \sum\sum_{\underset{\sim}{\ell}\underset{\sim}{\ell}'} \int_0^t dt_1 \int_0^t dt_2\; e^{i\underset{\sim}{k}\cdot\underset{\sim}{v}t_1} V_{\underset{\sim}{\ell}}\, \underset{\sim}{\ell}\cdot\frac{\partial}{\partial \underset{\sim}{p}}\, e^{-i(\underset{\sim}{k}+\underset{\sim}{\ell})\cdot\underset{\sim}{v}t_1} \times$$

$$\times\, e^{i(\underset{\sim}{k}+\underset{\sim}{\ell})\cdot\underset{\sim}{v}t_2} V_{\ell'}\, \underset{\sim}{\ell}'\cdot\frac{\partial}{\partial \underset{\sim}{p}}\, e^{-i(\underset{\sim}{k}+\underset{\sim}{\ell}+\underset{\sim}{\ell}')\cdot\underset{\sim}{v}t_2}\, \rho_{\underset{\sim}{k}+\underset{\sim}{\ell}+\underset{\sim}{\ell}'}(0) + O(\lambda^3). \tag{3.18}$$

Let us consider the solution not for the correlations $\rho_{\underset{\sim}{k}}$ but for the velocity distribution function $\rho_o(\underset{\sim}{p};t)$, we have (put $\underset{\sim}{k} = 0$ in 3.18):

$$\rho_o(t) = \rho_o(0) + (\tfrac{\lambda}{i})(\tfrac{8\pi^3}{\Omega}) \sum_{\underset{\sim}{\ell}} \int_0^t dt_1\, V_{\ell}\, \underset{\sim}{\ell}\cdot\frac{\partial}{\partial \underset{\sim}{p}}\, e^{-i\underset{\sim}{\ell}\cdot\underset{\sim}{v}t_1}\, \rho_{\underset{\sim}{\ell}}(0)$$

$$+ (\tfrac{\lambda}{i})^2(\tfrac{8\pi^3}{\Omega})^2 \sum_{\underset{\sim}{\ell},\underset{\sim}{\ell}'} \int_0^t dt_1 \int_0^{t_1} dt_2\, V_{\ell}\, \underset{\sim}{\ell}\cdot\frac{\partial}{\partial \underset{\sim}{p}}\, e^{i\underset{\sim}{\ell}\cdot\underset{\sim}{v}(t_2-t_1)}$$

$$V_{\ell'}\, \underset{\sim}{\ell}'\cdot\frac{\partial}{\partial \underset{\sim}{p}}\, e^{-i(\underset{\sim}{\ell}+\underset{\sim}{\ell}')\cdot\underset{\sim}{v}t_2}\, \rho_{\underset{\sim}{\ell}+\underset{\sim}{\ell}'}(0) + O(\lambda^3). \tag{3.19}$$

One can distinguish different contributions in (3.19): (1) the unperturbed term $\rho_o(0)$; (2) terms coming from the initial correlations with the scatterer (cf. $\rho_{\underset{\sim}{\ell}}(0)$ and $\rho_{\underset{\sim}{\ell}+\underset{\sim}{\ell}'\neq 0}(0)$ in (3.19)); (3) a λ^2 term relating $\rho_o(0)$ to $\rho_o(t)$ which we will show to represent the collision operator in the classical version of the Born approximation.

We will come closer to the scattering experiment by supposing the beam and the scatter to be uncorrelated at $t = 0$ ($\rho(0) = \rho_o(0)$) and inquiring for the form of $\rho_o(t)$ at large t ($\to\infty$). This is exactly the S-matrix point of view on a scattering experiment. We thus have from (3.19)

$$\rho_o(t) = \rho_o(0) + \lambda^2\, (\tfrac{8\pi^3}{\Omega})^2 \sum_{\underset{\sim}{\ell}} \int_0^t dt_1 \int_0^{t_1} dt_2 \Big[V_{\ell}\, \underset{\sim}{\ell}\cdot\frac{\partial}{\partial \underset{\sim}{p}}\, e^{i\underset{\sim}{\ell}\cdot\underset{\sim}{v}(t_2-t_1)}$$

$$V_{\ell}\, \underset{\sim}{\ell}\cdot\frac{\partial}{\partial \underset{\sim}{p}}\, \rho_o(0)\Big]. \tag{3.20}$$

First we have to compute the asymptotic form of the time integral. In the limit of an infinite volume we have:

$$\frac{8\pi^3}{\Omega}\sum_{\underset{\sim}{\ell}} \to \int d\underset{\sim}{\ell}. \tag{3.21}$$

Therefore we can write the second term of (3.20) in the form

$$I(t) = \int_0^t dt_1 \int_0^{t_1} dt_2 \int d\alpha \, e^{i\alpha(t_1-t_2)} \; f(\alpha) = \int d\alpha \; I(\alpha,t) \; f(\alpha). \tag{3.22}$$

We introduce the change of variables with the Jacobian $\frac{1}{2}$:

$$\tau = t_1 - t_2 \quad \text{and} \quad T = t_1 + t_2; \tag{3.23}$$

this yields:

$$I(\alpha,t) = \frac{1}{2}\int_0^t d\tau \, e^{i\alpha\tau} \int_\tau^{2t-\tau} dT = \int_0^t d\tau \, e^{i\alpha\tau}(t-\tau). \tag{3.24}$$

The first term of (3.24) grows systematically with time as t, while the second remains bounded. Indeed we have:

$$\int_0^t d\tau \, e^{i\alpha\tau} = \frac{\sin\alpha t}{\alpha} + i\,\frac{(1-\cos\alpha t)}{\alpha}, \tag{3.25}$$

and for large t:

$$\lim_{t\to\infty} \frac{\sin\alpha t}{\alpha} = \pi\delta(\alpha)$$

$$\lim_{t\to\infty} \frac{1-\cos\alpha t}{\alpha} = \mathcal{P}\left(\frac{1}{\alpha}\right) \equiv \begin{cases} \frac{1}{\alpha} & \text{if} \quad \alpha \neq 0 \\ 0 & \text{if} \quad \alpha = 0 \end{cases}. \tag{3.26}$$

Indeed (3.26) is nothing but the well known result:

$$\int_0^\infty d\tau\ e^{\pm i\alpha\tau} = \pi\delta_\pm(\alpha) = \pi\delta(\alpha) \pm i\mathcal{P}(\frac{1}{\alpha}) \qquad (3.27)$$

where $\delta_\pm$ is the so-called delta-plus or delta-minus function, δ is the Dirac function and $\mathcal{P}$ the Cauchy principal part. (All three are generalized functions.) We can thus write (3.20) asymptotically as:

$$\rho_o(t) = \rho_o(0) + \lambda^2 t\ \frac{8\pi^3}{\Omega} \int d\underset{\sim}{\ell}\ \underset{\sim}{\ell}\cdot\frac{\partial}{\partial \underset{\sim}{p}} |V_{\underset{\sim}{\ell}}|^2 \pi\delta(\underset{\sim}{\ell}\cdot\underset{\sim}{v})\underset{\sim}{\ell}\cdot\frac{\partial}{\partial \underset{\sim}{p}}\ \rho_o(\underset{\sim}{p};0) + O(\lambda^3), \qquad (3.28)$$

where we dropped the principal part term because it is odd in the $\underset{\sim}{\ell}$ integration. One can show (see Ref. 8) that Eq. (3.28) can be written:

$$\rho_o(\underset{\sim}{p};t) = \rho_o(\underset{\sim}{p};0) - \lambda^2 t\ \frac{8\pi^3}{\Omega}\ \frac{A}{v^3}\ \mathcal{L}^2\ \rho_o(\underset{\sim}{p};0) + O(\lambda^3), \qquad (3.29)$$

where A contains all characteristics of the potential and where the operator $\mathcal{L}^2$ (first introduced by DELCROIX (see Ref. 8)) is defined by:

$$\mathcal{L}^2 = -\ (\underset{\sim}{v} \times \frac{\partial}{\partial \underset{\sim}{v}})\cdot(\underset{\sim}{v} \times \frac{\partial}{\partial \underset{\sim}{v}}), \qquad (3.30)$$

and is the analogue in velocity space of the quantum mechanical angular momentum operator

$$\underset{\sim}{L}^2 = -\hbar^2(\underset{\sim}{x} \times \frac{\partial}{\partial \underset{\sim}{x}})\cdot(\underset{\sim}{x} \times \frac{\partial}{\partial \underset{\sim}{x}}). \qquad (3.31)$$

As is well known from quantum mechanics, the eigenfunctions of (3.30) are spherical harmonics in velocity space. From (3.29) it follows that the final distribution will be more spherically symmetric than the initial one. If $\rho_o(0)$ was already spherically symmetric it would remain unchanged. Summarizing, one can say that in this potential scattering experiment we have a contribution to $\rho_o(t)$ coming from $\rho_o(0)$ which (1) is linearly growing in time as it should be for the scattering of the incoming beam; (2) introduces a collision operator whose only effect (within the Born approximation) is to increase the spherical symmetry of the initial distribution; and (3) yields an irreversible behaviour, the stationary state being a spherically symmetric velocity distribution.

In this way we obtained the expected answer. However, is this the answer in general? To investigate this point we can go back to the general Eq. (3.17-18). Let us start with an homogeneous distribution:

$$\begin{cases} \rho_o(0) \neq 0 \\ \\ \rho_{\tilde{k}}(0) = 0 \, . \end{cases} \tag{3.32}$$

We can see on (3.18) that at time t we have:

$$\begin{cases} \rho_o(t) \neq 0 \\ \\ \rho_k(t) \neq 0 \, , \end{cases} \tag{3.33}$$

i.e., while the velocity distribution increases its symmetry (cf. (3.29)) there appears now the Fourier coefficients $\rho_{\tilde{k}}$. Therefore, the system at time t is no longer homogeneous. There exist space dependent correlations between the scattering center and the particles of the beam. Suppose now we invert the velocities at some time $t = t_1$, then the system will invert its history and come back in the initial situation after a time delay t_1, i.e. at time $t = 2t_1$. Thus in the time interval $0\text{-}t_1$ the symmetry of ρ_0 increased and a $\rho_{\underset{\sim}{k}}$ is created, i.e., spatial symmetry decreased. On the contrary during the time interval $t_1\text{-}2t_1$, $\rho_0(t)$ decreases its symmetry when reaching $\rho_0(0)$ while $\rho_{\underset{\sim}{k}}$ disappears, increasing the spatial symmetry. We are thus faced with two opposite types of evolution: 1) one in which the symmetry of ρ_0 increases and where ρ_0 determines $\rho_{\underset{\sim}{k}}$,* i.e. the "thermodynamic" evolution, 2) one in which the symmetry of ρ_0 decreases and where $\rho_{\underset{\sim}{k}}$ determines ρ_0, i.e. an "anti-thermodynamic" evolution. The basic difference between the two situations is that during $0\text{-}t_1$, one can assume the correlation to be of short range while during $t_1\text{-}2t_1$ the correlations are of a long range type because the particle will have to interact with the center in an arbitrarily removed future. This indicates that the domination of thermodynamic evolution in nature is simply an expression of our habit to encounter situations where short range correlations are dominant. We shall come back to these problems in Chap. 7.

After this example of potential scattering, which illustrates nicely the main features of the more complicated situations, we will generalize this to a real N-body system.

e) General Theory

In this case the Hamiltonian reads:

---

* This type of evolution was postulated by Bogoliubov to derive the kinetic equations.

$$H = \sum_i \frac{p_i^2}{2m_i} + \lambda \sum_{j<n} V_{jn}(|\underset{\sim}{x}_j - \underset{\sim}{x}_n|). \tag{3.34}$$

This yields the following Liouville operator:

$$L = L_o + \lambda\delta L,$$

$$L_o = -i \sum_j \underset{\sim}{p}_j \cdot \frac{\partial}{\partial \underset{\sim}{x}_j}, \tag{3.35}$$

$$\delta L = -i \sum_{j<n} \frac{\partial V_{jn}}{\partial \underset{\sim}{x}_j} \cdot \underset{\sim}{D}_{nj},$$

where we introduced the notation:

$$\underset{\sim}{D}_{nj} = \frac{\partial}{\partial \underset{\sim}{p}_n} - \frac{\partial}{\partial \underset{\sim}{p}_j}. \tag{3.36}$$

The matrix elements corresponding to (3.14) are now:

$$\langle\{\underset{\sim}{k}\}|\delta L|\{\underset{\sim}{k}'\}\rangle = \Omega^{-N} \int d\underset{\sim}{x}_1 \ldots d\underset{\sim}{x}_N \, e^{-i\sum_j \underset{\sim}{k}_j \cdot \underset{\sim}{x}_j} \, \delta L \, e^{+i\sum_j \underset{\sim}{k}'_j \cdot \underset{\sim}{x}_j} \tag{3.37}$$

$$= \frac{8\pi^3}{\Omega} \sum_{\underset{\sim}{\ell}} \sum_{j<n} V_{\ell} \underset{\sim}{\ell} \cdot \underset{\sim}{D}_{nj} \prod_{i \neq j,n} \delta_{\underset{\sim}{k}_i, \underset{\sim}{k}'_i} \, \delta_{\underset{\sim}{k}'_j, \underset{\sim}{k}_j - \underset{\sim}{\ell}} \, \delta_{\underset{\sim}{k}'_n, \underset{\sim}{k}_n + \underset{\sim}{\ell}}.$$

Eq. (3.37) expresses the selection rules on the matrix elements of δL: As we consider binary forces in (3.34), only two wave vectors can change in a single interaction; as, moreover, (3.34) is invariant for translations we have a conservation law:

$$\sum_i \underset{\sim}{k}_i = \sum_i \underset{\sim}{k}'_i. \tag{3.38}$$

The only-non vanishing matrix elements are thus given by:

$$\langle\{\underset{\sim}{k}\},\underset{\sim}{k}_j,\underset{\sim}{k}_n|\delta L|\{\underset{\sim}{k}\},\underset{\sim}{k}_j-\underset{\sim}{\ell},\underset{\sim}{k}_n+\underset{\sim}{\ell}\rangle . \tag{3.39}$$

We will picture this matrix element as:

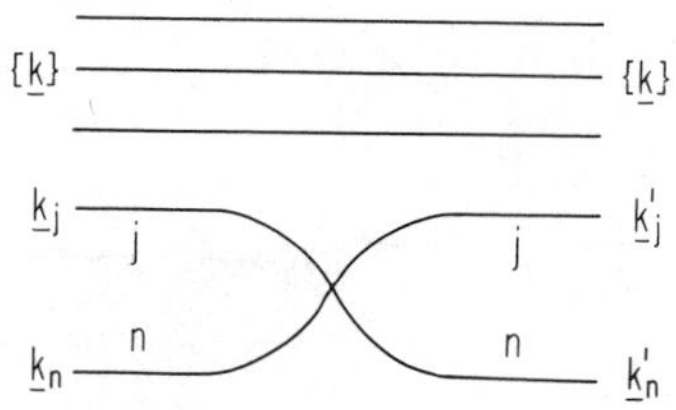

(3.40)

However, we will not draw lines corresponding to zero-wave vectors because we consider them to form our (correlational) vacuum state. In this sense (3.40) contains six different cases. First we draw lines only for the subset $\{\underset{\sim}{k}\}'$ of non-zero wave vectors of $\{\underset{\sim}{k}\}$, and then we distinguish all possible resonance conditions $\underset{\sim}{k}_j-\underset{\sim}{\ell} = 0,\ldots,$ i.e.,

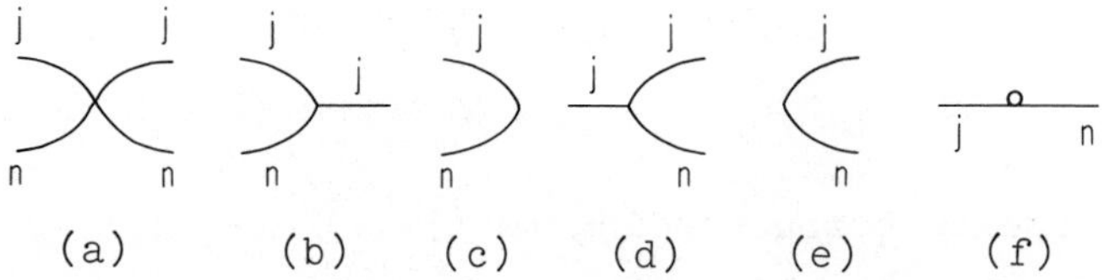

(3.41)

We will say that a transition of type (b)(c) corresponds to a creation, (d)(e) to a destruction and (a)(f) to a propagation of correlations. In the general case we consider now, we will not solve the Liouville equation in "interaction representation" by successive iteration as we did for the example of scattering but solve at once the Liouville equation with the aid of the resolvent formalism. Alternative methods can be found in the references. We start from the Liouville equation in the form given by (3.4) and introduce the resolvent operator R of the Liouville operator L:

$$R(z) = (L-zI)^{-1}, \tag{3.42}$$

where z is a complex variable. We have from (3.4)

$$\rho(t) = e^{-iLt}\rho(0). \tag{3.43}$$

Or, using the Cauchy formula for operators, we can write (3.43) as:

$$\rho(t) = -\frac{1}{2\pi i}\int_C dz\, \frac{e^{-izt}}{L-zI}\,\rho(0), \quad t > 0. \tag{3.44}$$

The integration contour C in (3.44) is a line antiparallel to the real axis lying above all singularities of the integrand. These singularities are real as they correspond to the eigenvalues of L which is Hermitian and thus C lies in the upper half z-plane (see Fig.3.4.)

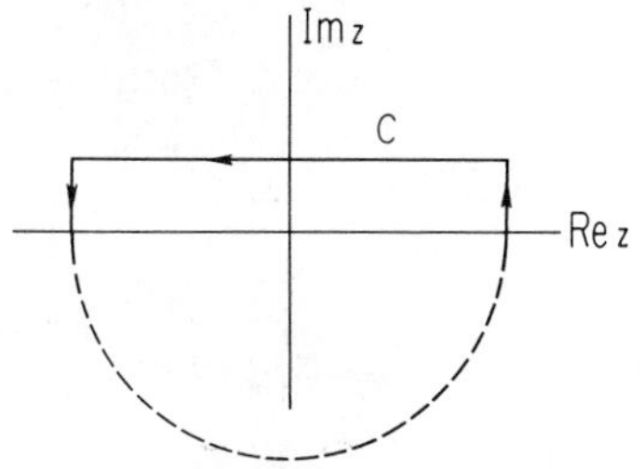

Fig. 3.4.

Integration contour in (3.44)

Introducing a formal expansion of $\rho(0)$ in the exact eigenfunctions of L in (3.44) one can check that (3.44) reduces to the usual Cauchy formula. We will assume here the validity of (3.44) in the limit of an infinite system.

Using the operator identity:

$$A^{-1}-B^{-1} = A^{-1}(B-A)B^{-1}, \tag{3.45}$$

for the perturbed and unperturbed resolvent:

$$A^{-1} = R(z) = (L-zI)^{-1}, \quad L = L_o + \lambda\delta L,$$
$$B^{-1} = R_o(z) = (L_o-zI)^{-1}, \tag{3.46}$$

we have:

$$R(z) = R_o(z) - \lambda R(z)\delta L\, R_o(z),$$
$$= \sum_{n=0}^{\infty} R_o(z) \left[-\lambda\delta L R_o(z)\right]^n, \tag{3.47}$$

where we formally solved the equation in R. Substituting (3.47) in (3.44) yields a perturbation expansion (λ) of the solution of the Liouville equation. We now perform a Fourier expansion of ρ with respect to the space variables:

$$\rho(t) = \Omega^{-N/2} \sum_{\{\underset{\sim}{k}\}} \tilde{\rho}_{\{\underset{\sim}{k}\}} e^{i\sum_j \underset{\sim}{k}_j \cdot \underset{\sim}{x}_j}. \tag{3.48}$$

Note that the Fourier coefficient $\tilde{\rho}_{\{\underset{\sim}{k}\}}$ appearing in (3.48) should not be confused with the one appearing in the interaction representation (3.13). The relation between the two is:

$$\tilde{\rho}_{\underset{\sim}{k}}(t) = \rho_{\underset{\sim}{k}}(t)\, e^{-i\underset{\sim}{k}\cdot\underset{\sim}{v}t}, \tag{3.49}$$

where $\underset{\sim}{k}$ stands for the whole set $\{\underset{\sim}{k}\}$. In the following we will, however, drop the tilde in (3.48). Using (3.49) and (3.47) together with (3.44) we get finally:

$$\rho_{\{\underset{\sim}{k}\}}(t) = -\frac{1}{2\pi i}\int_C dz\, e^{-izt} \sum_{\{\underset{\sim}{k}'\}} \sum_{n=0}^{\infty} \langle\{\underset{\sim}{k}\}|R_o(z)\left[-\lambda\delta L R_o(z)\right]^n|\{k'\}\rangle \rho_{\{\underset{\sim}{k}'\}}(0). \tag{3.50}$$

The matrix elements of δL are given in (3.37) and those of R_o are:

$$\langle\{\underset{\sim}{k}\}|R_o(z)|\{\underset{\sim}{k}'\}\rangle = \frac{1}{\sum_j \underset{\sim}{k}_j\cdot\underset{\sim}{v}_j - z} \prod_j \delta_{\underset{\sim}{k}_j,\underset{\sim}{k}'_j} \ . \qquad (3.51)$$

Thus R_o is diagonal in the plane wave representation. It propagates the correlation state $\{\underset{\sim}{k}\}$. We picture it as a superposition of lines, one line for each $\underset{\sim}{k}_j \neq 0$ in the set $\{\underset{\sim}{k}\}$. Thus, the succession of matrix elements of R_o and δL appearing in (3.50) can be pictured as a succession of lines (see (3.51))and vertices (see(3.40) and (3.41)). For example, the scattering operator in the previous example corresponds to the following matrix elements in (3.50):

$$\lambda^2\langle 0|R_o\delta LR_o\delta LR_o|0\rangle \ , \qquad (3.52)$$

and to the diagram:

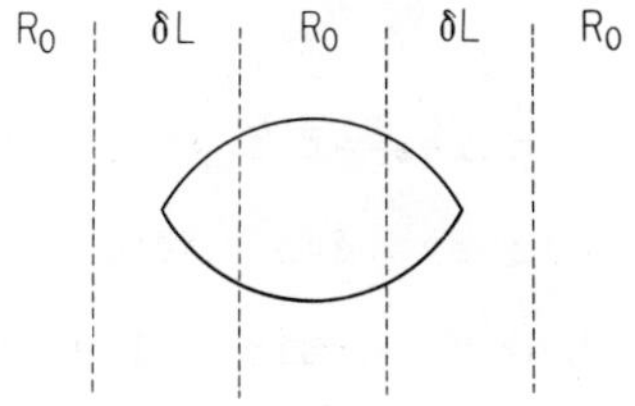

(3.53)

Let us investigate the importance of vacuum-vacuum transitions on the example (3.52). Using (3.51) we can write:

$$\langle 0|R_o\delta LR_o\delta LR_o|0\rangle = \frac{1}{-z}\langle 0|\delta LR_o\delta L|0\rangle\frac{1}{-z} \ , \qquad (3.54)$$

$$\equiv \frac{1}{z^2}\Psi(z),$$

where we introduce the finite frequency collision operator $\Psi(z)$.*

*The Laplace transform of $\Psi(z)$ corresponds to a time dependent collision operator. This is, of course, necessary if effects related to the finite duration of the collision are to be retained.

Consider the contribution of the second order pole at z = 0 of (3.54) to the integral (3.50). Taking the residue, assuming $\psi(z)$ regular at z = 0 (see below), we obtain a contribution of the form:

$$\left(-it\, e^{-izt}\, \psi(z) + e^{-izt} \frac{d\psi(z)}{dz}\right)_{z=0}. \qquad (3.55)$$

This consists of a term growing linearly with time and a term constant in time. In this way we come to the important conclusion that each vacuum-vacuum transition, that introducing a double pole at z = 0 in (3.50), yields a term proportional to t and thus describes a collision process. Similarly a triple pole at z = 0 will yield a t^2 factor and corresponds to two successive collisions. On the contrary, a term of the destruction (or creation) type corresponds to a single pole and yields thus a constant term:

$$< = \langle 0|R_o \delta L R_o|\underset{\sim}{k},-\underset{\sim}{k}\rangle \sim \frac{1}{z} \frac{1}{\underset{\sim}{k}\cdot\underset{\sim}{v}-\underset{\sim}{k}\cdot\underset{\sim}{v}'-z}. \qquad (3.56)$$

The contributions from the poles at $z \neq 0$, on the contrary, yield in general (except for unstable systems) damped oscillatory terms depending on the nature of the potential. Let us investigate these conclusions in a little more detail. From (3.54) we write:

$$\psi(z) = \sum_{\underset{\sim}{k}} \sum_{i<j} \langle 0|\delta L_{ij}|\underset{\sim}{k},-\underset{\sim}{k}\rangle \frac{1}{\underset{\sim}{k}\cdot\underset{\sim}{v}^{ij}-z} \langle \underset{\sim}{k},-\underset{\sim}{k}|\delta L_{ij}|0\rangle , \qquad (3.57)$$

$$\approx \underset{\sim}{D}_{ij}\cdot\left(\int d\underset{\sim}{k}\, \frac{|V_{\underset{\sim}{k}}|^2}{\underset{\sim}{k}\cdot\underset{\sim}{v}^{ij}-z}\, \underset{\sim}{k}\underset{\sim}{k}\right)\cdot\underset{\sim}{D}_{ij},$$

where $\underset{\sim}{v}^{ij} = \underset{\sim}{v}_i-\underset{\sim}{v}_j$ is the relative velocity. We have in (3.57) replaced the summation over $\underset{\sim}{k}$ by an integration. This means that according to (3.9) we have made the transition to an *infinite system*. This step is very important. As we shall see in Chap. 7, in this way we eliminate the quasiperiodic behavior characteristic of finite systems.

Let us at this point continue with systems so large that the transition from sums to integrals may be made. Then we obtain in (3.57) an integral which is of the form of a Cauchy integral

$$\Phi(z) = \frac{1}{2\pi i} \int_{-\infty}^{+\infty} d\omega \, \frac{f(\omega)}{\omega - z} \tag{3.58}$$

with $\omega = \underset{\sim}{k} \cdot \underset{\sim}{v}^{ij}$. This integral maps the function $f(\omega)$ defined on the real axis into the complex function $\Phi(z)$ defined in the complex z-plane except on the real axis. Since the properties of (3.58) are well known (see Ref. 8), we will only summarize them. For Im $z > 0$, (3.58) defines a function analytic in the upper half z-plane which we call $\Phi^+(z)$, and a function $\Phi^-(z)$ analytic in the lower half z-plane for Im $z < 0$. Let us investigate the value of $\Phi^\pm$ on the real axis. Remember that:

$$\lim_{\substack{\varepsilon \to 0 \\ \varepsilon > 0}} \frac{1}{a \mp i\varepsilon} = \pm i\pi\delta_{\mp}(a) = \pm i\pi\delta(a) + \mathcal{P}\left(\frac{1}{a}\right). \tag{3.59}$$

Approaching the real axis from above we have, using (3.58) and (3.59):

$$\Phi^+(x) \equiv \lim_{\substack{\varepsilon \to 0 \\ \varepsilon > 0}} \Phi^+(x + i\varepsilon) = \frac{1}{2\pi i} \int_{-\infty}^{\infty} d\omega i\pi\delta_-(\omega - x) \, f(\omega). \tag{3.60}$$

Approaching the real axis from below we have:

$$\Phi^-(x) \equiv \lim_{\substack{\varepsilon \to 0 \\ \varepsilon > 0}} \Phi^-(x - i\varepsilon) = \frac{1}{2\pi i} \int_{-\infty}^{\infty} d\omega(-i)\pi\delta_+(\omega - x) \, f(\omega). \tag{3.61}$$

Taking the difference of (3.60) and (3.61) we obtain:

$$\Phi^+(x) - \Phi^-(x) = \int_{-\infty}^{\infty} d\omega \; \delta(\omega - x) \; f(\omega) = f(x). \tag{3.62}$$

We can thus define $\Phi^\pm$ on the real axis (see (3.60-61)). Moreover, from (3.62) we see that they do not take the same value on the real axis but have a finite discontinuity (cut) given by f(x).From (3.62) we can define an analytic continuation of $\Phi^+(x)$ in the lower half plane using the analytic continuation of f(x):

$$\Phi^+(z) = \Phi^-(z) + f(z) \quad \text{Im } z < 0. \tag{3.63}$$

Indeed one can check that (3.63) approaches $\Phi^+(x)$ defined by (3.60), using (3.62), when x approaches the real axis from below. In this way (3.58) defines two functions $\Phi^+(z)$ and $\Phi^-(z)$, $\Phi^+(z)$ being regular for Im $z \geq 0$ and having singularities (coming from f(z)) for Im $z < 0$, and the opposite situation for $\Phi^-(z)$. The functions defined by R(z) and $\Psi(z)$ correspond thus to $R^+(z)$ and $\Psi^+(z)$ (see Fig. 3.4). The location of the poles depends strongly on the nature of the interactions. In the example (3.57) using a potential of range $\frac{1}{a}$:

$$V(r) \sim e^{-ar}, \tag{3.64}$$

one can show easily that this yields poles at $z = -iag$, and gives thus a time behaviour: exp (-agt) with g being the z-component of the relative velocity. Replacing g by an average velocity, we obtain the inverse of the collision or interaction time $t^{-1}_{coll.} = a\langle v\rangle$. Therefore one can say that in simple situations the poles of $\Psi^+(z)$ are located at a distance $t^{-1}_{coll.}$ from the real axis and yield thus strongly damped terms in $\exp(-t/t_{coll.})$. However, in more complicated situations, e.g., long range forces, this will no longer be the case. In a plasma, for example, the location of the poles depends on the dielectric constant and the collision becomes thus a statistical process.*

We will now reorder the formal perturbation solution of the Liouville equation (3.50) in a manner separating the collision processes which, as we know from the above examples, correspond to vacuum-vacuum transitions. Therefore consider Eq. (3.50) for $\{\underset{\sim}{k} = 0\}$ and separate the initial correlations $\{\underset{\sim}{k}' \neq 0\}$ from $\rho_o(0)$. The transition from $\rho_o(0)$ to $\rho_o(t)$ can be performed either without an intermediate vacuum state, or with one, two, ... intermediate vacuum states. We will write this as:

$$\rho_o(t) = \sum_{n=0}^{\infty} \Big(\;\; \Big)^n \rho_o(0), \qquad (3.65)$$

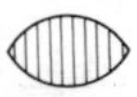

where we introduce the "irreducible diagonal fragment":

$$= \frac{\psi^+(z)}{-z} \; . \qquad (3.66)$$

Here, irreducible means that there is no intermediate vacuum state. We have:

$$\psi^+(z) = \sum_{n=1}^{\infty} \langle 0| -\lambda\delta L \left[R_o(-\lambda\delta L)\right]^n |0\rangle_{irr} . \qquad (3.67)$$

---

*Indeed the dielectric constant is then a functional of the distribution function $\rho_o(\underset{\sim}{p},t)$, expressing the fact that the collision is a collective process. The duration of the collision, therefore, also depends upon the distribution function (see Ref. 8).

On the other hand, the transition from $\rho_{\{\underset{\sim}{k}'\neq 0\}}(0)$ to $\rho_0(t)$ can be performed in an irreducible fragment which we call destruction fragment D(z):

$$D_{\{\underset{\sim}{k}'\}}(z) = $$ 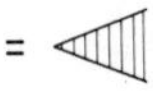 $$= \sum_n \langle 0|(-\lambda\delta LR_0)^n|\{\underset{\sim}{k}'\}\rangle_{irr} \quad , \qquad (3.68)$$

because it corresponds to a contribution in which the initial correlations are destroyed. This fragment (3.68) can then be followed by an arbitrary number of diagonal fragments (3.66), so that the perturbation series can be reordered as:

$$\rho_0(t) = \sum_{n=0}^{\infty} ($$ 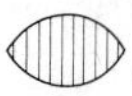 $$)^n \Big(\rho_0(0) + \sum_{\{\underset{\sim}{k}'\neq 0\}}$$ 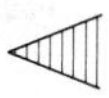 $$\rho_{\{\underset{\sim}{k}'\}}(0)\Big) \quad , \qquad (3.69)$$

or explicitly:

$$\rho_0(t) = -\frac{1}{2\pi i}\int_C dz\, e^{-izt}\,\frac{1}{-z}\sum_{n=0}^{\infty}\left(\frac{\psi^+(z)}{-z}\right)^n\left(\rho_0(0) + \sum_{\{\underset{\sim}{k}'\neq 0\}} D_{\{\underset{\sim}{k}'\}}\rho_{\{\underset{\sim}{k}'\}}(0)\right). \qquad (3.70)$$

Differentiating (3.70) with respect to t we obtain the general evolution equation:

$$\frac{\partial}{\partial t}\rho_0(t) \equiv \partial_t\rho_0(t) = -\frac{1}{2\pi}\int_C dz\, e^{-izt}\sum_{n=0}^{\infty}\left(\frac{\psi^+(z)}{-z}\right)^n(\rho_0(0) + F^+(z)) \quad , \qquad (3.71)$$

where we introduced the abbreviation:

$$F^+(z) = \sum_{\{\underset{\sim}{k}'\neq 0\}} D_{\{\underset{\sim}{k}'\}}(z)\,\rho_{\{\underset{\sim}{k}'\}}(0) \quad . \qquad (3.72)$$

Note that only the second term of the n=0 contribution has poles, i.e., survives. Separating this term we write from (3.71):

$$\partial_t \rho_o(t) = -\frac{1}{2\pi}\int_C dz\, e^{-izt}\left\{F^+(z)+\frac{\psi^+(z)}{-z}\sum_{n=0}^{\infty}\left(\frac{\psi^+(z)}{-z}\right)^n(\rho_o(0)+F^+(z))\right\}. \tag{3.73}$$

Introducing the Laplace transform:

$$f(t) = -\frac{1}{2\pi}\int_C dz\, e^{-izt} f(z) , \tag{3.74}$$

we have from (3.73), using again (3.70):

$$\partial_t \rho_o(t) = F(t) + \frac{-1}{2\pi}\int_C dz\, e^{-izt}\, \psi^+(z)\, i\, \rho_o(z) . \tag{3.75}$$

Using the convolution theorem for Laplace transforms we have:

$$\partial_t \rho_o(t) = \int_0^t d\tau\, G(\tau)\, \rho_o(t-\tau) + F(t) \tag{3.76}$$

where $G(t)$ is the Laplace transform of $i\, \psi^+(z)$. Note that we may write (3.75) in the alternative form:

$$\partial_t \rho_o(t) = i\, \psi^+(i\, \partial_t)\rho_o(t) + F(t) . \tag{3.77}$$

Eqs. (3.75) through (3.77) are the general evolution equations of the system. They are essentially a form equivalent to the Liouville equation ("projected" onto ρ_o) obtained by reordering the perturbation solution according to the number of collisions (diagonal fragments).* Similar equations for the correlation coefficients $\rho_{\{\underset{\sim}{k}\neq 0\}}$ can be deduced. We shall briefly consider these in section (h).

---

*The relation between the irreducibility condition in (3.67) and projection operators is briefly considered in Appendix I.

f) General Evolution Equation

We see that the general evolution Eq. (3.76) contains two contributions:

(1) a term $F(t)$ depending on the *initial* correlations and describing how the system forgets its initial correlations. In simple situations this term decays as $\exp(-t/t_{coll.})$ and can thus be neglected for long times. However in the case of long range correlations, as in turbulence, or for long range forces without screening, as for gravitation, one can generally not neglect this term;

(2) a term depending only on the velocity distribution at earlier times. This nonMarkoffian character expresses the finite duration of a collision process as described by the Laplace transform of the irreducible diagonal fragment (3.66).

In simple situations $G(\tau)$ is peaked around $\tau \sim t_{coll.}$. In this case one can simplify Eq. (3.76). Expanding in a Taylor series we have:

$$\int_0^t d\tau\; G(\tau)\rho_0(t-\tau) = \int_0^t d\tau\; G(\tau)\rho_0(t) - \int_0^t d\tau\; G(\tau)\tau\partial_t\rho_0(t) + \cdots. \qquad (3.78)$$

And for t large compared to $t_{coll.}$ we obtain:

$$\int_0^t d\tau\; \tau^n G(\tau) \simeq \int_0^\infty d\tau\; \tau^n G(\tau) = i^n \left[\frac{d^n}{dz^n}\, i\, \psi^+(z)\right]_{z\,=\,+i0} . \qquad (3.79)$$

We can rewrite (3.78) replacing the time derivatives by their value:

$$\int_0^\infty d\tau\; G(\tau)\rho_0(t-\tau) = i\,\Omega\,\psi^+(+i0)\rho_0(t) \qquad (3.80)$$

$$\Omega = 1 + \psi^{+\prime}(+i0) + \frac{1}{2}\psi^{+\prime\prime}(+i0)\;\psi^+(+i0) + (\psi^{+\prime})^2(+i0) + \cdots,$$

with

$$\Psi^{+'}(+i0) = \left(\frac{d\Psi^+}{dz}\right)_{z\to+i0} . \qquad (3.81)$$

There exist recurrence relations which permit us to express Ω explicitly in terms of $\Psi(+i0)$ and its derivatives to arbitrary order (References 7 and 9, and especially CL. GEORGE [10]).

Thus when $G(t)$ and $F(t)$ are peaked around $t \sim t_{coll.}$ then Eq. (3.76) simplifies to a "pseudoMarkoffian" form:

$$\partial_t \rho_o(t) = i\,\Omega(+i0)\,\Psi^+(+i0)\rho_o(t), \qquad (3.82)$$

where the Ω operator retains the asymptotic corrections due to the finite duration of a collision. $\Psi^+(+i0)$ corresponds to the asymptotic collision operator. Roughly speaking one can say Ω corresponds to a correction in $t_{coll.}/t_{rel.}$ where $t_{rel.}$ is the relaxation time:

$$\Omega\,\Psi^+ \simeq t^{-1}_{rel.}\left(1 + \frac{t_{coll.}}{t_{rel.}} + \cdots\right) . \qquad (3.83)$$

Indeed we have:

$$\Psi^+ \sim t^{-1}_{rel.} \qquad \Psi^{+'} \sim \frac{\Psi^+}{\langle \underset{\sim}{k}\cdot \underset{\sim}{v}\rangle} \sim \frac{t^{-1}_{rel.}}{t^{-1}_{coll.}} . \qquad (3.84)$$

On the other hand we have for a dilute gas, for example, $t^{-1}_{rel.} \sim a^{-2} c \langle v\rangle$, $t^{-1}_{coll.} \sim a\langle v\rangle$ where a^{-1} is the range of the potential and c the concentration. From this point of view (3.83) corresponds to density corrections similar to those appearing in the equation of state of an imperfect gas.* However, usually there exists a clear separation between the time scales $t_{rel.}$, $t_{coll.}$ and things are not so bad. It has been shown, for example, that for transport processes (steady state) Ω is not important and in fact drops out (see Refs. 8 and 11). Indeed (3.77) and (3.78) show that Ω is related to the nonstationarity of the distribution function during a collision.

Eq. (3.82) may be considered as a generalized Boltzmann kinetic equation. We may say therefore that we have a generalized Boltzmann situation when in the general Eq. (3.76) we may: (1) neglect the "memory" effects F(t), (2) retain only the asymptotic effects due to the finite duration of the collisions through operator Ω.

When, moreover, we retain only the lowest order contribution in terms of a suitable parameter, then the operator Ω cancels (even in the evolution equation and not only for steady states [7],[9] and we obtain the simplest and best known classes of transport equations (Fokker-Planck equation for weakly coupled systems, Boltzmann equation for dilute gases, Balescu-Lenard equation for plasmas). We may then say that we are in "strict" Boltzmann situations.

We shall in the next three chapters illustrate the general theory with examples taken from the three classes of situations: strict Boltzmann situations, generalized Boltzmann situations and nonBoltzmann situations described by the general Eq. (3.76).

Let us emphasize again how far reaching the steps are, which go from Eq. (3.76) to the generalized Boltzmann situations.

---

* The question of whether density corrections to transport coefficients arising from the kinetic equations may be expanded in powers of the concentration has been much discussed recently. We have summarized the situation in Appendix II.

The whole validity of phenomenological thermodynamics as discussed in Chap. 2 of these lectures depends on this step. Clearly, if memory effects cannot be neglected the system cannot evolve to the state of maximum entropy *whatever* the initial conditions.

If F(t) may be neglected but the complete nonMarkoffian form of the collision operator is retained, then the change of the distribution function in time is no longer a functional of its state at the *same* time. Therefore, even if the entropy is a well defined functional of the distribution function its change is no longer determined by the instantaneous state of the system.

Before we go to examples let us discuss in a little more detail some aspects of Eqs. (3.76) and (3.82).

g) Transformation Theory of the Kinetic Equation

By comparing the kinetic Eq. (3.82) with the integral representation (3.70) (in the case where $\rho_{\{\underset{\sim}{k}\}}(0) = 0$) we see that the following modified initial condition has to be used:

$$\bar{\rho}_o(0) = A\, \rho_o(0) , \tag{3.85}$$

where

$$A = \sum_{n=0}^{\infty} \frac{1}{n!} \left[\frac{d^n [\Psi^+(z)]^n}{dz^n}\right]_{z \to +i0} . \tag{3.86}$$

Therefore we have:

$$\bar{\rho}_o(t) = e^{i\Omega\Psi^+ t} A\, \rho_o(0) . \tag{3.87}$$

The solution involves therefore *two* noncommuting operators $\Omega\Psi^+$ and

A. The explicit form of A is derived directly from (3.86). For example, the first terms are:

$$A = 1+\psi^{+'}(+i0)+\frac{1}{2}\left(\psi^{+''}(+i0)\psi^{+}(+i0) +\psi^{+}(+i0)\psi^{+''}(+i0)\right) +\left(\psi^{+'}\right)^{2}(+i0) +\cdots \tag{3.88}$$

The operator A corresponds to a short time correction (on the scale of the relaxation time) which has to be applied to the initial distribution $\rho_o(0)$, because in the Eq. (3.82) short time transients have been neglected. This is illustrated schematically on Fig. 3.5. As we shall see the operator A is related to the "dressing" of the particles. On the other hand $\Omega\psi^+$ which is associated with time in (3.87) is a dynamical operator. Both are functionals of ψ^+ and its derivatives at z = +i0.

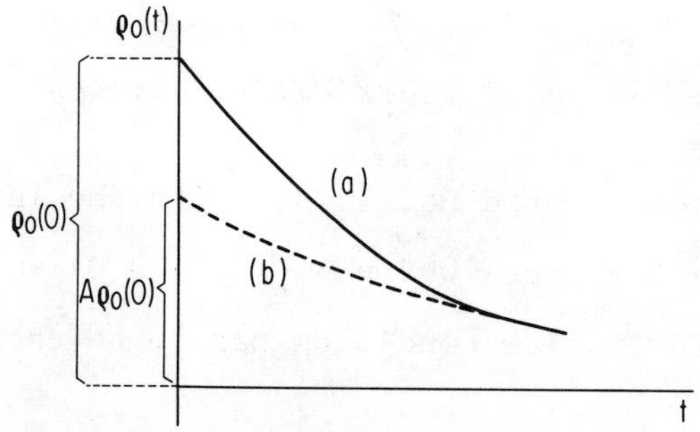

Fig. 3.5.

Meaning of operator A.

(a) exact evolution of distribution ρ_o;

(b) solution of pseudoMarkoffian equation (3.82)

These are alternative ways of representing the solutions (3.87) which are of great interest in connection with the renormalization problem we shall discuss in the next section.

Instead of (3.87) we may use the more symmetrical form:

$$\bar{\rho}_o(t) = \chi\, e^{it\tilde{\phi}}\chi\, \rho_o(0) \ , \tag{3.89}$$

where χ is a time independent linear operator (see Refs. 12 and 13). Through identification with (3.82) we obtain:

$$\chi \, \chi \;\; = A \quad , \qquad (3.90)$$

$$\tilde{\phi} = \chi^{-1} \, \Omega \, \Psi^{+} \, \chi . \qquad (3.91)$$

The characteristic feature of (3.89) is that the operator χ appears both at t = 0 and at t, while in (3.87) A acted only at t = 0. Using expressions for A and $\Omega\Psi$ (see (3.80) and (3.88)) in powers of the coupling constant λ (see (3.47)), we obtain through formulas (3.90) and (3.91) similar expansions for χ and $\tilde{\phi}$. The first terms are:

$$\chi = 1 + \frac{\lambda^2}{2} \Psi_2^{+\prime}(+i0) + \cdots \quad , \qquad (3.92)$$

$$\tilde{\phi} = \lambda^2 \Psi_2^{+}(+i0) + \lambda^4 \left\{ \Psi_4^{+}(+i0) + \frac{1}{2}[\Psi_2^{+\prime}(+i0), \Psi_2^{+}(+i0)]_{+} \right\} + \cdots , \qquad (3.93)$$

where $[\ ,\ ]_{+}$ is the symbol for an anticommutator; and where Ψ_2^{+} is the part of Ψ^{+} which is proportional to λ^2, and so on.

Let us now introduce the new distribution function:

$$\tilde{\rho}_0(t) = \chi^{-1} \, \bar{\rho}_0(t) \quad . \qquad (3.94)$$

We see immediately from (3.89) that $\tilde{\rho}_0(t)$ satisfies the kinetic equation:

$$\partial_t \tilde{\rho}_0 = i \, \tilde{\phi} \, \tilde{\rho}_0(t) \quad . \qquad (3.95)$$

The most remarkable feature of $i \, \tilde{\phi}$ is that it is a Hermitian operator. While the general proof has been given by GEORGE and is too involved to reproduce here, we may verify it easily in first order: $i\Psi^{+}(+i0)$ is a Hermitian operator (for the proof see Refs. 12 and 13); more generally $i\Psi^{+(n)}(+i0)$, where n is the order of derivative with respect to z, is Hermitian for n even and antiHermitian for n odd.

As a consequence (3.80) shows that in general $\Omega(i\Psi^{+})$ is not Hermitian as it contains the product:

$$i\Psi^{+''}(+i0)\, i\Psi^{+}(+i0) \quad , \tag{3.96}$$

of two noncommuting Hermitian operators (in a *nonsymmetrical* way). On the contrary $i\tilde{\phi}$ contains only symmetrized expressions as anticommutators (see (3.93)) and is Hermitian.

As for strictly Boltzmann situations ($\Omega = 1$) the collision operator $i\Psi^{+}$ is Hermitian, and we may say that (3.95) and *not* (3.82) is the natural generalization of the classical transport equations.

However, there remains still a degree of freedom open. Indeed we may introduce an operator corresponding to an arbitrary unitary transformation χ'' and write (3.89), more generally, as:

$$\bar{\rho}_0(t) = \chi\chi''\, e^{it((\chi'')^{-1}\tilde{\phi}\chi'')}\, (\chi\chi'')^{+}\rho_0(0) \quad . \tag{3.97}$$

If we define now a new distribution function $\tilde{\tilde{\rho}}_0$ and a new collision operator through:

$$\tilde{\tilde{\rho}}_0(t) = (\chi\chi'')^{-1}\, \bar{\rho}_0(t) \quad , \tag{3.98}$$

and

$$\tilde{\tilde{\phi}} = (\chi'')^{-1} \, \tilde{\phi} \, \chi'' \, , \tag{3.99}$$

we easily derive the kinetic equation:

$$\partial_t \tilde{\tilde{\rho}}_o = i \, \tilde{\tilde{\phi}} \, \tilde{\tilde{\rho}}_o(t) \, . \tag{3.100}$$

We see, therefore, that there exists an infinite number of kinetic equations with *Hermitian* collision operators differing through a unitary transformation.

This "degree of freedom" is very important (see Ref. 12 and Chap. 8b). We may use it to obtain a representation of the kinetic equation in which all macroscopic quantities (energy, entropy, ...) may be expressed in terms of $\tilde{\tilde{\rho}}_o$ alone, exactly as in the case of weakly coupled systems. But $\tilde{\tilde{\rho}}_o$ is now (in quantum mechanical language) a function of new entities, the "quasiparticles". In this way we obtain a remarkable, simple and transparent description of such systems. For example, the entropy has the usual combinatorial meaning; alternatively it is in this representation that the concept of physical, dressed particles becomes meaningful.

h) Correlations

Let us close this chapter dealing with the general theory with a few brief remarks about *correlations* (for more details see Refs. 7 and 9). Any given correlation $\rho_{\{\underset{\sim}{k}\}}$ may be split into two parts:

$$\rho_{\{\underset{\sim}{k}\}}(t) = \rho'_{\{\underset{\sim}{k}\}}(t) + \rho''_{\{\underset{\sim}{k}\}}(t) \, . \tag{3.101}$$

The evolution of the first part is given by an equation similar to the kinetic equation for $\rho_o(t)$ and describes the scattering of initially present correlations. This part vanishes for $t \to \infty$. It corresponds, briefly speaking, to memory effects and we shall neglect it here.

The second part corresponds to creation of correlations from $\rho_o(t)$. It is given by all creation fragments applied to $\rho_o(t)$:

$$\rho''_{\{\underset{\sim}{k}\}} \to$$ 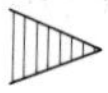 $$\rho_o(t) \ .$$

We may write asymptotically:

$$\rho''_{\{\underset{\sim}{k}\}}(t) = C_{\{\underset{\sim}{k}\}}(+i0)\ \rho_o(t) \qquad (3.102)$$

with (see for an analogous formula the *destruction* fragment (3.68)):

$$C_{\{\underset{\sim}{k}\}}(z) = \sum_{n=1}^{\infty} \langle\{\underset{\sim}{k}\}|(-\lambda R_o(z)\delta L)^n|0\rangle_{irr.} \qquad (3.103)$$

where again "irr." means that no intermediate state is equal to the vacuum of correlation.

The Eqs. (3.102) and (3.103) will be very important for us in Chap. 8 when we shall discuss the problem of entropy. An explicit example for the evaluation of the diagonal fragment as well as the creation and destruction fragments is given in Chap. 7, Sec. (b).

The new distribution function $\tilde{\rho}_o$ (or $\tilde{\tilde{\rho}}_o$) introduced in (3.94) (see also (3.100)) is a combination of the distribution function ρ_o and of the part $\rho''_{\{\underset{\sim}{k}\}}$ of the correlations which are a functional of ρ_o (see GEORGE [15]). As the existence of correlations is therefore taken into account by the very definition of $\tilde{\tilde{\rho}}_o$, it is not astonishing that it satisfies a much simpler evolution equation. From this point of view (3.95) or (3.100) appears as a kind of generalization of the well-known random phase approximation.

References

[1] A. KHINCHIN: Mathematical Foundations of Statistical Mechanics, Dover, New York, 1949.

[2a] I. E. FARQUHAR: Int. Conf. Stat. Mech., Copenhagen, 1966.

[2b] IA. SINAI: Izv. A.N. URSS. (Math.) 25, 899 (1961); 30, 15 (1966); Proc. Int. Math. Congress, Stockholm, 1962.

[3] A. BELLEMANS and J. ORBAN: private communication.

[4] B.J. ALDER and T.E. WAINWRIGHT: J. Chem. Phys. 33, 1439 (1960); Phys. Rev. 127, 359 (1962).

[5] J. LOSCHMIDT: Wiener Berichte 73, 139 (1876); 75, 67 (1877).

[6] S. CHANDRASEKHAR: Rev. Mod. Phys. 15, 1 (1943).

[7] I. PRIGOGINE: Non-Equilibrium Statistical Mechanics, Wiley-Interscience, New York, 1962.

[8] R. BALESCU: Statistical Mechanics of Charged Particles, Wiley-Interscience, New York, 1963.

[9] P. RÉSIBOIS: Irreversible Processes in Classical Gases, in: Many Particle Physics, ed. E. Meeron, Gordon and Breach, 1967.

[10] CL. GEORGE: Physica 30, 1513 (1964).

[11] R. BALESCU: Physica 31, 1599 (1965).

[12] I. PRIGOGINE and F. HENIN: Proc. of the I.U.P.A.P. Meeting, Copenhagen, 1966; p. 421, Benjamin, N.Y., 1967.

[13] CL. GEORGE: Physica 37, 182 (1967)..

[14] A. MANGENEY: Physica 30, 461 (1964).

[15] CL. GEORGE: Physica 36, 678 (1967); Bull. Cl. Sc. Acad. Belg.: to appear 1967.

4. Boltzmann Situations

As illustration of the general theory we consider first two examples which belong to what we called "strict" Boltzmann situations. Not only do we have the asymptotic Eq. (3.82) but we may even take $\Omega = 1$.

The two examples we shall treat are anharmonic solids and Brownian motion.

a) Anharmonic Solids

In general a solid behaves like a collection of oscillators. In equilibrium these oscillators can be considered as independent harmonic oscillators. However, the approach to equilibrium of the solid necessitates anharmonic forces coupling the different harmonic oscillators. This situation is typical for many fields in physics. For example, in the case of an electromagnetic field interacting with matter it is the presence of charged particles which introduces anharmonicity. This anharmonicity is responsible for the approach of the field variables to equilibrium. The situation has a nice symmetry: alternatively it is the presence of the field which permits matter to reach equilibrium. M. BAUS will in his lectures deal with these problems. Here, we will consider the example of a linear chain of N atoms of equal mass m. One can extend this easily to a three-dimensional lattice (see Ref. 1). The potential energy U of the system is assumed to have a minimum U_o when all atoms are regularly spaced, separated by a distance a. Denoting the displacement of the nth atom from this equilibrium position by the real number u_n we can expand the potential energy in a Taylor series:

$$U = U_o + \frac{1}{2}\sum_{n,n'} A_{nn'}\, u_n u_{n'} + \frac{1}{6}\sum_{n,n',n''} B_{nn'n''} u_n u_{n'} u_{n''} + \cdots , \quad (4.1)$$

where the coefficients $A_{nn'}$, $B_{nn'n''}$ characterize the harmonic and anharmonic forces of the solid. Their symmetry properties can be found for example in Ref. 1. We introduce the oscillator description through the so-called normal coordinate transformation defined as:

$$u_n = \sum_k q_k e^{ikna} \quad . \tag{4.2}$$

Using periodic boundary conditions $u_n = u_{n+N}$ we obtain the k-spectrum:

$$k = m\left(\frac{2\pi}{Na}\right) , \quad m = \text{integer} \quad . \tag{4.3}$$

The Hamiltonian of this system reads:

$$H = \sum_n \frac{\dot{u}_n^2}{2m} + (U - U_o) \quad . \tag{4.4}$$

If we use the action and angle variables (J_k, α_k) to describe the oscillator, the Hamiltonian will take a simple form. We have:

$$q_k = (2\ Nm)^{-1/2}\left\{(J_k/\omega_k)^{1/2} e^{i\alpha_k} + (J_{-k}/\omega_{-k})^{1/2} e^{-i\alpha_{-k}}\right\} , \tag{4.5}$$

$$\dot{q}_k = i\ \omega_k\ q_k ,$$

where $\omega_k = \omega_{-k}$ is the frequency of normal mode k, defined by (see Ref. 1)

$$m\omega_k^2 = \sum_{n,n'} A_{nn'} e^{ik(n-n')a} \quad . \tag{4.6}$$

The Hamiltonian reads, retaining only the first anharmonic term in

(4.1), i.e., the cubic term:

$$H = H_o + \lambda V$$

$$H_o = \sum_k \omega_k J_k$$

$$V = \sum_{kk'k''} \sum_{\varepsilon\varepsilon'\varepsilon''} V_{\varepsilon k,\varepsilon' k',\varepsilon'' k''} \left(\frac{J_k J_{k'} J_{k''}}{\omega_k \omega_{k'} \omega_{k''}}\right)^{1/2} e^{i(\varepsilon\alpha_k + \varepsilon'\alpha_{k'} + \varepsilon''\alpha_{k''})}, \tag{4.7}$$

where $\varepsilon, \varepsilon', \varepsilon''$ equal +1 or -1, and the strength of the anharmonicity is determined by $V_{k,k',k''}$:

$$V_{k,k',k''} = (2\ Nm)^{-3/2} \sum_{nn'n''} \frac{1}{6} B_{nn'n''}\ e^{i(kn+k'n'+k''n'')a}. \tag{4.8}$$

From (4.7), we deduce the Liouville operator $L = L_o + \lambda\delta L$, where:

$$L_o = -i \sum_k \omega_k \frac{\partial}{\partial\alpha_k} \quad , \tag{4.9}$$

$$\delta L = i \sum_{k,k',k''} \sum_{\varepsilon,\varepsilon',\varepsilon''=\pm 1} V_{\varepsilon k,\varepsilon' k',\varepsilon'' k''} \left(\frac{J_k J_{k'} J_{k''}}{\omega_k \omega_{k'} \omega_{k''}}\right)^{1/2}$$
$$e^{i(\varepsilon\alpha_k + \varepsilon'\alpha_{k'} + \varepsilon''\alpha_{k''})} \left\{ \frac{1}{2}\left(\frac{1}{J_k}\frac{\partial}{\partial\alpha_k} + \frac{1}{J_{k'}}\frac{\partial}{\partial\alpha_{k'}} + \frac{1}{J_{k''}}\frac{\partial}{\partial\alpha_{k''}}\right)\right.$$
$$\left. -i\left(\varepsilon\frac{\partial}{\partial J_k} + \varepsilon'\frac{\partial}{\partial J_{k'}} + \varepsilon''\frac{\partial}{\partial J_{k''}}\right)\right\} .$$

The matrix elements of the perturbation δL are defined as:

$$\langle\{n\}|\delta L|\{n'\}\rangle = \int_0^{2\pi} \cdots \int_0^{2\pi} \left(\frac{d\alpha}{2\pi}\right) e^{-i\sum n\alpha}\, \delta L\, e^{+i\sum n'\alpha} , \tag{4.10}$$

and the only nonvanishing elements are:

$$\langle\{n\}, n_k, n_{k'}, n_{k''}|\delta L|\{n\}, n_k+\varepsilon, n_{k'}+\varepsilon', n_{k''}+\varepsilon''\rangle . \tag{4.11}$$

Let us make the following comments on the above expressions. In these variables the action J_k is an invariant of the *harmonic* (linearized) motion, α_k being cyclic (see H_o and L_o). The anharmonicity is reflected in the nonlinearity of V and δL in the action variable. Moreover the α-dependence of V reflects the phase relations between the coupled harmonic oscillators. The matrix element (4.11) which is still a first order operator in the action variables describes a three-phonon interaction corresponding to the cubic term in Eq. (4.1). One can now derive an evolution equation (master equation) for the energy distribution of the oscillators $\rho_o(\{J\};t)$ just as we did for the velocity distribution of a gas in Chap. 3 (d). The weak coupling contribution (see (3.57)) to the asymptotic collision operator reads (see Ref. 1):

$$\lambda^2 i \Psi_2^+(+i0) = -\sum_{\{n\}} \langle 0|\delta L|\{n\}\rangle \pi\delta_-(\sum_k n_k\omega_k)\langle\{n\}|\delta L|0\rangle$$

$$= 6\pi \sum_{k,k',k''} \delta(\omega_k+\omega_{k'}-\omega_{k''}) \frac{|V_{k,k',-k''}|^2}{\omega_k\omega_{k'}\omega_{k''}} \left(\frac{\partial}{\partial J_k} + \frac{\partial}{\partial J_{k'}} - \frac{\partial}{\partial J_{k''}}\right)$$

$$J_k \, J_{k'} \, J_{k''} \left(\frac{\partial}{\partial J_k} + \frac{\partial}{\partial J_{k'}} - \frac{\partial}{\partial J_{k''}}\right) . \qquad (4.12)$$

In analogy with (3.53), contribution (4.12) will be represented by the diagram:

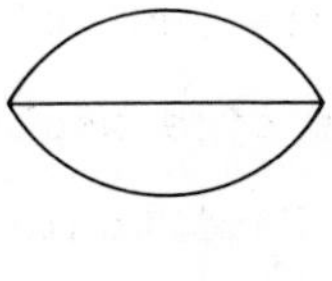 , (4.13)

corresponding to the simplest vacuum-vacuum transition for three-phonon interactions. The operator (4.12) is Hermitian, as can be verified. Neglecting the destruction fragment and all nonMarkoffian effects, we obtain the following asymptotic evolution equation:

$$\partial_t \rho_o(\{J\};t) = \lambda^2 \, i \, \Psi_2^+(+i0) \, \rho_o(\{J\};t). \qquad (4.14)$$

This equation describes the approach to equilibrium of a system of weakly coupled harmonic oscillators and was first obtained by PEIERLS as a Boltzmann equation for phonons using arguments similar to those BOLTZMANN used for gases.

The evolution Eq. (4.14) is as indicated by (4.12) a diffusion type of equation of second order in the action variables J_k. It is a special case of the Fokker-Planck stochastic equation (see Ref.6 of Chap. 3) which has been extensively studied in the theory of random processes.

From Eq. (4.14) using expression (4.12) one can easily derive the following H-theorem:

$$\frac{d}{dt} \int (dJ) \, \rho_o \ln \rho_o \leq 0 \, . \qquad (4.15)$$

The equilibrium state will be reached when:

$$\int (dJ) \, \rho_o \ln \rho_o = 0 \, . \qquad (4.16)$$

The solution of (4.16) is $\rho_o = f(H_o)$, i.e., an arbitrary function (determined by the initial conditions) of the unperturbed Hamiltonian H_o. The equilibrium properties of a weakly coupled system are thus determined by the unperturbed Hamiltonian only. These results are valid only for a large system (see Ref. 1). From the irreversible Eq. (4.14) for ρ_o (not the total ρ) one can derive an equation for the energy distribution of a single oscillator. Suppose, for example, $\rho_o(\{J\})$ is factored; then the evolution equation for $\rho_o(J_k)$ will be nonlinear (see (4.12)). In simple cases, as for example the evolution of a single mode, all other modes being in equilibrium (in the weakly coupled case they will remain so), this equation will become linear. In Ref. 1, one can find a complete treatment of the Brownian motion of a single normal mode of energy E. The distribution function is of the type:

$$\rho_o(E,t) = e^{-E/kT} \sum_{m=0}^{\infty} \alpha_m L_m (E/kT) e^{-mt/t_r} , \qquad (4.17)$$

where $L_m(x)$ is the mth Laguerre polynomial. In this way, for $t \to \infty$, only the Laguerre polynomial L_m with $m = 0$ survives and (4.17) goes over into the canonical equilibrium distribution.

b) Brownian Motion

A classical problem in nonequilibrium statistical mechanics is the approach to equilibrium of a heavy particle moving in an equilibrium fluid of light particles, i.e., the so-called Brownian motion. It was first studied by EINSTEIN and SMOLUCHOWSKI using a stochastic process method. In this type of problem the expansion parameter is the mass ratio m/M of the light (m) and heavy (M) particles, independently of the particle-particle coupling so that this process can occur in any fluid (gas, liquid). A second type of Brownian motion was studied by a semiphenomenological method by CHANDRASEKHAR and VON NEUMANN. In this type the Brownian particle has the same mass as the medium particles but is only weakly coupled to them so that in both problems each interaction is associated with only a small momentum transfer.

The weak coupling version is discussed in detail in Refs. 1 and 3. The case of the slightly anharmonic lattice considered in the preceding paragraph is an example of such a "weakly coupled"

Brownian motion. Another well known example is the so-called Landau equation for plasmas.

Here we shall study the Brownian motion of a heavy particle. In recent papers LEBOWITZ and RUBIN [2] as well as RÉSIBOIS and DAVIS [3],[6] have derived a theory of Brownian motion starting from mechanics. It is remarkable that complete agreement is obtained between the equations of motion as derived from mechanics and from stochastic considerations. We will only briefly sketch the result. The Hamiltonian of the system is

$$H = \frac{MV^2}{2} + \sum_{j=1}^{N} \frac{mv_j^2}{2} + \sum_{i<j} V(r_{ij}) + \sum_{j=1}^{N} U(r_j - R). \tag{4.18}$$

It contains the kinetic energy of the Brownian particle (M), of the fluid particle (m), the fluid-fluid particle interaction energy (V) and the fluid-Brownian particle interaction energy (U). From (4.18) one can deduce a Liouville operator L in the usual manner using a Poisson bracket in the complete phase space of the fluid particles and the Brownian particle. As usual L consists of an unperturbed (L_o) and perturbed (δL) part, but the important splitting here is not in powers of the coupling constant but in powers of the mass ratio $\gamma = (\frac{m}{M})^{1/2}$. The square root is introduced in order to study Brownian motion in the thermal region where $\gamma \sim V/v$. We then have $L_o = L_o^{fluid} + \gamma L_o'$, $\delta L = \delta L^{fluid} + \gamma \delta L'$. Following the general method of Chap. 3 (d) one can derive a master equation and integrating over the fluid particles an evolution equation for the Brownian particle distribution function $f(R,V;t)$. In lowest order, γ^2, (the zero and first order contributions in γ vanish) one obtains the following equation for f :

$$\partial_t f + V \cdot \frac{\partial}{\partial R} f = \xi \frac{\partial}{\partial V} \cdot (V + (\beta M)^{-1} \frac{\partial}{\partial V}) f , \tag{4.19}$$

where β is the Boltzmann factor $(kT)^{-1}$ and ξ the friction coefficient which can be written as an autocorrelation of forces:

$$\xi = \frac{\beta}{3M} \lim_{t_0 \to \infty} \int_0^\infty dt\, e^{-t/t_0} < \mathfrak{f}(0)\, \mathfrak{f}(t) >_{eq} \tag{4.20}$$

where $\mathfrak{f}$ is the total force exerted by the fluid on the Brownian particle. For more details see Ref. 4.

The most important consequence of the mechanical derivation of Brownian motion theory is that now one can discuss also Brownian motion in situations to which the classical stochastic theory does not apply. We may quote Brownian motion in nonuniform systems (for example in which there exists a thermal gradient, see NICOLIS [5]) or quantum theory of Brownian motion [6].

References

[1] I. PRIGOGINE: see Ref. 7 of Chap. 3.

[2] J. L. LEBOWITZ and E. RUBIN: Phys. Rev. 131, 2381 (1963).

[3] P. RÉSIBOIS and H. T. DAVIS: Physica 30, 1077 (1964).

[4] J. L. LEBOWITZ and P. RÉSIBOIS: Phys. Rev. 139, 1101 (1965).

[5] G. NICOLIS: J. Chem. Phys. 43, 1110 (1965).

[6] H. TED DAVIS and R. DAGONNIER: J. Chem. Phys. 44, 4030 (1966).
P. RÉSIBOIS and R. DAGONNIER: Bull. Cl. Sc. Acad. Belgique 52, 299 (1966); 52, 1475 (1966); Phys. Lett. 22, 252 (1966).
H. T. DAVIS, K. HIROÏKE and S.A. RICE: J. Chem. Phys. 43, 2633 (1965).
J. Mc KENNA and H. L. FRISCH: Phys. Rev. 145, 93 (1966).

5. Generalized Boltzmann Situations

a) The Three-Body Problem

We want now to go beyond the lowest order contributions (in the relevant parameter). The operator Ω (see Chap. 3(f)) can therefore no longer be taken as equal to one. We meet then a difficulty which appears already in the important work on three-body contributions to transport equations due to CHOH and UHLENBECK [1],[2]. We have indeed to distinguish between "genuine" three-body collisions and a succession of two two-body collisions (see Fig. 5.1).

Fig. 5.1.

a) a genuine three-body collision; b) a succession of two two-body collisions

As binary collisions are already taken into account through the usual Boltzmann collision operator, we have to calculate separately the genuine three-body term which gives the first concentration correction to the Boltzmann operator.

It is interesting to write down explicitly the Choh-Uhlenbeck formula for this correction term (for more details see Refs. 1,2):

$$\lim_{t\to\infty} \int d\underset{\sim}{x}_2\, d\underset{\sim}{x}_3 \int d\underset{\sim}{p}_2\, d\underset{\sim}{p}_3\, \frac{\partial V}{\partial \underset{\sim}{x}_{12}} \cdot \frac{\partial}{\partial \underset{\sim}{p}_1}\left(\mathscr{S}_t^{(123)} - \mathscr{S}_t^{(12)}\mathscr{S}_t^{(23)} - \mathscr{S}_t^{(12)}\mathscr{S}_t^{(13)} + \mathscr{S}_t^{(12)}\right), \tag{5.1}$$

where V is the interaction potential and $\mathscr{S}_t^{(12\ldots j)}$ the unitary operator of motion for j particles, defined in terms of the Liouville

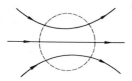

operator for these j particles, $\mathcal{L}^{(1\ldots j)}$ as:

$$\mathcal{S}_t^{(12\ldots j)} = \exp\left[-i\mathcal{L}^{(12\ldots j)}t\right]. \qquad (5.2)$$

Eq. (5.1) shows explicitly how one has to *subtract* successive binary collisions (as well as the single binary collision) from the complete three-body operator. Successions of two two-body collisions would give asymptotic contributions ~ t^2, and it is only by subtracting such terms that we obtain asymptotic contributions ~ t and that we may at all define transition probabilities per unit time.

In our formalism the subtraction procedure is *automatically performed* through the *irreducibility condition* (see Chap. 3 (e) and (f)). As a result of this condition there is no intermediate state corresponding to the vacuum of correlations which would give a 1/z factor and therefore a supplementary power for the time dependence.

It is clear that in scattering experiments such considerations are in general irrelevant. Indeed a "re-scattering" (Fig. 5.1b) introduces a supplementary Ω^{-1} factor (Ω = volume) which goes to zero in the limit of a large system. However, in many-body systems this is compensated by a supplementary factor N and we obtain a finite contribution.

It is interesting to consider the quantum mechanical version of this problem [3]. The standard tool for quantum scattering theory is the scattering T-matrix defined through the integral equation:

$$T_{pp'}(z) = V_{pp'} + \sum_{p''} V_{pp''} \frac{1}{z-E_{p''}} T_{p''p'}(z) \qquad (5.3)$$

(for more details see Ref. 4). Here we use the momentum (p) representation, E_p is the unperturbed energy. In terms of the T-matrix one has the usual expression,

$$P(i \to f) = |\langle i|T(E_i)|f\rangle|^2 \quad 2\pi\delta(E_i - E_p), \qquad (5.4)$$

for the transition probability per unit time between an initial state i with energy E_i and a final state E_f. RÉSIBOIS has shown that this expression is divergent when three — body contributions are calculated. The physical reason is precisely that sequences of two-body collisions are still included in (5.4). It may then be shown that the scattering matrix has poles on the real axis and we are not allowed to take the square (5.4) (all details can be found in the paper already mentioned, Ref. 3).

Higher order terms in the kinetic equation have been analyzed in other cases as well as for charged particles interacting with an electromagnetic field [5] or strongly coupled anharmonic lattices [6]. The situation is always the same: higher-order terms are *excess quantities* corresponding to excess transition probabilities and have no simple relation with T- (or S-)matrix theory.

b) Strongly Coupled Systems

The necessity to subtract lower-order terms is not the only complicating feature. When we consider higher-order terms we shall always have more than one intermediate state between the initial and final state of vacuum of correlations. As a result we shall have products of propagators and the structure of $\psi(z)$ is no longer given by a simple Cauchy integral of the form (3.58). As an example let us consider the four-phonon process represented in Fig. 5.2.

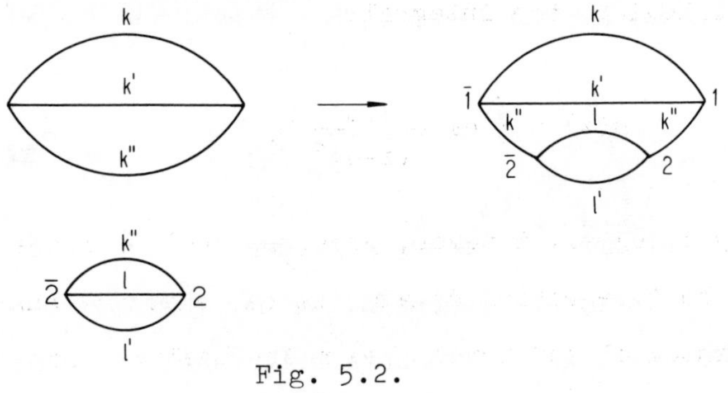

Fig. 5.2.

Four-Phonon Process

We have four vertices, each corresponding to a three-phonon process of the type discussed in Chap. 4 (a). After vertex 1 we have a three-phonon correlation between phonons k, k' and k"; at vertex 2 phonon k" is "destroyed" and we have a four-phonon correlation between k, k', l, l', and as Fig. 5.2 indicates this process may also be thought of as resulting from a superposition of two three-phonon processes of the type considered in Chap. 4 (a), occurring simultaneously.

We have here three intermediate states (two of them involve phonons k, k' and k" and are identical). Instead of a simple Cauchy integral we now have to consider the double integral (for more details see Ref. 6):

$$I(z) = \int d\omega \int d\omega' \frac{1}{(z-\omega)(z-\omega-\omega')(z-\omega)} f(\omega,\omega') \quad . \qquad (5.5)$$

We have to study carefully the analytic continuation of (5.5) for $z \to 0$ as each propagator may become singular on the real axis.

This may be done in a rather straightforward way. Through elementary transformations one reduces the integrals one is interested in to a Cauchy integral or to a succession of Cauchy integrals to which one may then apply the theory of Chap. 3 (e). A trivial example is the following: the integral

$$C(z) = \int d\omega \frac{f(\omega)}{(z-\omega)^2} , \tag{5.6}$$

is *not* a Cauchy integral. However, provided $f(\omega)$ vanishes at the boundaries of the integration domain, we can easily transform (5.6) into a Cauchy integral and hence obtain its analytic continuation. Indeed we have

$$C(z) = \int d\omega \frac{1}{z-\omega} \frac{d}{d\omega} f(\omega) , \tag{5.7}$$

and hence for $z \to +i0$ (see Chap. 3 (e))

$$C(+i0) = \int d\omega \; i\pi\delta_-(\omega) \frac{df(\omega)}{d\omega} . \tag{5.8}$$

It is worthwhile to notice that *no ambiguities* related, for example, to products of distributions are ever met in our formalism, as we deal with a *single* time variable and consequently with a single Laplace variable z (for the discussion of the ambiguities met in the usual formulation of quantum mechanics see GÜTTINGER [7], BOGOLIUBOV and SHIRKOV [8] and especially SCHIEVE [9]).

Using our technique of reduction to Cauchy integrals we obtain for (5.5) (see Ref. 6), if f is an even function of its arguments:

$$I(+i0) = -i\pi \int d\omega \int d\omega' \Big\{ \mathcal{P}\Big(\frac{1}{\omega}\Big) \delta(\omega+\omega') \Big(\frac{\partial}{\partial\omega} - \frac{\partial}{\partial\omega'}\Big) - \mathcal{P}\Big(\frac{1}{\omega'}\Big) \delta'(\omega)$$

$$- \delta(\omega) \mathcal{P}\Big(\frac{1}{\omega'}\Big) \frac{\partial}{\partial\omega'} \Big\} f(\omega,\omega') . \tag{5.9}$$

The contribution involving $\delta(\omega+\omega')$ expresses energy conservation for the four phonons involved in the process described in Fig. 5.2. This is the type of term we expected. But expression (5.9) contains also contributions involving energy conservation for fewer phonons (the term with $\delta(\omega)$) or no δ-function at all but its derivative.

It is then natural to consider "renormalizations." It is indeed natural to ask the following question: Perhaps this large collection of unexpected terms (one has to add all four phonon processes of the type described in Fig. 5.2) is due to the fact we use a "bare phonon" description? This is a very interesting problem which we shall discuss in the next paragraph.

c) The Renormalization Program - Strongly Coupled Anharmonic Oscillators

The importance of renormalization *both* in field theory and in statistical physics has been repeatedly emphasized (see e.g. DE DOMINICIS [10]). It is clear that the question of observation may, however, be quite different in both cases. The "bare" electron is to some extent a mathematical fiction useful as a starting point for a perturbational approach. There, at least as long as we deal with stable particles, one considers collisions between free particles and quanta. As emphasized by HEITLER [11] the important simplification is that in this case, in the limit of a large volume, the transition probability per unit volume vanishes as does, therefore, also the line breadth of the initial state. Neglecting the question of ultraviolet divergences which is irrelevant here, one may say that the problem of renormalization is then solved through an appropriate unitary transformation (an excellent survey is given by HEITLER [11]).

However, the problems occurring in statistical physics do not belong to this class. Here there is a finite transition probability whenever the concentration is considered to be finite in the limit of a large system. The same situation occurs for all problems involving discrete bound states, resonances and unstable particles.

For example in the case of unstable particles, the method generally used in such cases as due originally to PEIERLS [12] is to define the mass and life-time of an unstable particle through the real and the imaginary parts of a complex pole appearing in the "second Riemann sheet" of the analytic continuation of the propagator. One of the main advantages of this definition is that it is independent of the production mechanism (see the excellent discussion by LÉVY [13]).

However, even in simple cases this method may lead to ambiguities (see LÉVY, *loc. cit.*). Our kinetic method leads to a basically different approach to this problem. We now start with the asymptotic form (3.82) of the general evolution equation. As this equation is already obtained by neglecting the memory effects, we obtain also a theory which is independent of the production mechanism.

Let us, however, not go into details related to questions of unstable particles and illustrate the renormalization problem in statistical mechanics through the example of strongly coupled anharmonic oscillators (see also Sec.(e)).

In order to include dressing effects in the energy distribution function we use formula (3.94). We have now to take into account all four-phonon processes of the type represented in Fig. 5.2. As the result of elementary but lengthy calculations given in Ref. 6 the kinetic Eq. (3.95) takes the following form:

$$\frac{\partial}{\partial t} \tilde{\rho}_o(t) = i \, \tilde{\phi}_2 \, \tilde{\rho}_o(t) + i \, \tilde{\phi}_4 \, \tilde{\rho}_o(t) \quad . \tag{5.10}$$

The most interesting features are the following:

(a) The operator $\tilde{\phi}_2$ differs from the weak coupling operator (4.12) only by the fact that renormalization of the energy (or of the frequency ω) and of the interactions (V) is taken into account to order λ^2. We may write

$$\tilde{\phi}_2 = \phi_2 + \lambda^2(\Delta\phi_2)_\omega + \lambda^2(\Delta\phi_2)_V \quad . \tag{5.11}$$

Explicit expressions for the renormalized frequencies and interaction energies are given in our paper [6] and will not be repeated here. We may note only that the renormalized frequency is a linear functional of *all the occupation numbers*;

(b) Only energy conserving contributions remain in the collision operator. The whole collision operator therefore now lies on the "renormalized energy shell." This justifies the name "dressed distribution function" we gave to $\tilde{\rho}_o$. No anomalous terms subsist in the collision operator;

(c) The operator $i \, \tilde{\phi}_4$ while Hermitian is *not* positive definite. This is related to the fact that $\tilde{\phi}_4$ is an excess term corresponding to the difference between all four-phonon processes and *repeated* three-phonon interactions. The situation is here exactly the same as in the case of three-body collisions discussed in Sec. (a) of this chapter.

Many other examples may be discussed along similar lines (Lee model [14], dissipative processes in normal Fermi system [15],...).

It should also be noticed that the renormalizations we obtain to order λ^2 are similar to those which may be deduced by other techniques (see Ref. 20). This no longer occurs in higher orders in λ (see Ref. 21). For example, the mass of a V-particle when *unstable*, differs from that obtained by GLASER and KÄLLEN using the usual Green's function technique. We want to state explicitly that we do not consider Glaser and Källen's result as "wrong" and ours as "exact". We claim that thermodynamic quantities (such as the partition function) as well as the cross sections can be expressed in terms of one-particle states using our expressions. An essential point is that in our kinetic equation all processes are now energy conserving. No virtual processes appear at all. This is the basic justification for considering $\tilde{\rho}_0$ as the distribution function of the *physical* (interacting) particles.

Let us now consider another type of problem.

d) Pion Production in Baryon-Pion Scattering with a ρ-Meson in the Intermediate Stage

In order to illustrate how the kinetic equation may be useful in the discussion of high-energy problems, we shall now consider a problem which has been discussed in a qualitative way by MAYNÉ [16]. This is the problem of pion production in a baryon-pion collision with the creation of a metastable ρ-meson at the intermediate stage:

$$B_1 + \pi \rightarrow B_2 + \rho \rightarrow B_2 + \pi_1 + \pi_2 , \qquad (5.12)$$

where B denotes a baryon, π a π-meson and ρ a ρ-meson. The interaction Hamiltonian has been taken as a sum of two terms:

$$V = g_1 V_1 + g_2 V_2 \quad , \tag{5.13}$$

$$= g_1 \; B^+B(\pi^+\rho + \pi\rho^+) + g_2(\rho^+\pi\pi + \rho\,\pi^+\pi^+) \quad ,$$

where V_1 describes the process $(B\pi) \leftrightarrow (B\rho)$, while V_2 describes the disintegration of the ρ-meson $\rho \leftrightarrow (\pi\pi)$. The initial condition is such that we have one baryon with momentum p and a finite concentration of π-mesons with momentum k:

$$\rho_o(0) = \delta_{N_{B_p},1}\,\delta_{N_{\pi_k},n} \tag{5.14}$$

which corresponds for the dressed particle distribution to the initial condition (see (3.85) and (3.94)):

$$\tilde{\rho}_o(0) = \chi^{-1} A\, \rho_o(0) = \chi\, \rho_o(0) \quad . \tag{5.15}$$

Therefore the solution of the kinetic equation is:

$$\tilde{\rho}_o(t) = e^{it\tilde{\Phi}}\, \chi\, \rho_o(0) \quad . \tag{5.16}$$

The diagrams will be of three types: those which involve only V_1 or V_2, and those which involve vertices of both kinds. The lowest order diagrams of each type are represented on Fig. 5.3.

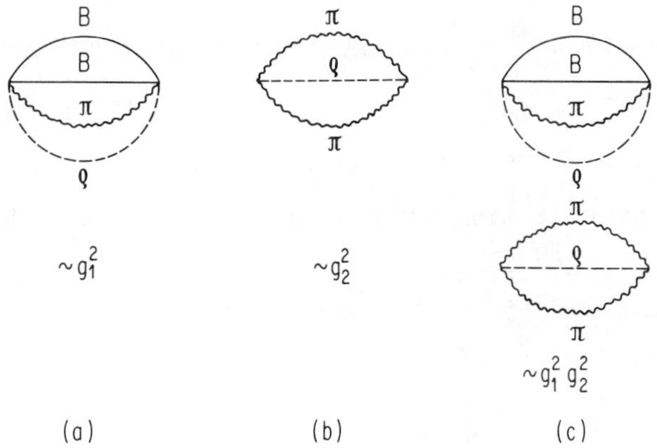

Fig. 5.3. Contributions to the baryon-pion scattering

The diagram (c) of Fig. 5.3 is very similar to the four-phonon process of Fig. 5.2. It corresponds to a V_1 process occurring "simultaneously" with a V_2 process and having one element (a ρ-meson or a π-meson) in common.

To the three types of processes (involving V_1, V_2 or both) correspond three types of contributions to the kinetic equation:

$$\Phi = \Phi^{(a)} + \Phi^{(b)} + \Phi^{(c)} \quad . \tag{5.17}$$

If we restrict our analysis to fourth order, both $\Phi^{(a)}$ and $\Phi^{(b)}$ will consist of a second-order renormalized contribution and a fourth order contribution similar to the ϕ_4 operator for anharmonic oscillators. In a first qualitative analysis, the latter have been omitted. The only contribution to $\Phi^{(c)}$ is of fourth order and has been kept. The computation of the various terms in the r.h.s. of (5.16) could be done easily because of the simplicity of the initial condition. Assuming a low concentration for pions, all terms of order $C^n (n>1)$ may be neglected. One then finds:

$$\tilde{\rho}_o(t) = a_1 + a_2 e^{-g_2^2 \beta t} + t \sum_{q,k_1,k_2} \gamma_{qk_1k_2} \tag{5.18}$$

$$\left\{ \delta_{N_{B_p},1} \, \delta_{N_{\pi_k},n} - \delta_{N_{B_q},1} \, \delta_{N_{\pi_k},n-1} \, \delta_{N_{\pi_{k_1}},1} \, \delta_{N_{\pi_{k_2}},1} \right\} ,$$

where a_1, a_2 are independent of time and $g_2^2 \beta$ is the inverse lifetime of the ρ-meson. The second term may then be dropped for times much longer than the lifetime of the ρ. The transition probability per unit time is then given by $\gamma_{qk_1k_2}$, which has been shown to be of the form:

$$\gamma_{qk_1k_2} = g_1^2\, g_2^2\, \sigma_1 + g_1^2\, g_2^2\, \sigma_2 / g_{2\beta}^2 \ . \qquad (5.19)$$

The quantities σ_1 and σ_2 differ essentially through the kind of delta-functions for energy conservation which appear, σ_1 involves a single delta-function $\delta(E_{B_1} + E_{\pi_k} - E_{B_2} - E_{\pi_{k_1}} - E_{\pi_{k_2}})$ which expresses the equality of the initial and final energies: in these processes the ρ-meson is virtual. The function σ_2 on the other hand involves a product of delta-functions $\delta(E_{B_1} + E_{\pi_k} - E_{B_2} - E_{\pi_{k_2}} - E_{\pi_{k_1}})\cdot \delta(E_\rho - E_{\pi_{k_1}} - E_{\pi_{k_2}})$. Here the ρ-meson is "real" and the energy is conserved in the disintegration process. A very interesting feature of this second contribution in the r.h.s. of (5.19) is that it is proportional to the lifetime of the ρ.

If one plots the transition probability as a function of $\left\{(E_{\pi_{k_1}} + E_{\pi_{k_2}})^2 - (\underset{\sim}{k}_1 + \underset{\sim}{k}_2)^2\right\}^{1/2}$, the first contribution gives a continuous background while the second one gives the characteristic peak at the mass of the ρ-meson. The experimental features are thus reproduced at least qualitatively. A more precise analysis of the shape of the curve is now going on. Similar problems are involved in resonance scattering of photons by atoms, in sequential decay processes, ... (see Refs. 22 and 23).

e) Relation with S-Matrix Theory

We have already noticed in Sec. (a) of this chapter that the application of S-matrix formalism leads to problems when one is concerned with higher order contributions to the kinetic equation. The easy elimination of such difficulties in our formalism through the use of the irreducibility condition is ultimately due to the fact that we work in *product* space corresponding to the density matrix. We have a single time sequence, and a simple ordering of the events along a *single* time coordinate may be performed.

Let us again consider the baryon-pion problem treated in Sec. (d). Because of the occurrence of the ρ-meson in energy conserving steps, the overall transition:

$$B_1 + \pi \rightarrow B_2 + \pi_1 + \pi_2 \quad , \tag{5.20}$$

may now involve an arbitrary number of t-factors. Moreover, diagram (c) of Fig. 5.3 now appears as a "correction" to the lowest order diagrams (a) and (b) exactly as the four-phonon process of Fig.5.2 is a correction to the three-phonon processes. For this reason its contribution, which as we have seen is essential for the "background" scattering cross section, is again an excess quantity which has no simple relation to S-matrix theory.

The relation between kinetic theory and S-matrix formalism appears to us as follows: In all problems of interest asymptotic elements are involved (limit of large volume, large number of degrees of freedom, long times ...). Essentially, in the S-matrix method, these asymptotic elements are introduced first and then the square is taken. This procedure is appropriate for simple cases such as two-body potential scattering. However, in more complicated situations it is not clear if a procedure based directly on the study of quadratic functionals of the wave functions would not be necessary to obtain a more physical and intuitive description of the time sequences involved. This is precisely what the kinetic theory aims to do. When we say that in Fig. 5.3 the process (c) corresponds to processes (a) and (b) occurring "simultaneously" on the scale of the duration of the processes, we mean a time ordering in the kinetic equations which has no analogue in the time ordering of the

S-matrix theory.

In picturesque terms, the S-matrix method (or the wave functions) introduces a "microscopic time" and the "kinetic time" (which is by construction the time of the observables) is a superposition of two microscopic times. We want only to formulate this problem here. F. HENIN and I are studying the example of sequential radiation decay, and a paper is in preparation.

References

[1] CHOH and UHLENBECK: see Ref. 5, Chap. 1.

[2] P. RÉSIBOIS: J. Math. Phys. 4, 166 (1963); Phys. Letters 9, 139 (1964).

[3] P. RÉSIBOIS: Ref. 6, Chap. 1.

[4] M. L. GOLDBERGER and K. M. WATSON: Collision Theory, J. Wiley, New York, 1964.

[5] A. MANGENEY: Physica 30, 461 (1964).

[6] F. HENIN, I. PRIGOGINE, CL. GEORGE and F. MAYNÉ: Physica 32, 1828 (1966).

[7] W. GÜTTINGER: Progr. Theor. Phys. 13, 612 (1955).

[8] N. N. BOGOLIUBOV and D. V. SHIRKOV: Introduction to the Theory of Quantized Fields, Wiley-Interscience, New York-London, 1959.

[9] W. C. SCHIEVE: Physica 34, 81 (1967).

[10] C. DE DOMINICIS: Suppl. Physica 26, 594 (1960).

[11] W. HEITLER: Quantum Theory of Radiation, 3d ed., Oxford University Press, 1953.

[12] R. E. PEIERLS: Proc. of the 1954 Glasgow Conference, London, Pergamon Press, p. 246.

[13] M. LÉVY: Nuovo Cim. 13, 115 (1959); 14, 612 (1959).

[14] I. PRIGOGINE and F. MAYNÉ: Phys. Letters 21, 42 (1966); F. MAYNÉ: Thèse, Bruxelles, 1968.

[15] P. RÉSIBOIS: Phys. Rev. 138, B 281 (1965); M. WATABE and F. DAGONNIER: Phys. Rev. 143, 110 (1966).

[16] F. MAYNÉ: To appear These, Bruxelles, 1968.

[17] J. SCHWINGER: Ann. Phys. 9, 169 (1960).

[18] G. HÖHLER: Z. Phys. 152, 546 (1958).

[19] F. HENIN: Physica (to appear 1968).

[20] D. C. WALLACE: Phys. Rev. 152 (2nd series), 247, 261 (1966).

[21] CL. GEORGE: To appear Physica 1968 (see also Ref. 1 of Chap. 9).

[22] I. PRIGOGINE: Quantum Theory of Dissipative Systems and Scattering Processes, Nobel Symposium V, Stockholm, 1967.

[23] I. PRIGOGINE, F. HENIN and CL. GEORGE: Dissipative Processes, Quantum States and Entropy, presented for publication in the Proc. Nat. Acad. Sci., U.S.A.

6. NonBoltzmann Situations

a) General Remarks

What is probably the most important single characteristic feature of Boltzmann situations is the existence of a *double* time scale: a short time t_{int} related to the duration of the interactions and a much longer time t_r related to the relaxation process. In the generalized Boltzmann situations studied in the previous sections the ratio t_{int}/t_r can no more be neglected. But even then these two time scales must remain separated. However, there exist much more extreme cases in which two time scales do not exist or play a completely different role.

The first case is realized when the Hamiltonian can no longer be split into an unperturbed part H_o and a perturbation λV, as we have always supposed till now. The existence and the meaning of irreversible processes is then a completely different problem. Contrary to the case of gases or lattices, we do not have even at the start a continuous spectrum associated with the eigenvalues of H_o (see Chap. 3 (c)).The method we followed has to be revised as we no longer have Cauchy integrals (see Chap. 3 (e)) to be used in conjunction with the resolvent technique. We shall study such a situation first in the somewhat simplified case of harmonic lattices (without anharmonic terms), then in the most interesting problem of Heisenberg spin systems analyzed recently by RÉSIBOIS and DE LEENER [1].

The second case corresponds to situations where the order of t_{int} with respect to t_r is "abnormal". We shall consider as an illustration the case of a gravitational plasma. As the result of the long range gravitational forces (gravitational screening is not known) the duration of an interaction is essentially infinite!

Here again we may expect a typical nonBoltzmann behavior.

b) Harmonic Lattices

For comparison with the example of spin systems that we shall consider later, let us briefly discuss time dependent processes in purely harmonic lattices. We therefore neglect completely the anharmonic terms in (4.1). Clearly, such a system can no longer reach equilibrium as the energy of each normal mode remains constant. No Boltzmann equation may therefore be expected.

Still there are some interesting features (see Ref. 2). We consider here only the case of one-dimensional lattices with nearest neighbor interactions. Let us call y_{2n} the velocity of the nth particle, y_{2n+1} the deviation of the distance between particle n and n+1 from its equilibrium value. All quantities are normalized so as to be dimensionless. The initial values of the y_n are y_n^o. Then, as indicated by SCHRÖDINGER [3] a long time ago (in 1914!) the solution of the equation of motion for harmonic forces takes the relatively simple form:

$$y_n(t) = \sum_{\nu} y_\nu^o \, J_{n-\nu}(t) \quad , \tag{6.1}$$

where $J_{n-\nu}$ is the Bessel function of order $n-\nu$. We may now consider an ensemble for which we know the probability of the initial data y_n^o, and we may then use (6.1) to calculate the probability distribution at any time. This is probably the first nonequilibrium many-body problem which has been solved exactly.

If we start, for example, with the initial ensemble:

$$y_\nu^o \, y_\mu^o = \delta_{\mu\nu}, \tag{6.2}$$

then Eq. (6.1) gives us:

$$\overline{y_n\ y_n^o} = J_0(t) \quad . \tag{6.3}$$

This is the autocorrelation function corresponding to y_n. Its behavior is indicated in Fig. 6.1

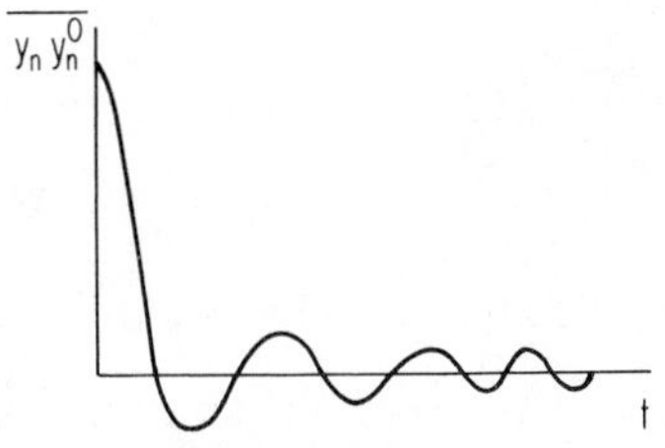

Fig. 6.1.

Schematic representation of the autocorrelation function (6.3).

There is nothing like a relaxation time. The approach to the equilibrium value is oscillating with an amplitude which decreases as

$$t^{-1/2} \quad . \tag{6.4}$$

At first it seemed to us that this behavior was essentially due to the nonergodic character of this system (invariance of all the energies of the normal modes). But we shall see that a rather similar behavior is found in the approach to equilibrium of spins interacting through a Heisenberg Hamiltonian. There also the spin autocorrelation function, while decreasing exponentially in time (in contrast with (6.4)), continues to oscillate for all times. The basic characteristic of these situations is that the Hamiltonian cannot be split as in (3.34) into two parts, one of which is the unperturbed part and the other part is the perturbation responsible for the approach to equilibrium. As a consequence, mechanical and dissipative behavior can no longer be separated.

The results we have described in this paragraph are not in contradiction with the results obtained in Chap. 4 (a) for anharmonic lattices. But we are here interested in the short time behavior (for times ~ 10^{-12} sec.) while the relaxation time introduced through anharmonicity generally corresponds to a much longer time scale (i.e., ~ 10^{-7} sec.).

c) Heisenberg Spin Systems

Let us consider N spins fixed at the sites of a three-dimensional lattice and interaction through the Heisenberg spin Hamiltonian ($\hbar = 1$).

$$H = -\sum_{i \neq j} J_{ij}\, \underset{\sim}{s}_i \cdot \underset{\sim}{s}_j \qquad (6.5)$$

where J_{ij} is the exchange integral between the spins situated at the lattice points i and j. The spin operator s obeys the usual commutation relations (see Ref. 1). A characteristic feature of (6.5) is that the Hamiltonian cannot be split into some unperturbed part plus a perturbation. Similarly, the corresponding Liouville operator is *not* the sum of an unperturbed operator L_o and a perturbation δL. Therefore, as in the case of the purely harmonic oscillators treated in the preceding paragraph we have a single time scale (which may be absorbed into the definition of the unit of time). However, here the corresponding equations of motion are *nonlinear* and we may expect that irreversible processes, driving the system to equilibrium, will occur. As has been shown by RÉSIBOIS and DE LEENER this evolution to equilibrium has most interesting features (see Ref. 1, where references to earlier work in this field can be found), which we shall briefly survey here.

In spite of the fact that we cannot use the perturbational approach, we can still classify the dynamic processes into processes corresponding to a creation, a propagation and a destruction of correlations. For example, we may consider the vacuum of correlations to vacuum of correlations process represented in Fig. 6.2.

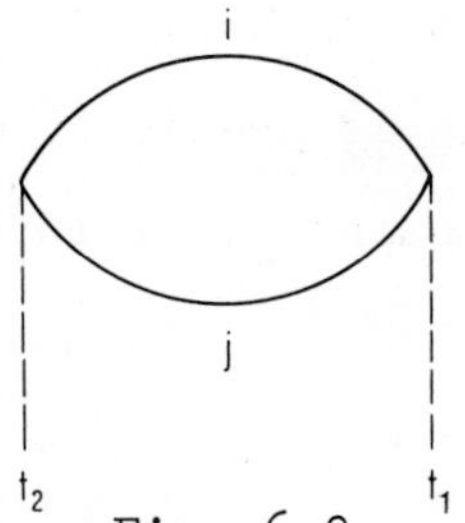

Fig. 6.2.

Diagram representing the creation at time t_1 of a binary correlation between spins i and j, destroyed at time t_2

But the basic difference with what we have seen before is that such a process gives now an asymptotic contribution to $\rho_0(t) \sim t^2$, instead of t. Indeed going back to Chap. 3 (d) we see that we have again to consider a formula of the type of (3.22) but now with $\alpha = 0$ in $\exp i\,\alpha(t_1 - t_2)$, which is the propagator coming from the unperturbed part of the Liouville operator which vanishes here identically. Therefore:

$$I(t) = \int_0^t dt_1 \int_0^{t_1} dt_2 \int d\alpha\ f(\alpha) \sim t^2 \quad . \tag{6.6}$$

Physically this difference has a very simple origin. In the usual gas-kinetic case the two particles "move" and the interaction time is finite (the origin of free motion is precisely H_0). Here, on the contrary, the spins are localized on the lattice sites and they are, so to say, in *permanent* interaction. This leads then to the divergences of the collision operator (6.6).

On the other hand, this continuous interaction leads us immediately to suppose that it will not be allowed to neglect multiple interactions with other spins. That is, each of the two spins, i and j, will at all times interact with the "bath" formed by the remaining spins. It is this flow of interactions involving a larger and larger number of spins which should introduce the necessary element for dissipation. This is precisely what RÉSIBOIS and DE LEENER have shown. The main quantity of interest here is the spin autocorrelation function

$$\underset{\approx}{\Gamma}(t) = \langle \underset{\sim}{s}(t)\, \underset{\sim}{s}(0) \rangle \ , \tag{6.7}$$

where $\underset{\sim}{s}(t)$ denotes the Heisenberg representation of the spin operator and where the average is over a canonical equilibrium distribution. RÉSIBOIS and DE LEENER have shown that (6.7) satisfies a nonMarkoffian kinetic equation of the following type:

$$\frac{\partial}{\partial t} \Gamma(t) = - \int_0^t d\tau \; G_o(\tau|\Gamma) \; \Gamma(t-\tau) \ , \tag{6.8}$$

Here $\Gamma(t)$ is the z-z component of the tensor (6.7). The kernel G_o is still a functional of Γ. In first approximation one has:

$$G_o(t|\Gamma) \sim (\Gamma(t))^2 \ , \tag{6.9}$$

so that the kernel in(6.8) is the square of the quantity itself. The Eq. (6.8) may be solved numerically (as well as analytically in the case of very short and very long times). The autocorrelation function $\Gamma(t)$ starts as a Gaussian for short times and dies out asymptotically with damped oscillations. Qualitatively the behavior is very similar to that of harmonic oscillators (see Fig. 6.1).

The square in (6.9) has a simple topological meaning: it corresponds to the replacement of each line in Fig. 6.2 by a "renormalized" line including interaction with the bath.

The results obtained by RÉSIBOIS and DE LEENER seem to me of fundamental importance. It gives a first example of a system in which the approach to equilibrium cannot be split into processes involving a few degrees of freedom only (no "cluster expansion" of the cross section, as a consequence no hierarchy with closure). In the kernel (6.9) contributions of arbitrary order are included.

From the practical point of view a rather similar situation is often realized in dense systems (Einstein model of solids, cell model) and similar methods should be applicable there.

d) Gravitational Plasmas

We now consider a gravitational plasma, i.e., an isolated system of N identical stars of mass m and velocities v_i, described by the Hamiltonian:

$$H = \sum_{i=1} \frac{m}{2} v_i^2 + \sum_{i<j} \frac{-Gm^2}{|\underset{\sim}{r}_i - \underset{\sim}{r}_j|} , \tag{6.10}$$

where G is the constant of gravitation. Contrary to the previous example where the Hamiltonian did non contain any H_0 and where there was only one time scale, the Hamiltonian (6.10) is of the usual plasma type so that one can again introduce two time scales t_{int} and t_r (see Chap. 3 (d)). However, whereas the ratio t_{int}/t_r can be considered small for short-range forces we have in our case $t_{int} \sim \infty$ for gravitational (long-range) forces, independently of t_r, so that the two time scales are no longer widely separated. Following Ref. 8 of Chap. 3 we will consider the gravitational potential as the limit of a screened potential:

$$\frac{\lambda}{r} = \lim_{\kappa \to 0} \lambda \frac{e^{-\kappa r}}{r} , \quad \lambda = -Gm^2 \quad . \tag{6.11}$$

The only non-dimensional parameter proportional to the coupling constant λ is $\Gamma = \lambda\, d^{1/3}(m\langle v^2\rangle)^{-1}$ when using as characteristic parameters for the systems : λ, m, the average star velocity $\langle v\rangle$ and the number density d. An evaluation of Γ for our galaxy in the vicinity of the sun yields $\Gamma \sim 10^{-6}$ so that one can restrict to a weak coupling approximation. However, we have to introduce this approximation into the general evolution Eq. (3.77) and not into any Boltzmann approximation of this equation, because none of the memory effects (nonMarkoffian, initial correlations) are negligible in our case. In the weak coupling case we have (see (3.57)):

$$\Psi(z) = 32\pi^3 \frac{\lambda^2 d}{m^2} \frac{\partial}{\partial \underset{\sim}{g}} \cdot \underset{\approx}{T} \cdot \frac{\partial}{\partial \underset{\sim}{g}} \qquad \underset{\sim}{g} = \underset{\sim}{v}_1 - \underset{\sim}{v}_2 \tag{6.12}$$

$$\underset{\approx}{T} = - i \int d\underset{\sim}{k} \quad |V_{\underset{\sim}{k}}|^2 \frac{\underset{\sim}{k}\,\underset{\sim}{k}}{\underset{\sim}{k}\cdot\underset{\sim}{g}-z} \, .$$

The evaluation of the tensor $\underset{\approx}{T}$ can be found in Balescu's book (Ref. 8, Chap. 3). After decomposition into transverse and longitudinal components with respect to the relative velocity $\underset{\sim}{g}$, the result is:

$$T_{\parallel} = \lim_{\kappa \to 0} \frac{1}{4\pi^2 i}\left(\frac{\kappa}{z+i\kappa g} - \frac{\mu}{z+i\mu g}\right) \tag{6.13}$$

$$T_{\perp} = \lim_{\kappa \to 0} \left(\frac{1}{8\pi^2 i\mu R^2} \cdot \frac{1}{z+i\mu g} - \frac{1}{4\pi^2 g} \cdot \ln \frac{z+i\kappa g}{z+i\mu g}\right)$$

where $\mu = (\kappa^2 + R^{-2})^{1/2}$ and R is the usual distance of closest approach (cut-off) of the two-body scattering problem. One can see in

(6.13) that if $\kappa \neq 0$ the collision operator $\Psi(z)$ will have an asymptotic limit $\Psi(+i0)$ corresponding, for example, to a series summation in the electrostatic plasma case. In our case ($\kappa=0$), $\Psi(z)$ contains a term $\sim \ln z$, so that one cannot define an asymptotic collision operator. This expresses the infinite duration of a collision in a gravitational plasma. In the electrostatic case the asymptotic collision operator has a meaning (but not the weak coupling approximation, as screening has to be taken into account). In the gravitational case the weak coupling approximation has a meaning ($\Gamma \ll 1$) but not the asymptotic approximation ($t_{int} \sim \infty$). We have thus to retain the nonMarkoffian evolution equation which will be of the form:

$$\partial_t \rho_o(t) = \Phi\rho_o(t) + [\ln(\frac{R}{g}\partial_t)\hat{\Psi}]\rho_o(t) + D(t), \tag{6.14}$$

where Φ and $\hat{\Psi}$ are differential operators (Ref. 8, Chap. 3). The term in $\ln \partial_t$ in (6.14) comes from the $\ln z$ contribution in $\Psi(z)$ and $D(t)$ from the destruction fragment. The main differences between (6.14) and the Boltzmann situation are the following:

1) In the Boltzmann case the destruction fragment $D(t)$ vanishes for $t \gg t_{int}$ and the canonical equilibrium distribution ρ_o^{eq} is a solution of the asymptotic equation:

$$\Psi(+i0)\ \rho_o^{eq} = 0 \quad . \tag{6.15}$$

In the gravitational case the memory effects are crucial. As $D(t)$ does not vanish for long times, the system cannot forget its initial conditions which in turn will determine the solutions of (6.14). Moreover, (for more details Ref. 4) while the equilibrium distribution satisfies:

$$\left(\ln \frac{R}{g} \partial_t\right) \Phi \rho_o^{eq} = 0 \quad , \tag{6.16}$$

one now has

$$\Phi \rho_o^{eq} \neq 0 \quad . \tag{6.17}$$

Therefore the solution ρ_o^{eq} can only be maintained through a compensation of dynamical and memory effects

$$\Phi \rho_o^{eq} + D^{eq} = 0 \quad . \tag{6.18}$$

This shows that for negligible initial correlations, i.e. when $D = 0$, the canonical equilibrium distribution cannot be reached by the system.

2) Until now it has not been possible to deduce an H-theorem from (6.14). If entropy is defined in terms of the usual H-quantity, it can decrease or increase. All proofs of the H-theorem are based on the properties of the asymptotic collision operator $\psi(+i0)$ and are not applicable here.

The possibility that the second principle of thermodynamics is valid only for short-range forces for a given state of gravitation cannot be discarded. It may well be that the condition of a large number of degrees of freedom is not sufficient to justify the second law and it may also be essential to consider the type of interaction (see also Chap. 7 (d)).

It is really a fascinating possibility that gravitation provides us with the "driving force" which keeps the universe "moving". This problem deserves future study.

References

[1] P. RÉSIBOIS and M. DE LEENER: Phys. Rev. 152, 305, 318 (1966); P. RÉSIBOIS, P. BORCKMANS and D. WALGRAEF: Phys. Rev. Lett. 17, 1290 (1966); P. BORCKMANS and D. WALGRAEF: Physica 35, 80 (1967).

[2] G. KLEIN and I. PRIGOGINE: Physica 19, 1053 (1953).

[3] E. SCHRÖDINGER: Ann. der Physik 44, 916 (1914).

[4] I. PRIGOGINE: Nature 209, 602 (1966); I. PRIGOGINE and G. SEVERNE: Physica 32, 1376 (1966).

7. Irreversibility

a) Introduction

During these lectures we have already made some comments about irreversibility and entropy (see Chap. 3 (a) and (c) and Chap. 6 (d)). We want now to come back to these questions in a more systematic fashion.

It is useful to distinguish the general problem of irreversibility from a discussion of the meaning of entropy. The first problem is the most general one: under what circumstances may we expect a system to reach a final equilibrium state described by equilibrium statistical mechanics? This is of course related to the ergodic problem discussed in Chap. 3 (a) (see also Ref. 7 of Chap. 3).

The introduction of entropy in classical thermodynamics is a most elegant way to *describe* both the approach to equilibrium and the final equilibrium properties on a phenomenological level. But the very idea of irreversibility may be formulated in terms of molecular distribution functions (for example, through the statement that the velocity distribution approaches the Maxwellian distribution) independently of the idea of entropy.

Now the whole concept of irreversibility is faced with two objections which have been raised in the form of paradoxes. These paradoxes due to ZERMELO [1] and LOSCHMIDT [2] are based on classical mechanics, but they are not modified in their essential features by quantum mechanics (see Ref. 6 of Chap. 3 for the original references).

As we shall discuss them, it is useful to restate them briefly:

1) Zermelo's Recurrence Paradox. This paradox is based on Poincaré's [3] recurrence theorem which states: "For almost all initial states an arbitrary function of phase space will infinitely often assume its initial value within arbitrary error, provided the system remains in a finite part of the phase space".

As a result it seems that irreversibility is incompatible with the validity of this theorem.

2) Loschmidt's Reversibility Paradox. As the laws of mechanics are symmetrical with respect to the time inversion

$$t \to -t ,$$

to each process corresponds a time reversed process. This again seems to be in contradiction with the existence of irreversible processes.

There exists an enormous literature on this subject. One may say that a clear understanding of the way in which microscopic reversibility and macroscopic irreversibility may be reconciled is due to SMOLUCHOWSKI [4] (see CHANDRASEKHAR, Ref. 6, Chap. 3 for an excellent review). However, Smoluchowski's fundamental work was based on a *stochastic theory of molecular processes* (similar to the Brownian motion theory we mentioned in Chap. 3 (b)).

It is therefore interesting to follow Smoluchowski's discussion but starting now with a mechanical description of the system.

b) Discussion of Zermelo's Paradox

Let us consider a system of interacting particles in a finite volume Ω using periodic boundary conditions. It is now essential to consider a finite system to which Poincaré's theorem, stated above, may be applied.

The permissible values of the wave vector $\underset{\sim}{k}$ are given by (3.9):

$$\underset{\sim}{k} = \frac{2\pi}{\Omega^{1/3}} \underset{\sim}{n} \quad . \tag{7.1}$$

We may still define the collision operator $\Psi(z)$,but we have to be careful not to replace sums over $\underset{\sim}{k}$ by integrals. As a result we now have in (3.58) (omitting unnecessary factors):

$$\Psi(z) \sim \sum_{\underset{\sim}{k}} \frac{|V_{\underset{\sim}{k}}|^2 \underset{\sim}{k}\,\underset{\sim}{k}}{\underset{\sim}{k} \cdot \underset{\sim}{v} - z} \quad . \tag{7.2}$$

This defines a function which is regular everywhere except at a denumerable number of poles *situated on the real axis* for values of z given by (we suppose that $\underset{\sim}{v}$ is taken as the polar axis):

$$z_n = n_z v \frac{2\pi}{\Omega^{1/3}} \tag{7.3}$$

We no longer have a Cauchy integral of the type (3.58) and there do not exist *two* distinct functions Φ^+ and Φ^-, as in the continuous case, but only one function defined in the complete z plane.

We see how deeply the topology of the problem is altered. By going back to the t-description we shall still have Eq. (3.77-78). However, because of the position of the poles of $\Psi(z)$, $G(\tau)$ will be a sum of periodic functions in τ and we have to expect a quasi-periodic behavior for $\rho_0(t)$ in agreement with Poincaré's theorem.* As the poles (7.3) give for $G(\tau)$ contributions proportional to

$$\sim \; e^{iz_n t} \; , \tag{7.4}$$

we may expect that the "fine structure" of $\Psi(z)$ will become apparent for long times (for which $(z_n - z_{n-1})t$ gives appreciable contributions).

We see that the theory we have developed in these lectures bypasses the difficulties associated with Zermelo's paradox by considering directly the case of an *infinite system* (for which the sum in (7.2) may be replaced by an integral). But this means that at present we cannot apply this theory to long times with respect to Poincaré's time. In other words our theory applies to young systems which have been prepared and observed over relatively short periods.

For most systems of interest this is hardly a limitation (because for a system of say 10^{23} particles Poincaré's time is probably by orders of magnitude larger than the presumed age of the universe), but it would be interesting to develop a theory valid at all stages of the evolution in order to see how irreversibility arises when the size of the system increases [10].

c) Discussion of Loschmidt's Paradox

We want to show that the complete evolution Eq. (3.77-78) satisfies exactly the conditions imposed by mechanical reversibility. It may therefore describe both evolution towards equilibrium as represented on Fig. 3.1 and evolution away from equilibrium as in

*It would even be very interesting to use in detail this method to obtain at least in simple cases of interacting systems an estimate of Poincaré's recurrence time.

Fig. 3.2.

The situation is basically the same as the one discussed in Chap. 3 (d) in connection with the scattering problem.

Let us directly consider a simple example (see PRIGOGINE and RÉSIBOIS [5], BALESCU [6]). We follow exactly Balescu's recent work [6]. We consider the example of a classical, nonrelativistic gas of particles interacting through an exponential potential:

$$V(r) = V_o\, e^{-\kappa r} \quad , \tag{7.5}$$

whose Fourier transform is:

$$V_{\underset{\sim}{k}} = \frac{8\pi\kappa V_o}{(\underset{\sim}{k}^2+\kappa^2)^2} \quad . \tag{7.6}$$

We shall study the evolution approximately, to the lowest nontrivial order in V_o, i.e. up to order V_o^2. We shall, however, make no other approximation; in particular the time dependence will be studied exactly.* The diagonal fragment operator to this approximation is given by

$$\Psi(z) = \sum_{i<j} \frac{1}{8\pi^3\Omega}\, \underset{\sim}{\partial}_{ij} \cdot \int d\underset{\sim}{k}\; \underset{\sim}{k}\, V_{\underset{\sim}{k}}\, \frac{1}{i(\underset{\sim}{k}\cdot\underset{\sim}{g}_{ij}-z)}\, \underset{\sim}{k}\, V_{\underset{\sim}{k}} \cdot \underset{\sim}{\partial}_{ij} , \tag{7.7}$$

where Ω is the volume, $\underset{\sim}{\partial}_{ij} \equiv \frac{\partial}{\partial \underset{\sim}{p}_i} - \frac{\partial}{\partial \underset{\sim}{p}_j}$ and $\underset{\sim}{g}_{ij} = \underset{\sim}{v}_i - \underset{\sim}{v}_j$.

We now pose our problem in the following way:

1.) We start at time $t = 0$ with a system without correlations:

---

* This is the main difference with respect to the treatment of Ref.5 where an asymptotic long-time approximation was used.

$$\rho_0(v;0) \neq 0$$

$$\rho_k(v;0) = 0 \quad . \tag{7.8}$$

In order to avoid later confusion, we call (7.8) the "*preinitial condition.*"

2.) We let the system evolve freely up to time $t_0 > 0$. During this time, the velocity distribution has changed (up to order V_0^2) to:

$$\rho_0^{(D)}(v;t)=\rho_0(v;0)+\frac{1}{2\pi}\int_0^t dt'\int_0^{t'} d\tau\int dz\ e^{-iz(t'-\tau)}\ \Psi(z)\ \rho_0(v;0) \ , \tag{7.9}$$

for $t \leq t_0$.

On the other hand, correlations have appeared by the action of the creation fragment on the initial $\rho_0(v;0)$:

$$\rho_{\underset{\sim}{k},-\underset{\sim}{k}}^{(D)}(v;t) = \frac{e^{-i\underset{\sim}{k}\cdot\underset{\sim}{g}_{ij}}-1}{i\underset{\sim}{k}\cdot\underset{\sim}{g}_{ij}}\ \underset{\sim}{k}\ V_{\underset{\sim}{k}} \cdot \underset{\sim}{\partial}_{ij}\ \rho_0(v;0) \ , \tag{7.10}$$

for $t \leq t_0$.

3.) In the *direct case*, we consider t_0 as a new initial time, (7.9) and (7.10) giving the new initial condition (for $t = t_0$), and we study the evolution for times $t > t_0$. It is given, up to second order, by:

$$\rho_0^{(D)}(v;t) = \rho_0^{(D)}(v;t_0)+ \int_{t_0}^{t} dt'\int_{t_0}^{t'} d\tau\ \frac{1}{2\pi}\int dz\ e^{-iz(t-\tau)}\ \Psi(z)\ \rho_0(v;0)$$

$$+ \frac{1}{2\pi}\int dz\ e^{-iz(t-t_0)} \sum_{i<j} \int d\underset{\sim}{k}\ D_{ij}(\underset{\sim}{k};z)\ \rho_{\underset{\sim}{k},-\underset{\sim}{k}}^{(D)}(v;t_0), \tag{7.11}$$

for $t \geq t_0$,

with

$$D_{ij}(\underset{\sim}{k};z) = \underset{\sim}{\partial}_{ij} \cdot \left\{ \underset{\sim}{k}\, V_{\underset{\sim}{k}} \frac{1}{i(\underset{\sim}{k}\cdot \underset{\sim}{g}_{ij} - z)} \right\} . \tag{7.12}$$

4.) To study the *reversed case* we change abruptly all velocities at time t_o; $\underset{\sim}{v} \to -\underset{\sim}{v}$. This will be called the (nonrelativistic) T-transformation:

$$\rho_o^{(T)}(v;t_o) = \rho_o(-v;0) + \frac{1}{2\pi} \int_0^{t_o} dt' \int_0^{t'} d\tau \int dz\, e^{-iz(t'-\tau)} \Psi(z) \rho_o(-v;0), \tag{7.13}$$

$$\rho_{\underset{\sim}{k},-\underset{\sim}{k}}^{(T)}(v;t_o) = \frac{e^{i\underset{\sim}{k}\cdot \underset{\sim}{g}_{ij} t_o} - 1}{-i\underset{\sim}{k}\cdot \underset{\sim}{g}_{ij}}\, \underset{\sim}{k}\, V_{\underset{\sim}{k}} \cdot (-\underset{\sim}{\partial}_{ij})\, \rho_o(-v;0) . \tag{7.14}$$

Considering now (7.13) with (7.14) as the new initial condition, its normal evolution for $t > t_o$ is given again by (7.11), where the superscript (D) is simply replaced by (T).

We now make an additional simplifying assumption, namely that $\rho_o(v;0)$ is an isotropic distribution:*

$$\rho_o(v;0) = \rho_o(-v;0) . \tag{7.15}$$

It is then immediately seen, from (7.7), (7.8) and (7.14), that

$$\rho_o^{(T)}(v;t_o) = \rho_o^{(D)}(v;t_o) . \tag{7.16}$$

This fact already points out the important conclusion that the whole difference between "direct" and "reversed" evolution lies in

---

* This assumption is not essential: all the conclusions of this paper can be trivially generalized to include the case where (7.15) is not realized.

the destruction fragment.

Let us now perform the integrations explicitly. We shall first write out explicitly the master equation in the two cases considered.*

We shall simplify the notations as much as possible. Substituting (7.6) into (7.7), we perform the integration over $\underset{\sim}{k}$ in a system of cylindrical coordinates illustrated in Fig. 7.1.

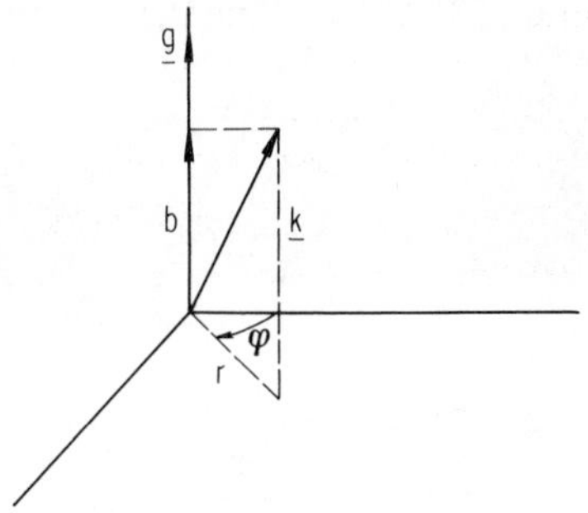

Fig. 7.1. Coordinate system for the integration over $\underset{\sim}{k}$

The integrations over r and ϕ are trivial, and we are left with an expression of the form

$$\Psi(z) = \sum A \; \partial_r \, \mathscr{C}_{rs}(z) \; \partial_s \qquad (7.17)$$

where ∂_r $(r = x,y,z)$ is the rth component of the gradient ∂_{ij} and A is a numerical constant, proportional to V_o^1, whose value is of no interest to us. The summation sign $\sum$ over the particles will also be omitted henceforth. We shall moreover introduce nondimensional variables:

$$\tilde{b} = b/\kappa$$
$$\tilde{z} = z/\kappa g \qquad (7.18)$$
$$\tilde{t} = \kappa g t \quad ,$$

* The destruction fragment being an explicit functional of the initial condition, the equation will have different forms in the two cases.

(the tildes being omitted after substitution). In terms of these variables the x-x component of the "diffusion coefficient" $\mathscr{C}_{rs}(z)$ is given by:

$$\mathscr{C}_{xx}(z) = \int_{-\infty}^{\infty} db \frac{1}{(b^2+1)^2} \frac{1}{b-z} \quad , \quad z \in S_+ \quad , \tag{7.19}$$

where S_+ denotes the upper half-plane of the complex variable z. The destruction fragment is treated in a similar way using (7.6), (7.10), (7.11), (7.12) and (7.14). Finally, the master equation to order V_o^2 is written as:

$$\partial_t \rho_o^{(D,T)}(v;t) = A\, \partial_r \left\{ \int_{t_o}^{t} d\tau \frac{1}{2\pi} \int dz\, e^{-iz(t-\tau)} \mathscr{C}_{rs}(z)\, \partial_s \rho_o(v;\tau) \right.$$
$$\left. + \frac{1}{2\pi} \int dz\, e^{-iz(t-t_o)} \mathscr{D}_{rs}^{(D,T)}(z)\, \partial_s \rho_o(v:0) \right\} . \tag{7.20}$$

The x-x component of $\mathscr{D}_{rs}(z)$ is given, in the two cases, by:

$$\mathscr{D}_{xx}^{(D)}(z) = \int_{-\infty}^{\infty} db \frac{1}{(b^2+1)^2} \frac{1}{b-z} \frac{e^{-ibt_o}-1}{b} \quad , \quad z \in S_+ \quad , \tag{7.21}$$

$$\mathscr{D}_{xx}^{(T)}(z) = \int_{-\infty}^{\infty} db \frac{1}{(b^2+1)^2} \frac{1}{b-z} \frac{e^{ibt_o}-1}{b} \quad , \quad z \in S_+ \quad . \tag{7.22}$$

Alternatively, Eq. (7.20) can be written, after the Laplace transformations have been carried out explicitly, as:

$$\partial_t \rho_o^{(D,T)}(v;t) = B\, \partial_r \left\{ \int_{t_o}^{t} d\tau\, \mathscr{C}_{rs}(t-\tau)\, \partial_s \rho_o(v;\tau) + \mathscr{D}_{rs}^{(D,T)}(t)\, \partial_s \rho_o(v;0) \right\} \tag{7.23}$$

with obvious definition for $\mathscr{C}_{rs}(t)$, $\mathscr{D}_{rs}(t)$. Let us now discuss in some detail the various functions appearing here.

We first note (as was pointed out already) that the diagonal fragment operator $\mathscr{C}_{rs}(z)$ (it will be called briefly "the cycle") is invariant under the velocity inversion. Its explicit form is easily calculated from (7.19) by the method of residues:

$$\mathcal{C}_{xx}(z) = \frac{1}{z+i} + \frac{i}{(z+i)^2} \quad . \tag{7.24}$$

As expected, $\mathcal{C}_{xx}(z)$ has the usual properties of a Cauchy integral (Chap. 3 (e)). It is regular in the upper half-plane and has a double pole in S_- at $z = -i$. Its Laplace transform is

$$\mathcal{C}_{xx}(t) = e^{-t} + te^{-t} \quad . \tag{7.25}$$

Hence the "memory kernel" is a function decaying to zero after a time of the order of the duration of a collision ($t = 1$ in our units).

Consider now the *direct destruction fragment* (7.21). It is a Cauchy integral of a function obeying the Lipschitz condition and can be explicitly calculated, with the result:

$$\mathcal{D}^{(D)}_{xx}(z) = \frac{t_o e^{-t_o}}{i+z} + \left(\frac{2}{i+z} + \frac{i}{(i+z)^2}\right) \quad (e^{-t_o}-1). \tag{7.26}$$

Its Laplace transform is

$$\mathcal{D}^{(D)}_{xx}(t) = (t+2)\ e^{-t} - (t-t_o+2)\ e^{-(t-t_o)} \quad , \quad t \geq t_o \quad . \tag{7.27}$$

This function has all the "normal" properties postulated in the general theory: it is a monotonically decreasing function of time which vanishes after a few collision times. This function is shown in curve (a) of Fig. 7.2.

Let us now study the *reversed destruction fragment* (7.22). Just like (7.21), it is a Cauchy integral of a Lipschitz function. But because of the changed sign in the exponential, it must be evaluated by completing the contour of integration by a semicircle at infinity in the upper half-plane. The result is

$$\mathcal{D}^{(T)}_{xx}(z) = \frac{t_o e^{-t_o}}{z-i} + \left[\frac{2}{z-i} - \frac{i}{(z-i)^2}\right](e^{-t_o}-1) + 4\,\frac{e^{izt_o}-1}{z(z+i)^2(z-i)^2} . \quad (7.28)$$

This function has a peculiar structure, which is responsible for its behavior. It can be written in the form:

$$\mathcal{D}^{(T)}_{xx}(z) = A(z) + e^{izt_o} B(z) \quad . \qquad (7.29)$$

Both A(z) and B(z) have poles in the upper half-plane S_+ at $z = i$ and at $z = 0$, and one in the lower half-plane at $z = -i$. Hence, neither of the two terms is regular in S_+. However, the residues of A(z) at $z = i$ and at $z = 0$ equal minus the residues of $e^{izt_o} B(z)$ in the corresponding poles. Therefore, the complete function $\mathcal{D}^{(T)}_{xx}(z)$ is a Cauchy integral and is regular in the (finite) upper half-plane. However, because of the exponential character of the second term, $\mathcal{D}^{(T)}_{xx}(z)$ will have a peculiar behavior under Laplace transformation:

$$\mathcal{D}^{(T)}_{xx}(t) = \frac{1}{2\pi}\int_C dz\; e^{-iz(t-t_o)}\,\{A(z) + e^{izt_o} B(z)\} \quad , \qquad (7.30)$$

where, to make everything definite, the contour C must lie above $z = i$. Consider first $t > 2t_o$; then the contour C can be completed by a semicircle in S_- for both terms. According to the previous discussion, the only singularity of the complete integrand is a pole in $z = -i$ and hence $\mathcal{D}^{(T)}_{xx}(z)$ has an exponentially decaying behavior, just like in the direct case. But for $t_o \leq t \leq 2t_o$, the contour must be closed below for the first term and above for the second term. By this process the two terms are separated and as a result the contributions of the singularities in $z = 0$ and $z = i$ no longer cancel. Actually the second term gives zero. Therefore, $\mathcal{D}_{xx}^{(T)}(t)$ is the sum of the residues of $A(z)$ at $z = -i$, $z = 0$ and $z = +i$ and hence it is the sum of a decaying exponential, a constant and a growing exponential. Explicitly, the result of the calculation is:

$$\mathcal{D}_{xx}^{(T)}(t) = \begin{cases} -(t-2t_o-2)\, e^{t-2t_o} - 4 + (t-t_o+2)\, e^{-(t-t_o)}, & t_o \leq t \leq 2t_o, \\ -(t-2t_o+2)\, e^{-(t-2t_o)} + (t-t_o+2)\, e^{-(t-t_o)}, & t \geq 2t_o. \end{cases} \tag{7.31}$$

This function is shown in curve (b) of Fig. 7.2.

Hence, the reversed initial condition leads to a destruction fragment which, during a time t_o, has an abnormal behavior in the sense that it contains an exponentially growing function. Now remember that under normal conditions (such as in the direct case) after a few collision times the destruction fragment can be neglected and the master equation goes over to its asymptotic form which describes the "*kinetic regime*" of approach to equilibrium. We now see that if the initial condition is of the type $\rho_k^{(T)}(t_o)$, the destruction fragment cannot be neglected, because during a time t_o it becomes more and more important and only after a time t_o does the evolution again become normal. It is true that the "antikinetic" behavior is only transient, but on the other hand t_o is as long as one desires. We shall resume the general discussion below.

Let us now consider the effect of the antikinetic behavior on the solution of the master equation. Eq. (7.9) can be rewritten in our condensed notation as:

$$\rho_o(t) = \rho_o(0) + B\,\partial_r\,\Gamma_{rs}(t)\,\partial_s\,\rho_o(0)\;, \qquad t\leq t_o, \tag{7.32}$$

with

$$\Gamma_{rs}(t) = \int_0^t dt' \int_0^{t'} d\tau\; \mathscr{E}_{rs}(t'-\tau)\;, \qquad t\leq t_o\;. \tag{7.33}$$

For times longer than t_o we must distinguish the direct and reversed evolution:

$$\rho_o^{(D,T)}(t) = \rho_o(0) + B\,\partial_r\,\{\Gamma_{rs}(t) + \Delta_{rs}^{(D,T)}(t)\}\,\partial_s\rho_o(0)\;, \qquad t\geq t_o, \tag{7.34}$$

where

$$\Gamma_{rs}(t) = \int_0^{t_o} dt' \int_0^{t'} d\tau\; \mathscr{E}_{rs}(t'-\tau) + \int_{t_o}^{t} dt' \int_{t_o}^{t'} d\tau\; \mathscr{E}_{rs}(t'-\tau)\;, \tag{7.35}$$

and

$$\Delta_{rs}^{(D,T)}(t) = \int_{t_o}^{t} dt'\; \mathscr{D}_{rs}^{(D,T)}(t')\;. \tag{7.36}$$

The calculations are quite simple and yield the following results:

$$\Gamma_{xx}(t) = 2t + (t+3)\,e^{-t} - 3\;, \qquad t\leq t_o\;, \tag{7.37}$$

$$\Gamma_{xx}(t) = 2t + (t-t_o+3)\,e^{-(t-t_o)} + (t_o+3)\,e^{-t_o} - 6\;, \quad t\geq t_o. \tag{7.38}$$

The first term, linear in t, is characteristic of the diagonal fragment. It is well known that the apparent divergence in time to which it leads is suppressed by summing a partial perturbation series containing all powers of $V_o^2 t$, which finally leads to an exponential decay on the long relaxation time scale. Here, however, we are only considering terms up to order V_o^2 and are not interested in this extra summation.

The direct destruction fragment contributes the following term to the solution:

$$\Delta_{xx}^{(D)}(t) = -(t-t_o+3)\, e^{-(t-t_o)} + (t+3)\, e^{-t} - (t_o + 3)\, e^{-t_o} + 3, \qquad t \geq t_o \,. \tag{7.39}$$

The complete direct solution for $t \geq t_o$ is given by the sum of (7.38) and (7.39):

$$\sum_{xx}^{(D)}(t) \equiv \Gamma_{xx}(t) + \Delta_{xx}^{(D)}(t) = 2t + (t+3)\, e^{-t} - 3 \;, \quad t \geq t_o \,. \tag{7.40}$$

It is identical in form with (7.37), as it should be. Indeed, in the direct case, nothing new has happened physically at time t_o. The splitting of the solution into a cycle contribution and a destruction term is a purely mathematical artifact and must lead to the same solution as the one obtained by starting at $t = 0$ with the cycle alone. In particular, the total solution must be independent of t_o.

Consider now the reversed destruction fragment contribution:

$$\Delta_{xx}^{(T)}(t) = \begin{cases} -4(t-t_o)-(t-2t_o-3)\, e^{t-2t_o}-(3+t_o)\, e^{-t_o}-(t-t_o+3)e^{-(t-t_o)}+3, & t_o \leq t \leq 2t_o \,, \\ (t-2t_o+3)\, e^{-(t-2t_o)} -(t-t_o+3)\, e^{-(t-t_o)}-(t_o+3)e^{-t_o}-4t_o+3, & t \geq 2t_o \,. \end{cases} \tag{7.41}$$

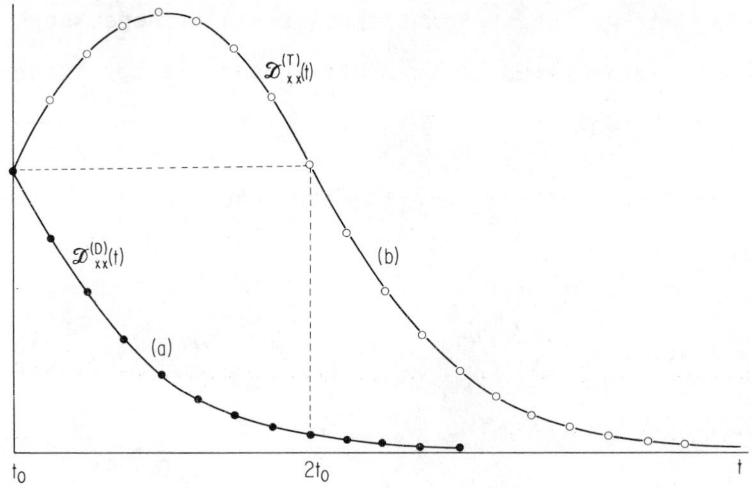

Fig. 7.2.

Behavior of the direct and reversed destruction fragments

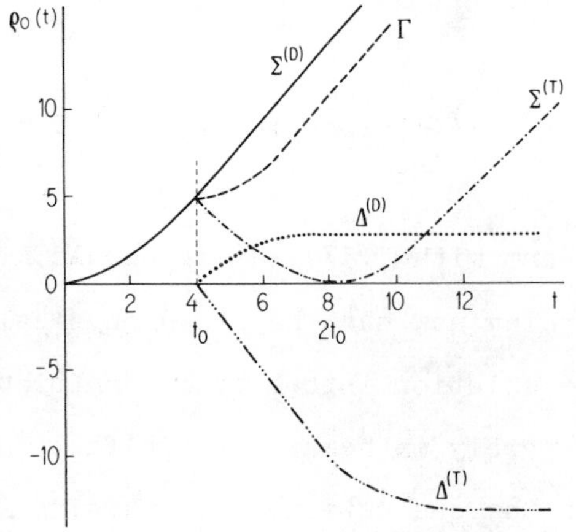

Fig. 7.3.

Solutions of the master equation (see text)

Whereas the direct destruction contribution rapidly tends to a constant, as can be seen in Fig. 7.3, the reversed destruction fragment grows monotonically in absolute value during a time t_o, until it finally cancels exactly the cycle contribution at $t = 2t_o$. After that time the growth is stopped and $\Delta^{(T)}_{xx}$ also becomes constant. Adding (7.41) to (7.38), we obtain the total reversed solution:

$$\Sigma_{xx}^{(T)}(t) = \begin{cases} -2(t-2t_o) - (t-2t_o-3)\, e^{t-2t_o} - 3 \ , & t_o \le t \le 2t_o \ , \\ 2(t-2t_o) + (t-2t_o+3)\, e^{-(t-2t_o)} - 3 \ , & t \ge 2t_o \ . \end{cases} \tag{7.42}$$

The reversed solution has remarkable features, which one expects intuitively.

1.) For $t_o \le t \le 2t_o$ the solution reverses step by step the evolution which had occurred between 0 and t_o. More precisely, we have the relation:

$$\Sigma_{xx}^{(T)}(t) = \Sigma_{xx}^{(D)}(2t-t) \ , \quad t_o \le t \le 2t_o \ . \tag{7.43}$$

2.) As a consequence,

$$\Sigma_{xx}^{(T)}(2t_o) = 0 \ . \tag{7.44}$$

Hence, at time $t = 2t_o$ the reversed system has reached exactly its pre-initial starting point:

$$\rho_o^{(T)}(2t_o) = \rho_o(0) \ . \tag{7.45}$$

3.) From there on, of course, the evolution must continue as in the direct case, because we are starting afresh with the same initial conditions. Hence,

$$\Sigma_{xx}^{(T)}(t) = \Sigma_{xx}^{(D)}(t-2t_o) \ , \quad t \ge 2t_o \ . \tag{7.46}$$

This is exactly the behavior expected from purely mechanical arguments.

This example (for the general case see Ref. 6) shows clearly how a single equation (the master equation studied in Chap. 3) describes correctly both the kinetic evolution towards equilibrium and the antikinetic evolution away from equilibrium. The type of behavior realized depends on the initial correlations. If the initial correlations are created by the interactions alone, as is the predominant case for many-body systems in nature, the evolution will be of the kinetic type.

d) Meaning of Irreversibility

As the example we have studied has clearly indicated, a fundamental role is played by the memory effects represented by the destruction fragment in our general evolution Eq. (3.77). It is only when this memory effect can be neglected for times small with respect to the total time necessary for the system to reach equilibrium, that we speak in general about approach to equilibrium.

As we have seen in the Liouville formulation of dynamics we have transitions, between different types of correlations, induced by the interactions (see Chap. 3 (d) and (e)). Schematically we have:

$$\rho_0 \rightleftarrows \rho_2 \rightleftarrows \rho_3 \rightleftarrows \cdots \quad , \tag{7.47}$$

where ρ_n represents a correlation between n particles and ρ_0 as usual the vacuum of correlations (this is the velocity or energy distribution function). When the destruction fragment may be neglected, ρ_0 satisfies a closed equation and the ρ_n become functionals of ρ_0 (see also Ref. 5):

$$\rho_0 \rightarrow \rho_2 \rightarrow \rho_3 \rightarrow \cdots \ . \tag{7.48}$$

On the contrary when the destruction fragment is dominant the correlations ρ_n determine the behavior of ρ_o:

$$\rho_o \leftarrow \rho_2 \leftarrow \rho_3 \leftarrow \dots \quad (7.49)$$

These two directions of flow of correlations correspond respectively to thermodynamic and "anti thermodynamic" behavior.

The case which will be realized depends on the nature of the *initial correlations* (as well as on the type of interactions, see below).

The empirically observed dominance of (7.48) over (7.49) can certainly not be explained in terms of measure theory alone. It would be incorrect to assume that behavior (7.49) is realized only by initial states of measure zero. Indeed as the example studied in Sec. (c) has shown, it is sufficient to invert the velocities to go from (7.48) to (7.49). However, our theory gives us at least some indication about this problem. If we start with an initial condition in which the particles were uncorrelated and use the creation fragment (see Chap. 3 (h)) to calculate correlations at a later moment, we obtain a new ensemble which, as verified explicitly in the example of Sec. (c), evolves again to equilibrium.

Now, our theory, as we have seen in Sec. (b) is applicable to systems young with respect to Poincaré's time. In other words, correlations formed spontaneously (through material interactions) in *young systems* give rise to destruction fragments which vanish rapidly (on the scale of the relaxation time) and do not prevent the system from going to equilibrium. This statement follows directly from what we have seen in Sec. (c) and can be verified in other simple cases.

In other words the dominance of thermodynamic evolution would again be a consequence of the fact that we deal with systems over time scales much shorter than Poincaré's time.

If we would use as the initial state a statistical state of a many-body system left in isolation over a period comparable to Poincaré's time, we could say nothing about the type of evolution the system would have to follow afterwards.

This is in agreement with the conclusions reached by different people such as PENROSE and PERCIVAL [7] or DE BEAUREGARD [8].

The only property we have to require from the destruction fragment (and therefore from the initial state of correlations) is its disappearance over times sufficiently short, but further indications about its value or form are generally of no great interest. Properties such as relaxation times and transport coefficients are independent of the destruction fragment. Therefore, a new nondynamical principle which would give the "most probable" initial conditions as proposed by some authors [9] is not necessary and would be of no utility so far as we can see.

However, there are also other remarks which should be made. The destruction fragment depends, of course, not only on the initial correlations but also on the type of interactions which are considered. In the case of gravitational forces, memory effects persist at least in some cases (see Chap. 4 (d)). The whole concept of irreversibility is then to be reconsidered very carefully. Our experience with irreversible processes refers of course only to the effect of short range forces.

In problems involving particles and fields or interacting fields similar considerations are valid. The number of degrees of freedom is then infinite (if we do not introduce some cut-off parameter). Therefore, strictly speaking, Poincaré's recurrence time becomes infinite. Moreover, the distinction between reversible microscopic phenomena and macroscopic irreversibility loses much of its meaning in modern theory of elementary particles because most of the elementary particles are unstable (see Chap. 5 (d) and Refs. 6 and 14 of Chap. 5). For this reason the study of irreversibility is today more interesting and important than ever.

References

[1] E. ZERMELO: Ann. Physik 57, 585 (1896); 59, 793 (1896).

[2] J. LOSCHMIDT: see Ref. 5, Chap. 3.

[3] H. POINCARÉ: Méthodes Nouvelles de la Mécanique Céleste, Gauthier-Villars, 1892, reprinted by Dover, New York, 1957.

[4] M. VON SMOLUCHOWSKI: see Ref. 7, Chap. 1.

[5] I. PRIGOGINE and P. RÉSIBOIS: Estratto dagli Atti del Simposio Lagrangiano, Acad. della Scienze di Torino (1964).

[6] R. BALESCU: Physica 36, 433 (1967).

[7] O. PENROSE and I. C. PERCIVAL: Proc. Phys. Soc. 79, 605 (1962).

[8] C. DE BEAUREGARD: Le Second Principe de la Science du Temps, Ed. du Seuil, Paris, 1963.

[9] E. T. JAYNES: Brandeis Lectures, 1962.

[10] Such a theory is being developed at present by WALGRAEF and BORCKMANS for paramagnetic systems in an external field (Physica 35, 80 (1967)). The splitting, due to the field, introduces a discrete spectrum which has to be made "continuous" through local interactions to obtain irreversible processes (which satisfy a δ-function conservation). In such cases, there exists a critical value of the strength of the interactions at which irreversibility starts. The same situation

is expected to exist when the splitting is due to the finite size of the system. There also the irreversibility would "start" at a critical value. As a consequence the analytic behavior of irreversibility as a function of the coupling constant λ would be completely altered.

8. Entropy

a) Thermodynamic Entropy and Statistical Mechanics

Before we discuss the statistical mechanical interpretation of entropy, let us first summarize the main properties of thermodynamic entropy:

1.) It has to increase with time (at least in a time-averaged sense) for an isolated system:

$$\frac{dS}{dt} \geq 0 \; . \tag{8.1}$$

If the system is not isolated we have to be able to split the time evolution $\frac{dS}{dt}$ into an entropy flow and into an entropy production term such that in agreement with Chap. 2 we have for the entropy production:

$$\frac{d_i S}{dt} \geq 0 \; . \tag{8.2}$$

2.) For long times we have to require that:

$$S \underset{t\to\infty}{\to} S_{eq} \; , \tag{8.3}$$

where S_{eq} is the equilibrium value of entropy. Its relation with equilibrium statistical mechanics is well known since the work of Gibbs.

In this chapter we are not concerned with the more general problem of the relation between dynamics and irreversibility which has been discussed in Chap. 7 but only with the problem of expressing S in terms of distribution functions.

The classical dilemma which is faced here is the following (see Ref. 1-4): the Boltzmann definition of the $\mathcal{H}$-quantity in terms of the velocity (or energy) distribution ρ_o:

$$\mathcal{H}_B = \int \rho_o(\underset{\sim}{p}) \log \rho_o(\underset{\sim}{p}) \, d\underset{\sim}{p} \quad , \tag{8.4}$$

leads as well known (see, e.g., Ref. 2 of Chap. 2) for dilute or weakly coupled systems both to the $\mathcal{H}$-theorem and to the correct value of entropy at equilibrium if the identification:

$$S = - k \, \mathcal{H}_B \tag{8.5}$$

is made. Therefore, both conditions (8.1) and (8.3) are satisfied. The difficulty lies in the extension of Boltzmann's functional to strongly coupled (or dense) systems for which correlations can no longer be neglected.

On the other hand the Gibbs definition of $\mathcal{H}$ in terms of the complete distribution function ρ:

$$\mathcal{H}_G = \int \rho(\underset{\sim}{q},\underset{\sim}{p}) \log \rho(\underset{\sim}{q},\underset{\sim}{p}) \, d\underset{\sim}{p} \, d\underset{\sim}{q} \quad , \tag{8.6}$$

leads to the correct thermodynamic value for *all systems* (in the case of quantum systems minor changes as, for example, replacement of the integral by sums have to be introduced, but we do not go into details here). In other words, at equilibrium, Gibbs' expression takes into account, in a correct and complete way, all correlations. However, out of equilibrium, for *isolated* systems which satisfy the laws of classical or quantum mechanics, expression (8.6) cannot represent correctly the entropy as $\mathcal{H}_G$ is *under these circumstances* an invariant. This point was noticed many years ago by the

EHRENFESTS [3] and is a straightforward consequence of the fact that as a consequence of Liouville's theorem (or its quantum mechanical version), the complete phase distribution function ρ is an invariant. This of course is no longer so if a more general class of systems is considered (for example, if stochastic boundary conditions are introduced or if the system is temporarily brought into contact with a heat reservoir, etc. (see Ref. 4)). But the consideration of more general situations does not solve the basic problem posed in statements (1) or (2) of the second law referring to *isolated* systems. What we want is to understand the thermodynamic behavior in terms of statistical mechanics. To overcome the difficulty various coarse grainings have been proposed (see Ref. 4). This essentially replaces $\mathcal{H}_G$ by some new functional. While interesting properties may be obtained in this way (see, for example, TOLMAN [4]), as far as we know, no proof that such a coarse grained quantity would lead to an $\mathcal{H}$-theorem has been given. As pointed out by PAULI [1], it is easy, through coarse graining, to derive an $\mathcal{H}$-quantity $\bar{\mathcal{H}}_G$ such that:

$$\bar{\mathcal{H}}_G(t_1) \leq \bar{\mathcal{H}}_G(t_o) \quad \text{for } t_1 \geq t_o \text{ ,}^* \tag{8.7}$$

but we can say nothing about the relative values of $\bar{\mathcal{H}}_G(t_1)$, $\bar{\mathcal{H}}_G(t_2)$, when both t_1 and t_2 are larger than t_o (for example, $t_2>t_1>t_o$). This is, of course, much less than can be derived from Boltzmann's $\mathcal{H}$-theorem.

Therefore the problem remains, is it possible to construct a functional which includes the effect of correlations but still satisfies an $\mathcal{H}$-theorem (in the Boltzmann sense) and gives at

---

* This is true even for $t_1 \neq t_o$, which indicates that this has nothing to do with irreversibility in the thermodynamic sense.

equilibrium the correct thermodynamic entropy? As we shall see, this is closely connected to a basic dynamical problem. How does it happen that correlations which at equilibrium play a minor role (at least in some systems) are *always* essential out of equilibrium and in fact so important that they reduce the Gibbs $\mathcal{H}_G$ to a constant?

As we shall see we only begin to understand the answer to these fascinating but difficult questions.

b) Entropy in Strongly Coupled Anharmonic Oscillators - Disorder and Entropy

Let us come back to the problem of strongly coupled anharmonic oscillators that we briefly analyzed in Chap. 5 (c).

We have seen that the dressed distribution function $\tilde{\rho}_0$ satisfies a simpler kinetic equation without any "anomalous terms." It is therefore natural to try to define a generalized Boltzmann $\mathcal{H}$-quantity through the relation:

$$\mathcal{H}_B = \int \tilde{\rho}_0(J) \log \tilde{\rho}_0(J)\, dJ \quad , \qquad (8.8)$$

(we use action-angle variables (see Ref. 7 of Chap. 3) as well as notations appropriate to classical system). It was shown in detail in a recent paper [5] that both properties (8.1) and (8.3) are verified when $\mathcal{H}_B$ is defined by (8.8). We shall not reproduce the proofs here.

It is therefore reasonable to identify (8.8) with the thermodynamic entropy of a strongly coupled lattice. Going back from $\tilde{\rho}_0$ to the bare phonon or normal mode distribution ρ_0, we obtain the following expression (see Ref. 5 for more details):

$$\mathcal{H}_B = \int dJ \, \{\rho_o \log \rho_o + \frac{\lambda^2}{2\rho_o} \sum_{\{k\varepsilon\}} \mathcal{P}\left(\frac{1}{[\varepsilon_k \omega_k]}\right) \frac{\partial}{\partial [\varepsilon_k \omega_k]}$$

$$\times \frac{|V_{\{k\varepsilon\}}|^2}{\omega_k \omega_{k'} \omega_{k''}} \, J_k \, J_{k'} \, J_{k''} \, [\varepsilon_k \frac{\partial \rho_o}{\partial J_k}]^2 \} \; , \tag{8.9}$$

where the following abbreviations have been used:

$$\sum_{\{k\varepsilon\}} \equiv \sum_{k\, k'\, k''} \; \sum_{\varepsilon_k \varepsilon_{k'} \varepsilon_{k''} = \pm 1} \; ,$$

$$[\varepsilon_k A_k] \equiv \varepsilon_k A_k + \varepsilon_{k'} A_{k'} + \varepsilon_{k''} A_{k''},$$

$$V_{\{k\varepsilon\}} = V_{\varepsilon_k k, \; \varepsilon_{k'} k', \; \varepsilon_{k''} k''} \; . \tag{8.10}$$

We now have two terms. One is the usual Boltzmann term which gives a *positive* contribution to entropy and can be understood easily in terms of the "disorder" interpretation of entropy. On the contrary, as can be easily verified, the second term gives at least at equilibrium a *negative* contribution to entropy. The correlations between normal modes order the system. This is, we believe, a very important remark to keep in mind to avoid misinterpretation of the entropy concept. It is only for weakly coupled or dilute systems that entropy may be identified with "disorder". A strongly interacting equilibrium system is *highly organized.*

This is already apparent in simple cases. Molecular chaos is an appropriate expression for the space relations in a gas, certainly neither for a liquid nor a solid. (This seems to be overlooked in all popular discussions in which, invariably, entropy is explained as a measure of disorder.)

Expression (8.9) is valid till order λ^2. In higher orders in λ the situation is not so simple. We may take advantage of the unitary operator χ'', which we mentioned in Chap. 3 (g), to obtain a distribution function $\tilde{\tilde{\rho}}_0$ for which the Boltzmann functional (8.8) correctly represents the entropy both at equilibrium and out of equilibrium. We have shown that at least up to order λ^4 inclusive, this is possible in a unique way. We now see more clearly the importance of the transformation theory we have introduced in Chap. 3. It leads to a representation of the kinetic equation in which all properties have a "*particle*" interpretation and do not depend explicitly on the correlations (which enter of course in the definition of the distribution functions).

Let us now discuss more closely the relation of (8.9) with the Gibbs entropy.

c) Canonical Entropy and Resonance Effects

To the Gibbs $\mathcal{H}$-quantity (8.6) corresponds what we may call the "canonical" entropy:

$$S_{can} = - k \int dJ \, d\alpha \, \rho \log \rho . \tag{8.11}$$

We may separate S_{can} into two parts, one related to the energy distribution and the other to correlations:

$$S_{can} = S_{kin} + S_{corr} , \tag{8.12}$$

with

$$S_{kin} = - k \int dJ \, \rho_0 \log \rho_0 . \tag{8.13}$$

This is simply the usual Boltzmann contribution. To order λ^2 we have:

$$S_{corr} = - k \int dJ \frac{1}{2\rho_o} \sum_{\{k,\varepsilon\}} \rho_{\{\varepsilon k\}} \rho_{\{-\varepsilon k\}} \tag{8.14}$$

where $\rho_{\{\varepsilon k\}}$ is a three-phonon correlation. We now want to analyze more carefully this term. To do this we go back to Chap. 3 (h). We have seen that an arbitrary correlation may be split into two parts:

$$\rho_{\{\varepsilon k\}} = \rho'_{\{\varepsilon k\}} + \rho''_{\{\varepsilon k\}} . \tag{8.15}$$

$\rho'_{\{\varepsilon k\}}$ corresponds to memory effects and vanishes asymptotically; it is of no interest here. On the contrary $\rho''_{\{\varepsilon k\}}$ is a functional of ρ_o. The basic quantity to be discussed is:

$$\sum_{\{k,\varepsilon\}} \rho''_{\{\varepsilon k\}} \rho''_{\{-\varepsilon k\}} . \tag{8.16}$$

Great care has to be used to discuss (8.16) as this is a *nonlinear* functional of the correlation. If we use for each of the two correlations formula (3.102) we obtain, using Laplace transform, an expression of the form:

$$\int dk\ f(k) \int dz_1 \frac{1}{z_1 - \omega_k} \int dz_2 \frac{1}{z_2 + \omega_k} . \tag{8.17}$$

If now we go to the limit $z_1 \to +i0$, $z_2 \to +i0$ without care, we obtain, following the general theory of Chap. 3 (e), a product of two δ-functions, which is, strictly speaking, meaningless. As a result we have to perform carefully the operations involving the *two* Laplace variables z_1 and z_2. When this is done, one obtains two types of contributions:

1.) a contribution which grows systematically with time and vanishes at equilibrium (i.e., which contains a t factor and an operator which vanishes identically when applied on the equilibrium distribution ρ_o^{eq});

2.) a contribution which does not depend *explicitly* on time and is identical to the second term of (8.9).

The existence of the first term is due to constructive interference (or resonance) between $\rho''_{\{\varepsilon k\}}$ and $\rho''_{\{-\varepsilon k\}}$ in (8.16). Such effects are well known when going from the wave function to the density matrix (i.e. the square of a wave function) in quantum mechanics (see VAN HOVE [8]).

The canonical entropy thus contains three types of terms:

$$S_{can} = S^{I}\,[\rho_o(t);\rho'_{\{\varepsilon k\}}] + S^{II}[\rho_o(t);t] + S^{III}[\rho_o(t)]\,. \tag{8.18}$$

Here S^I is a memory effect (due to $\rho'_{\{\varepsilon k\}}$); S^{II} depends on time both implicitly through $\rho_o(t)$ *and* explicitly; finally S^{III} depends on time *only* implicitly through $\rho_o(t)$. The third term is in our case identical to (8.9).

At equilibrium both S^I and S^{II} vanish. We see therefore that the canonical entropy contains different types of contributions, one of which survives only at equilibrium.

Moreover, out of equilibrium, we see what is the compensation mechanism between kinetic effects and correlations. In fact,for an anharmonic solid, it is possible to show [9] that the $\lambda^2 t$ contribution to (8.14) (involved in S^{II}) cancels exactly the systematic evolution of the kinetic entropy due to the change in ρ_o. Of course, the compensation of the kinetic term cannot be due only to the effect of binary correlations. For longer times, the whole expression (8.14) will take its equilibrium value; therefore, the corresponding S^{II} will vanish and the compensation in the canonical entropy will be due to the effect of nonequilibrium correlations of higher order. This flow of correlations is of course in complete agreement with the qualitative picture presented in Chap. 7 (d).

d) Numerical Calculations

The conclusion we have reached in the preceding paragraph may be verified both on other simple examples and by computer calculations.

For instance NICOLIS [10] (see also, for the example of spin systems, PHILIPPOT [11]) has considered a classical homogeneous gas. Up to order λ^2 the correlation entropy may be written:

$$- \frac{S_{corr}}{k} = \int d\underset{\sim}{v}_1 \, d\underset{\sim}{v}_2 \, d\underset{\sim}{r}_{12} \, \frac{g_{12}^2(t)}{\rho_1(\underset{\sim}{v}_1,t)\rho_2(\underset{\sim}{v}_2,t)} \; . \qquad (8.19)$$

The notations are self-evident. Using a Gaussian interaction potential:

$$V = V_o \, e^{-\alpha r_{12}^2} \quad , \qquad (8.20)$$

and the weak coupling approximation, the correlation function can be computed exactly. The important feature is that out of equilibrium this function contains long-range contributions. Again (8.19) may be split into different contributions, one of which may be proved to be the thermodynamic contribution which satisfies an $\mathcal{H}$-

theorem and gives rise to the complete thermodynamic entropy (including here terms of order λ^2).

In order to check these conclusions, BELLEMANS and ORBAN are at present performing computer experiments [12]. They use a two-dimensional system of hard discs (up to 1,225), initially located at the sites of a square lattice, with all velocities being equal in magnitude (see Chap. 3 (a)). The kinetic entropy has the well-known behavior (see, e.g., ALDER and WAINWRIGHT (Ref. 4 of Chap. 3)): it remains constant until the first collision occurs and then it increases until the equilibrium value is reached (see Fig. 3.1). As for the r.h.s. of (8.19), they obtain the behavior in Fig. 8.1.

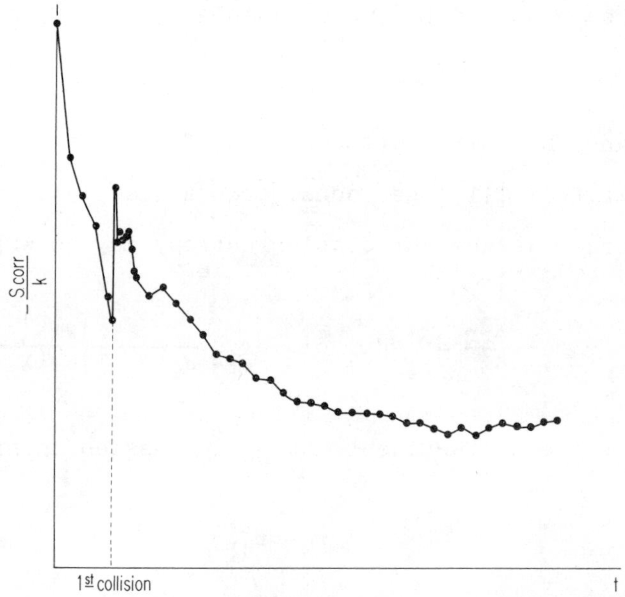

Fig. 8.1. Correlational entropy computed for 1,225 discs

Before the first collision, it decreases. This effect is not yet completely understood. The most interesting feature for our present discussion is that it increases rapidly after the first collision. Later it starts to decrease.

The calculations confirm, at least qualitatively, our discussion of the time evolution of the canonical entropy. The abnormal increase of the correlational entropy (8.19) after the first collision tends to compensate the effect of collisions on the kinetic entropy (8.13). A quantitative discussion of the calculations is unfortunately difficult. It would require a large number of particles (of the order of 5,000) to cut down sufficiently the effect of fluctuations.

Thus, the time behavior of the correlational entropy shows that out of equilibrium the effect of correlations on entropy is quite different from at equilibrium. This corresponds precisely to the supplementary terms S^{II} in the expression (8.18) of the canonical entropy. *The Gibbs expression for entropy cannot be continued out of equilibrium to obtain the thermodynamic entropy.* The abnormal contribution S^{II}, as well as the memory term S^{I}, which vanish at equilibrium have to be subtracted. This discussion shows the complexity of the dynamic effects which have to be considered to understand the microscopic meaning of entropy.

This complexity is due to the fact that the entropy is a *nonlinear functional* of the correlations and that its discussion is of an order of magnitude more difficult than the discussion of physical quantities as energy, specific heat, equation of state, which depend linearly on correlations (i.e., on the Fourier coefficients ρ_k of the distribution).

References

[1] W. PAULI: Int. Conf. on Statistical Mechanics, Florence (1949).

[2] J. YVON: L'Entropie, Dunod, Paris (1965).

[3] P. and T. EHRENFEST: Encyclopedie der Mathematischen Wissenschaften, vol. 4, p. 4 (1911); reprinted by Cornell University Press, Ithaca, N.Y. (1959).

[4] R. C. TOLMAN: The Principles of Statistical Mechanics, Clarendon Press, Oxford (1959).

[5] I. PRIGOGINE, F. HENIN and CL. GEORGE: Physica 32, 1873 (1966).

[6] J. W. GIBBS: The Collected Works, vol. 2, Longmans, Green and Co., London, New York (1928).

[7] A. I. KHINCHIN: Mathematical Foundations of Information Theory, Dover, New York (1957).

[8] L. VAN HOVE: Physica 21, 512 (1955).

[9] F. HENIN: to appear Physica, 1968 .

[10] G. NICOLIS: J. Chem. Phys. 46, 702 (1967).

[11] J. PHILIPPOT and D. WALGRAEF: Proc. of the I.U.P.A.P. Meeting, Copenhagen, 1966, p. 414, Benjamin, N.Y. (1967).

[12] A. BELLEMANNS and J. ORBAN: Phys. Lett. 24 A, 620 (1967).

9. Concluding Remarks

The initial purpose of nonequilibrium statistical mechanics was very modest: its aim was to provide us with a better physical and mathematical understanding of the approximations involved in the derivation of the traditional kinetic equations associated with the names of BOLTZMANN, PAULI, and FOKKER-PLANCK.

To-day the situation is quite different. We now understand in much more detail the assumptions necessary to derive such kinetic equations but we see also that these classical equations are only valid for very specific, rather exceptionally simple situations (see Chap. 6). A wide new field of research is now open for detailed investigations.

It seems probable that methods of nonequilibrium statistical mechanics will throw some light on basic unsolved questions of quantum theory.

Indeed the similarity of some of the problems which arise in field theory and in statistical mechanics is rather striking. In both cases we deal with situations involving an infinite number of degrees of freedom, and some kind of asymptotic procedure is necessary to extract physical information.

For example, what is the relation between particles and fields? For free fields this is trivial, but it is no longer so for interacting fields. If some canonical transformation could reduce the Hamiltonian to a sum of independent terms there would be no problem, but that seems out of the question. Therefore, we have to involve other considerations to introduce the particles associated with the interacting fields. The type of transformation theory of Chap. 3 and Chap. 8 may be useful here. If we may realize a representation in which the entropy at equilibrium and out of equilibrium has a purely combinatorial meaning in terms of particles, we may consider these particles as the "physical" particles associated with the initial Hamiltonian.

One could say that we use the classical argument about entropy in a reverse form: one proves usually that particles, when *weakly* coupled, have a purely combinatorial entropy. We put the entropy into the combinatorial form and conclude that the particles are then well defined, physical entities.

We cannot go into more detail here. Suffice it to say that in this way we obtain a generalization of the usual quantum rules to states with a *finite* life time (see Refs. 1 and 2).

Whatever the future of this specific approach may be, I am convinced that the association between quantum physics and statistical physics can only become closer and will lead ultimately to a formulation of the laws of motion in which the statistical features of quantum mechanics will be even more accentuated.

References

[1] I. PRIGOGINE: Dissipative Processes, Quantum States and Field Theory, XIVe Conseil de Physique Solvay, Octobre 1967, Bruxelles (to appear).

[2] I. PRIGOGINE, F. HENIN and CL. GEORGE: Dissipative Processes, Quantum States and Entropy, presented for publication to the Proc. Nat. Acad. of Sci., U.S.A.

Appendix I.* The General Evolution Equation and Projection Operators

In Chap. 3 we have derived the general evolution Eq. (3.76). In this derivation we introduced the different steps which are used in further applications, namely, the Laplace transform (3.44), the perturbation expansion (3.47) and the Fourier transforms (3.48). However, as was first shown by ZWANZIG [1], the general evolution equation by itself can be obtained without reference to these Fourier-Laplace transforms, using only abstract operator techniques. As this derivation of Eq. (3.76) is very compact we will summarize it here.

In Chap. 3, we stressed repeatedly the importance of irreducible vacuum-to-vacuum transitions. The important point herein, with respect to Eq. (3.76), is the distinction between a state corresponding to nonvanishing correlations ρ_c and a state corresponding to the vacuum of correlations ρ. This distinction can be exhibited formally through the introduction of an operator projecting the total distribution function or state vector ρ onto these two subspaces.

$$\rho_0 = P_0 \rho ,$$
$$\rho_c = P_c \rho , \qquad \text{(A I.1)}$$

where P_0, P_c are the "projection operators" onto ρ_0 and ρ_c, respecttively, i.e.,

$$P_0^2 = P_0 = P_0^+ ,$$
$$P_c^2 = P_c = P_c^+ . \qquad \text{(A I.2)}$$

*This Appendix is due to M. BAUS.

(Mathematically (A I.2) expresses the fact that we have two idempotent $P_0^2 = P_0$ and Hermitian $P_0^+ = P_0$ operators. For more detail see Ref. 2.) As a matter of fact we have:

$$\rho = \rho_0 + \rho_c \quad , \tag{A I.3}$$

i.e., these operators project onto complementary subspaces:

$$P_0 + P_c = I \quad . \tag{A I.4}$$

Moreover, they project onto orthogonal subspaces:

$$P_0 P_c = P_c P_0 = 0 \; . \tag{A I.5}$$

One easily can check properties (A I.1-5) by using the Fourier representation of P_0 and P_c used in Chap. 3, namely:

$$P_0 = |0\rangle\langle 0| \quad ,$$
$$P_c = \sum_\phi{}' |\phi\rangle\langle\phi| \quad , \tag{A I.6}$$

where the prime on $\sum$ means that the Fourier state $\phi = 0$ (vacuum) is excluded in the summation.

We will now write the complete evolution operator e^{-iLt} (see (3.4)) in such a way as to exhibit the irreducible transitions. We apply the operator identity (3.45) to the resolvent for L and to the resolvent for LP_c to obtain the following identity:

$$e^{-iLt} = e^{-iLP_c t} + \int_0^t d\tau \; e^{-iLP_c \tau}(-iLP_0) \; e^{-iL(t-\tau)} \; . \tag{A I.7}$$

(An alternative method using the Laplace transform of the projections of the Liouville equation can be found in Ref. 1.)
Multiplying (A I.7) by P_c we obtain:

$$P_c\, e^{-iLt} = e^{-iP_cLt}\, P_c + \int_0^t d\tau\, e^{-iP_cL\tau}\, P_c(-iLP_0)\, e^{-iL(t-\tau)} \; . \qquad \text{(A I.8)}$$

Indeed we have:

$$P_c\, e^{-iLP_ct} \equiv P_c\, e^{-iLP_ct}\, P_c^{-1}\, P_c \equiv e^{-iP_c(LP_c)P_c^{-1}t}\, P_c \equiv e^{-iP_cLt}\, P_c .$$

We now write the Liouville Eq. (3.4) as:

$$i\, \partial_t\, \rho(t) = L\, P_0\, \rho(t) + L\, P_c\, \rho(t) \; . \qquad \text{(A I.9)}$$

Rewriting the second term as:

$$L\, P_c\, \rho(t) = L\, P_c\, e^{-iLt}\, \rho(0) \quad , \qquad \text{(A I.10)}$$

and substituting (A I.8) we obtain for (A I.9):

$$i\, \partial_t\, \rho(t) = L\, P_0\, \rho(t) + \int_0^t d\tau\, L\, e^{-iP_cL\tau}\, (-iP_cL)P_0\, \rho(t-\tau)$$
$$+ L\, e^{-iP_cLt}\, P_c\, \rho(0). \qquad \text{(A I.11)}$$

Projecting (A I.11) onto the two subspaces we obtain:

$$i\, \partial_t\, P_0\, \rho(t) = P_0\, L\, P_0\, \rho(t) + \int_0^t d\tau\, P_0\, L\, e^{-iP_cL\tau}(-iP_cL)\, P_0\, \rho(t-\tau)$$
$$+ P_0\, L\, e^{-iP_cLt}\, P_c\, \rho(0) \quad , \qquad \text{(A I.12)}$$

$$i\,\partial_t\, P_c\,\rho\,(t) = P_c\, L\, P_0\, \rho(t) + \int_0^t d\tau\; P_c\, L\, e^{-iP_cL\tau}(-iP_cL)\; P_0\, \rho(t-\tau)$$

$$+ P_c\, L\, e^{-iP_cLt}\; P_c\, \rho(0). \qquad \text{(A I.13)}$$

The identity (A I.8) shows that the solution of (A I.13) is simply

$$P_c\, \rho(t) = e^{-iP_cLt}\; P_c\, \rho(0) + \int_0^t d\tau\; e^{-iP_cL\tau}(-iP_cL)\; P_0\, \rho(t-\tau)\;. \qquad \text{(A I.14)}$$

Eqs. (A I.12-13) are the formal evolution equations for the vacuum of correlations and for the correlations, respectively. Eq. (A I.14) expresses the correlations in terms of the initial correlations and of the solution of(A I.12). If we now use the explicit representation of P_0 and P_c given in (A I.6), Eqs. (A I.12,13,14) become identical with those discussed in Chap. 3. Indeed, consider for example Eq. (A I.12). In the representation (A I.6) we have:

$$P_0\; L\; P_0 = 0 \quad , \qquad \text{(A I.15)}$$

because only L_0 is diagonal in this representation and, moreover, it has a zero eigenvalue when acting on a vacuum state. Taking now the scalar product of the remainder of (A I.12) with $\langle 0|$ we have ($\rho_0(t) = \langle 0|P_0\; \rho(t)\rangle$):

$$i\,\partial_t\; \rho_0(t) = \int_0^t d\tau\; G(\tau)\; \rho_0(t-\tau) + F(t) \quad , \qquad \text{(A I.16)}$$

where:

$$G(\tau) = -\, i\, \langle 0|L\; e^{-iP_cL\tau}\; P_c\; L|0\rangle \quad , \qquad \text{(A I.17)}$$

$$F(t) = \sum_\phi{}' \langle 0|L\; e^{-iP_cLt}|\phi\rangle\;\; \rho_\phi(0) \;, \qquad \text{(A I.18)}$$

with:

$$\rho_\phi = \langle\phi|\rho\rangle \ .$$

If we Laplace transform (A I.17 and 18) and use a perturbation expansion (see (3.47)) for $(P_cL-zI)^{-1}$, one immediately verifies the identity of (A I.16) and (3.76). The irreducibility condition of Chap.3 is expressed here by the appearance of P_cL instead of L in the resolvent. A more detailed discussion of the connection between (A I.13 and 14) and Chap. 3, as well as the extension of the projection operator techniques to inhomogeneous systems where (A I.6) is no longer a good representation, can be found in Ref. 3.

This method shows that the physical idea of "dynamics of correlations" used in Chap. 3 may be expressed in an elegant and concise way through a projection operator technique. However, in every concrete application it is essential to have a good representation of the abstract operators P_0 and P_c. Recent applications of this technique [4] show that this is not always easy. There appear confusions mainly with respect to the irreducibility condition ($L \neq P_cL$) (for a detailed discussion see [3]) which already shows how delicate these formal manipulations become when separated from their physical context.

References

[1] R. ZWANZIG: (a) J. Chem. Phys. 33, 1338 (1960); (b) Lectures in Theoretical Physics (Boulder) 3, 106 (1960); (c) Physica 30, 1109 (1964).

[2] A. MESSIAH: Mécanique Quantique, vol. I, Dunod, Paris (1962).

[3] M. BAUS: Acad. Roy. Belg., Bull. Cl. Sciences 53, 1291, 1332, 1352 (1967).

[4] (a) J. BIEL and L.M. GARRIDO: Nuovo Cim. 40 B, 197 (1965); (b) J. BIEL: Nuovo Cim. 40 B , 213 (1965); (c) P. MAZUR and J. BIEL: Physica 32, 1633 (1966); (d) R.H. TERWIEL and P. MAZUR: Physica 32, 1813 (1966).

Appendix II. Nonanalytic Density Behavior of Transport Coefficients*

Recently, a great deal of effort has been devoted to the calculation of transport coefficients in moderately dense gases. These calculations are always confronted, in one way or another, with the explicit evaluation of collision operators involving three, four, ... n particles. In principle, the formal expression (3.67) (when expanded according to the number of colliding particles) or compact expressions like (5.1) may be used for such calculations. However, neither of these expressions is very adequate for explicit manipulations. Indeed, in the former case, we have an expansion in the coupling parameter λ, which is not convergent for realistic strong repulsive forces, while the Choh-Uhlenbeck type of formula (5.1) requires, in principle, the solution of the exact n-body motion, which is not available in analytical form.

Let us indicate how these difficulties may be bypassed. We consider for definiteness the three-body operator. From (3.67) we have:

$$\Psi^{+}_{(123)}(+i0) = \sum_{n=1}^{\infty} \langle 0|-\lambda\delta L\,[R_o(-\lambda\delta L)^n]|0\rangle_{irr(123)}\,, \tag{A II.1}$$

where the subscript (123) means that we consider only interactions between the triplet of particles 1,2 and 3. We may rearrange this expression as an infinite sum of sequences of two-body collision operators (*binary collision expansion*). To accomplish this we introduce the operator $T_{(12)}$, defined by the following integral equation:

$$T_{(12)} = -\lambda\delta L_{(12)} - \lambda\delta L_{(12)}R_o(123)T_{(12)}\,. \tag{A II.2}$$

*This Appendix is due to Dr. P. RÉSIBOIS.

Clearly $T_{(12)}$ describes the exact motion of particles 1 and 2 while particle 3 is moving freely. There is no difficulty in principle in calculating this operator exactly.

It is then rather easy to show that (A II.1) may be rewritten as (see Ref. 1):

$$\Psi^{+}_{(123)}(+i0) = \sum_{\substack{(\alpha\beta)\\(\gamma\delta)\\ \vdots}} \langle 0| [T_{(\alpha\beta)}R_oT_{(\gamma\delta)}R_oT_{(\varepsilon\rho)} + T_{(\alpha\beta)}R_oT_{(\gamma\delta)}R_oT_{(\varepsilon\rho)}R_oT_{(\mu\nu)} + \ldots]|0\rangle_{irr} , \qquad (A\ II.3)$$

where the couples $(\alpha\beta)$ $(\gamma\delta)$... are chosen from the set (123) in all possible ways, the only restriction being that the same pair should not be repeated in succession. For instance, one of the terms in (A II.3) will be:

$$\bar{\Psi}^{+}_{(123)}(+i0) = \langle 0|T_{(12)}R_oT_{(13)}R_oT_{(12)}|0\rangle_{irr} , \qquad (A\ II.4)$$

while a contribution of the type:

$$\bar{\bar{\Psi}}^{+}_{(123)}(+i0) = \langle 0|T_{(12)}R_oT_{(12)}R_oT_{(23)}\ |0\rangle_{irr} , \qquad (A\ II.5)$$

does not appear.

The physical interpretation of Eq. (A II.3) is quite clear: the three-body collision is described in terms of all possible sequences of two-body motions, the third particle being considered as provisionally free.

Although it is very difficult to show that the series (A II.3) converges, explicit calculations indicate that, in three dimensions, the binary expansion of the three-particle collision operator is a very powerful method for approximate calculation. However, when the same method is applied to the four-body operator $\Psi_{(1234)}$, it can be shown that the result is infinite! Worse than that, this divergence is not a consequence of the particular type of expansion used here but is really a genuine physical property of the dynamics of four particles (see Ref. 2). As the same type of difficulty already occurs with the three-body operator (A II.1) in *two* dimensions, we shall discuss this simpler case here.

Let us write down more explicitly the particular term (A II.4). With the help of (3.46), we readily obtain:

$$\Psi^{+}_{(123)}(+i0) \propto \int d^2\underset{\sim}{k} \; \langle 0|T_{(12)}|\,\underset{\sim}{k},-\underset{\sim}{k}\rangle \frac{1}{\underset{\sim}{k}\cdot\underset{\sim}{v}_{12}-i0} \langle\underset{\sim}{k},-\underset{\sim}{k}|T_{(13)}|\,\underset{\sim}{k},-\underset{\sim}{k}\rangle$$

$$\frac{1}{\underset{\sim}{k}\cdot\underset{\sim}{v}_{12}-i0} \langle\underset{\sim}{k},-\underset{\sim}{k}|T_{(12)}|0\rangle \quad , \qquad \text{(A II.6)}$$

where we have left out unimportant factors. Let us stress that wavenumber conservation (see (3.37)) forces us to take the same wavenumber $\underset{\sim}{k}$ in the two propagators.

In order to investigate the behavior of (A II.6) we look at the integrand for small wavenumber $\underset{\sim}{k}$. Clearly, we have from (A II.2):

$$\lim_{k\to 0} \langle 0|T_{(12)}|\underset{\sim}{k},-\underset{\sim}{k}\rangle = \lim_{k\to 0} \langle\underset{\sim}{k},-\underset{\sim}{k}|T_{(12)}|\underset{\sim}{k},-\underset{\sim}{k}\rangle = \Psi^{+}_{(12)}(+i0) \quad , \qquad \text{(A II.7)}$$

where $\Psi^{+}_{(12)}(+i0)$ denotes the *two particle Boltzmann operator*. This operator is, of course, bounded and when it acts on a function of the velocities v_1 and v_2, it gives the following result:

$$\Psi^{+}_{(12)}(+i0)\; \phi(\underset{\sim}{v}_1,\underset{\sim}{v}_2) = \int d\Omega\; \sigma(\Omega,v_{12})\; v_{12}[\phi(\underset{\sim}{v}_1',\underset{\sim}{v}_2'') - \phi(\underset{\sim}{v}_1,\underset{\sim}{v}_2)], \qquad \text{(A II.8)}$$

where $\sigma(\Omega, v_{12})$ is the scattering cross-section at angle Ω, and $\underset{\sim}{v}'_1$, $\underset{\sim}{v}'_2$ are the velocities resulting from $\underset{\sim}{v}_1$, $\underset{\sim}{v}_2$ after the collision process.

When (A II.8) is inserted into (A II.6), it may be shown by a rather delicate but rigorous calculation that

$$\overline{\Psi}^{+}_{(123)}(i0) \propto \int \frac{dk}{k} \propto \lim_{k\to 0} (\ln k) \quad . \qquad \text{(A II.9)}$$

Let us point out that the naive point of view, which would simply replace the Boltzmann operator (A II.7) by a constant (relaxation time τ_r^{-1}) or by its lowest-order Born approximation:

$$\Psi^{+}_{(12)}(i0) \simeq \lambda^2 \langle 0 | \delta L_{(12)} R_o(12)\; \delta L_{(12)} | 0 \rangle \ , \qquad \text{(A II.10)}$$

does not lead to the correct divergent form (A II.9). Indeed, in these two cases, one gets:

$$\overline{\Psi}^{+}_{(123)}(i0) \propto \lim_{\varepsilon\to 0} \int \frac{d^2\underset{\sim}{k}}{(\underset{\sim}{k}\cdot\underset{\sim}{v}_{12} - i\varepsilon)^n} \quad \text{(n finite)} \ , \qquad \text{(A II.11)}$$

which is a well-defined and finite quantity (see Ref. 3). Thus, the divergence appears only when the exact Boltzmann collision operator (A II.8) is considered. We shall see presently the reason for this result.

In order to understand the physical meaning of this divergence, let us consider the simpler problem of one light particle 1, interacting on a plane with two fixed scattering centers (2-3) (corresponding to the two-dimensional Lorentz gas). We can easily compute the order-of-magnitude of the cross-section for a scattering process corresponding to the sequence (12), (13), (12) (cf.Fig. A II.1).

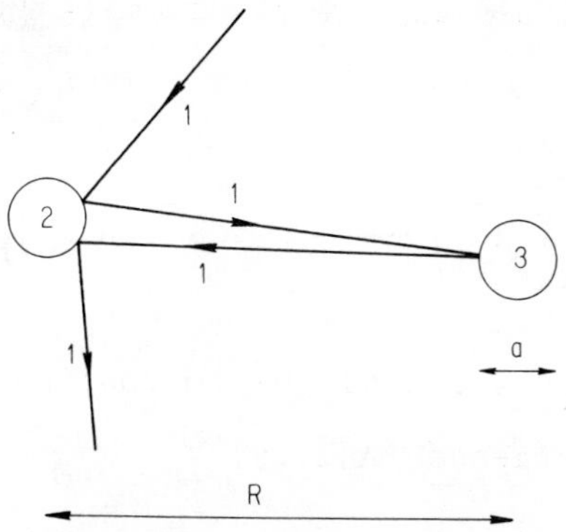

Fig. A II.1

Triple collision in a Lorentz gas

Indeed, the probability per unit time for the collision process (12) is

$$p_{(12)} \sim \frac{av_1}{S} , \tag{A II.12}$$

where v_1 is the velocity of particle 1 and S the total surface area of the system. The conditional probability for the later process(13) is obtained by a simple geometrical argument:

$$p'_{(13)} \sim \frac{a}{R} . \tag{A II.13}$$

Similarly we get for the probability of recollision (12):

$$p''_{(12)} \sim \frac{a}{R} . \tag{A II.14}$$

The total scattering cross-section is thus:

$$\sigma \sim \frac{a^3 v}{SR^2} , \tag{A II.15}$$

which decreases very slowly with the distance R. If we now suppose that the fixed spheres are distributed randomly on the surface S (with density $n = \frac{N}{S}$), we may compute from (A II.15) an *average cross-section:*

$$\bar{\sigma} \sim n\, v\, a^3 \int \frac{d^2R}{R^2} \sim n\, v\, a^3 \ln R \Big|_{R\,=\,\infty} \sim \infty \,. \qquad \text{(A II.16)}$$

This result is just the analogue in configuration space of the small wavenumber divergence (A II.9).

The picture shown in Fig. A II.1 makes it clear that the diverging contributions come from collision processes in which the light particle suffers a "head-on" ($\sim 180^o$) scattering on particle 3. Thus, it should not be surprising that any finite order Born approximation of the type (A II.10) does not lead to the correct divergence because, as is well known, such an approximation does not describe correctly large-angle deflections (see, for instance, Ref. 4).

We shall not enter here into the details of the various theories (see Ref. 2) which have been developed in order to introduce a cut-off in the diverging cross-sections (A II.9), (A II.10). Let us simply mention that the situation is very similar to that of plasmas and electrolytes, because important contributions can come from particles 2 and 3 even when separated by large distance, i.e., we are not allowed to neglect the presence of the other particles in the system. These considerations lead to a natural cut-off at the mean free-path $\Lambda \sim (an)^{-1}$ and one obtains, instead of (A II.16),

$$\bar{\sigma} \sim n\, v\, a^3 \ln (an)^{-1} \,, \qquad \text{(A II.17)}$$

which is a nonanalytic function of the density n. This in turn gives rise to a nonanalytic density behavior of the observable transport coefficients.

Let us conclude with one remark of general interest.

A great deal of work in nonequilibrium statistical mechanics has been based on the Bogoliubov theory (see Ref. 5) which rests heavily upon the assumption of a density expansion for transport coefficients. The nonanalyticity we have described in this section has led some people to cast doubts on the validity of the Bogoliubov method and indeed there are, *sensu stricto*, serious fundamental difficulties. We would, however, like to stress that, in our opinion, the type of theory we have developed in this course does not lead to these problems. Although our theory leads to results identical to Bogoliubov's in the limit of long-time *when the density is chosen as the expansion parameter* (see Ref. 6), the generalized master Eq. (3.76) and its asymptotic limit (3.82) result from *identities* derived from the Liouville equation, and are valid under very general conditions on the initial state of the system independently of any specific expansion parameter. The fact that a nonanalytical behavior results when the density is chosen as the expansion parameter does not seem to us to be of more fundamental importance than the similar difficulties found in the case of plasmas and electrolytes. However, it shows the great care that should be exercised in applying the general theory to specific situations and also indicates the need for rigorous proofs of the existence of the various operators, like $\Psi^{+}(z)$ which we have introduced.

References

[1] See, for instance: J. BROCAS and P. RÉSIBOIS: Bull. Ac. Sc. Belgique 47, 226 (1961). K. KAWASAKI and L. OPPENHEIM: Phys. Rev. 136 A, 1519 (1964).

[2] See the papers by: E.G.D. COHEN, I. OPPENHEIM, J. VAN LEEUWEN, at the Copenhagen Meeting on Statistical Mechanics (1966).

[3] J. STECKI: Phys. Letters 19, 123 (1965).

[4] D. BOHM: Quantum theory (Constable, 1954, London).

[5] See, for instance, E.G.D. COHEN: Fundamental Problems in Statistical Mechanics (North Holland Pub. Co., 1962, Amsterdam).

[6] J. BROCAS and P. RÉSIBOIS: Physica 32, 1050 (1966).

Interactions in a Classical Relativistic Plasma

M. Baus*

We will review in four lectures the work done on classical relativistic plasmas at the Brussels school of Professor PRIGOGINE. An introduction to the methods used together with applications to other fields of physics can be found in the lecture notes of Prigogine's course at this summer school.

This review will be, in some sense, systematic. As a result it will contain topics already some years old together with topics some weeks old. These topics are considered here because the problems related to interactions in plasmas are always nonlinear, as can be seen from the microscopic Maxwell-Lorentz equations of motion. Nevertheless we will follow a method which proved to be very fruitful in theoretical physics of this century; namely, we will first put forward as many linear features as possible, for the very simple reason that linear mathematics is a beautiful, extremely well developed subject. The theoretical reason for doing this is that from the start it allows one to distinguish the different basic physical situations with respect to irreversibility on a formal linear equation. Of course, the equations describing real situations are highly nonlinear, but at least one knows what situation one is describing. This is usually not so clear when doing approximations on nonlinear equations.

---

*Aspirant au Fonds National Belge de la Recherche Scientifique.

1. Introduction

We will describe and investigate the physics of the following system: (1) an assembly of N charged particles, i.e., a fully ionized plasma; (2) these particles are at least of two species $\left(N = \sum_\alpha N_\alpha\right)$, say electrons and ions ($\alpha = 1,2$); and (3) the particles are considered as point charges without internal degrees of freedom and are thus completely characterized by an electric charge e_α and a rest mass m_α. The system is assumed to be electrically neutral $\left(\sum_\alpha e_\alpha N_\alpha = 0\right)$. This plasma will be described in the classical relativistic region. By classical we mean that we neglect all quantum effects ($\hbar \to 0$). By relativistic we mean that:

1) We retain relativistic dynamics for particles and will thus obtain significant departures from the usual laws for sufficiently high particle velocities, that is, high temperatures. This is justified by the fact that high temperature is the natural condition for most of the known plasmas.

2) The particles interact through a field with finite propagation velocity, namely, the electromagnetic field they create plus, eventually, an initial external field incorporated in the initial condition (we will not consider here external fields present at $t \neq 0$). This means that in some way we have to deal with the electromagnetic field and thus incorporate a theory of radiation. As, moreover, this interaction introduces the Maxwell-Lorentz equations we will have to replace the classical or Galilean relativity principle by the relativistic or Lorentz relativity principle.

3) Thus, we have to guarantee that the theory is invariant under Lorentz transformations.

The joint condition, classical and relativistic, necessitates some special comment. Indeed, we have to keep in mind that whereas relativistic means high temperatures, at very high temperatures creation-annihilation phenomena may play an important role which we cannot describe. The real region of description is thus the intermediate one in which the kinetic energy of the electrons (kT_α) is not yet comparable with their rest mass energy. Moreover, the classical description of radiation is accurate only for high frequencies (ν_λ) (compared to the field energy kT_λ). Finally, we have $kT_\alpha < 2m_\alpha c^2$ and $kT_\lambda << \hbar\nu_\lambda < 2m_\alpha c^2$. Although this program corresponds to the experimental situation, it was considered only in recent years, mainly because of its complexity and because of the computational difficulties involved.

2. Lorentz-Invariant Statistical Mechanical Formalism for a Classical Relativistic Plasma Interacting with an Electromagnetic Field

a) *Introduction*

This relativistic program started only some years ago; nevertheless there exists already a spectrum of formalisms, that can be divided into two types: those using the BBKGY hierarchy or those using directly the Liouville equation for the coupled system of particles and fields with or without explicit introduction of field variables. We will use here the methods of the Prigogine school for studying directly the Liouville equation. These different methods agree on the results for problems in common, but it is too early to say what are their advantages.* Strictly speaking there is a third

*For the nonrelativistic case see, however, Prigonine's course.

kind of approach in which one tries to describe the particle interactions without the explicit use of a field. We think, however, that the theorem of D.G. CURRIE [7] (showing that if a description of a two-particle system is to be Lorentz-invariant, then the particles have constant velocity and thus no interaction) puts a point after these attempts; so that if we want interactions and Lorentz invariance then we have to face the difficulties introduced by the use of a field. A recent attempt to eliminate the self-field from the start by using the Lorentz-Dirac equation of motion together with an enlarged phase space including the accelerations has been made by R. HAKIM.[5] However, the intrinsic difficulties of such a procedure do not cancel its necessity at the present stage of development. For the relativistic extension of the classical methods presented in Refs. 1, 2, and 3 we will take as a guide the work by A. MANGENEY.[4] The Lorentz-invariant form of this theory was set up by R. BALESCU and others in Ref. 6.

b) Lorentz-Invariant Formalism

In order to realize the above relativistic program, one usually takes over the methods of relativistic mechanics, collecting all equations, making 4-vectors and scalar products out of it, so as to guarantee the explicit invariance under Lorentz transformations through the covariance properties of the equations. For the one-body equations of mechanics this works very well, but for an N-body problem one is faced with the difficulty that the fourth components of the particle positions introduce N-time components, which in one way or another have to be related to the time of the observer, i.e., the irreversible time. It has been indicated by P.A.M. DIRAC that since the relation between the accelerated particles and the observer is not a Lorentz transformation, the relation between these N + 1 times is not given by dynamics but introduces N arbitrary functions. In order to avoid such an arbitrariness for the basic concept of irreversibility theories, namely, time, we will use another method developed by DIRAC [8] and CURRIE, JORDAN and SUDARSHAN [7] in order to set up a relativistic Hamiltonian quantum theory avoiding the unsatisfactory features of all present-day covariant quantum field theories, but which can as well be fruitful in the field of classical relativistic mechanics. It is this method which was adopted by BALESCU and collaborators [6] to our problem. The main difference from previous methods is that in this method we do not use a tensorial representation of the Lorentz group, so that we can still use the usual 3-vectors and the usual phase space. Nevertheless, in any representation of the Lorentz group each quantity has a well defined transformation law. In other words, we do not restrict the Lorentz invariance to its covariant expression. We will only summarize this method here; the details can be found in the original literature. Let us separate the different levels of the requirement of relativistic invariance.

First, in physical space $x = (\underset{\sim}{x},t)$ we have to require that the laws of nature are invariant under the inhomogeneous Lorentz group of coordinate transformations (with evident notations):

$$x' = \mathcal{L}x, \quad x'_{\mu} = a_{\mu} + \Lambda^{\gamma}_{\mu}x_{\gamma}, \quad \begin{bmatrix} \mu = (0,k) \\ \mu = 0,1,2,3 \end{bmatrix}; \tag{2.1}$$

$$\mathcal{L} = (a,\Lambda), \quad \Lambda^{\mu}_{\gamma}\Lambda^{\gamma}_{\lambda} = \delta^{\mu}_{\lambda} .$$

As a change of coordinate system is equivalent to a change of reference system or a change of observer, (2.1) establishes a group of equivalent observers, i.e., observers having the same physics. More concretely this means that the laws of nature are independent of the position and the clock of the observer and thus invariant in a transformation:

$$\underset{\sim}{x} \rightarrow \underset{\sim}{x} + \underset{\sim}{a},$$
$$t \rightarrow t + \tau, \qquad (2.2)$$

expressing homogeneity of space and time (all space-time points have the same properties). Moreover, space is isotropic, i.e., $(\underset{\sim}{x},t)$ and $(R\underset{\sim}{x},t)$ with $R = (\Lambda^1_k)$ are equivalent observers. And last but not least, the laws of nature are independent of the state of motion of a system as long as it leaves Maxwell's equations invariant, i.e., observers rotated in space-time (Λ^0_k) are equivalent. All this is summarized in the requirement of invariance under the inhomogeneous Lorentz group of coordinate transformations. We will not consider here the discrete elements of this group (space or time inversion). A discussion of these can be found in a forthcoming paper by R. BALESCU.[9]

The second level is the description of the system. We will consider a classical Hamiltonian description. Then we have at our disposal the phase space Γ of the system, a vector space R of sufficiently regular functions on Γ whose elements are the dynamical quantities (particle current, field intensity,...), and a subspace S of R of density functions (particle distribution function,...). As usual, an average is defined by a bilinear functional (A,F); $A \varepsilon R$, $F \varepsilon S$. In each coordinate transformation of (2.1) a dynamical quantity will change because each dynamical quantity is defined with respect to a given frame of reference. For example, the position of a particle, $\underset{\sim}{x}_j$, is defined with respect to a given frame of reference. Each coordinate transformation in physical space introduces therefore an application of R into R (inner automorphism) of the dynamical variable in the old system to the dynamical variable in the new system. As the Lorentz group (2.1) can be built up by infinitesimal Lorentz transformations $\delta\mathcal{L} = (\delta a, \delta\Lambda)$ we have to consider only these. Each infinitesimal Lorentz transformation $\delta\mathcal{L}$ in physical space induces, therefore, an infinitesimal transformation in phase space, thus in R. In order to preserve the Hamiltonian character of the theory this infinitesimal transformation has to be an infinitesimal canonical transformation of Γ. Thus, for any A of R we have:

$$A \to A + [A,G], \qquad (2.3)$$

when

$$x \to x + \delta\mathcal{L}(x).$$

Here we expressed the canonical transformation (2.3) corresponding to $\delta\mathcal{L}$ with the help of a generating function G and the Poisson bracket (see Ref. 3). This defines a Lie-bracket operation in R which is a linear, antisymmetric, and nonassociative inner composition law. In a classical theory we have to identify it with the Poisson bracket (P.B.) associated with our phase space and its well known properties, but which in other theories can be another operation (commutator, Moyal bracket,...) having the properties mentioned and called in general a Lie bracket. To the full Lorentz group, which is a ten parameter group (see Ref. 1), will thus correspond ten generating functions. Moreover, it is easy to show that they form a subspace of R, and that the generating function of the commutator of two Lorentz transformations in physical space is the P.B. of the corresponding generating functions, so that this subspace is closed under the P.B. operation. This is mathematically summarized by saying that the canonical transformations generated by these generating functions form a Lie group and that these generating functions form an algebra (vector space plus P.B.) which is the Lie algebra of the Lie group. So that once more this beautiful chapter of linear mathematics arises in context with physics. Let us now call (for reasons which will soon become clear) H, $\underset{\sim}{P}$, $\underset{\sim}{J}$ and $\underset{\sim}{K}$ the generators of the infinitesimal canonical transformations associated with the following infinitesimal Lorentz transformations: time translation; space translation; space rotation; and space-time rotation (proper Lorentz transformation). Since the commutator of two Lorentz transformations is a Lorentz transformation, the P.B. of two generators is a generator, and one easily checks that this yields the following algebra:

$$[H,\underset{\sim}{P}] = 0; \quad [H,\underset{\sim}{J}] = 0; \quad [P_i,P_j] = 0;$$

$$[J_i,J_j] = \varepsilon_{ijk}J_k; \quad [J_i,P_j] = \varepsilon_{ijk}P_k; \quad [J_i,K_j] = \varepsilon_{ijk}K_k; \qquad (2.4)$$

$$[K_i,K_j] = -\varepsilon_{ijk}J_k; \quad [K_i,P_j] = \delta_{ij}H; \quad [\underset{\sim}{K},H] = \underset{\sim}{P};$$

where ε_{ijk} is the usual antisymmetric unit tensor. Let us call the set of these generators G. With each element T of G we can generate a one-parameter group of finite canonical transformations:

$$A \to e^{[T]s}(A) = A + [A,T]s + \ldots \; (\forall A \varepsilon R, \forall T \varepsilon G). \qquad (2.5)$$

This corresponds to the Lorentz transformation to a new reference system: which is translated an amount s in time when T = H; which is displaced by s in the j direction when $T = P_j$; which is rotated in space by an angle s about the j axis when $T = J_j$; and which is moving uniformly in the j direction with a velocity v_j = th s (in American notation th s = tanh s).

Moreover, as the canonical transformations preserve measure in phase space (they have unit Jacobian), the average value of a dynamical variable is conserved by G, i.e.:

$$(A,F) = 1\cdot(e^{[T]s}(A),e^{[T]s}(F)),(\forall A\varepsilon R,\forall F\varepsilon S,\forall T\varepsilon G), \qquad (2.6)$$

as is the subset S of density functions. Property (2.6) introduces at this stage the well known existence of different but equivalent dynamical pictures, the extremes being the Schrödinger and Heisenberg pictures. Indeed, if for a change of reference (or observer) characterized by $e^{[T]s}$, a dynamical variable A goes over into:

$$A' = e^{[T]s}(A), \qquad (2.7)$$

for a given state F of the system, then the average value $\langle A\rangle = (A,F)$ goes over into $\langle A\rangle'$:

$$\langle A\rangle' = (A',F). \qquad (2.8)$$

Then using (2.6) we have:

$$\langle A\rangle' = (A,F'), \quad F' = e^{-[T]s}(F). \qquad (2.9)$$

Thus, $A,F \to A',F$ or A,F' are equivalent descriptions of the transformation, namely its Heisenberg and Schrödinger pictures. This expresses an inner symmetry principle contained in any relativity principle, namely, that if a transformation leads from the description of a system by the original observer to the description of the same system by a new observer (for example, translated an amount s), then this transformation cannot be distinguished from the transformation leading from the original description to a description by the same observer, but of a new system (translated an amount -s). That is, if there exists a set of equivalent observers then there exists a set of equivalent systems. The usual expression of this is that the function H which generates the description of observers displaced in time is the same as the one which generates descriptions of the system displaced in time with respect to the original observer, i.e., the dynamical evolution of the system. From this point of view the content of the canonical equations, i.e., the equations of motion in phase space, is that the point representing the system at time t of the physical space moves during the time interval dt along the natural trajectory following an infinitesimal canonical transformation of generating function H and parameter t. This was first noticed by SOPHUS LIE studying infinitesimal "restricted contact transformations". A similar interpretation holds for the other elements of G. Again from this point of view relations (2.4) imply that H, $\underset{\sim}{P}$, $\underset{\sim}{J}$ are invariant along the natural trajectory, expressing thus the conservation of energy, momentum and angular momentum of the system during the motion. This already indicates what quantities we have to build in order to find the generators satisfying (2.4) and deduce $\underset{\sim}{K}$ from it. With the aid of these tools we will now indicate how the theory satisfies the Einstein-Lorentz relativity principle:

1) If A represents a dynamical quantity, F the state of a system, and (A,F) the average value of A for the system in state F (for a given reference frame), then there always exist elements A' of R and F' of S representing the same quantity (a possible state of the same system), with respect to another reference frame, such that:

$$(A,F) = (A',F') \tag{2.10}$$

(see (2.6)). Thus, to a possible average value of some dynamical quantity for a system in a given state with respect to a given reference frame, there corresponds a possible state of the same system giving the same average value for the same dynamical variable described in any other relativistic equivalent reference frame. In other words, with respect to any equivalent reference frame we can describe a system in terms of the same set of variables, states and measurements.

2) A dynamical description is transformed into a dynamical description with respect to the new reference frame. Let us introduce an abbreviation for time evolution and Lorentz transformation:

$$A(t) = e^{[H]t}(A), \qquad B' = e^{[K_1]s}(B), \tag{2.11}$$

then we have:

$$(e^{[K_1]s}e^{[H]t}(A)) = A'(t) = (e^{[K_1]s}e^{[H]t}e^{-[K_1]s}).(e^{[K_1]s}(A)), \tag{2.12}$$

or

$$A'(t) = (e^{[H]t}(A))' = e^{[H']t}(A'). \tag{2.13}$$

That is, the evolution of A under H is transformed into the evolution of the transformed value of A under the transformed Hamiltonian H':

$$H' = e^{[K_1]s} H e^{-[K_1]s} = e^{[K_1]s}(H). \tag{2.14}$$

3) The equation of motion is:

$$\partial_t A(t) = [A(t),H]. \tag{2.15}$$

Since P.B.'s are preserved by canonical transformations:

$$([A(t),H])' = [A'(t),H'], \tag{2.16}$$

we can Lorentz transform the equation of motion (2.15) and obtain:

$$\partial_t A'(t) = [A'(t),H']. \tag{2.17}$$

That is, we recover the same equation of motion as (2.15) but involving the transformed quantities, thereby proving the Lorentz invariance of the theory if (2.4) is satisfied. For a realization of (2.4), see Ref. 6 . Summarizing, we can say that with this formalism we realize a theory which is globally Lorentz-invariant and in which we know how to transform to any other equivalent observer.

The interesting feature of this formalism is that we have only one time t for all observers. As can be seen from (2.1) this implies a synchronization of the different observers, i. e., an implicit use of dynamics. This is expressed by the fact that H and $\underset{\sim}{K}$ both involve time and depend on the coupling constant (dynamics), whereas $\underset{\sim}{P}$ and $\underset{\sim}{J}$ are purely geometric transformations. This clearly shows that a Lorentz transformation for an N-body system is a dynamical problem of the same complexity as solving the Liouville equation. Thus, when the Liouville equation is not trivial to solve in some approximation, then it is not trivial to show the Lorentz invariance of this approximation. In other words, whereas the theory is globally invariant (exact H and exact $\underset{\sim}{K}$), an approximation on the dynamics (H) will in general be invariant only for an approximate Lorentz transformation. An investigation of this on the example of the relativistic Landau equation, which a 4-vector formalism proves to be invariant, is going on at present. However, we will not discuss Lorentz transformations here, but instead develop the description of our system for a given observer.

c) *Perturbation Treatment of the Liouville Equation*

In the previous section we have introduced a formalism by means of which we can study the evolution of any distribution function (or any dynamical variable) with respect to any of the ten parameters of the Lorentz group. In doing this we met two "dynamical problems", i.e., problems for which we need the solution of the canonical equations of motion and which can thus be solved for an interacting N-body system only by a perturbation treatment. These two problems were the variation of the distribution function, say ρ, with respect to time t and with respect to the (constant) average velocity $\underset{\sim}{v} = (v_1,v_2,v_3) = c(\text{ths}_1,\text{ths}_2,\text{ths}_3)$, or in other words the time evolution and the proper Lorentz transformations of ρ. For the other six variations one can solve exactly the corresponding equations of "motion" (translation and rotation). We will now introduce a perturbation treatment for the time evolution of ρ. The second dynamical problem, the proper Lorentz transformation, has different features and needs a different perturbation treatment, which is still being studied.

The evolution in time of our system within a given reference frame (for which we take all parameters zero corresponding to the identity transformations) is generated by H, the Hamiltonian of the system. We have for any density function ρ:

$$\rho(t) = e^{-[H]t}\rho(0), \tag{2.18}$$

where $\rho(0) \equiv \rho(0,\ldots,0)$ is the initial condition in our reference frame (where the parameters were zero) and $\rho(t,0,\ldots,0)$ is the value of ρ in this frame at time t of this reference frame. In what follows we will leave out the nine zero parameters and always use the Schrödinger picture. Differentiating Eq. (2.18) with respect to time we obtain the Liouville equation:

$$\partial_t\rho(t) = -[\rho(t),H] \equiv [H,\rho(t)]; \qquad \rho(t=0) = \rho(0). \tag{2.19}$$

The system we will describe is governed by the following Maxwell-Lorentz equations of motion:

$$\dot{\underset{\sim}{p}}_j = e_j \underset{\sim}{e}(j) + \frac{e_j}{c} \underset{\sim}{v}_j \times \underset{\sim}{b}(j); \qquad j = 1,\ldots,N;$$

$$\underset{\sim}{p}_j = m_j \gamma_j \underset{\sim}{v}_j; \qquad \gamma_j = (1-v_j^2/c^2)^{-(1/2)}; \tag{2.20}$$

$$\underset{\sim}{\nabla}\times\underset{\sim}{e}(\underset{\sim}{r}) + \frac{1}{c}\,\partial_t \underset{\sim}{b}(\underset{\sim}{r}) = 0;$$

$$\underset{\sim}{\nabla}\times\underset{\sim}{b}(\underset{\sim}{r}) - \frac{1}{c}\,\partial_t \underset{\sim}{e}(\underset{\sim}{r}) = \frac{4\pi}{c} \sum_{j=1}^{N} e_j \underset{\sim}{v}_j \delta(\underset{\sim}{r}-\underset{\sim}{x}_j) \equiv \frac{4\pi}{c}\,\underset{\sim}{j}(\underset{\sim}{r}); \tag{2.21a}$$

$$\underset{\sim}{\nabla}\cdot\underset{\sim}{e}(\underset{\sim}{r}) = 4\pi \sum_{j=1}^{N} e_j \delta(\underset{\sim}{r}-\underset{\sim}{x}_j) \equiv 4\pi q(\underset{\sim}{r});$$

$$\underset{\sim}{\nabla}\cdot\underset{\sim}{b}(\underset{\sim}{r}) = 0; \tag{2.21b}$$

where Eq. (2.20) is the equation of motion of the particles interacting with the electromagnetic field (without radiation reaction), $\underset{\sim}{v}_j$ is the velocity, e_j the charge, m_j the rest mass, p_j the relativistic mechanical momentum of particle j, c the vacuum phase velocity of light, and $\underset{\sim}{e}(j)$, $\underset{\sim}{b}(j)$ the microscopic electric and magnetic fields evaluated at the position $\underset{\sim}{x}_j$ of particle j. Eqs. (2.21a) are the equations of motion of the fields, $\underset{\sim}{\nabla}$ is the gradient with respect to the spatial point $\underset{\sim}{r}$ and δ is a Dirac δ-function. Eqs. (2.21b) are the usual constraints on the longitudinal fields. It is implicitly understood that all quantities $(\underset{\sim}{x}_j, \underset{\sim}{v}_j, \underset{\sim}{e}, \underset{\sim}{b})$ are taken at time t in Eq. (2.20) and Eq. (2.21).

This Newtonian description can be translated into the following Hamiltonian one:

$$\dot{\underset{\sim}{x}}_j = \frac{\partial H}{\partial \underset{\sim}{\mathscr{P}}_j} \;; \quad \dot{\underset{\sim}{\mathscr{P}}}_j = - \frac{\partial H}{\partial \underset{\sim}{x}_j} \;; \quad \dot{\xi}_\lambda = \frac{\partial H}{\partial \eta_\lambda} \;; \quad \dot{\eta}_\lambda = - \frac{\partial H}{\partial \xi_\lambda} \,. \tag{2.22}$$

Here we describe the coupled system of particles and field oscillators by the following set of canonical variables defining our phase space Γ: (1) the canonical variables for the particles $\{\underset{\sim}{x}_j, \underset{\sim}{\mathscr{P}}_j\}$ i.e., the positions $\underset{\sim}{x}_j$ and canonical momenta $\underset{\sim}{\mathscr{P}}_j = \underset{\sim}{p}_j + \frac{e_j}{c}\, \underset{\sim}{a}(\underset{\sim}{x}_j)$, where $\underset{\sim}{a}$ is the vector potential of the fields $\underset{\sim}{e}$ and $\underset{\sim}{b}$; (2) the angle and action variables $\{\xi_\lambda, \eta_\lambda\}$ for the infinite set of field oscillators. These are defined from the potentials by a Fourier expansion:

$$\underset{\sim}{a}(\underset{\sim}{r},t) = \sum_\lambda \underset{\sim}{\varepsilon}_\lambda (q_\lambda(t) \cdot e^{i\underset{\sim}{k}_\lambda \cdot \underset{\sim}{r}} + \text{c.c.}),$$

$$a_0(\underset{\sim}{r},t) = \sum_{\underset{\sim}{k}_\lambda} (q_{\underset{\sim}{k}_\lambda,0}(t) \cdot e^{i\underset{\sim}{k}_\lambda \cdot \underset{\sim}{r}} + \text{c.c.}), \tag{2.23}$$

$$q_\lambda = \left(\frac{c^2 n_\lambda}{\Omega+\nu_\lambda}\right)^{1/2} e^{-i2\pi\xi_\lambda},$$

where c.c. denotes complex conjugate, λ denotes the set of the corresponding wave vector $\underset{\sim}{k}_\lambda$ and corresponding polarization state μ_λ, ν_λ is the vacuum frequency c $[\underset{\sim}{k}_\lambda]$, Ω the normalization volume, and $\underset{\sim}{\varepsilon}_\lambda$ the polarization vectors:

$$\varepsilon_{\underset{\sim}{k},\mu} \cdot \varepsilon_{\underset{\sim}{k},\mu'} = \delta_{\mu,\mu'}, \quad \varepsilon_{\underset{\sim}{k},3} = \frac{\underset{\sim}{k}}{[\underset{\sim}{k}]} \equiv \hat{\underset{\sim}{k}}, \quad \varepsilon_{\underset{\sim}{k},0} = 1, \tag{2.24}$$

where $\mu,\mu' = (1,2,3)$. The Hamiltonian H is given by the usual expression:

$$H = \sum_j (p_{0j}+e_j a_0(j)) + \sum_\lambda \frac{\tilde{\nu}_\lambda}{2\pi} n_\lambda,$$

$$p_{0j} = (\mu_j^2 + c^2 \underset{\sim}{p}_j^2)^{1/2}, \quad \mu_j = m_j c^2, \quad \underset{\sim}{p}_j = \mathscr{P}_j - \frac{e_j}{c} a(j), \tag{2.25}$$

$$\tilde{\nu}_\lambda = \begin{bmatrix} \nu_\lambda & (\mu_\lambda = 1,2,3) \\ -\nu_\lambda & (\mu_\lambda = 0) \end{bmatrix} .$$

The Hamiltonian (2.25) is consistent with a realization of the Lie algebra of Eq. (2.4) (see Ref. 6). H represents a set of harmonic oscillators coupled through the particle term, i.e. the interaction of the field oscillators with matter gives rise to an anharmonicity. One can choose a gauge for the potentials, for example the Lorentz gauge $\partial_t a_0 + \underset{\sim}{\nabla}\cdot\underset{\sim}{a} = 0$ or the Coulomb gauge $\underset{\sim}{\nabla}\cdot\underset{\sim}{a} = 0$ (then $a_0(\underset{\sim}{r}) = \sum_j \frac{e_j}{|\underset{\sim}{r} - \underset{\sim}{x}_j|}$ and $\mu_\lambda = 1,2$). Note that in the formalism of (IIb) we do not need a Lorentz-invariant gauge (as, for example, is the Lorentz gauge) in order to obtain Lorentz-invariant results. Indeed, one can always include gauge transformations into the Lie group and obtain gauge preserving solutions. In the following we will specify the gauge according to the problem we treat.

Eq. (2.18) is a compact way of writing the exact dynamics of our system. We now introduce a statistical element into the theory. We do not specify rigorously the huge number of initial conditions of our system, which is supposed to be macroscopic. Instead we represent the initial state - not by a single point in phase space - but by a measurable set of possible points distributed according to $\rho(P;0)$. Thus $\rho(P;0)d\Gamma_P$ is the probability that the system is represented at t = 0 by a point P' belonging to the volume element $d\Gamma_P$ around P in Γ. We suppose $\rho(P;0)$ to be a real, nonnegative function, normalized to one on Γ, periodic in $\{\underset{\sim}{x}_j\}$ on the boundaries of the box in which we enclose our system and of period one in $\{\xi_\lambda\}$ (the angle variable) according to the definition of Eq.(2.23). The time evolution of this distribution function is given by Eq.(2.19) which we rewrite as:

$$i\partial_t\rho(t) = \tilde{L}\rho(t); \qquad \tilde{L} = i\,[\,H,\,] \equiv -i\,[\,H,\,], \qquad (2.26)$$

where we have introduced the Liouville operator $\tilde{L}$. This operator is Hermitian in the space of ρ-functions for which the product is integrable on Γ and which vanish at the boundaries of Γ. Physically we have to require, however, that ρ decrease sufficiently rapidly when the energy in any degree of freedom becomes infinite (i.e., when $|\underset{\sim}{v}_j| \to c$ or $\eta_\lambda \to \infty$). The operator $\tilde{L}$ will then be Hermitian for any ρ-function which represents a physical state of the system.

Finally, we will make a change of variables $\underset{\sim}{\mathscr{P}}_j \to \underset{\sim}{p}_j = \underset{\sim}{\mathscr{P}}_j - \frac{e_j}{c}\underset{\sim}{a}(j)$, thereby introducing a new phase space. Since the Jacobian of this transformation is one (a necessary but not a sufficient condition for this transformation to be a canonical transformation), the new distribution function is numerically equal to the old one and we can still denote it by ρ. As this transformation is noncanonical, the Poisson bracket defining $\tilde{L}$ is not conserved and we obtain thus a new Liouville operator which we denote by L. We perform this transformation because the perturbation theory which we will now introduce reduces to the familiar nonrelativistic one only in the mechanical momentum variables $\underset{\sim}{p}$. This follows because the transformation introduces the coupling constant and because the velocity is a function only of the mechanical momentum in both cases but a function of all canonical variables in the relativistic case.

Since we are using field variables (which permits us to do a relativistic theory of an interacting system), our phase space is infinite-dimensional. This should not bother us because in any physical problem we will use only a reduced finite-dimensional subspace.

We will now introduce a perturbation theory of the transformed Liouville equation:

$$i\partial_t \rho(t) = L\rho(t). \tag{2.27}$$

Since (2.27) is *linear* we can solve it formally by the resolvent method:

$$\rho(t) = e^{-iLt}\rho(0). \tag{2.28}$$

Using Cauchy's formula we write:

$$e^{-iLt} = \frac{1}{2\pi i}\int_C dz\, e^{-izt}\mathcal{R}(z) \equiv \mathcal{L}\left(\frac{\mathcal{R}(z)}{i}\right), \quad \mathcal{R}(z) = (L-zI)^{-1}, \qquad (2.29)$$

where $\mathcal{L}$ defines our Laplace transform, C is a contour parallel to the real axis in the upper half of the complex z plane, and I is the identity operator. If we split L and $\mathcal{R}$ into perturbed and unperturbed parts:

$$L = L_0 + \delta L, \qquad \mathcal{R} = \mathcal{R}^0(z) + (\mathcal{R}-\mathcal{R}^0),$$

where $\mathcal{R}^0(z) = (L_0-zI)^{-1}$, and then use the operator identity $A^{-1} - B^{-1} = -B^{-1}(A-B)A^{-1}$ for $\mathcal{R}$ and $\mathcal{R}^0$, we obtain the equation $\mathcal{R} - \mathcal{R}^0 = -\mathcal{R}^0\delta L\mathcal{R}$. The formal solution for $\mathcal{R}$ can be written as a perturbation series:

$$\mathcal{R}(z) = \sum_{n=0}^{\infty} \mathcal{R}^0(z)\left(-\delta L\mathcal{R}^0(z)\right)^n . \qquad (2.30)$$

Using Eqs. (2.28-2.30) we can write the formal perturbation solution of the Liouville equation as:

$$\rho(t) = \frac{1}{2\pi i}\int_C dz\, e^{-izt}\sum_{n=0}^{\infty} \mathcal{R}^0(z)(-\delta L\mathcal{R}^0(z))^n \rho(0). \qquad (2.31)$$

We now represent ρ by a Fourier representation in the space of eigenfunctions $|\phi\rangle$ of L_0:

$$|\rho\rangle = \sum_{\phi} |\phi\rangle\langle\phi|\rho\rangle, \qquad (2.32)$$

and we obtain:

$$L_0|\phi> = E_\phi|\phi> \tag{2.33}$$

where

$$\phi = \{\underset{\sim}{k}_{j_1},\ldots,\underset{\sim}{k}_{j_n};\, n_{\lambda_1},\ldots,n_{\lambda_{j_s}}\}$$

$$|\phi> = \exp i\,\Big(\sum_{j\varepsilon\phi} \underset{\sim}{k}_j\cdot\underset{\sim}{x}_j + \sum_{\lambda\varepsilon\phi} 2\pi n_\lambda \xi_1\Big)$$

$$E_\phi = \sum_{j\varepsilon\phi} \underset{\sim}{k}_j\cdot\underset{\sim}{v}_j + \sum_{\lambda\varepsilon\phi} n_\lambda \tilde{\nu}_\lambda .$$

As the oscillators are already described in Fourier space, this choice symmetrizes the situation, but what is more important, it allows us to express in a simple way the two basic assumptions of the theory. First, the initial condition for the solution of the Liouville equation $\rho(0)$ (see (2.31)) has to be such that any reduced distribution function of a finite number of degrees of freedom remains finite in the limit:

$$N \to \infty, \qquad \Omega \to \infty, \qquad \frac{N}{\Omega} \to d = \text{finite number density.}$$

This limit expresses mathematically that the system is macroscopic ($N \sim N_A$, $\Omega \sim N_A$ atomic volumes, where N_A is Avogadro's number, $N_A \sim 10^{23}$). In this limit the $\underset{\sim}{k}$ spectrum becomes continuous and one can replace everywhere $\frac{8\pi^3}{\Omega}\sum_{\underset{\sim}{k}}$ by $\int d\underset{\sim}{k}$. This is possible when one has the right number of $8\pi^3/\Omega$ factors. Therefore, we define new Fourier components as:

$$\langle\phi|\rho\rangle = \Omega^{-N} \left(\frac{8\pi^3}{\Omega}\right)^{|\phi|} \rho_\phi, \qquad |\phi| = m_\phi + \sum_{\lambda\epsilon\phi} \frac{1}{2}|n_\lambda| \tag{2.34}$$

where m_ϕ is the number of independent wave vectors $\underset{\sim}{k}_j$ in ϕ. In this limit, ρ_ϕ will be a function of N and Ω only through the finite ratio d. This property is conserved in time when assumed at t = 0 (see Ref. 2).

For the second basic assumption we neglect, at the initial time, correlations between: particle momenta (i.e., velocity correlations); action variables; and momentum-action variables. The Fourier representation introduces two types of Fourier components which evolve separately. The components ρ_ϕ with total wave vector $\underset{\sim}{k}_\phi = \sum_{j\epsilon\phi} \underset{\sim}{k}_j + \sum_{\lambda\epsilon\phi} n_\lambda \underset{\sim}{k}_\lambda$ equal zero. These components correspond to homogeneous correlations that depend only on the position differences and contain as a special member the distribution function ρ_0 of particle momenta and action variables which we will call the vacuum state of correlations. The components with $\underset{\sim}{k}_\phi \neq 0$ correspond to inhomogeneities and inhomogeneous correlations. The second hypothesis can now be written as:

$$\rho_\phi = \rho_{\phi'} \prod_{j\notin\phi'} \rho_0(\underset{\sim}{p}_j) \prod_{\lambda\notin\phi'} \rho_0(n_\lambda), \tag{2.35}$$

where ϕ' is the subset of ϕ for which $\underset{\sim}{k}_j \neq 0$ and $n_\lambda \neq 0$. In equilibrium this condition is satisfied and out of equilibrium one can show [2] that it is a consequence of a finite correlation range for particles. However, for the oscillator part it remains an assumption. This property is again conserved in time. [2]

We can now transform Eqs. (2.27) and (2.31) into Fourier space and obtain:

$$\rho_\phi(t) = \frac{1}{2\pi i}\int_C dz\, e^{-izt} \sum_{n=0}^{\infty} \sum_{\psi} \langle\phi|\mathcal{R}^0(z)(-\delta L\mathcal{R}^0(z))^n|\psi\rangle \left(\frac{8\pi^3}{\Omega}\right)^{|\psi|-|\phi|} \rho_\psi(0), \tag{2.36}$$

$$i\partial_t\rho_\phi(t) = E_\phi\rho_\phi(t) + \sum_\psi \langle\phi|\delta L|\psi\rangle\left(\frac{8\pi^3}{\Omega}\right)^{|\psi|-|\phi|} \rho_\psi(t). \tag{2.37}$$

Eqs. (2.36) and (2.37) clearly indicate how in this representation we really have a dynamics of correlations, the interactions inducing transitions between different correlation states.

The perturbation solution (2.37) can be computed with the aid of the following matrix elements (see, for example, Ref. 4):

$$\langle\phi|L_0|\phi'\rangle = E_\phi\delta_{\phi,\phi'}$$

$$\langle\phi|\mathcal{R}^0(z)|\phi'\rangle = \frac{1}{E_\phi - z}\delta_{\phi,\phi'} \tag{2.38}$$

$$\langle\phi|\delta L|\phi'\rangle = \sum_{j,\lambda} \langle\phi|\delta L_{j\lambda}|\phi'\rangle$$

$$\langle\phi|\delta L_{j\lambda}|\phi'\rangle = e_j\nu_\lambda^{-1/2}\Omega^{-1/2} \sum_{\varepsilon=\pm1} \varepsilon\left(n_\lambda^{1/2}A(\lambda,j) - \frac{\varepsilon n_\lambda}{4\pi n_\lambda^{1/2}}\Delta(\lambda,j)\right)$$

$$\delta_{\underset{\sim}{k}_j,\underset{\sim}{k}'_j-\varepsilon\underset{\sim}{k}_\lambda}\,\delta_{n_\lambda,n'_\lambda+\varepsilon} \prod_{\ell\neq j}\delta_{\underset{\sim}{k}_\ell,\underset{\sim}{k}'_\ell} \prod_{\omega\neq\lambda}\delta_{n_\omega,n'_\omega}\,.$$

For the details of the calculation see Ref. 4. The terms in the matrix element of interaction $\delta L_{j\lambda}$ are defined by:

$$\Delta(\lambda,j) = \begin{cases} 2\pi \underset{\sim}{v}_j \cdot \underset{\sim}{\varepsilon}_\lambda, & \mu_\lambda = 1,2,3 \\ -2\pi c, & \mu_\lambda = 0 \end{cases};$$

$$A(\lambda,j) = \Delta(\lambda,j) \frac{\partial}{\partial n_\lambda} + \underset{\sim}{\mathcal{F}}(\lambda,j) \cdot \underset{\sim}{\partial}_j; \tag{2.39}$$

$$\underset{\sim}{\mathcal{F}}(\lambda,j) = \begin{bmatrix} \underset{\sim}{v}_j \times (\underset{\sim}{\varepsilon}_\lambda \times \underset{\sim}{k}_\lambda) - \nu_\lambda \underset{\sim}{\varepsilon}_\lambda, & \mu_\lambda = 1,2,3 \\ c\underset{\sim}{k}_\lambda, & \mu_\lambda = 0 \end{bmatrix}.$$

In the Coulomb gauge, the scalar and longitudinal polarizations $\mu_\lambda = 0,3$, have to be replaced by the matrix elements of the Coulomb potential as given by:

$$\langle \phi | \delta L^c | \phi' \rangle = \sum_{j,n} \frac{4\pi e_j e_n}{\Omega} \frac{\underset{\sim}{k}_j' - \underset{\sim}{k}_j}{|\underset{\sim}{k}_j' - \underset{\sim}{k}_j|^2} \cdot \underset{\sim}{\partial}_{jn} \Big[\delta_{\underset{\sim}{k}_j + \underset{\sim}{k}_n, \underset{\sim}{k}_j' + \underset{\sim}{k}_n'} \prod_{\ell \neq j,n} \delta_{\underset{\sim}{k}_\ell, \underset{\sim}{k}_\ell'} \prod_\lambda \delta_{n_\lambda, n_\lambda'} \Big], \tag{2.40}$$

where

$$\underset{\sim}{\partial}_{jn} \equiv \frac{\partial}{\partial \underset{\sim}{p}_j} - \frac{\partial}{\partial \underset{\sim}{p}_n} .$$

To obtain an insight into the perturbation expression (Eq. (2.37)), which is an infinite series, it is useful to "picture" each term of the series with the help of diagrams constructed as follows:

To each matrix element $\langle \phi | \mathcal{R}^0(z) | \phi' \rangle$ or "propagator" we associate a number of superposed lines, a full line for each particle with nonvanishing wave vector $\underset{\sim}{k}_j$, a dotted line for each oscillator with nonvanishing n_λ, for example:

$$\langle \underset{\sim}{k}_j, \underset{\sim}{k}_n, \pm 1_\lambda, \pm 2_\omega | \mathcal{R}^0 | \underset{\sim}{k}_j, \underset{\sim}{k}_n, \pm 1_\lambda, \pm 2_\omega \rangle \equiv \tag{2.41}$$

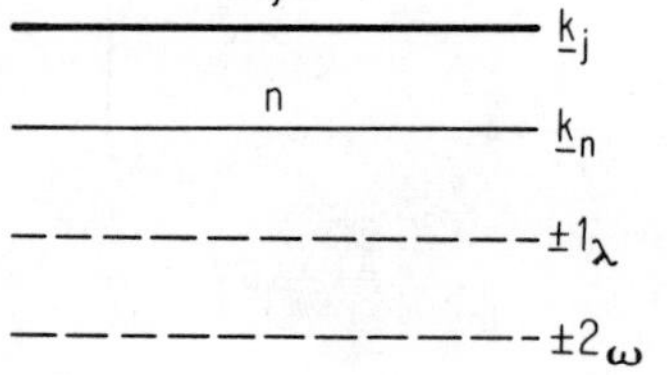

To each matrix element $\langle \phi | \delta L | \phi' \rangle$ or "vertex" we associate a vertex on the interacting lines of (2.41). The elementary vertices by means of which one can build up any vertex are:

$$\langle \underset{\sim}{k}_j | \delta L_{j\lambda} | \underset{\sim}{k}'_j, \pm 1_\lambda \rangle \equiv$$

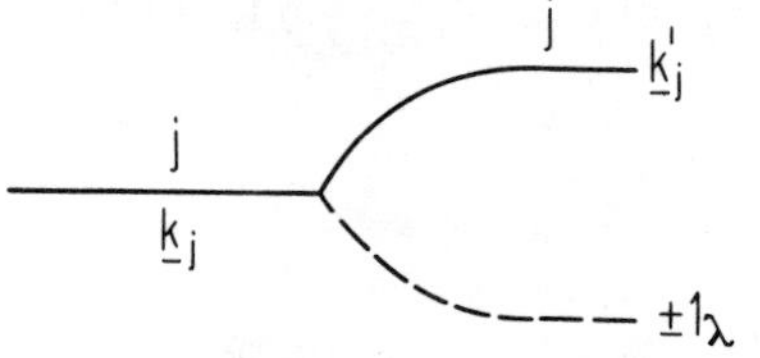

$$= \Omega^{-1/2} e_j \nu_\lambda^{-1/2} (\mp) A(\lambda, j) n_\lambda^{1/2} \delta_{\underset{\sim}{k}_j - \underset{\sim}{k}'_j \mp \underset{\sim}{k}_\lambda};$$

$$\langle \underset{\sim}{k}_j, \pm 1_\lambda | \delta L_{j\lambda} | \underset{\sim}{k}'_j \rangle \equiv \tag{2.42}$$

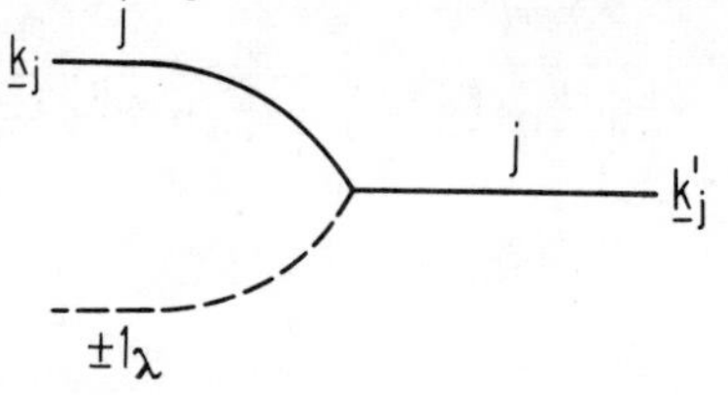

$$= \Omega^{-1/2} e_j \nu_\lambda^{-1/2} (\pm) n_\lambda^{1/2} A(\lambda, j) \delta_{\underset{\sim}{k}_j - \underset{\sim}{k}'_j \pm \underset{\sim}{k}_\lambda},$$

and the special cases in which $\underset{\sim}{k}_j$ or $\underset{\sim}{k}'_j$ vanish are:

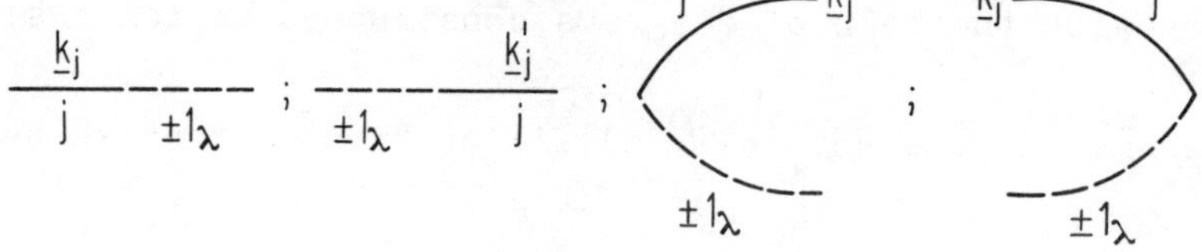

Similarly, in the Coulomb gauge we have to replace $\mu_\lambda = 0,3$ by:

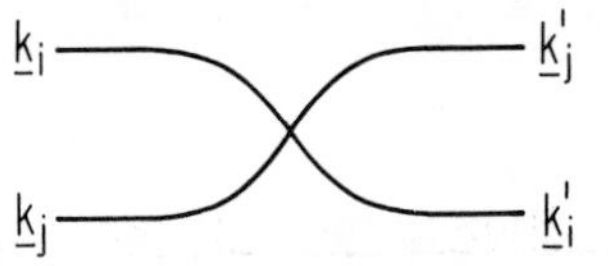

$$= \langle \underset{\sim}{k}_i, \underset{\sim}{k}_j | \delta L^c_{i,j} | \underset{\sim}{k}_i', \underset{\sim}{k}_j' \rangle$$

$$\equiv \frac{4\pi e_i e_j}{\Omega} \frac{\underset{\sim}{k}_j' - \underset{\sim}{k}_j}{|\underset{\sim}{k}_j' - \underset{\sim}{k}_j|^2} \cdot \underset{\sim}{\partial}_{ji} \delta_{\underset{\sim}{k}_i + \underset{\sim}{k}_j - \underset{\sim}{k}_i' - \underset{\sim}{k}_j'} , \qquad (2.43)$$

and the special cases with one or two vanishing wave vectors,

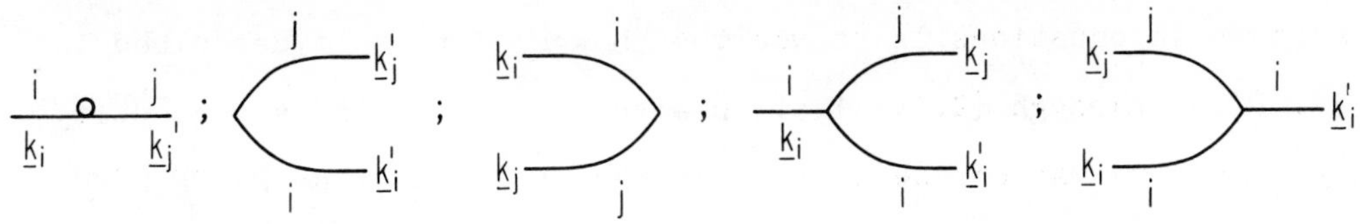

Finally, one connects the lines from the vertices to the lines from the propagators with the same label, for example:

$$\langle 0 | \delta L_{j\lambda} | \underset{\sim}{k}_j, \pm 1_\lambda \rangle \langle \underset{\sim}{k}_j, \pm 1_\lambda | \mathcal{R}^0(z) | \underset{\sim}{k}_j, \pm 1_\lambda \rangle \langle \underset{\sim}{k}_j, \pm 1_\lambda | \delta L_{j\lambda} | 0 \rangle \equiv$$

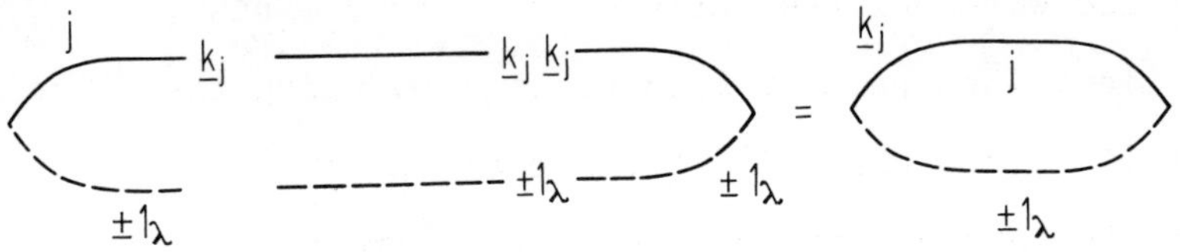

(2.44)

or a less simple one,

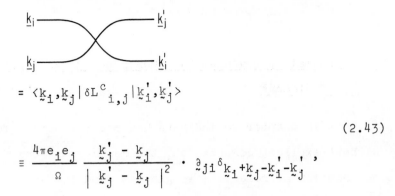

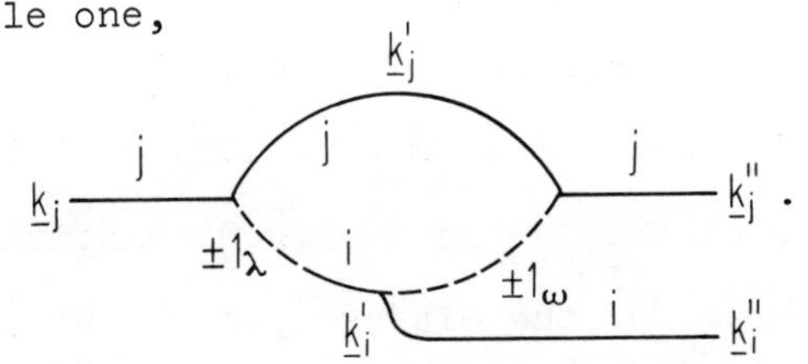

(2.45)

In interpreting these diagrams time runs from right to left. At constant time all degrees of freedom which are in an inhomogeneous or fluctuating state are indicated. These degrees of freedom are correlated at a time instant, and going from right to left one can follow the dynamics of these correlations in time, as they are created out of, or destroyed into, the vacuum of correlations.

3. Formal Approach to the Phenomenological Electrodynamics of Plasmas

In this chapter we indicate how one can derive formally, i.e., without explicit computation, the basic equations for a macroscopic description of relativistic plasmas (and in fact for more complicated media) and see how far they coincide with those used in phenomenological electrodynamics. The simplest procedure for obtaining macroscopic equations is to use the change of picture described in Eqs. (2.6) through (2.9), i.e., use the fact that for any quantity A the average $\langle A\rangle$ can be written in the SCHRÖDINGER or HEISENBERG picture, i.e.:

$$\langle A\rangle(t) = \int d\Gamma\, A(0)\rho(t) \equiv \int d\Gamma\, A(t)\rho(0). \qquad (3.1)$$

This is valid whenever the surface terms in Γ vanish, i.e., whenever the Liouville operator L is Hermitian. Using the Heisenberg picture (A(t)) (see Chap. 1), the evolution equation for $\langle A\rangle$ is seen to be the average of the microscopic evolution equation of A(t). Thus, the average or macroscopic fields $\underset{\sim}{E} \equiv \langle\underset{\sim}{e}\rangle$, $\underset{\sim}{B} \equiv \langle\underset{\sim}{b}\rangle$ will obey Maxwell's equations of the form of (2.21) with the microscopic density and current replaced by the averaged quantities $\langle\rho\rangle$ and $\langle j\rangle$. That is, the average of the microscopic electric and magnetic fields $\underset{\sim}{e}$ and $\underset{\sim}{b}$ is interpreted as the electric field $\underset{\sim}{E}$ and the magnetic induction $\underset{\sim}{B}$ of the usual Maxwell equations. The same conclusion can be derived in the Schrödinger picture where the hermiticity condition of L appears explicitly [10]. Now one can do the same for the equations of motion of particles, for the energy-momentum balance equations, as well as for charge conservation, and we obtain:

$$\partial_t \langle \sum_j p_{0j} \delta(\underset{\sim}{r}-\underset{\sim}{x}_j)\rangle + \underset{\sim}{\nabla} \cdot \langle \sum_j \underset{\sim}{p}_j \delta(\underset{\sim}{r}-\underset{\sim}{x}_j)\rangle = \langle \underset{\sim}{e}(\underset{\sim}{r}) \cdot \underset{\sim}{j}(\underset{\sim}{r})\rangle; \qquad (3.2)$$

particle momentum balance equation:

$$\partial_t \langle \sum_j \underset{\sim}{p}_j \delta(\underset{\sim}{r}-\underset{\sim}{x}_j)\rangle + \underset{\sim}{\nabla} \cdot \langle \sum_j \underset{\sim}{v}_j \underset{\sim}{p}_j \delta(\underset{\sim}{r}-\underset{\sim}{x}_j)\rangle = \langle q(\underset{\sim}{r}) \underset{\sim}{e}(\underset{\sim}{r}) + \frac{1}{c} \underset{\sim}{j}(\underset{\sim}{r}) \times \underset{\sim}{b}(\underset{\sim}{r})\rangle; \qquad (3.3)$$

field energy balance equation:

$$\frac{1}{8\pi} \partial_t \langle \underset{\sim}{e}^2(\underset{\sim}{r}) + \underset{\sim}{b}^2(\underset{\sim}{r})\rangle + \frac{c}{4\pi} \underset{\sim}{\nabla} \cdot \langle \underset{\sim}{e}(\underset{\sim}{r}) \times \underset{\sim}{b}(\underset{\sim}{r})\rangle = -\langle \underset{\sim}{e}(\underset{\sim}{r}) \cdot \underset{\sim}{j}(\underset{\sim}{r})\rangle; \qquad (3.4)$$

field momentum balance equation:

$$\frac{1}{4\pi c} \partial_t \langle \underset{\sim}{e}(\underset{\sim}{r}) \times \underset{\sim}{b}(\underset{\sim}{r})\rangle + \frac{1}{4\pi} \underset{\sim}{\nabla} \cdot \langle \frac{\underset{\sim}{e}^2(\underset{\sim}{r}) + \underset{\sim}{b}^2(\underset{\sim}{r})}{2} \underset{\approx}{1} - \underset{\sim}{e}(\underset{\sim}{r}) \underset{\sim}{e}(\underset{\sim}{r}) - \underset{\sim}{b}(\underset{\sim}{r}) \underset{\sim}{b}(\underset{\sim}{r})\rangle = -\langle q(\underset{\sim}{r}) \underset{\sim}{e}(r) + \frac{1}{c} \underset{\sim}{j}(\underset{\sim}{r}) \times \underset{\sim}{b}(\underset{\sim}{r})\rangle; \qquad (3.5)$$

charge conservation:

$$\partial_t \langle q(\underset{\sim}{r})\rangle + \underset{\sim}{\nabla} \cdot \langle \underset{\sim}{j}(\underset{\sim}{r})\rangle = 0; \qquad (3.6)$$

where all quantities are evaluated at time t. The macroscopic Lorentz force can be obtained from Eq. (3.3) by integrating over $\underset{\sim}{r}$ and putting in evidence the summation sign on j. The characteristic feature of Eqs. (3.2)-(3.5) with respect to the Maxwell equations is that they are quadratic in the averaging procedure, i.e., they contain field-matter correlations. However, the balance equations for the field are peculiar in the sense that they are in fact a superposition of balance equations for the average and fluctuating fields:

$$\underset{\sim}{e} = \langle\underset{\sim}{e}\rangle + \delta\underset{\sim}{e}; \quad \langle\underset{\sim}{e}^2\rangle = \langle\underset{\sim}{e}\rangle^2 + \langle\delta\underset{\sim}{e}^2\rangle; \quad \text{etc.} \tag{3.7}$$

Since $\underset{\sim}{E} = \langle\underset{\sim}{e}\rangle$ and $\underset{\sim}{B} = \langle\underset{\sim}{b}\rangle$ obey the Maxwell equations, we can use these to derive the equation:

$$\frac{1}{8\pi}\,\partial_t(\underset{\sim}{E}^2(\underset{\sim}{r})+\underset{\sim}{B}^2(\underset{\sim}{r})) + \frac{c}{4\pi}\,\underset{\sim}{\nabla}\cdot(\underset{\sim}{E}(\underset{\sim}{r})\times\underset{\sim}{B}(\underset{\sim}{r})) = -\underset{\sim}{E}(\underset{\sim}{r})\cdot\langle\underset{\sim}{j}(\underset{\sim}{r})\rangle. \tag{3.8}$$

Subtracting Eq. (3.8) from Eq.(3.4), we obtain a balance equation for the energy of the fluctuating fields:

$$\frac{1}{8\pi}\,\partial_t\langle\delta\underset{\sim}{e}^2(\underset{\sim}{r})+\delta\underset{\sim}{b}^2(\underset{\sim}{r})\rangle + \frac{c}{4\pi}\,\underset{\sim}{\nabla}\cdot\langle\delta\underset{\sim}{e}(\underset{\sim}{r})\times\delta\underset{\sim}{b}(\underset{\sim}{r})\rangle = -\langle\delta\underset{\sim}{e}(\underset{\sim}{r})\cdot\delta\underset{\sim}{j}(\underset{\sim}{r})\rangle, \tag{3.9}$$

and similarly for the momentum balance equation (3.5).

Thus, as a direct consequence of the linear character of Maxwell's equations and the quadratic nonlinearity of the balance equations with respect to the averaging procedure, we obtain the peculiar situation where the energy and momentum of the average fields depend only on the average sources and not on the correlations at time t. Furthermore, the energy and momentum of the fluctuating fields depend only on the fluctuating sources and the correlations at time t and not on the average quantities, i.e., there is a superposition at time t of correlations and average quantities. However, for the particle balance equations, (3.2) and (3.3), the situation is not so favorable because the microscopic equations of motion already contain correlations (one has to take the fields at the position of the particle). Therefore, we have a correlational contribution to the macroscopic density of the Lorentz force. For example, from Eq.(3.3):

$$\langle q(\underset{\sim}{r})\underset{\sim}{e}(\underset{\sim}{r})\rangle = \langle q(\underset{\sim}{r})\rangle\underset{\sim}{E}(\underset{\sim}{r}) + \langle \delta q(\underset{\sim}{r})\delta\underset{\sim}{e}(\underset{\sim}{r})\rangle. \qquad (3.10)$$

In phenomenological electrodynamics one uses Eq. (3.8) and the first term of Eq. (3.10) and similar terms for the other balance equations. While Eq. (3.8) is an exact equation it does not represent the energy in the field, but only that part of it which is independent of the correlations at time t. It can therefore be used only for a collisionless inhomogeneous plasma (in a homogeneous plasma there are only fluctuating fields). On the other hand, equations containing only the first term of (3.10) are not exact, but contain an approximation of the same kind as using the exact Eq. (3.8) seperately from Eq. (3.4) of which it is a part.

The connection with phenomenological electrodynamics can be pushed further by introducing polarizations. This is performed by splitting the sources into external or given sources and "media" or induced sources ($\underset{\sim}{j} = \underset{\sim}{j}^{ext} + \underset{\sim}{j}^{ind}$, etc. ...). The induced sources are then eliminated from the macroscopic Maxwell equations by introducing new fields $\underset{\sim}{P}$, $\underset{\sim}{M}$, usually called electric polarization and magnetization. However, this four-field formulation is redundant and for theoretical purposes it is often useful to introduce only one new field $\underset{\sim}{D}$, the generalized electric induction, not to be confused with the usual electric induction defined in terms of $\underset{\sim}{P}$ (see Refs. 10 and 11). From the statistical point of view these systems of equations are closed. In a phenomenological theory one tries to solve the field equations without having to solve the particle equations.

This is done by assuming some relations between the fields; e.g., in linear electrodynamics we have in the Fourier-Laplace space $\underset{\sim}{D}_{\underset{\sim}{k}}(\omega) = \underset{\approx}{\varepsilon}_{\underset{\sim}{k}}(\omega)\cdot\underset{\sim}{E}_{\underset{\sim}{k}}(\omega)$ which is the most general linear relation between these two fields for a medium whose properties are homogeneous in space and time and which exhibits temporal and spatial dispersion. In a statistical theory, however, one can compute these coefficients ($\underset{\approx}{\varepsilon}$ as well as the nonlinear ones). One can now rewrite the balance equation (3.8) for the average fields so as to include the polarization fields, i.e., by transforming from the average vacuum normal modes to the real dispersive normal modes of the medium, with the restriction that they have to exist (as they do in a "weakly dispersive" medium). This is done in detail in Refs. 11 and 12. This is as far as one can go in order to connect the microscopic description to the macroscopic one without using analytic methods. For any treatment of the fluctuating fields as well as for any explicit evaluation of the different quantities one has to use perturbation theory explicitly. We will illustrate this for some processes in the next chapter.

4. Kinetic Approach to the Electrodynamics of Homogeneous Systems

In this and subsequent chapters we will develop a kinetic approach for *homogeneous* systems, thus excluding all hydrodynamic aspects of matter and radiation. In such a system there exist no average fields ($\langle\underset{\sim}{e}\rangle = 0 = \langle\underset{\sim}{b}\rangle$) or densities ($\langle j_\mu\rangle = 0$), so that only the fluctuating fields will act during a collision process. Our system will thus be completely described by the subclass of Fourier components ρ_ϕ for which $\underset{\sim}{k}_\phi = 0$, i.e., the distribution functions for particle momenta $\rho_0(\underset{\sim}{p}_j)$ and action variables $\rho_0(\eta_\lambda)$ and the correlations $\rho_\phi(\underset{\sim}{k}_\phi = 0)$. In analogy with the Boltzmann equation, one can derive an equation governing the general time evolutions of ρ_0 (see also Prigogine's lecture notes).

a) Kinetic Equations

Consider Eq. (2.37) for $\phi \equiv 0$ and use Eq. (2.36) to obtain:

$$i\partial_t \rho_0(t) = \sum_{\phi} \langle 0|\delta L|\phi\rangle \left(\frac{8\pi^3}{\Omega}\right)^{|\phi|} \frac{1}{2\pi i}\int_C dz\, e^{-izt} \sum_{n=0}^{\infty} \sum_{\psi}$$

$$\langle \phi|\mathscr{R}^0(-\delta L\mathscr{R}^0)^n|\psi\rangle \left(\frac{8\pi^3}{\Omega}\right)^{|\psi|-|\phi|} \rho_\psi(0),$$

or

$$i\partial_t \rho_0(t) = -\frac{1}{2\pi i}\int_C dz\, e^{-izt} \sum_{n=1}^{\infty} \sum_{\phi}$$

$$\langle 0|(-\delta L\mathscr{R}^0)^n|\phi\rangle\left(\frac{8\pi^3}{\Omega}\right)^{|\phi|} \rho_\phi(0), \qquad (4.1)$$

where all distribution functions are nonreduced ones. In terms of diagrams one can represent a general contribution to this perturbation series as:

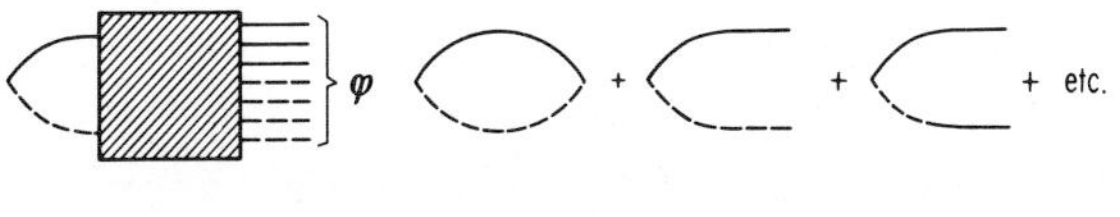

(4.2)

where the box in (4.2) can contain any contribution. We now split these contributions into two parts:

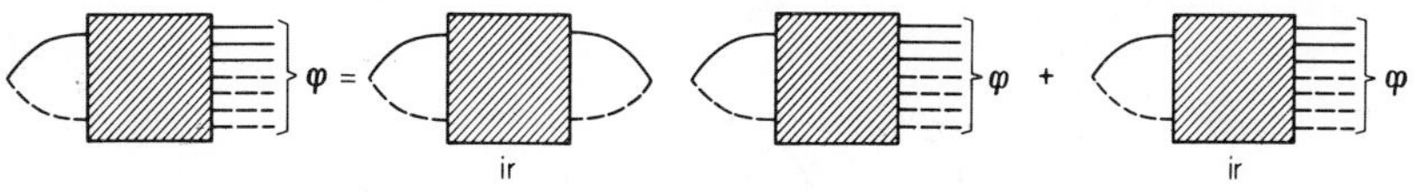

(4.3)

where we have separated out all "irreducible" contributions (the second term of (4.3), denoted by ir), i.e., all the contributions from transitions from ϕ at t = 0 to the correlational vacuum state at time t without ever passing again through a correlational vacuum state at some intermediate time. In the reducible contributions (first term of (4.3)) we have indicated the last (in time) vacuum state, putting thus in evidence an irreducible vacuum-vacuum transition as indicated in (4.3). Substituting (4.3) in (4.1), we obtain:

$$i\partial_t\rho_0(t) = -\frac{1}{2\pi i}\int_C dz\, e^{-izt}\Bigg[\sum_{n=1}^{\infty}\langle 0|(-\delta L\rho^0)^n\delta L|0\rangle_{ir} \cdot \sum_{\substack{n'=0\\ \phi}}^{\infty}\langle 0|\mathcal{R}^0(-\delta L\mathcal{R}^0)^{n'}|\phi\rangle\left(\frac{8\pi^3}{\Omega}\right)^{|\phi|}\rho_\phi(0) + \sum_{\substack{n=1\\ \phi}}^{\infty}\langle 0|(-\delta L\mathcal{R}^0)^n|\phi\rangle_{ir}\left(\frac{8\pi^3}{\Omega}\right)^{|\phi|}\rho_\phi(0)\Bigg], \tag{4.4}$$

or in compact form

$$i\partial_t\rho_0(t) = \frac{-1}{2\pi i}\int_C dz\, e^{-izt}\,\{\tfrac{1}{i}\,\Psi(z)i\rho_0(z) + D(z)\}\ . \tag{4.5}$$

Using our definition (2.29) of the Laplace transform, we obtain from Eq. (4.5):

$$\begin{aligned}\partial_t\rho_0(t) &= \mathcal{L}(\psi(z)\rho_0(z)+D(z))\\ &= \int_0^t d\tau\ \Psi(\tau)\rho_0(t-\tau) + D(t),\end{aligned} \tag{4.6}$$

where $\Psi(t)$ and $D(t)$ are the Laplace transforms of $\Psi(z)$ and $D(z)$ defined from Eqs. (4.4) and (4.5) as:

$$\Psi(z) = i \sum_{n=1}^{\infty} \langle 0|\delta L(-\mathcal{R}^0(z)\delta L)^n|0\rangle_{ir}$$

$$D(z) = \sum_{\phi\neq 0} \sum_{n=1}^{\infty} \langle 0|(-\delta L\mathcal{R}^0(z))^n|\phi\rangle_{ir} \left(\frac{8\pi^3}{\Omega}\right)^{|\phi|} \rho_\phi(0) \qquad (4.7)$$

$$= \sum_{\phi\neq 0} D_\phi(z).$$

Eq. (4.6), usually called "master equation", has several remarkable features:

1) It shows that the time evolution of ρ_0 consists of two additive parts. The first part depends only on ρ_0 and not on the correlations. It contains an integration over the past history of the system, and yields a nonMarkoffian contribution to $\partial_t\rho_0(t)$. The second part depends on the initial value of the correlations and not on ρ_0. This strongly contrasts with the BBKGY hierarchy (in our case the hierarchy is given by Eq. (2.37) without making use of Eq. (2.36)) where $\partial_t\rho_0$ depends only on the binary correlations, but at time t. For a given approximation, Eq. (4.6) remains but simplifies (see below), whereas the hierarchy has to be truncated, assuming that correlations are asymptotically a functional of ρ_0 and estimating the order of the various terms of the hierarchy with respect to the given approximation. In the oscillator formalism of the relativistic case, these estimates proved to be delicate, leading to errors (see Ref. 13). Note, moreover, that it will not be allowed in the future to derive a hierarchy from:

$$\frac{\partial}{\partial s}\,\rho(s) = [\rho(s), K_1], \tag{4.8}$$

which is the analogue of the Liouville equation for Lorentz transformations (see Chap. 2b). This follows, because in this case one cannot develop an asymptotic hierarchy theory in s that can be truncated (except for the ultra-relativistic case: $v \to c$, $s \to \infty$).

2) In order to obtain known results (for example, the Boltzmann equation) from Eq. (4.6), it is known by detailed investigation (see Refs. 1, 2 and 3) that one has to assume:

$$\left.\begin{array}{l} \Psi(t) \approx 0 \\ \\ D(t) \approx 0 \end{array}\right\} \quad \text{for} \quad t >> t_c. \tag{4.9}$$

For systems to which the results apply, (4.9) can be shown to be valid in the lowest order in the perturbation expansion. For our case, the collision time, t_c, is of the order of the inverse plasma frequency ω_p^{-1} for a stable plasma with a correlation range of the order of the inverse Debye length L_D^{-1}.

3) For $t >> t_c$, when (4.9) is fulfilled one can write:

$$\begin{aligned} \partial_t \rho_0(t) &\approx \int_0^t d\tau\ \Psi(\tau)\rho_0(t-\tau) \approx \int_0^\infty d\tau\ \Psi(\tau)\rho_0(t-\tau) \\ &= \Psi(+i0)\rho_0(t) + \ldots = \Omega(+i0)\Psi(+i0)\rho_0(t). \end{aligned} \tag{4.10}$$

Thus, ρ_0 satisfies a closed equation containing an asymptotic collision operator $\Psi(+i0)$ and the asymptotic effects of the nonMarkoffian character, i.e., the finite duration of the collisions, through the

Ω operator ($\Omega = 1+i\Psi'+\ldots$).

4) Eq. (4.6) is reversible because it is strictly equivalent to a projection onto $|0>$ of the Liouville equation. Now the main goal of statistical mechanics is to explain the nonreversible character of macroscopic systems. It is a nice feature of the master equation that the property which introduces irreversibility is separated out in D(t). Namely, if for sufficiently long times $D(t)\to 0$, then the equation becomes irreversible, because for such times one can use $\Psi(+i0)$ (see Eq. (4.10)) and show that this term drives ρ_0 to the canonical distribution. Moreover, for such times the correlations become functions of ρ_0 only, and one obtains the thermodynamic situation. It is important to note that this crucial property is by no means a necessary property of a physical system. Roughly speaking, D(t) depends on the range of the potential (t_c) and on the range of the initial correlations (t_{corr}), whereas $\Psi(t)$ depends only on the potential. Thus, correcting (4.9), the crucial properties are:

$$
\begin{aligned}
\Psi(t) &\approx 0 \qquad t >> t_c \\
D(t) &\approx 0 \qquad t >> t_c, t_{corr}.
\end{aligned}
\tag{4.11}
$$

These conditions can be violated in two extreme situations. In the case of a long-range potential (for the Coulomb potential we know the effective potential is short-range due to screening but for gravitation there is no screening) we have $t_c \approx \infty$ and in the case of long-range correlations (for example, some kind of turbulent state) we have $t_{corr} \approx \infty$. Finally, we can also violate this condition by a special preparation of the system, for example, take $\rho_\phi(0) = 0$ and let the system evolve an arbitrarily long time t_1;

reverse the velocities and magnetic fields and take this state as a new initial condition. Now as the correlations in this new initial condition were created out of ρ_0 on a macroscopic time t_1, D(t) will decay on a macroscopic time $t_1 >> t_c, t_{corr}$ and $D(2t_1) = 0$. So one can say that irreversibility is in some sense a property true "on the average"; in other words it refers to a certain class of systems.

5) Reducing the master equation to one particle, say j, or one oscillator, say λ, we obtain the coupled system of equations:

$$\partial_t \rho_0(\underset{\sim}{p}_j;t) = C_j \rho_0(\underset{\sim}{p}_j;t) + D_j(t), \qquad C_\alpha = \int_\alpha d\Gamma \int_0^t d\tau \, \Psi(\tau) e^{-\tau \partial_t} \rho_0^{(\alpha)}(t),$$

$$\partial_t \rho_0(n_\lambda;t) = C_\lambda \rho_0(n_\lambda;t) + D_\lambda(t), \qquad D_\alpha(t) = \int_\alpha d\Gamma \, D(t), \tag{4.12}$$

where C_α (α= j or λ) contains an integration over the whole phase space except α and where $\rho_0^{(\alpha)}$ is α independent. If for sufficiently long times, say $t >> t_c$, one can neglect $D_\alpha(t)$, then the system (4.12) becomes closed, yielding a coupled system of general kinetic equations. We will investigate this system using an expansion procedure in a small parameter.

b) General Description of a Plasma Within the Ring Approximation

In the case of plasmas, the parameter which is small for most of the experimental conditions is the inverse of the number of particles in a Debye sphere n_D^{-1}:

$$n_D = dL_D^3 = \left(\frac{kT}{e^2 d^{1/3}}\right)^{3/2} ; \quad L_D^{-1} = \left(\frac{e^2 d}{kT}\right)^{1/2}$$

where e is the charge, d is the number density, T is the temperature, k is the Boltzmann constant and L_D is the Debye length. For a multiple species plasma, this parameter, n_D^{-1} is small provided the ion terms are of the same order as the electron terms. It is also small for low density, high temperature plasmas, and thus especially for low density relativistic plasmas. The expansion parameter n_D^{-1} is a power series in the coupling constant e^2:

$$n_D^{-1} \sim e^2 \sqrt{e^2 d} \sim e^2(e^2 d)^m, \quad \text{all } m.$$

As we will see, this indicates the importance of collective processes within this approximation. The pth order approximation consists in a summation of all diagrams of order $n_D^{-p} \sim e^{2p}(e^2 d)^q$, all q. In the first approximation, i.e., for ring diagrams (p = 1), we will obtain the relativistic extension of the Balescu type of equations. In the next paragraph we will consider some typical n_D^{-2} contributions. In this paragraph we will treat the n_D^{-1} approximation completely, starting, however, with the first density corrections (p = 1,q = 0,1) before including the collective effects (p = 1, all q). Note finally that $\mathcal{O}(C_\lambda) = d\,\mathcal{O}(C_j)$ for the same diagram; therefore we always give the order $\mathcal{O}$ of the more familiar C_j when speaking of the order of a contribution.

b.1 Self-energy. Consider the zero order density correction (p = 1,q = 0). The corresponding contribution is (see 2.44):

$$\Psi(z) = -i\langle 0|\delta L \mathcal{R}^0(z)\delta L|0\rangle_{ir} =$$ 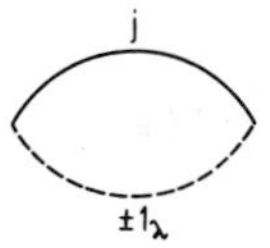

. (4.13)

This diagram has no classical analogue because a classical vertex by itself is already of order e^2 because of the Coulomb potential.

It corresponds to the contribution to the evolution of $\rho_0(j)$ and $\rho_0(\lambda)$ due to a transitory correlation between the particle (j) and the field (oscillator λ). The interactions create a fluctuation in the particle density and a fluctuating field, thus correlating the motion of the particle to the motion of the field determined by this particle. This one-body process corresponds to the self-interaction or self-energy of the particle. Indeed, let us write the contribution of this diagram to the particle equation (see IIc):

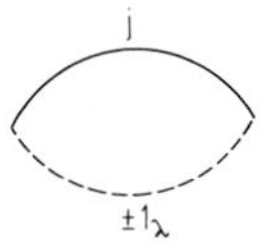

$$= -i \sum_{\lambda} \sum_{\pm} (e_j \Omega^{-1/2} \nu_\lambda^{-1/2} (\mp) A(\lambda,j) \eta_\lambda^{1/2})$$

$$\frac{1}{E_\lambda + E_j - z} ((\pm)\eta_\lambda^{1/2} A(\lambda,j) e_j \nu_\lambda^{-1/2} \Omega^{-1/2}) \, \delta_{\underset{\sim}{k}_j, \mp \underset{\sim}{k}_\lambda}$$

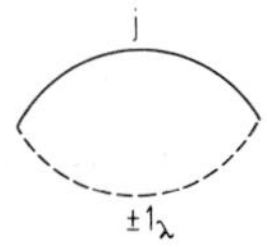

$$= -i \sum_{\lambda} \sum_{\pm} e_j^2 \Omega^{-1} \nu_\lambda^{-1} A(\lambda,j) \frac{\eta_\lambda}{\pm\tilde{\nu}_\lambda \mp \underset{\sim}{k}_\lambda \cdot \underset{\sim}{v}_j - z} A(\lambda,j). \tag{4.14}$$

For the contribution to C_λ, replace $\sum_\lambda$ by $\sum_j$. We see that for a finite duration of the field-particle collision ($z \neq 0$) and a continuous spectrum, we have a $\underset{\sim}{k}_\lambda$ integration of the form $\int_0^\infty dk_\lambda \, k_\lambda \ldots$ (where $k_\lambda = |\underset{\sim}{k}_\lambda|$), and thus a divergence for large k_λ as characteristic for the self-energy problem. For the long-time limit $z \to +i0$ we have:

$$\lim_{z \to +i0} \frac{1}{a-z} = i\pi\delta_-(a) = \mathscr{P}\left(\frac{1}{a}\right) + i\pi\delta(a), \tag{4.15}$$

and in our case: $a = \pm(\tilde{\nu}_\lambda - \underset{\sim}{k}_\lambda \cdot \underset{\sim}{v}_j)$. But as $|\underset{\sim}{v}_j| < c$, $\delta(a)$ cannot be satisfied, ($\rho_0 = 0$ for $|\underset{\sim}{v}_j| = c$), and there remains:

$$\sum_{\pm} \frac{1}{\pm(\tilde{\nu}_\lambda - \underset{\sim}{k}_\lambda \cdot \underset{\sim}{v}_j)} \equiv 0. \tag{4.16}$$

As this discussion is based only on the property of the propagator, the same will be true for D(t). Asymptotically, in this approximation one concludes, the particle momenta and the field energy remain constant. A more detailed investigation would show that these contributions vanish for times long compared to the Lorentz time t_L, which is the time necessary for light to cross the classical electron radius, $ct_L = \frac{e^2}{mc^2}$. As $t_L \ll t_c \ll t$ this term does not contribute at the Markoffian kinetic level, which is the level of interest when speaking of kinetic equations in the usual sense. However, this divergence will appear each time we consider the motion of a specific particle. One can introduce a Lorentz-invariant cut-off in the k_λ integral. This leads to the theory of extended particles with a renormalized mass and as first approximation the Lorentz-Dirac equation of motion[14].

b.2 Relativistic Landau Equation. Consider now the first density correction (q = 1) to the first order approximation (p = 1), i.e., all diagrams of order $e^2(e^2d)$ containing four vertices (e^4) and two particles (d). The irreducible diagrams contributing to the collision operator at this order are:

(I) 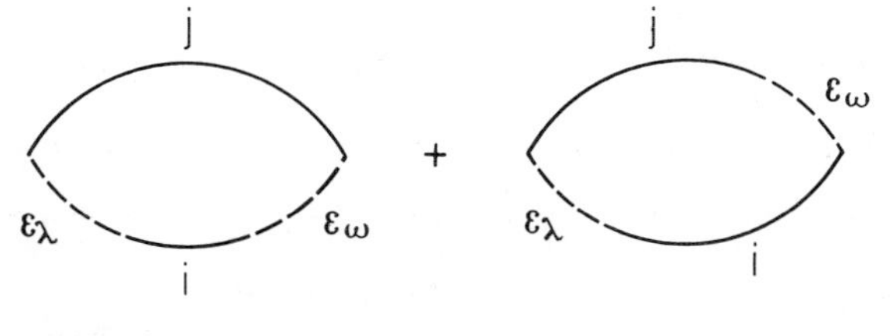 (4.17)

(II) 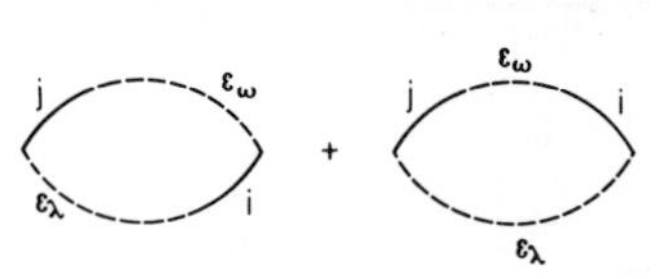 (4.18)

(III) 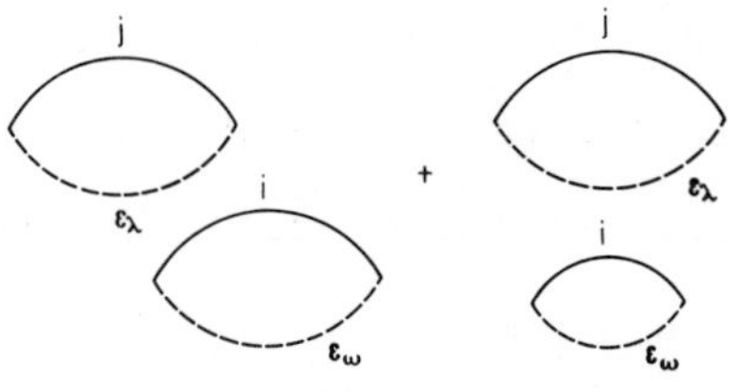 (4.19)

We split the diagrams into three classes corresponding to three different processes. Now we will have to sum diagrams and the problem becomes nontrivial. We focus our attention only on the asymptotic contributions to the nonMarkoffian kinetic equation, i.e., we are interested in times $t >> t_c = \omega_p^{-1}$ and suppose $\psi(t)$ and $D(t)$ to vanish on such time scales, so that the quantity of interest is $\psi(+i0)$. Class III consists of nonconnected diagrams and will thus vanish, except if $\lambda = \omega$ ($i \neq j$, otherwise we leave our approximation $\sim d$). Indeed if $i \neq j$, $\lambda \neq \omega$ we obtain a vanishing surface term at the (i,ω) vertex:

$$\ldots \int dp_i \int d\eta_\omega \, A(i,\omega)\eta_\omega^{1/2} \ldots \rho_0(i,\omega) \equiv 0. \qquad (4.20)$$

When $\lambda = \omega$, $\varepsilon_\lambda = -\varepsilon_\omega$, class III reduces to class I because one can add the field lines algebraically. The remaining case $\lambda = \omega$, $\varepsilon_\lambda = \varepsilon_\omega$ will vanish asymptotically because the central propagator will yield a delta-function of positive argument. Class III corresponds thus only to transient (short-time) processes (in fact,twice a self-energy diagram). Similarly, the central propagator of the diagram of (4.18) will yield a vanishing delta-function or a reducible contribution. This indicates that there is no (asymptotic) two-mode coupling.

The remaining diagrams of (4.17) correspond to the establishment of a correlation between the two particles, each particle interacting with the electromagnetic field created by the other one. They correspond thus to the classical "cycle" diagram  but with the Coulomb interaction  replaced by a retarded interaction  through the field. Their summation is straightforward[4,15], yielding an asymptotic collision operator (see Eqs. (4.4) and (4.10)):

$$\Psi_I(+i0) = \sum_{i,j,\lambda} 16\pi^3 \left(\frac{e_i e_j}{\Omega}\right)^2 \mathbf{k}\cdot\boldsymbol{\partial}_{ij}\sigma_{ij}^2\delta(\mathbf{k}\cdot\mathbf{v}_{ij})\mathbf{k}\cdot\boldsymbol{\partial}_{ij}, \tag{4.21}$$

from which we obtain the kinetic equations:

$$\partial_t\rho_0(\mathbf{p}_j;t) = \sum_{\alpha=1,2} 2e_j^2e_\alpha^2 d_\alpha \int d\mathbf{p}_\alpha \int d\mathbf{k}\; \mathbf{k}\cdot\boldsymbol{\partial}_j\sigma_{j\alpha}^2\delta(\mathbf{k}\cdot\mathbf{v}_{\alpha j})$$

$$\mathbf{k}\cdot\boldsymbol{\partial}_{j\alpha}\rho_0(\mathbf{p}_\alpha;t)\rho_0(\mathbf{p}_j;t), \tag{4.22}$$

$$\partial_t\rho_0(\eta_\lambda;t) = 0,$$

with the cross section:

$$k^2\sigma_{ij}^2 = k^2\left[\frac{c^2 - \mathbf{v}_i\cdot\mathbf{v}_j}{(ck)^2 - (\mathbf{k}\cdot\mathbf{v}_j)^2}\right]^2. \tag{4.23}$$

In the first equation of (4.22) one recognizes the relativistic extension of the Landau equation to which it reduces in the non-relativistic limit:

$$\mathbf{p}_j \to \mathbf{p}_j^{NR} = m_j\mathbf{v}_j; \qquad \gamma_j \to 1;$$

$$\sigma \to \sigma^{NR} = k^{-2}. \tag{4.24}$$

The Landau equation describes the irreversible approach to equilibrium of a plasma through Coulomb collisions. Here the difference from the Landau equation is that instead of the Fourier transform of the Coulomb potential (σ^{NR}) there appears a velocity dependent potential (σ_{ij}) which is nothing but the Fourier transform of $A_\mu(i)v^\mu(j)$ where $v^\mu(j) = (\underset{\sim}{v}_j, c)$ and $A_\mu(i) = (\underset{\sim}{A}, A_0)$ is the Lienard-Wiechert retarded potential of particle i. This equation presents the same divergence as the classical Landau equation, i.e., it is meaningful only in the k-region $L_D^{-1} << | \underset{\sim}{k} | << L_m^{-1}$, between the inverse of the Debye length and the inverse of the minimum impact parameter. The second equation of (4.22) indicates that the distribution of each normal mode of the field remains constant, i.e., there is no radiation in this approximation. Notice finally that the delta-function in the Landau equation expresses the collisional invariance of the energy of the two colliding particles, i.e.,

$$\delta(\underset{\sim}{k}\cdot\underset{\sim}{v}_{ij})\underset{\sim}{k}\cdot\underset{\sim}{\partial}_{ij}(p_{0i}+p_{0j}) \sim \delta(\underset{\sim}{k}\cdot\underset{\sim}{v}_{ij})\underset{\sim}{k}\cdot\underset{\sim}{v}_{ij} \equiv 0, \qquad (4.25)$$

and is thus the classical analogue of the quantum mechanical energy-conserving delta-function.

b.3 General Ring Equations. After these two introductory examples we will consider the exact first order solution, i.e., n_D^{-1} (or p = 1, all q) corresponding to the collective interactions. All diagrams contributing to the finite frequency collision operator $\Psi(z)$ within the n_D^{-1} approximation can be written

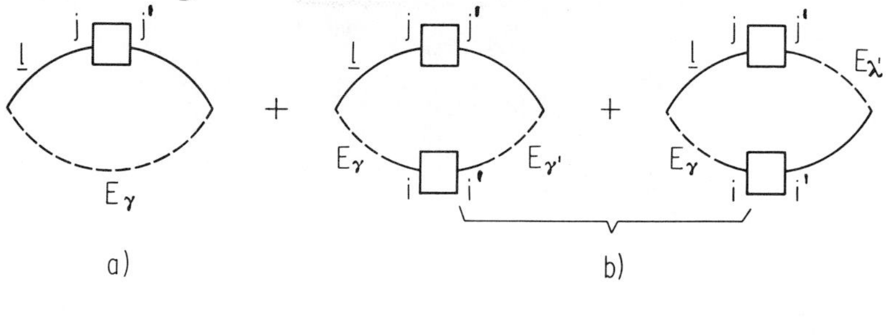

(4.26)

where the box is defined as:

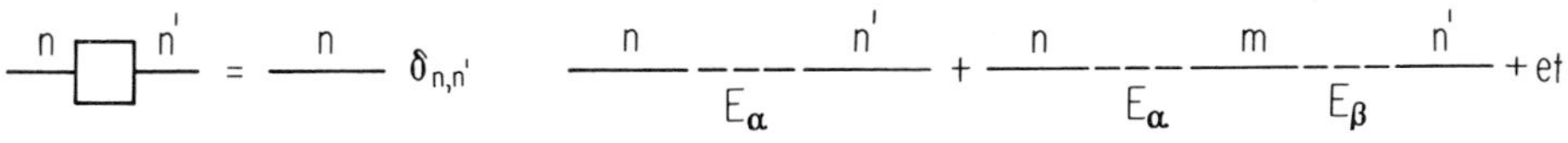

$$\underset{n}{\rule{0pt}{0pt}}\,\square\,\underset{n'}{\rule{0pt}{0pt}} = \underset{n}{\rule{0pt}{0pt}}\,\delta_{n,n'} \qquad \underset{n}{\rule{0pt}{0pt}}\underset{E_\alpha}{\text{-----}}\underset{n'}{\rule{0pt}{0pt}} + \underset{n}{\rule{0pt}{0pt}}\underset{E_\alpha}{\text{----}}\underset{m}{\rule{0pt}{0pt}}\underset{E_\beta}{\text{----}}\underset{n'}{\rule{0pt}{0pt}} + \text{etc.} \qquad (4.27)$$

Indeed, the first and the last vertex are determined by the homogeneity condition, and in between one can use only vertices which introduce a new particle for each e^2 factor so as to obtain powers of $(e^2 d)$. The only vertices which satisfy this condition are obviously ... ——— and ——— ... . Diagrams (4.26) represent exactly the relativistic extension of the ring diagrams of the Balescu-Lenard equation; therefore, we will still call them rings and also all equations deriving from (4.26) are called ring equations. Note that the diagrams of (4.26) correspond to the self-energy diagram (a) and the Landau diagrams (b) (see (4.13) and (4.17)-(4.18)) but with the single particle propagator (———) replaced by a screened (Vlasov) propagator (—□—). Note also that the a-rings have no classical analogue.

Finally, it is understood that one has to consider all relative time orders of the boxes in (4.26). Clearly the processes described by these diagrams are the self-interaction of a screened particle (a) and a Landau type of collision between two screened particles, i.e., two particles interacting through a screened field. Bearing this picture in mind one can derive some exact results in an *ad hoc* way, especially for stable plasmas. However, in the spirit of the preceding sections we will use an exact method and sum up the diagrams. As a result we will obtain all possible information within this approximation. For the details of the tricky but elementary algebra, see Ref. 16. The form of the diagrams suggests the use of the relativistic extension of a factorization theorem proven by

RESIBOIS in Ref. 17. We will derive this theorem in a simple way as follows. Rewrite (4.7) as:

$$\Psi(z) = \frac{1}{i}\langle 0|\delta L \frac{1}{L-z} \delta L|0\rangle_{ir}, \qquad (4.28)$$

(which shows that the collision operator is always an autocorrelation of forces δL). Let us split L into two groups according to the splitting of the degrees of freedom $L = L_1 + L_2 + L_{12}$, where L_{12} contains the interactions between the two groups and define an approximation in which you neglect the nonzero eigenvalues of L_{12}, i.e.:

$$\langle 0|\delta L \frac{1}{L-z} \delta L|0\rangle_{ir} \overset{\text{approx}}{=} \langle 0|\delta L \frac{1}{L_1 + L_2 - z} \delta L|0\rangle_{ir} . \qquad (4.29)$$

Using Cauchy's formula we can write for the collision operator, in this approximation and for any subdivision $L = L_1 + L_2 + L_{12}$ such that $[L_1,L_2]_- = 0$, the formula:

$$\Psi_{app}(z) = \frac{1}{2\pi i} \int_{C'} dz' \ \langle 0|\delta L \frac{1}{L_1-z'} \frac{1}{L_2+z'-z} \delta L|0\rangle_{ir} , \qquad (4.30)$$

where C' is a contour enclosing all eigenvalues of L_1 and excluding all those of $z - L_2$, i.e., a line antiparallel to the real axis such that Im z > Im z' > 0 and closed in the lower halfplane. One can rewrite (4.30) as:

$$\Psi_{app}(t) = \langle 0|\delta L\, e^{-iL_1 t} e^{-iL_2 t} \delta L|0\rangle_{ir} . \qquad (4.31)$$

We will apply this theorem to the summation of the b-rings with L_1, L_2 being the upper and lower branch (line), the extreme vertices being excluded. This theorem states that for two commuting branches, the relative order of the vertices is irrelevant and the sum over all relative orders equals the product of the two branches evaluated at the final time. The use of this theorem makes the summation of the diagrams straightforward, yielding collision operators containing field independent as well as field dependent contributions from the a- and b-rings. We will omit here the details of the calculations (see Ref. 16) and review only the qualitative aspects. We can obtain general expressions giving all the information on the system in a situation in which we can neglect: 1) terms of higher order in n_D^{-1}, 2) ternary and higher order initial correlations (then the destruction fragment can be obtained from $\Psi(z)$ disregarding the last vertex). Since these expressions are in general too complicated to be manageable, we will specify the desired information to be retained for further approximations.

1) The particle equation is coupled to $\langle\eta_\lambda\rangle$ so that we need only the first moment of the field equation. When taking this moment all the contributions depend linearly on the tensors $\underset{\approx}{D}^{\pm} = \underset{\approx}{D}(\pm\underset{\sim}{k},\omega_\pm)$.

$$\underset{\approx}{D}(k,\omega) = \left[\omega^2\underset{\approx}{\varepsilon}(\underset{\sim}{k},\omega)-\nu^2\underset{\approx}{T}\right]^{-1}; \quad \nu = c|\underset{\sim}{k}|, \quad \underset{\approx}{T} = \underset{\approx}{1} - \underset{\approx}{L}, \quad \underset{\approx}{L} = \hat{\underset{\sim}{k}}\,\hat{\underset{\sim}{k}}, \tag{4.32}$$

where $\omega_+ = z'$, $\omega_- = z - z'$, $\underset{\approx}{T}$ and $\underset{\approx}{L}$ are the projectors on transverse and longitudinal k-space, $\hat{\underset{\sim}{k}} = \frac{\underset{\sim}{k}}{|\underset{\sim}{k}|}$, and $\underset{\approx}{\varepsilon}$ is the generalized dielectric tensor defined as:

$$\underset{\approx}{\varepsilon}(\underset{\sim}{k},\omega) = \underset{\approx}{1} + \sum_\alpha \frac{4\pi e_\alpha^2 d_\alpha}{\omega} \int d\underset{\sim}{p}_\alpha\, \underset{\sim}{v}_\alpha \frac{1}{\omega - \underset{\sim}{k}\cdot\underset{\sim}{v}_\alpha}$$

$$(\underset{\sim}{\partial}_\alpha - \frac{1}{\omega}(\underset{\sim}{v}_\alpha\cdot\underset{\sim}{k}\;\underset{\approx}{T}\cdot\underset{\sim}{\partial}_\alpha - \underset{\sim}{v}_\alpha\cdot\underset{\approx}{T}\;\underset{\sim}{k}\cdot\underset{\sim}{\partial}_\alpha))\rho_0(\underset{\sim}{p}_\alpha;t). \tag{4.33}$$

This tensor expresses the dispersive properties of the plasma within this approximation.

2) We will neglect the variation of ρ_0 during a collision, i.e., $\rho_0(t-\tau) \approx \rho_0(t)$. This is correct only when there are no strong instabilities in the system.

3) Although the destruction fragment can be important in unstable plasmas, here it will usually be similar to the collision operator into which one can incorporate it formally. In the following we will neglect it. This approximation implies an assumption on the range of the initial correlations.

Now we still have to specify what contributions we retain in the z and z' integration. In the z' integration we will have contributions from poles from free propagators and poles from the dispersion relation, det $\underset{\approx}{D}^{-1} = 0$, where $\underset{\approx}{D}^{-1}$ is the inverse of the tensor given in Eq. (4.32). For the z integration we have:

$$\int_0^t d\tau \, \frac{1}{2\pi} \int_C dz \, e^{-iz\tau} \, C(z) = \frac{-1}{2\pi i} \int_C dz \, \frac{e^{-izt}}{z} \, C(z)$$
$$= C(+i0) + \sum_{z_j} \frac{e^{-iz_j t}}{z_j} \operatorname*{Res}_{z=z_j} C(z), \tag{4.34}$$

where z_j are the poles of the reduced collision operator C(z). Note that in Eq. (4.34) C(+i0) depends on time only through $\rho_0(t)$, whereas the second term, say C(t), depends explicitly and implicitly on time. The nature of the time dependence of C(t) depends strongly on the position of the roots of the dispersion relation. For roots in the upper halfplane (unstable plasma waves) C(t) grows, for roots in the lower halfplane (weakly stable plasma waves) C(t) decays, and for roots far down in the lower halfplane (damped plasma waves) C(t) is negligible for long times. Roughly speaking, the damping is proportional to $c|\underset{\sim}{k}|$, and one can define two complementary regions in the $\underset{\sim}{k}$-integral contained in the collision operator for the particle equation. The radiation region corresponds to small k's, say from 0 to L_D^{-1}, and the collision or opacity region runs from L_D^{-1} to some inverse minimum impact parameter L_m^{-1}. In the radiation region the dominant contributions will come from the emission of (longitudinal and transverse) plasma waves, i.e., C(t), whereas this emission will be strongly damped in the collision region where C(+i0) will be dominant.

b.4 Ring Equations for Stable Systems. To obtain a more manageable expression let us consider a stable system, i.e., we suppose the dispersion relation yields no unstable modes and only strongly damped stable ones. Then a fantastic simplification occurs - only one term survives, namely, the contribution to C(+i0) of the residue at $z' = \underset{\sim}{k}\cdot\underset{\sim}{v}_j$. The corresponding kinetic equations are:

$$\partial_t \rho_0(\underset{\sim}{p}_j;t) = \sum_\alpha 2e_j^2 e_\alpha^2 d_\alpha \int d\underset{\sim}{k} \int d\underset{\sim}{p}_\alpha \, \underset{\sim}{k}\cdot\underset{\sim}{\partial}_j \delta(\underset{\sim}{k}\cdot\underset{\sim}{v}_{j\alpha}) Q^+ Q^- \underset{\sim}{k}\cdot\underset{\sim}{\partial}_{j\alpha}$$

$$\rho_0(\underset{\sim}{p}_j;t)\rho_0(\underset{\sim}{p}_\alpha;t), \tag{4.35}$$

$$\partial_t \rho_0(n_\lambda;t) = 0,$$

where $Q^\pm$ is defined as:

$$Q^\pm = \underset{\sim}{v}_j\cdot\underset{\approx}{D}(\pm\underset{\sim}{k}, \pm\underset{\sim}{k}\cdot\underset{\sim}{v}_i)\cdot\underset{\sim}{v}_i \equiv (Q^\mp)^*, \tag{4.36}$$

and $\underset{\approx}{D}$ was defined in (4.32). The product $Q^+Q^- = | Q^+ |^2$ appearing in Eq. (4.35) is k^{-2} times the scattering cross section. Before reviewing the properties of Eq. (4.35) we notice the following special cases:

1) When the medium is isotropic in $\underset{\sim}{r}$ space, there is only one independent vector in $\underset{\approx}{\varepsilon}(\underset{\sim}{k},\omega)$, namely $\underset{\sim}{k}$, and one can write

$$\begin{aligned} \underset{\approx}{\varepsilon}(\underset{\sim}{k},\omega) &= \ldots \underset{\approx}{1} + \ldots \underset{\approx}{L}, \\ &= \varepsilon_L(\underset{\sim}{k},\omega)\underset{\approx}{L} + \varepsilon_T(\underset{\sim}{k},\omega)\underset{\approx}{T}. \end{aligned} \tag{4.37}$$

This yields:

$$\delta(\underset{\sim}{k}\cdot\underset{\sim}{v}_{j\alpha})\,|\underset{\sim}{v}_j\cdot\underset{\approx}{D}\cdot\underset{\sim}{v}_\alpha|^2 = \delta(\underset{\sim}{k}\cdot\underset{\sim}{v}_{j\alpha})\left|\frac{1}{k^2\varepsilon_L} + \frac{(\hat{\underset{\sim}{k}}\times\underset{\sim}{v}_j)\cdot(\hat{\underset{\sim}{k}}\times\underset{\sim}{v}_\alpha)}{(\underset{\sim}{k}\cdot\underset{\sim}{v}_\alpha)^2\varepsilon_T - c|\underset{\sim}{k}|^2}\right|^2, \tag{4.38}$$

where ε_L is the usual or longitudinal dielectric constant and where ε_T is the transverse dielectric constant. The magnetic permeability μ is defined by the relation:

$$1 - \mu^{-1}(\underset{\sim}{k},\omega) = \frac{\omega^2}{(ck)^2}(\varepsilon_T - \varepsilon_L). \tag{4.39}$$

This shows explicitly that supposing $\underset{\approx}{\varepsilon}$ to be a scalar (this is consistent only when neglecting spatial dispersion) is equivalent to the neglect of magnetic properties, i.e., $\mu = 1$.

2) In the extreme case $\underset{\approx}{\varepsilon} = \underset{\approx}{1}$, Eq. (4.38) reduces to the relativistic Landau cross section ($\sigma^2_{j\alpha}$ of Eq. (4.23)).

3) When the system is supposed isotropic in velocity space, for all t, i.e., $\rho_0(\underset{\sim}{p};t) \equiv \rho_0(|\underset{\sim}{p}|;t)$, one obtains:

$$|\underset{\sim}{v}_j \cdot \underset{\approx}{D} \cdot \underset{\sim}{v}_\alpha|^2 = \frac{1}{\left| k^2 \varepsilon_L \right|^2} + \left| \frac{(\hat{\underset{\sim}{k}} \times \underset{\sim}{v}_j) \cdot (\hat{\underset{\sim}{k}} \times \underset{\sim}{v}_\alpha)}{(\underset{\sim}{k} \cdot \underset{\sim}{v}_\alpha)^2 \varepsilon_T - (ck)^2} \right|, \qquad (4.40)$$

where, furthermore, the magnetic terms in ε_T vanish.

Note that it is only in two cases (Eqs. (4.38, 4.40)) that one can distinguish between longitudinal and transverse plasma waves, i.e., plasmons and radiation, whereas the general case (Eq.(4.35)) still contains a coupling between the transverse and longitudinal fields.

The particle equation of (4.35) was first obtained in the special form (4.40) for isotropic systems by KLIMONTOVITCH [12] using his formalism. SILIN [19] derived the particle equation of (4.35) using an heuristic method. The nonrelativistic equation obtained by DUPREE [20] does not agree with our equation. The equation obtained by ROSTOKER, AAMODT and ELDRIDGE [21] is of the same form but contains a different dielectric tensor $\underset{\approx}{D}$.

The equation for the field variables (4.35) states that the energy distribution of each normal mode remains constant within this approximation. The first moment of this equation could be inferred from the fact that the particle's energy is conserved and that the different normal modes are independent in this approximation. With our assumptions, most of the effects disappear as short-time effects. Note, for example, that there is no contribution from the a-rings. In this approximation the whole behavior of the system is described by the particle equation whose properties we now briefly review (see also Refs. 2 and 18):

1) The equation strongly reminds one of the Landau equation in the sense that the velocity dependent potential now becomes screened by the dispersion tensor $\underset{\approx}{D}$ containing the complete dielectric tensor

$\underset{\approx}{\varepsilon}(\underset{\sim}{k},\omega;t)$. The electromagnetic field still plays the role of a catalyst in the formation of the polarization cloud leading to a screened potential. There is no radiation field in the system, only a local induction field, and the particles interact only through virtual plasma waves.

2) Eq. (4.35) is still of the Fokker-Planck type, i.e., a diffusion equation in momentum space:

$$\partial_t \rho_0(j) = \underset{\sim}{\partial}_j \cdot \underset{\approx}{d} \cdot \underset{\sim}{\partial}_j \rho_0(j) + \underset{\sim}{\partial}_j \cdot (\underset{\sim}{a} \rho_0(j)), \tag{4.41}$$

containing a diffusion $\underset{\approx}{d}$ and a friction $\underset{\sim}{a}$"constant".

3) It is a "long-time" equation valid for $t >> \omega_p^{-1}$ when $n_D^{-1} << 1$.

4) The Maxwell momentum distributions for electrons and ions are solutions of (4.35) when taken at the same temperature.

5) In analogy with the Boltzmann equation, one can show it conserves the number, the momentum and the energy of the particles.

6) One can show an H-theorem for Eq. (4.35), i.e., it drives the system to a canonical equilibrium state.

7) Eq. (4.35) forms a very complicated system of integro-differential equations for the different particle species. It is highly nonlinear because the unknown functions $\rho_0(t)$ appear in the denominator through $\underset{\approx}{\varepsilon}$. This indicates the constant adaption of the polarization cloud to the state of the system, yielding finally in equilibrium a Debye 4-potential.

8) When the polarization of the medium is neglected the relativistic Landau equation is recovered. This is meaningful only for states sufficiently close to equilibrium so that there are only

few particles having velocities large compared to the thermal velocity and that the emission (at $\underset{\sim}{k}\cdot\underset{\sim}{v}$) of high frequency plasma waves is improbable. For weakly stable states, for example, this assumption is not permitted.

9) The relaxation time is of the order $n_D\omega_p^{-1}$, and thus widely separated from ω_p^{-1} for large n_D.

10) This equation can be extended to weakly inhomogeneous systems.[22]

Finally, it should be noted that although one can write equations of the type of (4.35) with four-vectors,[18] their Lorentz invariance is not evident because neither the factorization, $\rho_0(j)\rho_0(\alpha)$, nor the concept of weak inhomogeneity are invariant concepts.

b.5 Ring Equations for Unstable Systems. In the approximation complementary to the one considered in b.4 , the dominant contribution to the collision operators will come from the radiation region, i.e., from C(t) (see b.3). In this region the behavior of the system depends very strongly on the nature of the proper modes. When, for example, there are only weakly unstable modes, then the separation into two regions has a definite meaning. One can first study the stabilization (see below) of these modes in the radiation region, and,once stabilized, the system will be governed by collisions, i.e., one can use the collision region only. As a matter of fact, when there are only strongly damped modes present, the radiation region will be irrelevant. However, when different types of modes are possible, especially when weakly stable modes exist, a separation into regions is not possible and the original complicated structure remains. Moreover, in some situations the destruction fragment will not be negligible (it will grow or decay slowly).

However, when restricting to binary initial correlations, one can incorporate its contributions into the collision operator as done, for example, in the classical case, by NICHIKAWA and OSAKA [23] (this paper contains misconceptions, some of which were corrected in a paper by MATSUDAIRA [24]). Here we will describe only the qualitative aspects of the kinetic equations within the radiation region; details will be found in Ref. 16.

The particle equation (4.35) of b.4 will now be replaced by a FOKKER-PLANCK equation whose diffusion and friction coefficients will depend on the unstable modes and will have an explicit exponential time dependence (as in Eq. (4.34)), corresponding to all different possible modes. Moreover, they will depend linearly on $\langle \eta_\lambda \rangle$, i.e., on the energy in these modes. The particle equation is now coupled to:

$$\partial_t \langle \eta_\lambda \rangle = e_\lambda - a_\lambda \langle \eta_\lambda \rangle - \sum_\mu P^+_{\lambda,\mu} \langle \eta_{\underset{\sim}{k}_{\lambda,\mu}} \rangle - \sum_\mu P^-_{\lambda,\mu} \langle \eta_{-\underset{\sim}{k}_{\lambda,\mu}} \rangle . \tag{4.42}$$

This equation describes all radiation processes possible within the n_D^{-1} approximation, i.e., the emission and absorption (e_λ, a_λ) of longitudinal and transverse modes and the scattering between the $(\pm \underset{\sim}{k}_\lambda, \mu)$ modes. There is no nonlinear mode coupling in Eq. (4.42); this process will appear only in the n_D^{-2} approximation. The redistribution of energy described by (4.42) is governed by the particle distribution $\rho_0(\underset{\sim}{p}_j;t)$, i.e., the coefficients e_λ, a_λ, $p^\pm_{\lambda,\mu}$ are functionals of $\rho_0(\underset{\sim}{p}_j;t)$ but have no definite sign, thus mixing the real and induced processes.

One can also take over from the nonrelativistic case the interpretation of the instability problem, i.e., the collision term depending on the unstable mode will drive the plasma to a stable state, i.e., will push the pole which is a functional of $\rho_0(j;t)$ into the lower halfplane, as the relative drift energy (for the two stream instability) is converted into thermal energy (see Refs. 2 and 25). However, in the relativistic case there will exist supplementary stabilization mechanisms, namely, not only redistribution of the particle energy, but also of the field energy, and proper radiation will become possible as described by (4.42). Finally, it should be noticed that the equations are of the so-called quasilinear form, i.e., the nonlinear particle equation is coupled to a linear equation (4.42) (when neglecting the time dependence of $\rho_0(j)$ in the coefficients) for the energy of the (unstable) modes. This should not be confused with the quasilinear equations for an inhomogeneous collisionless plasma as derived from the nonlinear Vlasov equation.[26] However, the interpretation of the stabilization mechanism is the same. When there is initially an unstable mode it will grow according to Eq. (4.42) ($a_\lambda < 0$), but when $\langle\eta_\lambda\rangle$ increases, the diffusion and friction coefficients of the particle equation will grow, yielding a more effective redistribution of the energy between the particles by diffusion and a more effective slowing down of the rapid particles by increased friction. This in turn will reshape the momentum distribution from an unstable to a stable one. It should be noticed, however, that our quasilinear equations are long-time equations, whereas the quasilinear equations for collisionless plasmas are short-time equations. The approximations, such as replacing $\rho_0(j;t)$ by $\rho_0(j;0)$, should therefore be handled more carefully in our case.

c) Typical Radiation Processes

In the approximation (n_D^{-1}) of the preceding section (b.4) the weakness of the particle collisions was such that no direct radiation was produced (soft collisions). All radiation was produced by plasma wave excitations. To increase the strength of the coupling (e) between the particles and the field we will now consider the second order, n_D^{-2} or $e^4(e^2d)^m$, all m. Now there will be a collisional energy exchange between the field and the particles. However, whereas the first-order could be considered a single process, an extremely large number of different processes can now occur. Therefore, we will single out two complementary radiation phenomena within the low density corrections m = 0,1. For m = 0 we obtain Thomson scattering, and for m = 1 a variety of phenomena can occur, such as Coulomb and retarded Bremsstrahlung, "double" Thomson scattering, Boltzmann contributions and screening corrections to lower order processes. We will single out the classical Bremsstrahlung process. Because of the restriction on m we disregard all effects of screening on the radiation. Therefore, our treatment will be meaningful only for frequencies $\nu_\lambda >> \omega_p$, because for $\nu_\lambda \approx \omega_p$ the screening becomes so important as to make the refractive index vanish for long wavelengths (i.e. for radiation) and no radiation can propagate in the system at this frequency.

c.1 Thomson Scattering. Consider all diagrams of order e^4 (m = 0) containing one particle and four vertices. The only such diagrams are

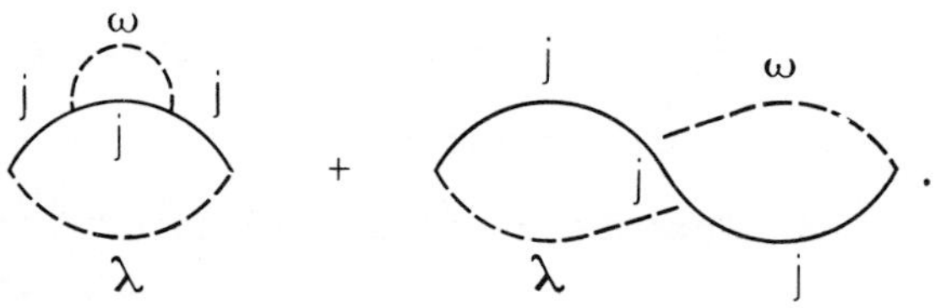

(4.43)

They represent the creation and destruction of a correlation between a particle and an oscillator, during which the creation and destruction of a correlation with another oscillator is established. In other words, an electromagnetic wave interacts with a particle, the particle is accelerated by the field of the wave and emits another wave. This corresponds to the scattering of electromagnetic waves on a particle. As it involves only one particle, it is an incoherent scattering, i.e., the scattered intensity is equal to the sum of the intensities scattered by each particle. It is high-frequency scattering ($\nu_\lambda >> \omega_p$) and cooperative phenomena will not contribute (hence incoherence). Such an incoherent, high-frequency scattering is called Thomson scattering or the classical Compton effect. We can now compute the asymptotic collision operator $\Psi(+i0)$ corresponding to (4.43), where the field has to be restricted to the radiation field, i.e., μ_λ, μ_ω = 1,2. This calculation is straightforward[4,15] yielding:

$$\Psi(+i0) = \sum_{j,\lambda,\omega} \pi e_j^4 \frac{(\nu_\lambda \nu_\omega)^{-1}}{\Omega^2} (2\pi\partial_{\lambda\omega} - \underset{\sim}{k}_{\lambda\omega}\cdot\underset{\sim}{\partial}_j)\eta_\lambda\eta_\omega \qquad (4.44)$$

$$(\alpha^j_{\lambda,\omega})^2 \, \delta(b^j_\lambda - b^j_\omega)\,(2\pi\partial_{\lambda\omega} - \underset{\sim}{k}_{\lambda\omega}\cdot\underset{\sim}{\partial}_j)$$

where

$$\partial_{\lambda\omega} = \partial_\lambda - \partial_\omega, \quad \partial_\lambda = \frac{\partial}{\partial\eta_\lambda}, \quad \underset{\sim}{k}_{\lambda\omega} = \underset{\sim}{k}_\lambda - \underset{\sim}{k}_\omega, \quad b^j_\lambda = \nu_\lambda - \underset{\sim}{k}_\lambda\cdot\underset{\sim}{v}_j,$$

and

$$\alpha^j_{\lambda,\omega} = \left[\underset{\sim}{\mathcal{F}}(j,\lambda)\cdot\underset{\sim}{\partial}_j, \frac{\underset{\sim}{v}_j\cdot\underset{\sim}{\varepsilon}_\omega}{b^j_\omega}\right]_- = \underset{\sim}{\mathcal{F}}(j,\lambda)\cdot\underset{\sim}{\partial}_j\left(\frac{\underset{\sim}{v}_j\cdot\underset{\sim}{\varepsilon}_\omega}{b^j_\omega}\right) \equiv \alpha^j_{\omega,\lambda}, \qquad (4.45)$$

where the anticommutator symbol [,]_ is not to be confused with the Poisson bracket [,].

From (4.44) we obtain the kinetic equations:

$$\partial_t \rho_0(n_\lambda;t) = \frac{c}{2\pi} \sum_\alpha d_\alpha \int d\nu_\omega d\Omega_{\hat{\underset{\sim}{k}}_\omega} \sum_{\mu_\omega} \int d\underset{\sim}{p}_\alpha \int dn_\omega n_\omega$$

$$\partial_\lambda n_\lambda \sigma^\alpha_{\lambda,\omega} \, \delta(b^\alpha_\lambda - b^\alpha_\omega) \, (2\pi \partial_{\lambda\omega} - \underset{\sim}{k}_{\lambda\omega} \cdot \underset{\sim}{\partial}_\alpha)$$

$$\rho_0(\underset{\sim}{p}_\alpha, n_\lambda, n_\omega;t), \tag{4.46}$$

$$\partial_t \rho_0(\underset{\sim}{p}_j;t) = \frac{c^4}{8^2\pi^5} \sum_{\mu_\lambda,\mu_\omega} \int d\underset{\sim}{k}_\lambda d\underset{\sim}{k}_\omega \int dn_\lambda dn_\omega$$

$$\underset{\sim}{k}_{\omega\lambda} \cdot \underset{\sim}{\partial}_j n_\lambda n_\omega \sigma^j_{\lambda,\omega} \, \delta(b^j_\lambda - b^j_\omega) \, (2\pi \partial_{\lambda\omega} - \underset{\sim}{k}_{\lambda\omega} \cdot \underset{\sim}{\partial}_j)$$

$$\rho_0(\underset{\sim}{p}_j, n_\lambda, n_\omega;t), \tag{4.47}$$

where $\rho_0(\underset{\sim}{p}, n_\lambda, n_\omega;t) \equiv \rho_0(\underset{\sim}{p};t)\rho_0(n_\lambda;t)\rho_0(n_\omega;t)$ and where $\sigma^\alpha_{\lambda,\omega}$, the scattering cross section by a particle of species α from mode ω into mode λ, is defined as:

$$\sigma^\alpha_{\lambda,\omega} = \frac{e^4_\alpha}{c^4} \frac{\nu_\omega}{\nu_\lambda} (\alpha^\alpha_{\lambda,\omega})^2 \; . \tag{4.48}$$

The coupled system of Eqs. (4.46) and (4.47) describes the approach to equilibrium of the system of charged particles and the radiation field through Thomson scattering. This process induces an energy transfer between the normal modes and the particles. The total energy of the particles and the two oscillators is conserved, as expressed by the delta-function in (4.46) and (4.47) (see Eq.(4.25)). This delta-function shows that the frequencies of the incident and scattered wave are the same in the reference system of the particle, i.e., the Doppler-shifted frequencies are equal. In the nonrelativistic limit the Doppler shift vanishes and the scattering becomes elastic:

$$\delta(b_\lambda^\alpha - b_\omega^\alpha) \overset{\text{N.R.}}{\rightarrow} \delta(\nu_\lambda - \nu_\omega), \tag{4.49}$$

and the Thomson formula is obtained :

$$\sigma_{\lambda,\omega}^\alpha \overset{\text{N.R.}}{\rightarrow} r_0^2(\underset{\sim}{\varepsilon}_\lambda \cdot \underset{\sim}{\varepsilon}_\omega)^2, \tag{4.50}$$

where $r_0 = e_\alpha^2/m_\alpha c^2$ is the classical electron radius. As $r_0 \sim 10^{-12}$cm for electrons, the cross section (4.50) is very small and the scattering will thus be quite ineffective (large relaxation time). However, when screening effects are included, (4.50) will become:

$$\sigma \sim \frac{r_0^2}{|\varepsilon(\underset{\sim}{k},\omega)|^2}, \tag{4.51}$$

which will be large in the neighborhood of propagating modes. Finally, it should be noticed that one can include the other polarizations in the above treatment so as to obtain scattering between different types of modes.

c.2 Normal and Anomalous Bremsstrahlung. The Bremsstrahlung process, i.e., radiation produced by a particle-particle collision or a free-free transition in the quantum mechanical sense, is very important in plasma physics. It is one of the major energy losses in fusion experiments and one of the main diagnostic tools. Through the study of its spectra, one can determine such plasma parameters as the density, temperature, etc. The most effective collision is the e-i (electron-ion) collision in which the e and the i are accelerated, each particle emitting radiation. As the average acceleration ratio equals the inverse mass ratio (which is small) the major part of the radiation comes from the electrons, that is

$$m_e \underset{\sim}{a}_e = \underset{\sim}{F} = m_i \underset{\sim}{a}_i \,, \quad \frac{\langle a_i \rangle}{\langle a_e \rangle} = \frac{m_e}{m_i} \ll 1. \tag{4.52}$$

The e-e, i-i collisions produce some quadrupole radiation, but it becomes appreciable only for very high temperatures. The characteristic frequency of the radiation produced by an e-i collision corresponds to the inverse of the collision time:

$$\omega = \frac{\langle v_e \rangle}{b} \,, \tag{4.53}$$

where b is a typical impact parameter and $\langle v_e \rangle$ some average electron velocity. For the minimum impact parameter ($kT = e^2/b_{min}$) we obtain the maximum of the spectrum:

$$\omega_{max} = \frac{\langle v_e \rangle}{b_{min}} \approx \left(\frac{kT_e}{m_e}\right)^{1/2} \cdot \left(\frac{kT_e}{e^2}\right) \approx 10^9 \, T_e^{3/2}. \tag{4.54}$$

Thus, the range of validity of our calculation will be:

$$\omega_p \ll \nu_\lambda \ll 10^9 \, T^{3/2}. \tag{4.55}$$

For electrons in a typical thermonuclear plasma we have $10^{12} \ll \nu_\lambda \ll 10^{21}$ (in sec^{-1}), thus quite an important frequency domain. In our system Bremsstrahlung consists of the usual Bremsstrahlung when the particle collision is a Coulomb collision, and of relativistic Bremsstrahlung when the collision proceeds via the retarded field (cf. classical and relativistic Landau equation). We will consider here only Coulomb Bremsstrahlung, which is dominant for moderately high temperatures. The diagrams describing this process are (in Coulomb gauge):

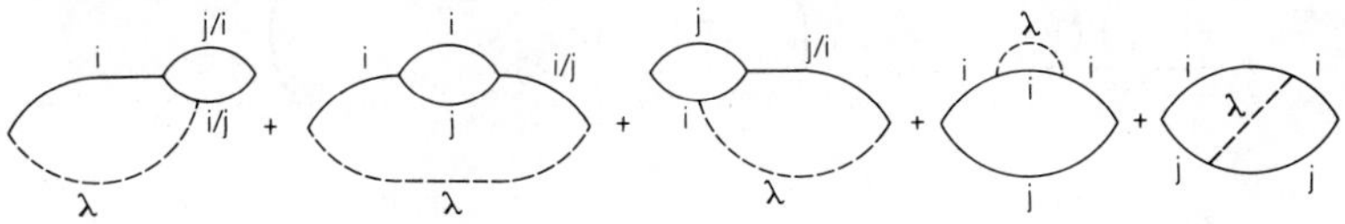

(4.56)

i.e., a Coulomb collision,  and a "photon" emission (λ). The sign i/j indicates two possibilities. We will omit here the lengthy calculations of the asymptotic collision operator corresponding to (4.56) (see Refs. 4 and 15). However, we will describe some of the complications of (4.56) with respect to the previous calculations. Mathematically the asymptotic limit ($z \to +i0$) for (4.56) has to be taken carefully, because some of the diagrams contain the same propagator twice, and the limit of the product of propagators is not equal to the product of the limits. Physically, the asymptotic value of (4.56) contains different contributions. Indeed, consider the contribution to $\Psi(+i0)$ of the first diagram of (4.56). In the asymptotic limit the three propagators become three δ_- functions. For the reason given in b.1 only the principal part of the first propagator remains and for parity reasons we are left with the two possibilities: $\mathcal{P}\delta\mathcal{P}$ and $\mathcal{P}\mathcal{P}\delta$. In the first possibility, the δ_- function will conserve the energy of the colliding system, and we call this contribution to $\Psi(+i0)$, Φ. In the second possibility the photon does not contribute to the energy conservation and we call this type of contribution χ. The result can then be written as:

$$\Psi_{Br}(+i0) = \Phi + \chi, \tag{4.57}$$

$$\Phi = \sum_{i,j,\underset{\sim}{\ell},\lambda} \frac{(4\pi e_i e_j)^2}{\Omega^3} \nu_\lambda^{-1} (2\pi\partial_\lambda - \underset{\sim}{k}_\lambda \cdot \underset{\sim}{\partial}_i + \underset{\sim}{\ell} \cdot \underset{\sim}{\partial}_{ji})$$

$$\pi n_\lambda \, \delta(b_\lambda^i + \underset{\sim}{\ell} \cdot \underset{\sim}{v}_{ji}) \left(e_i \alpha_{\lambda,i}^{\underset{\sim}{\ell}} + e_j \alpha_{\lambda,j}^{-(\underset{\sim}{\ell} + \underset{\sim}{k}_\lambda)} \right)^2$$

$$(2\pi\partial_\lambda - \underset{\sim}{k}_\lambda \cdot \underset{\sim}{\partial}_i + \underset{\sim}{\ell} \cdot \underset{\sim}{\partial}_{ji}),$$

$$\chi = -\frac{i}{2}\left[\Psi_2', \Psi_4^c\right]_+ + \Delta\Phi^c,$$

where we have used the symbols of the preceding section and where the cross section is defined in terms of $\alpha_{\lambda,i}^{\underset{\sim}{\ell}}$:

$$\alpha_{\lambda,i}^{\underset{\sim}{\ell}} = \frac{1}{(b_\lambda^i)^2 \ell^2} \left[A(\lambda,i), \underset{\sim}{\ell} \cdot \underset{\sim}{v}_i\right]_- = \left[-\underset{\sim}{\ell} \cdot \underset{\sim}{\partial}_i , \frac{\underset{\sim}{v}_i \cdot \underset{\sim}{\varepsilon}_\lambda}{\ell^2 b_\lambda^i} \right]_-$$

$$= \frac{1}{(b_\lambda^i)^2 \ell^2 p_{0i}} (c^2 \underset{\sim}{\ell} \cdot \underset{\sim}{\mathcal{F}}(\lambda,i) + \nu_\lambda (\underset{\sim}{\ell} \cdot \underset{\sim}{v}_i)(\underset{\sim}{v}_i \cdot \underset{\sim}{\varepsilon}_\lambda)). \tag{4.58}$$

We also wrote the χ part of (4.57) in a form putting in evidence an anticommutator of Ψ_4^c (the pure Coulombic part of the Landau operator) and Ψ_2' the asymptotic value of $\frac{d}{dz}\Psi_2(z)$, $\Psi_2(z)$ being defined in Eq. (4.13). Let us first consider the properties of Φ: (1) Φ conserves the energy of the colliding system, i.e.,

$$\Phi(p_{0j} + p_{0i} + \frac{\nu_\lambda}{2\pi} n_\lambda) \equiv 0; \tag{4.59}$$

(2) there is no small $\underset{\sim}{\ell}$ divergence. This was already noticed by OSTER [27] and shows that the individual Bremsstrahlung has a definite meaning even without taking collective effects into account. However, as usual, the large $\underset{\sim}{\ell}$ divergence has to be cut off at the inverse of the minimum impact parameter; (3) it contains as a special case the results of OSTER and SCHEUER.[27] This contribution corresponds, therefore, to the normal Bramsstrahlung. The second contribution, χ, has the following unexpected features: (1) χ only conserves the particle energy, i.e., the λ oscillator is virtual; (2) it is a higher order (compared to Φ) differential operator with a cross section which is not positively defined; (3) unlike Φ, there is a small $\underset{\sim}{\ell}$ divergence; (4) it is a nonequilibrium effect, vanishing at equilibrium when $T_e = T_i$; (5) there appears a remarkable structure - an anticommutator of a lower order collision operator with $i\Psi'$. The existence of these contributions, which constitute the "anomalous Bremsstrahlung" terms, was confirmed in a quantum mechanical treatment by CHAPPELL.[28] A general approach to the study and the interpretation of these terms was proposed by PRIGOGINE and a group of collaborators (see PRIGOGINE's lecture notes). The basic idea is that they correspond to a dynamical renormalization process. An application of this approach to Bremsstrahlung can be found in Ref. 28, where it is shown that one can define new, screened oscillators radiating only normal Bremsstrahlung.

5. Kinetic Approach to Phenomenological Radiation Laws for Homogeneous Systems

In Chapter 4 we considered some microscopic electrodynamic processes and then derived their observable aspect by means of statistical mechanics. That is, in some sense, a finite temperature field theory. In this chapter, however, we will take the opposite point of view. We will focus our attention on relations which were obtained by phenomenological considerations and try to determine their validity within the statistical electrodynamics of homogeneous systems. We will pursue this in the same way as we investigated the basic laws of macroscopic electrodynamics in Chap. 3, but here we will have to use perturbation theory, at least implicitly. We will consider two cases, the radiation transfer equation and the relation between material and radiative transport coefficients such as the relation between the absorption coefficient and the electrical conductivity.

a) *General Radiation Transfer Equation*

From phenomenological considerations one obtains (see CHANDRASEKHAR, Ref. 29) the following transport equation for $\langle u_\lambda \rangle$, the local intensity of radiation of wavelength and polarization λ, in a medium of density d:

$$\frac{1}{d}\, \hat{\underset{\sim}{k}}_\lambda \cdot \underset{\sim}{\nabla} \langle u_\lambda \rangle = e_\lambda - a_\lambda \langle u_\lambda \rangle, \qquad (5.1)$$

where e_λ and a_λ are the (spontaneous) emission and absorption coefficients per unit volume for radiation of frequency ν_λ, polarization μ_λ, propagating in $d\Omega_{\underset{\sim}{k}_\lambda}$ (for our purpose we denoted the intensity I_λ

as an average $I_\lambda \equiv \langle u_\lambda \rangle$). Equation (5.1) is derived for stationary systems. For nonstationary systems one expects the addition of $\frac{1}{c}\,\partial_t\langle u_\lambda\rangle$ to the left-hand side of Eq. (5.1). Thus, for nonstationary homogeneous systems one expects an equation of the form:

$$\frac{1}{c}\,\partial_t\langle u_\lambda\rangle = e_\lambda - a_\lambda\langle u_\lambda\rangle, \tag{5.2}$$

where u_λ will now be the density of states $\nu_\lambda^2/8\pi^3c^3$ times the energy $\frac{\nu_\lambda}{2\pi}\,\eta_\lambda$ of normal mode λ, i.e., $d_\lambda\eta_\lambda$, with d_λ a numerical factor $(\nu_\lambda^3/16\pi^4c^3)$. We will show that Eq. (5.2) is valid to all orders in the coupling constant e. It is thus the analogue for fluctuating fields of the more usual equation derived from Eq. (3.8) for weakly dispersive media (while Eq. (5.2) is exact, in general). It should be realized that writing Eq. (5.2) with a term proportional to $\langle u_\lambda\rangle$ and grouping everything else in e_λ, as in the phenomenological Eq. (5.1), is unphysical. Indeed all contributions corresponding to mode scattering, i.e., proportional to $\langle u_\lambda'\rangle$, $\lambda' = (\pm \underset{\sim}{k}_\lambda, \mu_\lambda' \neq \mu_\lambda)$, are hidden in e_λ (see Eq. (4.42) which is a special case of Eq. (5.2)), as are also all nonlinear coupling terms in (5.2). We will now prove Eq. (5.2). Consider the exact reduced master equation:

$$\partial_t\rho_0(\eta_\lambda;t) = \int_\lambda d\Gamma \int_0^t dz\;\Psi(\tau)\rho_0(t-\tau) + D_\lambda(t), \tag{5.3}$$

or its usual asymptotic form,

$$\partial_t\rho_0(\eta_\lambda;t) = \int_\lambda d\Gamma\;\Psi(+i0)\rho_0(t). \tag{5.4}$$

The most general contribution to $\Psi(z)$ has the form:

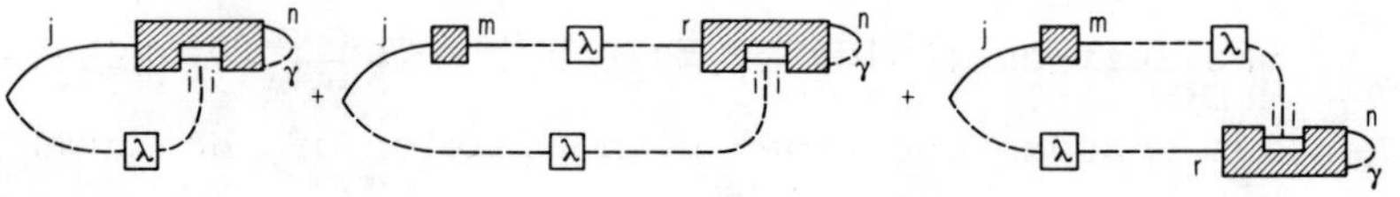

(5.5)

where the black box ▨ can contain everything except the oscillator λ, while the "modified λ-line" is defined as:

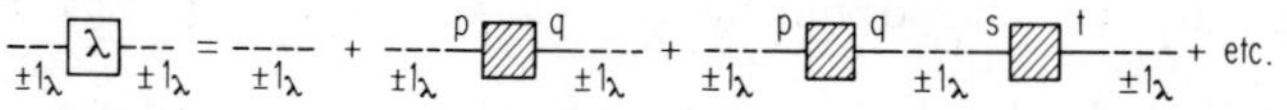

(5.6)

It is easy to check that all other possible structures with respect to the λ oscillator introduce a supplementary relation between the wave vectors, yielding a contribution in Ω^{-1} which will vanish in the limit of a large volume (see also Ref. 4). Furthermore, the irreducibility condition requires the λ-lines of the λ-boxes to be nonoverlapping; therefore, the λ-vertices will always follow in the order

$$(A(\lambda,j_1)\eta_\lambda A(\lambda,j_2))(A(\lambda,j_3)\eta_\lambda A(\lambda,j_4)) \; \ldots \; , \tag{5.7}$$

(see Eqs. (2.38)-(2.42)), where the vertex operator $A(\lambda,j)$ is the sum of an operator acting on $\lambda(A_\lambda(j))$ and an operator acting on $j(A_j(\lambda))$ (see (2.39)). The structure of Eq. (5.2) with respect to the η_λ integration is thus:

$$\partial_t \langle u_\lambda \rangle = d_\lambda \partial_t \langle \eta_\lambda \rangle \sim \int d\Gamma \eta_\lambda \Big[(A(\lambda,J)\eta_\lambda A(\lambda,j_1)) \; \ldots$$

$$(A(\lambda,j_2)\eta_\lambda A(\lambda,j_3)) \quad \ldots \Big] \rho_0 . \tag{5.8}$$

Integration by parts yields:

$$\sim \int d\eta_\lambda ((-)\Delta_{j\lambda})^2 (A_i(\lambda)(-)\Delta_{i\lambda})^n (A_j(\lambda)(-)\Delta_{\lambda n} + A_j(\lambda) A_n(\lambda) \langle \eta_\lambda \rangle), \quad (5.9)$$

where $\Delta_{j\lambda}$ is the coefficient of ∂_λ in $A(j,\lambda)$ (see (2.39)). Eq. (5.9) means that one can replace the first two λ-vertices by $(-)\Delta_{\lambda j_1}$ and $(-)\Delta_{\lambda j_2}$, the $(2n+1)^{th}$ vertex by $A_{j_{2n}}(\lambda)$, the $(2n+2)^{th}$ vertex by $(-)\Delta_{\lambda j_{2n+1}}$ $(n = 1,2,\ldots)$, the last λ-vertex by $(-\Delta_{\lambda i} + A_i(\lambda)\langle \eta_\lambda \rangle)$. This proves, then, that the sum of all contributions is the sum of an η_λ independent one and a contribution proportional to $\langle \eta_\lambda \rangle$, i.e., Eq. (5.2). As this proof is independent of the propagators, one can generalize Eq. (5.2) to the nonMarkoffian case (finite z) or include the destruction fragment. In fact such generalizations are quite formal because e_λ and a_λ considered as observable quantities have usually only an asymptotic meaning. We will not consider them here. From the general transfer Eq. (5.2) we can deduce Kirchoff's law. Indeed for a stationary state we have:

$$\frac{1}{c}\,\partial_t \langle u_\lambda \rangle = 0 = e_\lambda - a_\lambda \langle u_\lambda \rangle ; \quad \frac{e_\lambda}{a_\lambda} = \langle u_\lambda \rangle , \quad (5.10)$$

and for an equilibrium state:

$$\frac{e_\lambda^{eq}}{a_\lambda^{eq}} = \langle u_\lambda \rangle^{eq} = \frac{\nu_\lambda^2}{8\pi^3 c^3}\, kT_\lambda . \quad (5.11)$$

It should be noticed that this is a powerful proof of phenomenological relations (5.2), (5.10), and (5.11) which, for example, were shown to be valid in lowest order (n_D^{-2}) by DUPREE,[30] whereas they are valid in general. However, for inhomogeneous systems this will no longer be the case. In particular, second order moments (i.e., quadratic in the field strengths) different from $<u_\lambda>$ will appear.

In phenomenological methods one usually has in mind a particular emission or absorption process which is computed for one electron and averaged over the particle distribution. It is clear that out of equilibrium, when the equations for the field and the particles are coupled, this is not allowed. Moreover, one usually computes a_λ and determines e_λ from Kirchoff's law (5.10), a procedure which again is not correct out of equilibrium. Note that our method does not suffer from these limitations.

b) Relation Between Absorption Coefficient and Electrical Conductivity. Phenomenological considerations give a relation between the absorption coefficient (K_ν) at frequency ν and the electrical conductivity at the same frequency (σ_ν):

$$K_\nu = \frac{4\pi}{\text{"c"}} \sigma_\nu , \tag{5.12}$$

where "c" has to be interpreted as the group velocity of energy, i.e., "c" = $cn(\lambda)$ (c is the vacuum phase velocity and $n(\lambda)$ is the refractive index at frequency ν). In LANDAU and LIFSCHITZ, [31] for example, one can find the improved relation:

$$K(\underset{\sim}{k},\omega) = \frac{4\pi}{c \text{ Re } n(\underset{\sim}{k},\omega)} \text{Re } \sigma^T(\underset{\sim}{k},\omega), \tag{5.13}$$

where Re denotes the real part and where the refractive index is defined as a function of the transverse (T) dielectric constant as:

$$n^2(\underset{\sim}{k},\omega) = \varepsilon^T(\underset{\sim}{k},\omega),$$

$$\underset{\approx}{\varepsilon}(\underset{\sim}{k},\omega) = \underset{\approx}{1} + i\,\frac{4\pi}{\omega}\,\underset{\approx}{\sigma}(\underset{\sim}{k},\omega). \tag{5.14}$$

On the other hand, from statistical considerations of energy losses from test particles, ROSTOKER [32] obtained the following relation for an equilibrium Coulomb plasma:

$$K(\underset{\sim}{k},\omega) = 4\pi \text{ Re } \sigma^L(\underset{\sim}{k},\omega), \tag{5.15}$$

where L denotes the longitudinal part and where K is defined by linearizing the power absorption in the energy density, i.e., $\mathrm{Re}(j_L E_L^*) = 2K\ \mathrm{Re}(\frac{E_L E_L^*}{8\pi})$.

We will now investigate this relation by a method independent of any phenomenological idea, making explicit only the considerations of the previous paragraph (a) (see also Ref. 33). Keeping in mind Eqs. (5.2) and (5.5) we write:

$$\alpha_\lambda^{e/a}(z;t) = \frac{d_\lambda}{ic}\int d\Gamma\eta_\lambda\ \langle 0|\delta L_\lambda(j)\,\frac{1}{L-z}\,\delta L^{e/a}(i,\lambda|0_\lambda\phi_{(\lambda)}\rangle_{ir}\ \rho_{\phi_{(\lambda)}}(z;t)\ \rho_0(\eta_\lambda;t)$$

$$\rho_{\phi_{(\lambda)}}(z;t)\ \rho_0(\lambda) = \langle 0_\lambda\phi_{(\lambda)}|\frac{1}{L_{(\lambda)} - z}\,\delta L(n,\gamma)|0\rangle_{ir}\rho_0(t) \tag{5.16}$$

where $\phi_{(\lambda)}$ and $L_{(\lambda)}$ denote the λ-independent state ϕ and Liouville operator L, respectively, where e/a means an emission/absorption contribution and $\delta L^e(i,\lambda) = \delta L_\lambda(i)$, $\delta L^a(i,\lambda) = \delta L_i(\lambda)$, and where (j,λ) and (i,λ) are the first and last λ-interactions (see Eq. (5.5)). Let us introduce the following definitions from Eq. (5.16):

$$\alpha_\lambda^e = e_\lambda(z;t), \qquad\qquad \alpha_\lambda^a = -a_\lambda(z;t)\langle u_\lambda\rangle(t), \tag{5.17}$$

together with the definition:

$$\sigma_\lambda^{e/a}(z;t) = \int d\Gamma \sum_{\substack{\phi,\underset{\sim}{k}_i\\ j,i}} \frac{(-)e_j\Delta_{j\lambda}}{\Omega} \langle \underset{\sim}{k}_j \left| \frac{1}{L-z} \right| \phi,\underset{\sim}{k}_i\rangle \; e_i A^{e/a}(i,\lambda)\nu_\lambda^{-1}(-)\rho_{\phi,\underset{\sim}{k}_i}(z;t)\rho_0(t) \tag{5.18}$$

$=$ 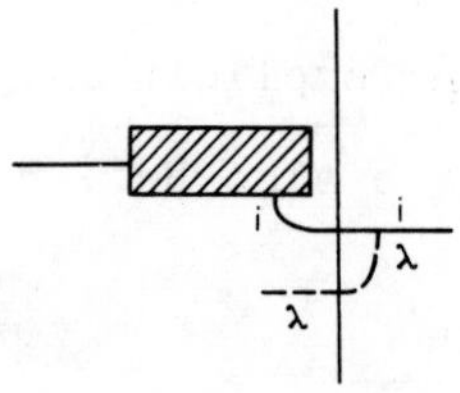

where $A^e_{(i,\lambda)} = (-)\Delta_{i\lambda}$, $A^a_{(i,\lambda)} = A_i(\lambda)$. Finally we define:

$$s_\lambda(z;t) = \int d\Gamma \sum_{m,m'} (-)e_m\Delta_{m\lambda}\Omega^{-1} \langle \underset{\sim}{k}_m \left| \frac{1}{L-z} \right| \underset{\sim}{k}'_m\rangle \; e_{m'}A_{m'}(\lambda)\nu_\lambda^{-1}(-)\rho_0(t) \tag{5.19}$$

$=$.

Moreover, for the modified λ-line, we introduce a renormalized propagator:

$$P_\lambda(z;t) = \frac{1}{E_\lambda - z}\left(1 + \frac{s_\lambda(z;t)}{E_\lambda - z} + \ldots\right) = \frac{1}{E_\lambda - z}\cdot\frac{1}{1 - \dfrac{s_\lambda}{E_\lambda - z}}$$

$$= \frac{1}{E_\lambda - s_\lambda(z;t) - z} \tag{5.20}$$

$=$ 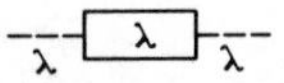 .

In writing Eqs. (5.16)-(5.20), we used the properties of the η_λ-integration of the preceding section and grouped the factors so as to put in evidence structures related to the electrical conductivity (see σ_λ, s_λ). We now use the factorization theorem (4.30) to obtain the sum given by Eq. (5.5):

$$\alpha_\lambda^{e/a}(z;t) = \frac{1}{2\pi i}\int_{C'} dz' \frac{d_\lambda}{ic}\Big[\sigma_\lambda^{e/a}(z-z';t)P_\lambda(z';t) + s_\lambda(z-z';t)$$

$$P_\lambda(z-z';t)\sigma_\lambda^{e/a}(z-z';t)P_\lambda(z';t)$$

$$+ P_\lambda(z-z';t)\sigma_\lambda^{e/a}(z-z';t)s(z';t)P_\lambda(z';t)\Big] \tag{5.21}$$

$$= \frac{1}{2\pi i}\int_{C'} dz' \frac{d_\lambda}{ic}\sigma_\lambda^{e/a}(z-z';t)P_\lambda(z';t)$$

$$\Big[1 + P_\lambda(z-z';t)\{s_\lambda(z';t) + s_\lambda(z-z';t)\}\Big].$$

One has to be careful when the dependence on $\pm 1_\lambda$ is made explicit. The $\pm$ signs will appear as follows: $P(z') = P^{\mp}$, $P(z-z') = P^{\pm}$, $s(z') = s^{\pm}$, $s(z-z') = s^{\mp}$, $\sigma(z-z') = \sigma^{\pm}$. The emission and absorption coefficients can now be written as:

$$
\begin{aligned}
e_\lambda(t) &= \alpha_\lambda^e(+i0;t), \\
a_\lambda(t) &= \frac{-1}{\langle u_\lambda\rangle(t)}\,\alpha_\lambda^a(+i0;t).
\end{aligned}
\tag{5.22}
$$

We now compare $a_\lambda(t)$ with the usual electrical conductivity defined as (see Ref. 10):

$$
\underset{\approx}{\sigma}_{\underset{\sim}{k}}(z;eq) = \frac{\delta}{\delta E^e_{\underset{\sim}{k}}(z)}\; j^{(1)}_{\underset{\sim}{k}}(z)\ , \tag{5.23}
$$

where $E^e_{\underset{\sim}{k}}(z)$ is the external electric field in Fourier-Laplace space and $j^{(1)}_{\underset{\sim}{k}}(z)$ is the first-order (linear conductivity) induced current in the same space, or:

$$
j^{(1)}_{\underset{\sim}{k}}(z) = -\int d\Gamma \sum_{\substack{\phi,\underset{\sim}{k}_i\\ i,j}} e_j \underset{\sim}{v}_j \Omega^{-1}\langle \underset{\sim}{k}^{(j)}|\frac{1}{L-z}|\phi,\underset{\sim}{k}_i\rangle \underset{\sim}{\mathcal{F}}^e_{\underset{\sim}{k}}(z)\cdot \underset{\sim}{\partial}_i \rho^{eq}_{\phi,\underset{\sim}{k}_i+\underset{\sim}{k}}\ , \tag{5.24}
$$

where $\underset{\sim}{\mathcal{F}}$ is the Lorentz force corresponding to the external fields, i.e., $\underset{\sim}{\mathcal{F}}(i) = e_i \underset{\sim}{E}(i) + \frac{e_i}{c}\underset{\sim}{v}_i\times \underset{\sim}{B}(i)$ (see Ref. 10). If we replace the equilibrium distribution ρ^{eq}_ϕ by $\rho_\phi(z;t)$ (see Eq. (5.16)) in Eq. (5.24) we obtain a nonequilibrium electrical conductivity $\underset{\approx}{\sigma}_{\underset{\sim}{k}}(z;t)$. If we neglect the screening of the λ oscillator in Eq. (5.21), i.e.,

$$
\sigma P(1+P(s+s)) \sim \sigma P \sim \sigma\,\frac{1}{E_\lambda - z'}, \tag{5.25}
$$

one obtains from Eqs. (5.22), (5.21), (5.18) and (5.24) the relation:

$$a_\lambda(t) = \sum_{\pm} \frac{1}{c} \sigma^{a}_{\lambda}(\pm\nu_\lambda + i0;t) = \frac{2\pi}{c} \sum_{I} \underset{\sim}{\varepsilon}_\lambda \cdot \underset{\approx}{\sigma}_{\pm \underset{\sim}{k}_\lambda}(\pm\nu_\lambda;t)\cdot \underset{\sim}{\varepsilon}_\lambda$$

$$= \frac{4\pi}{c} \operatorname{Re} \underset{\sim}{\varepsilon}_\lambda \cdot \underset{\approx}{\sigma}_{\underset{\sim}{k}_\lambda}(\nu_\lambda;t)\cdot \underset{\sim}{\varepsilon}_\lambda . \qquad (5.26)$$

If we assume the system in equilibrium, λ a longitudinal mode and the Vlasov approximation for $\underset{\approx}{\sigma}$, Eq. (5.26) yields the result of ROSTOKER (Eq. (5.15)). Comparing Eq. (5.26) to Eq. (5.13) we see that our neglect of the λ-screening (Eq. 5.25)) is equivalent to putting Re n = 1. We will not investigate here whether when retaining this screening we obtain a refractive index still defined as in Eq. (5.17). However, the question, "What is the microscopic definition of the refractive index?" is very important, as it may indicate, for example, whether there is a statistical "aging" of the photon which could be related to the cosmological red-shift.

The above formula may also be used to investigate relations between material and radiation transport coefficients out of equilibrium (see Refs. 10 and 34), especially nonequilibrium generalizations of Kirchoff's law (5.10). The strength of this type of derivation is that they do use perturbation theory only implicitly. However, it is expected that this will no longer be possible when extending to the systems of real interest, the inhomogeneous systems.

References

[1] PRIGOGINE, I.: Nonequilibrium Statistical Mechanics. New York: Interscience, 1962.

[2] BALESCU, R.: Statistical Mechanics of Charged Particles. New York: Interscience, 1963.

[3] RÉSIBOIS, P.: Irreversible Processes in Classical Gases. In: Many-Particle Physics, ed. E. Meeron, Gordon and Breach, 1967.

[4] MANGENEY, A.: a) Physica 30, 461 (1964); b) Annales de Physique 9, No. 1 (1964).

[5] HAKIM, R.: J. Math. Phys. 8, 1315, 1379 (1967).

[6] BALESCU, R., and T. KOTERA: Physica 33, 558 (1967); R. BALESCU, T. KOTERA, and E. PIÑA: Physica 33, 581 (1967); R. BALESCU, M. BAUS, and A. PYTTE: Bull. Cl. Sciences Acad. Royal Belg. 53, 1043 (1967); A. PYTTE, and R. BALESCU: To appear, Bull. Cl. Sciences Acad. Roy. Belg. (1967); R. BALESCU: To appear, Physica (1968); M. BAUS: To appear (1968).

[7] CURRIE, D.G., T.F. JORDAN, and E.C.G. SUDARSHAN: Rev. Mod. Phys. 35, 350 (1963).

[8] DIRAC, P.A.M.: Rev. Mod. Phys. 21, 392 (1949).

[9] BALESCU, R.: Physica 36, 433 (1967).

[10] BAUS, M.: Mémoire de licence, Université Libre de Bruxelles (1964).

[11] SILIN, V.P., and A.A. RUKHADZE: Electromagnetic Properties of Plasmas. Moscow: Atomizdat, 1961 (in Russian). Partially translated in Sov. Phys. Usp. 4, 459 (1961).

[12] KLIMONTOVITCH, Y.L.: Statistical Theory of Nonequilibrium Processes in a Plasma. Moscow: Univ. of Moscow, 1964 (in Russian).

[13] ROSTOKER, N., R. AAMODT, and O. ELDRIDGE: Ann. Phys. 31, 243 (1965).

[14] PRIGOGINE, I., and F. HENIN: Acad. Roy. Belg., mém. sc., 35, fascicule 7.

[15] BAUS, M.: unpublished, see thesis Université Libre de Bruxelles (1968).

[16] BAUS, M.: Ph. D. thesis, University of Brussels, Brussels 1968.

[17] RÉSIBOIS, P.: Phys. Fluids 6, 817 (1963).

[18] See Ref. 12

[19] SILIN, V.P.: Sov. Phys. JETP 13, 1244 (1961).

[20] DUPREE, T.H.: Phys. Fluids 6, 1714 (1963); Phys. Fluids 7, 923 (1964).

[21] See Ref. 13

[22] deGOTTAL, P.: Thèse, Université Libre de Bruxelles (1964).

[23] NISHIKAWA, K., and Y. OSAKA: Progr. Th. Phys. 33, 402 (1965).

[24] MATSUDAIRA, N.: Phys. Fluids 9, 539 (1966).

[25] ABRAHAM, B.W.: J. Math. Phys. 6, 630 (1965).

[26] Plasma Physics: Trieste Seminar of the International Centre for Theor. Phys., I.A.E.A., Vienna (1965), Chap. 4.

[27] OSTER, L.: Rev. Mod. Phys. 33, 525 (1961). P.A.G. SCHEUER: Monthly Notices Royal Astro. Soc. 120, 231 (1960).

[28] BAUS, M.: Phys. Fluids 9, 1427 (1966).

[29] CHANDRASEKHAR, S.: Stellar Structure, Dover 1954.

[30] See Ref. 20.

[31] LANDAU, L.D., and E.M.LIFSCHITZ: Electrodynamics of Continuous Media. New York, Pergamon 1960.

[32] ROSTOKER, N.: Nuclear Fusion 1, 101 (1961).

[33] SERGYSELS, R.: Mémoire de licence, Université Libre de Bruxelles (1965).

[34] MANGENEY, A.: Annales d'Astrophysique 30, 649 (1967).

Nonlinear Optics

N. Bloembergen

1. Phenomenological Survey of Nonlinear Optical Effects

By way of introduction to the field of nonlinear optics, we shall discuss some simple classical models of a nonlinear relationship between the field amplitudes and the current density or polarization induced in a medium by prescribed electromagnetic waves. It turns out that a simple perturbation theory is well suited to describe the nonlinear electromagnetic response of a material under practical nondestructive conditions.

The induced nonlinear source densities in turn create components of the electromagnetic field. This second step leads to a self-consistency condition between the sources and the fields. Typical nonlinear problems which arise will be discussed in the following chapters.

A phenomenological discussion of the first step - the nonlinear response of a medium to *prescribed* electromagnetic fields - will lead to a purely physical interpretation of the resulting nonlinear terms, which is directly related to some basic experimental situations. An understanding of the physical facts is essential to comprehend the similarities as well as the differences in the treatment of nonlinear optical and corresponding nonlinear mechanical problems, which will be the subject of the second and third chapters.

These brief notes of Chapter 1 are based upon and should be amplified by the following references:

1. N. BLOEMBERGEN: Nonlinear Optics (New York: Benjamin, 1965), chaps. 1 and 3.
2. P. N. BUTCHER: Nonlinear Optical Phenomena, Engineering Experiment Station, Bulletin 200, Ohio State University (Columbus, Ohio, 1965).
3. P. FRANKEN and J. F. WARD: Rev. Mod. Phys., 1963.

1.1 The Anharmonic Oscillator as a Classical Model of Nonlinearity

Consider a plane-polarized electromagnetic wave which drives a linear, slightly anharmonic oscillator of mass m and charge e. The incident wave propagates in the z-direction and contains two frequencies, ω_1 and ω_2, with corresponding wave vectors, $\underset{\sim}{k}_1$ and $\underset{\sim}{k}_2$. The equation of motion is

$$\ddot{x} + \Gamma\dot{x} + \omega_o^2 x + vx^2 = (e/m)[A_1 \exp(i\underset{\sim}{k}_1 \cdot \underset{\sim}{r} - i\omega_1 t) + A_1^* \exp(- i\underset{\sim}{k}_1 \cdot \underset{\sim}{r} + i\omega_1 t) + A_2 \exp(i\underset{\sim}{k}_2 \cdot \underset{\sim}{r} - i\omega_2 t) + A_2^* \exp(- i\underset{\sim}{k}_2 \cdot \underset{\sim}{r} + i\omega_2 t)]. \tag{1.1}$$

In the absence of the nonlinear term, vx^2/m, in the restoring force, one obtains the well known linear approximation,

$$x(\omega_1) = \frac{e}{m(- \omega_1^2 + \omega_o^2 - i\omega_1\Gamma)} A_1 \exp(i\underset{\sim}{k}_1 \cdot \underset{\sim}{r} - i\omega_1 t),$$

and corresponding terms for the Fourier components, $x(\omega_2)$, $x(-\omega_1)$ and $x(-\omega_2)$. It is useful to distinguish between positive and negative frequency components. Note that the amplitudes of the real waves are $2|A_1|$ and $2|A_2|$, respectively. In the presence of the nonlinear term, a solution of x in terms of a double Fourier series, containing all combination frequencies $\pm n_1\omega_1 \pm n_2\omega_2$ (n_1 and n_2 are integers), may be postulated. The terms can be found by the method of successive approximations, in ascending powers of the field amplitudes. They will be ordered according to the Fourier components involved.

In the lowest order nonlinear approximation one finds terms at the second-harmonic frequencies $2\omega_1$, $2\omega_2$, a term at zero frequency representing the rectification of light by the quadratic nonlinearity, vx^2, and terms at the sum and difference beats between the two light waves, $\omega_1 + \omega_2$ and $\omega_1 - \omega_2$. Only two Fourier components will be reproduced here; a second harmonic,

$$x(2\omega_1) = \frac{-(e^2/m^2)vA_1^2 \exp(2ik_1z - 2i\omega_1t)}{D^2(\omega_1)\, D(2\omega_1)} ; \qquad (1.2)$$

and the difference beat,

$$x(\omega_1 - \omega_2) = \frac{-2(e^2/m^2)vA_1A_2^*\exp[i(k_1 - k_2)z - i(\omega_1 - \omega_2)\, t]}{D(\omega_1)\, D^*(\omega_2)\, D(\omega_1 - \omega_2)} \qquad (1.3)$$

where the abbreviated notation

$$D(\omega) = \omega_o^2 - \omega^2 - i\Gamma\omega = D^*(-\omega)$$

for the denominators has been introduced. From Eq. (1.2) one derives immediately a nonlinear polarization and a nonlinear susceptibility describing second-harmonic production

$$P_x^{NL}(2\omega) = \chi_{xxx}(-2\omega, \omega, \omega)A_x^2(\omega) \exp\{2i\underset{\sim}{k}_1 \cdot \underset{\sim}{r} - 2i\omega t\}$$

$$\chi_{xxx}(-\,2\omega,\omega,\omega) = \frac{-N_o(e^3/m^2)v}{D^2(\omega)\;D(2\omega)} \;.$$

Note that the second-harmonic field from an incident fundamental wave with wave vector $\underset{\sim}{k}_1$ is generated by a phased array of dipoles, described by 2 $\underset{\sim}{k}_1$. Due to color dispersion, normally the fundamental and harmonic wave do not propagate with the same velocity, $2\underset{\sim}{k}_1(\omega_1) \neq \underset{\sim}{k}_2(2\omega_1)$. This leads to typical interference effects between the forced harmonic wave and the free harmonic wave. The linear dispersion plays a very important role in nonlinear optics and accounts for several differences with the problems normally encountered in nonlinear mechanics.

The indices xxx are of course superfluous in the one-dimensional example. They serve as a reminder of the fact that the susceptibility really connects three vectors and has the transformation properties of a third-rank tensor. The extension to a three-dimensional harmonic oscillator is straightforward.

Note that the dispersion of the lowest order nonlinear susceptibility is described with reference to a set of three frequencies rather than a single frequency. The first frequency refers to the Fourier component of polarization, the remaining frequencies to the ordered field components on the right. The dispersive properties are enhanced near resonance of one of the denominators. If, for example, the difference frequency is equal to the resonant frequency, $D(\omega_1 - \omega_2) = i\omega_o\Gamma$ for $\omega_1 - \omega_2 = \omega_o$, the nonlinear susceptibility at the difference frequency is far larger than all others.

When $\omega_1 - \omega_2$ is equal or nearly equal to the resonant frequency ω_o, it is permissible to retain only this Fourier component in the calculation of next higher order nonlinearities. It beats in vx^2

with linear terms to yield components at $2\omega_1 - \omega_2$, $\omega_1 - 2\omega_2$, in addition to terms at the original frequencies, ω_1 and $-\omega_2$. In this manner one finds, for example,

$$x^{NL}(\omega_2)^* = x^{NL}(-\omega_2) = \frac{4(e^3/m^3)\, v^2}{(D^*(\omega_2))^2 |D(\omega_1)|^2 D(\omega_1 - \omega_2)} A_2^* |A_1|^2$$

or

$$\chi_{xxxx}(-\omega_2, +\omega_2, +\omega_1, -\omega_1) = \frac{4\, N_0(e^4/m^3)v^2}{D^2(\omega_2)|D(\omega_1)|^2\, D^*(\omega_1 - \omega_2)} .$$

At resonance $\omega_1 - \omega_2 = \omega_0$ and $\omega_1 \gg \omega_0$, $D(\omega_1 - \omega_2)$ is purely imaginary; the other factors in the denominator are real and

$$\chi_{xxxx}(-\omega_2, \omega_2, +\omega_1, -\omega_1) = -\frac{4iN_0(e^4/m^3)\, v^2}{\omega_0\Gamma(\omega_1^2 - \omega_0^2)^2(\omega_2^2 - \omega_0^2)^2} .$$

In the same manner one finds

$$\chi_{xxxx}(-\omega_1, \omega_1, +\omega_2, -\omega_2) = \chi^*_{xxxx}(-\omega_2, \omega_2, +\omega_1, -\omega_1), \quad (1.4)$$

and

$$\chi_{xxxx}(-2\omega_1 + \omega_2, \omega_1, \omega_1, -\omega_2) = \frac{4N_0(e^4/m^3)v^2}{D^2(\omega_1)D^*(\omega_2)D(2\omega_1 - \omega_2)D(\omega_1 - \omega_2)} .$$

Smaller nonresonant terms should be added to obtain the general expression for these nonlinear susceptibilities. The physical interpretation of these terms will be postponed to the end of this chapter.

The following order of magnitude relationship exists between the first order nonlinear polarization and the linear polarization and between nonlinear polarizations of successive orders:

$$|P^{NL}/P^L| \approx |P^{NL}_{(n+1)}/P^{NL}_{(n)}| \approx \frac{e|A|}{mD}\frac{v}{D} .$$

It may be expected from the physical nature of electronic binding that if the deviation x is of the order of the radius a of the equilibrium orbital of the electron, the nonlinear force mvx^2 is of the same order as the linear force, $m\,\omega_o^2\,a = e|A_{at}|$, where A_{at} is the intra-atomic electric field binding the electron. Therefore, $v/D \approx \approx v/\omega_o^2 \approx a^{-1}$, and the ratio of the magnitudes of polarization in successive orders is $e|A|/m\,\omega_o^2\,a = |A|/|A_{at}|$. The electric field amplitude of the light wave must be compared with the electric field inside the atom, which is typically of the order 3×10^8 volts/cm. Therefore, even for the extreme power flux densities of 10^{10} watts/cm^2 in the focus of a Q-switched laser, the nonlinear response can still be treated as a small perturbation, since $|A/A_{at}| \sim 3 \times 10^{-3}$ in this extreme case. It should be noted that the ratio is enhanced by a factor, $Q = \omega_o/\Gamma$, whenever a resonance in one of the factors in the denominator occurs. It should also be noted that even if the magnitude of a nonlinear effect is small, it may still be detectable due to the excellent discrimination in optical experiments. Terhune has, for example, detected third-harmonic generation, even though only one out of every 10^{15} photons was converted in this manner. Since a typical ruby laser pulse contains 4×10^{18} photons representing one joule, a thousand ultraviolet photons per pulse at λ = 2313 Å are readily detectable. The polarization at the third-harmonic frequency 3ω is of course given by a similar susceptibility,

$$\chi\,(-\,3\omega,\ \omega,\ \omega,\ \omega) = \frac{N_o\,(e^4/m^3)\,v^2}{D^3(\omega_1)\,D(2\omega_1)\,D(3\omega_1)}\ .$$

It is important to note that the convergence of the perturbation procedure would be threatened at the highest attainable field strengths, when resonant denominators occur at the same time. Experimentally the critical power density in media, which are not absorbing at the laser frequency but do show stimulated Raman and

Brillouin scattering, occurs at power flux densities of about 10^{11} watts/cm^2. At these levels material damage, breakdown in gases and fluids, and damage tracks in glasses and crystals occur. In strongly absorbing materials, melting and evaporation occur with the formation of strongly absorbing ionized gas atmospheres. These physical manifestations may well be concomitant with, or antecede, the breakdown in mathematical convergence of the perturbation approach. Conversely, in the domain of nonlinear electromagnetic phenomena without material damage, to which the following discussions will be limited, the perturbation approach is valid.

1.2 Nonlinear Source Terms

The most general way to describe the nonlinear properties of a material is to expand the current density in terms of a power series of the vector potential,

$$j_i = \sigma^L_{ik} \frac{1}{c} \frac{\partial A_k}{\partial t} + \sigma^{NL}_{ikl} A_k A_l + \sigma^{NL}_{iklm} A_k A_l A_m + \ldots .$$

The summation convention over repeated indices is understood. Nonlinear conductivity tensors of successively higher rank have been introduced. The current density distribution, together with the vacuum field vectors, gives a complete description of the electromagnetic phenomena. Let us introduce the microscopic Maxwell equations

$$\nabla \times \underset{\sim}{h} = \frac{4\pi}{c} \underset{\sim}{j} + \frac{1}{c} \frac{\partial \underset{\sim}{e}}{\partial t}$$

$$\nabla \times \underset{\sim}{e} = - \frac{1}{c} \frac{\partial \underset{\sim}{h}}{\partial t}$$

$$\nabla \cdot \underset{\sim}{e} = 4\pi\phi = 0, \text{ and } \nabla \cdot \underset{\sim}{b} = 0.$$

The macroscopic Maxwell equations are obtained by averaging over volume elements large compared to the atomic dimensions but small compared to the optical wavelength. In terms of the macroscopic fields, Maxwell's equations become

$$\nabla \times \underset{\sim}{B} = \frac{1}{c}\frac{\partial \underset{\sim}{E}}{\partial t} + \frac{4\pi}{c}\underset{\sim}{J}$$

$$\nabla \times \underset{\sim}{E} = -\frac{1}{c}\frac{\partial \underset{\sim}{B}}{\partial t} .$$

The macroscopic current density is often given as a multipole expansion, after the convection current has been split off,

$$\underset{\sim}{J} = \underset{\sim}{J}_{cond} + \frac{\partial \underset{\sim}{P}}{\partial t} + c\,\nabla \times \underset{\sim}{M} - \frac{\partial}{\partial t}(\nabla \cdot \underset{\approx}{Q}) + \ldots .$$

The electric polarization $\underset{\sim}{P}$, the magnetization $\underset{\sim}{M}$, and the electric quadrupole density $\underset{\approx}{Q}$ may all be considered as a combined power series expansion in the electric and magnetic field amplitudes as well as their gradients.

When the medium is not dissipative, a time-averaged thermodynamic potential can be introduced. It is essential that a time interval T can be defined such that $(\omega_i - \omega_j)\,T \gg 1$, where ω_i and ω_j are any pair of frequency components occurring in the Fourier expansion of the field, while at the same time the Fourier amplitude,

$$E(\omega_i,t) = \frac{1}{2T}\int_{(t-T)}^{(t+T)} E(t')\exp(-i\omega_i t')dt' ,$$

is independent of T. This can be done only if the nonlinearity is small and the Fourier component $E(\omega_i,t)$ can be considered as a slowly varying amplitude. As explained in Chap. 3 of Nonlinear Optics, one retains only those terms in the free enthalpy which are very slowly varying in time, i.e., the sum of the frequencies of the Fourier components in each term must equal zero. One should also be careful in handling the linear dispersion associated with distant absorption bands. In this manner the following terms in the time-averaged free enthalpy may be considered:

$$- F = \alpha_{ij}E_iE_j + \chi_{ij}^{magn} H_iH_j + \chi_{ij}^{(1)}E_iH_j + \chi_{ij}^{(2)}E_i \frac{\partial H_j}{\partial t}$$

$$+ \chi_{ijk}^{(3)} E_iE_jE_k + \chi_{ijkl}^{(4)} E_iE_jE_kE_l + \chi_{ijk}^{(5)} E_iH_jH_k$$

$$+ \chi_{ijk}^{(6)} E_i \frac{\partial E_j}{\partial t} H_k + \chi_{ijkl}^{(7)} E_iE_j\nabla_kE_l + \chi_{ijkl}^{(8)}E_iE_jH_kH_l$$

$$+ \ldots.$$

If the fields are constant in time, these terms represent well known dc effects. The first two terms contain the well known linear electric and magnetic susceptibilities. The third term represents a magnetoelectric effect that occurs, for example, in antiferromagnetic Cr_2O_3. This crystal is invariant for simultaneous time reversal and spatial inversion, but not for either of these operations alone. The crystal acquires a magnetic moment by application of an electric field, etc. The term in $\chi^{(3)}$ can occur only in media which lack inversion symmetry. The term in $\chi^{(5)}$ occurs in piezoelectric paramagnetic crystals, such as $NiSO_4 \cdot 6H_2O$. The term with $\chi^{(4)}$ describes, for example, paraelectric saturation,etc.

For a light wave, $E_i = H_j = A_1 \exp(-i\omega t)$, the first term, $\alpha(\omega)A_1A_1^*$, describes the optical polarizability or the index of refraction. The term in $\chi^{(2)}$ describes natural optical activity. It vanishes in media with a center of inversion.

The time-independent term, $- \chi_{iik}^{(3)}(\omega,-\omega,0)A_iA_iE_k^{dc}(0)$, describes the linear electro-optic Kerr effect. In piezoelectric crystals the index of refraction is a linear function of an applied electric field. This same term describes the rectification of light. A dc polarization is developed in the crystal, proportional to the intensity of the light beam

$$P_k^{dc}(0) = -\frac{\partial F}{\partial E_k^{dc}} = \chi_{kii}^{(3)}(0,\omega,-\omega)A_i A_i^*.$$

One has the symmetry relation,

$$\chi_{kii}^{(3)}(0,\omega,-\omega) = \chi_{iik}^{(3)}(\omega,-\omega,0).$$

It is not necessary to restrict the third frequency to zero. The time-averaged free enthalphy,

$$\chi_{ijk}^{(3)}\{A_i(\omega_1)A_j(\omega_2)\, A_k^*(\omega_3) + A_i^*(\omega_1)A_j^*(\omega_2)A_k(\omega_3)\},$$

describes the generation of sum and difference frequencies in a nondissipative crystal, for which χ is real. Consider a cartesian coordinate system with $\underset{\sim}{A}_1$ in the j-direction, $\underset{\sim}{A}_2$ in the k-direction, and $\underset{\sim}{A}_3$ in the l-direction. One finds

$$P_{1,j} = -\frac{\partial F}{\partial A_1} = \chi_{jkl}(-\omega_1,-\omega_2,+\omega_3)A_2^*A_3\ ,$$

$$P_{2,k} = -\frac{\partial F}{\partial A_2} = \chi_{kjl}(-\omega_2,-\omega_1,\omega_3)A_1A_3 = \chi_{jkl}(-\omega_1,-\omega_2,+\omega_3)A_1A_3$$

where the convention is made that the first index on the nonlinear susceptibility refers to the component of polarization, the second to the first field amplitude, and the third to the last field amplitude. This leads to the permutation symmetry rule that the indices on the nonlinear susceptibility may be interchanged provided the corresponding frequencies are interchanged simultaneously.
If all fields are time independent (all frequencies equal to zero), this leads to a tensor symmetric in all three indices. For second harmonic generation the tensor is symmetric in two indices and has, therefore, the same symmetry properties as a piezoelectric

tensor. If all three frequencies lie in an interval in which the dispersion is negligible, the tensor may to a very good approximation be considered to be symmetrical in the indices as pointed out by KLEINMAN. For certain applications, such as the generation of sum or difference frequencies in optically active fluids, the lack of this symmetry due to dispersion is essential. The dispersion of the nonlinear susceptibility is also important in materials that are absorbing in or near the frequencies of interest.

In this case the nonlinear susceptibility is complex, as shown by the harmonic oscillator example. The imaginary part cannot be related immediately to an absorption process. In second harmonic generation, for example, the complex susceptibility just implies a change of phase of the harmonic polarization with respect to the incident field,

$$P(2\omega) = |\chi^{NL}| e^{i\phi_{NL}} |A_1|^2 e^{(2i\phi_1 + 2i\underset{\sim}{k}_1 \cdot \underset{\sim}{r})} .$$

The phase ϕ_{NL} can be determined by interference of the second harmonic radiation fields from an absorbing and a nonabsorbing piezoelectric crystal, excited by the same fundamental laser beam.

It can be shown that the time-averaged power absorbed by the medium, cubic in the field amplitudes, is given by

$$W = + \omega_3 [\mathrm{Im}\chi(\omega_1, \omega_2, -\omega_3)] [A_1 A_2 A_3^* + A_1^* A_2^* A_3] .$$

This work can still be positive or negative depending on the relative phases $\phi_1 + \phi_2 - \phi_3$. This work describes the interference between single and two photon quantum transitions. If the medium is, for example, absorbing near ω_2, one has a matrix element of the form

$$P_{gn}A_2 + \frac{P_{gn'}P_{n'n}}{(\omega_3 - \omega_{gn'})} A_3A_1^* .$$

Here P_{gn} is the electric dipole matrix element between quantum states $|g\rangle$ and $|n\rangle$ and $\hbar\omega_{gn}$ is the corresponding energy. In the square of this element, destructive or constructive interference of the one and two quantum photon transitions becomes evident. It is important to note that the physical interpretation of the imaginary part of the third-rank tensor can be given only by considering simultaneously the two photon absorption process which is described by a fourth-rank tensor susceptibility.

When the medium is dissipative and nonlinear no general symmetry relations have yet been derived. For the linear dissipative medium one has of course a dissipation function and the Onsager relations. For the nonlinear dissipative medium one must rely on special relations, derived for specific models. The relationship (1.4) which exists for the Raman type susceptibility is an example of a complex symmetry relation.

1.3 Nonlinearities in Media with Inversion Symmetry and in Isotropic Media

The free enthalpy term which is quartic in the electric field amplitudes is the dominant nonlinear term in media with inversion symmetry. The symmetry properties of the fourth-rank tensor will not be elaborated, but the variety of physical effects described by this term may be enumerated as follows.

a) The intensity dependent index of refraction is described by a polarization of the form

$$P_i(\omega) = \chi_{ijkl}(-\omega,\omega,-\omega,\omega)E_j(\omega)E_k^*(\omega)E_l(\omega).$$

The important effect of self-focusing is described by this term, with $i = j = k = 1$. The index can also be changed by the presence of another light beam,

$$P_i(\omega_1) = \chi_{iikk}(-\omega_1,\omega_1,-\omega_2,\omega_2)E_i(\omega_1)|E_k(\omega_2)|^2.$$

If one takes $\omega_2 = 0$, this term represents the well known quadratic Kerr effect. The birefringence induced by a laser pulse has been observed.

b) Two quanta absorption is described by the imaginary part of this same nonlinear susceptibility,

$$P_i(\omega_1) = i\,\chi''_{iikk}(-\omega_1,\omega_1,\omega_1,-\omega_1)|E_k(\omega_1)|^2E_i(\omega_1).$$

This term implies an intensity dependent absorption coefficient. From the anharmonic oscillator model it is clear that χ'' has a maximum, when either $2\omega_1 = \omega_o$ or $\omega_1 = \omega_o$. In the former case we have two photon absorption; the absorption increases with increasing intensity. When ω_1 is at a resonant frequency, this term describes the onset of saturation. The absorption decreases with increasing intensity. This effect occurs in bleachable filters.

c) The stimulated Raman effect is described by a polarization at the Stokes frequency, ninety degrees out of phase with the Stokes field and proportional to the laser intensity,

$$P_i(\omega_2) = i\chi''_{ijkl}(-\omega_2,\omega_2,\omega_1,-\omega_1)A_j(\omega_2)A_k(\omega_1)A_2^*(\omega_1).$$

When $\omega_2 = \omega_1 - \omega_o$ (ω_o is a resonant frequency of the material), $\chi'' < 0$. One has indeed an exponential gain at ω_2. Conversely, there is an exponential loss at ω_1 proportional to the intensity at ω_2. These properties are explicitly verified for the anharmonic oscillator model in the introductory section. The signs would be reversed, if the population difference between the resonant levels were reversed. These are clearly two photon processes, in which one photon is emitted and one absorbed.

d) Many combination frequencies can be created, and in the most general case, any (complex) scattering among four light waves is described by this nonlinearity. An important example is the creation of a polarization at the antiStokes frequency,

$$P_i(\omega_3 = 2\omega_1-\omega_2) = \chi_{ijkl}(-\omega_3,\omega_1,\omega_1,-\omega_2)A_j(\omega_1)A_k(\omega_1)A_l(\omega_2).$$

This susceptibility is also purely imaginary, if $\omega_1-\omega_2 = \omega_o$, but is real away from atomic resonances.

e) Third harmonic radiation is of course generated by the polarization

$$P_i(3\omega) = \chi_{ijkl}(-3\omega,\omega,\omega,\omega)A_j(\omega)A_k(\omega)A_l(\omega).$$

We shall not discuss all other possible combination frequencies, nor the detailed symmetry properties of these and other tensors occurring in the general expansion.

If the medium is isotropic, such as in a glass, liquid, or gas, the nonlinear tensors should reduce to a number of scalar and pseudoscalar quantities. For an isotropic medium we have the following nonvanishing terms. $\underset{\sim}{P}(\omega_3 = \omega_1 + \omega_2) = \chi(\underset{\sim}{E}_1 x \underset{\sim}{E}_2)$ exists when the fluid is optically active, i.e., when it has no center of inversion symmetry. Second harmonic generation is possible from a term with magnetic dipole character

$$\underset{\sim}{P}(2\omega) = \alpha(\underset{\sim}{E} x \underset{\sim}{H}) = \alpha'(\frac{\partial \underset{\sim}{E}}{\partial t} x \underset{\sim}{H}),$$

where α is a pseudoscalar, α' a scalar. A term with electric quadrupole character, $\underset{\sim}{Q}(2\omega) = \beta'\underset{\sim}{EE}$, also exists. This quadrupolarization is equivalent to a surface dipole source term, $\underset{\sim}{P}_{eff} = -\nabla \cdot \underset{\sim}{Q}$. Integration over a volume element containing the boundary of the nonlinear medium yields

$$\underset{\sim}{P}_{eff}(2\omega) = \beta'\underset{\sim}{E}_{ins}(\underset{\sim}{E}_{ins} \cdot \hat{n}),$$

where $\underset{\sim}{E}_{ins} \cdot \hat{n}$ is the normal component of $\underset{\sim}{E}$ inside the nonlinear dielectric. This component is related to the divergence $\underset{\sim}{E}$ at the surface and is equal to $(\nabla \cdot \underset{\sim}{E})_{surf}/(\varepsilon(\omega) -1)$, which is a δ-function at the surface, $\underset{\sim}{P}_{eff} = \beta \underset{\sim}{E} (\nabla \cdot \underset{\sim}{E})$. For a free electron plasma, one finds readily from the equations of motion for a free electron gas with N electrons per cm^3.

$$\alpha = \frac{-ie^2N}{4m^2c\omega^3}, \quad \beta = \frac{e^2N}{m^2\omega^2\omega_p^2},$$

where

$$\varepsilon(\omega) = 1 - \frac{\omega_p^2}{\omega^2} = 1 - \frac{4\pi N_e^2}{m\omega^2}.$$

Jha first called attention to the importance of the surface term, which contributes more to the second harmonic intensity than the magnetic dipole term.

When a dc magnetic field is present, the term in $\underset{\sim}{P}(\omega) = iV(\underset{\sim}{E}x\underset{\sim}{H}_o)$ describes the Faraday rotation. The constant V is proportional to the Verdet constant. There is a term in the free enthalpy,

$$F = iV \underset{\sim}{E} \cdot (\underset{\sim}{E} \times \underset{\sim}{H}_o) = iV \underset{\sim}{H}_o \cdot (\underset{\sim}{E} \times \underset{\sim}{E}^*) = 2V H_o\{|E_{rc}|^2 - |E_{1c}|^2\}.$$

When this free enthalpy is differentiated with respect to $\underset{\sim}{H}_o$, one gets the inverse Faraday effect, or magnetization of a nonabsorbing body by a circular polarized light beam.

The intensity dependent index of refraction in an isotropic medium is characterized by two constants,

$$\underset{\sim}{P}(\omega) = \chi_1 \underset{\sim}{E}(\omega) \ \{\underset{\sim}{E}(\omega) \cdot \underset{\sim}{E}^*(\omega)\} + \chi_2 \underset{\sim}{E}^*(\omega) \ \left(\underset{\sim}{E}(\omega) \cdot \underset{\sim}{E}(\omega)\right),$$

and the third harmonic polarization by one constant,

$$\underset{\sim}{P}(3\omega) = \chi_3 \underset{\sim}{E}(\omega) \ \{\underset{\sim}{E}(\omega) \cdot \underset{\sim}{E}(\omega)\}.$$

The stimulated Raman effect needs in principle three constants to describe the directional and polarization effects that may occur in an isotropic medium,

$$\underset{\sim}{P}(\omega_2) = \chi_4 \underset{\sim}{E} \ (\underset{\sim}{E}_1 \cdot \underset{\sim}{E}_1^*) + \chi_5 \underset{\sim}{E}_1^* \ (\underset{\sim}{E}_1 \cdot \underset{\sim}{E}_2) + \chi_6 \underset{\sim}{E}_1 \ (\underset{\sim}{E}_1^* \cdot \underset{\sim}{E}_2).$$

These number subscripts bear no relation to the bracketed superscripts in the tensors used in the general expressions for F, but just denote different elements in the tensor $\chi^{(4)}_{ijkl}$.

2. Wave Propagation in Nonlinear Media

2.1 The Nonlinear Wave Equation

The nonlinearities of the constitutive relationships must now be incorporated into Maxwell's equations. This leads to the nonlinear wave equation for the vector potential (we take the Coulomb gauge with $\nabla \cdot \underset{\sim}{A} = 0$ and $\emptyset = 0$),

$$\nabla^2 \underset{\sim}{A} - \frac{1}{c^2} \ddot{\underset{\sim}{A}} = - \frac{4\pi}{c} \underset{\sim}{j} \ (\underset{\sim}{A}).$$

It gives somewhat more physical insight if we consider only an electric dipole nonlinearity, so that $\tilde{P}(E) = \tilde{\tilde{\chi}}^{L}\tilde{E} + \tilde{P}^{NL}(\tilde{E})$. Maxwell's equations can then be written, if for the sake of simplicity the magnetic permeability is taken equal to the vacuum value of unity,

$$\nabla \times \tilde{E} = -\frac{1}{c}\frac{\partial \tilde{B}}{\partial t}$$

$$\nabla \times \tilde{B} = +\frac{1}{c}\frac{\delta \tilde{E}}{\delta t} + \frac{1}{c}\frac{\partial \tilde{E}}{\partial t} + \frac{4\pi}{c}\frac{\partial \tilde{P}(\tilde{E})}{\partial t} .$$

These equations are considerably more complicated than the corresponding equations occurring in the nonlinear mechanics of shock waves because of the vector character of the field quantities and time dependence of the polarization response to the electric field value. If the important linear dispersion is ignored, one may assume that P at any instant is only a function of E at that same instant. If a linear polarized wave is considered, the problem becomes formally equivalent to the shock wave problem and the method of characteristics of Courant and Friedrichs (Supersonic Flow and Shock Waves, New York, 1948) may be used. This has been done by BROER (see, for example, L.J.F. BROER, Z. angewandte Math. u. Physik 16, 18, 1965) and by ROSEN (Phys. Rev. 139, A 539, 1965).

A nonmagnetic medium (B = H) fills a half space, z > 0. It has an arbitrary functional relationship between D or P^{NLS} and E. Maxwell's equations for one linear polarization taken in the y-direction in the medium are

$$\frac{\partial E_y}{\partial z} - \frac{1}{c}\frac{\partial B_x}{\partial t} = 0, \quad \frac{\partial B_x}{\partial z} = \frac{1}{c}\frac{\partial D_y}{\partial t} .$$

Henceforth the subscripts x and y will be omitted. Define a function $v^{-2}(E) = (dD/dE) \cdot c^{-2}$. The equations can be written as

$$\frac{\partial B}{\partial t} = c\,\frac{\partial E}{\partial z}\,, \quad \frac{\partial B}{\partial z} = v^{-2}(E)c\,\frac{\partial E}{\partial t}\,.$$

A solution can be written in the differential form,

$$dE = -(v/c)\ dB \quad \text{for} \quad dz/dt = v\ ,$$

$$dE = +(v/c)\ dB \quad \text{for} \quad dz/dt = -v\ .$$

There exist solutions in the nonlinear medium in which the wave travels away from the boundary. If the time variation at the boundary is prescribed E_y $(z = 0) = G(t)$, then the solution in the medium is

$$E = G\,\{t - \frac{z}{v(E)}\}\ .$$

The function G(t) is of course in turn determined by the incident and reflected waves in the vacuum, $z < 0$. The former is assumed to be given $f_{inc}(t - z/c)$. If the reflected wave is denoted by $f_R(t + z/c)$, the vacuum fields $(z < 0)$ are

$$E_{vac} = f_{inc}(t - z/c) + f_R(t + z/c),$$

$$B_{vac} = -f_{inc}(t - z/c) + f_R(t + z/c).$$

The latter solution has to be matched to the magnetic field inside the medium at z = 0. Integration of the characteristic differential gives

$$B(z = 0,t) = -\int_0^{G(t)} (c/v)\ dE = -\int_0^{G(t)} \left(\frac{dD}{dE}\right)^{1/2} dE.$$

Matching the tangential components at z = 0, one finds the implicit relation between the reflected and the incident wave,

$$f_i - f_R = \int_0^{f_i + f_R} \left(\frac{dD}{dE}\right)^{1/2} dE \ .$$

An inherent limitation of these solutions is that they do not remain single valued for large values of z. The solution can be described physically by saying that a certain value of E propagates with a certain velocity v(E). If the point with E = 0 travels faster than the point with $E = E_{max}$, different values of E overtake each other. That something goes wrong can be seen by differentiating

$$E = G(t - p(E)z),$$

where we have written p(E) instead of $v^{-1}(E)$.

$$\frac{\partial E}{\partial z} = \left(-p(E) - z \frac{\partial p}{\partial E} \frac{\partial E}{\partial z}\right) G' \ ,$$

$$\frac{\partial E}{\partial z} = \frac{- p \ G'}{1 + z \frac{\partial p}{\partial E} G'} \ .$$

Thus, $\frac{\partial E}{\partial z}$ becomes infinite for large z, if $\frac{\partial p}{\partial E}$ G' is negative. This would imply the formation of a shock front, the steepness of the boundary being determined by some dissipative processes whose origin is not known in the electromagnetic case. In the gas shock wave the shock front structure is of course connected with the collisional mean free path between gas molecules, that is by the viscosity and heat conduction. These collision times are very short compared to the period of the acoustic waves. At any instant the pressure may be taken as a function of the acoustic amplitude. In practice, this situation will never arise in the EM case. For a very intense light flux ($\sim 10^{11}$ watts/cm^2) obtainable in a focused laser beam, the velocity in the wave maxima may be $1:10^6$ slower than in the nodes. Therefore, after about 10^6 wavelengths, a series of shock fronts would have developed in an originally sinusoidal wave. If the wave energy had been compressed by a small fraction, the shorter wavelength components would have produced a linear dispersion in the velocities, which would be very much larger than $1:10^6$. Therefore, this phenomenon will not occur. Another way of expressing this fact is that the optical period is short compared to damping times, etc. Therefore, the polarization at any instant is not a function of E alone at that instant, but depends through a response function on earlier values of the field. Furthermore, material breakdown, ionization, and evaporation would occur as soon as the electric field would develop frequency components in the steepened front that can be absorbed and produce carriers in the medium.

BROER has also solved the dispersionless case for oblique incidence. He finds that the direction of propagation in space depends on the local value of the field strength. This is not too surprising, since an intensity dependent index of refraction would have the same effect. In fact, if one takes for the incident wave a sine wave $f_i(z = 0) = \cos \omega_o t$, and expands the function $p(E) = v^{-1}(E)$ in an ascending power series in E, the term in E^3 will give a correction to the linear law of refraction. The deviation in angle would be about 10^{-6} radian in the strongest realizable field and would be very difficult to detect.

This same series expansion would give rise to harmonics in the reflected wave f_R and the transmitted solution, $G(t - p(E)z)$. The latter will contain terms that will continue to increase with z. These so-called secular terms cannot be avoided in the wave propagation problem without dispersion.

In the remainder of this lecture we shall reproduce a systematic method to handle these secular terms. This method was developed by BOGOLIUBOV in nonlinear mechanics (N. KRYLOV and N. BOGOLIUBOV, Introduction to Nonlinear Mechanics, translated by S. LEFSHETZ, Princeton University Press, Princeton, New Jersey, 1947). The method was extended to the partial differential wave equation by D. MONTGOMERY and D.A. TIDMAN (Phys. of Fluids 7, 242, 1964), where further references may be found. It should be noted at this time that the actual physical situation encountered with intense laser beams in fluids and crystals requires a somewhat different approach that will be discussed in the next lecture.

2.2 Bogoliubov's Method for a One-dimensional Nonlinear Mechanical System

Consider the equation

$$\left(\frac{d^2}{dt^2} + \omega_o^2\right) f(t) = -\varepsilon F(f , \frac{df}{dt}) ,$$

where ε is a formal expansion parameter and F an arbitrary non-linear function. For $\varepsilon = 0$ the solution is $f = a \sin(\omega_o t + \phi)$. For small $\varepsilon \neq 0$, a and ϕ may be considered "slowly varying" functions of the time. Their time derivatives are of order ε. If second order derivatives are ignored or required to vanish, substitution into the original equation leads to

$$\omega_o \frac{da}{dt} \cos(\omega_o t + \phi) - \omega_o a \frac{d\phi}{dt} \sin(\omega_o t + \phi)$$

$$= - \frac{1}{2} \varepsilon F \{a \sin(\omega_o t + \phi), a \omega_o \cos(\omega_o t + \phi)\}.$$

Expand F into a Fourier series, introduce $\psi = \omega_o t + \phi$,

$$F\{\sin\psi,\ a\,\omega_o\cos\psi\} = f_o(a) + \sum_{n=1}^{\infty}[f_n(a)\cos n\psi + g_n(a)\sin n\psi],$$

$$\frac{da}{dt} = -\frac{1}{2}\frac{\varepsilon}{\omega_o} f_1(a) + \varepsilon\ \text{(time varying terms)},$$

$$\frac{d\phi}{dt} = +\ \frac{1}{2}\ \frac{\varepsilon}{a\omega_o} g_1(a) + \varepsilon\ \text{(time varying terms)}.$$

The time varying terms give rise to a constant displacement and higher harmonics in f. This suggests attempting a better solution in the form

$$f = a\sin(\omega_o t + \phi) + \frac{\varepsilon}{\omega_o^2}\Big[-f_o(a)$$

$$+ \sum_{n=2}^{\infty}\frac{f_n(a)\cos n(\omega_o t + \phi) + g_n(a)\sin n(\omega_o t + \phi)}{n^2 - 1}\Big].$$

Substituting into the original equation and ignoring terms of order ε^2 gives

$$\frac{da}{dt} = -\frac{1}{2}\frac{\varepsilon}{\omega_o} f_1(a),$$

$$\frac{d\phi}{dt} = +\frac{1}{2}\frac{\varepsilon}{a\omega_o} g_1(a).$$

To order ε the second derivatives indeed vanish. These equations determine the secular perturbation. Note that $d\phi/dt$ represents the change in frequency, $\omega = \omega_o + \frac{\varepsilon}{2\omega_o a} g_1(a)$. If $F(f, df/dt)$ does not depend on df/dt, then all $f_n = 0$. The Fourier expansion contains only sine terms. Therefore, the amplitude is constant and there is a constant frequency shift determined by the above expression. If $F = f^2$ or $f\frac{df}{dt}$ or $\left(\frac{df}{dt}\right)^2$, the Fourier expansion does not contain the terms f_1 and g_1. In this case the nonlinear driving term has no Fourier component near the fundamental frequency, and no secular perturbation is present. For further examples and higher order approximations we refer to the textbook: N. BOGOLIUBOV and Y.A. MITROPOLSKI, Asymptotic Methods in the Theory of Nonlinear Oscillations (Gordon and Breach, New York, 1961).

2.3 The Secular and Nonsecular Behavior of the Wave Equation

MONTGOMERY and TIDMAN examine the equation

$$\left(\frac{\partial^2}{\partial t^2} - c^2 \frac{\partial^2}{\partial x^2} + \lambda^2 \right) f(x, t) = \varepsilon F\left(f, \frac{\partial f}{\partial t}, \frac{\partial f}{\partial x}\right),$$

where c^2 and λ^2 are positive real constants. Perturbations will be considered around the solution of a single monochromatic wave for $\varepsilon = 0$,

$$f = a \cos (k_o x - \omega_o t + \phi),$$

where $$\omega_o^2 = c^2 k_o^2 + \lambda^2.$$

Following the same procedure, we shall seek a solution to the non-linear problem, $\varepsilon \neq 0$, in which $f = a \cos \psi + \varepsilon u_1(a, \psi) + \varepsilon^2 u_2(a,\psi)$. Here $\psi = \omega_o t + \phi$, and $a(x, t)$ is a "slowly varying" amplitude which satisfies

$$\frac{\partial a}{\partial t} = \varepsilon A_1(a) + \ldots.,$$

$$\frac{\partial a}{\partial x} = \varepsilon D_1(a) + \ldots. \quad .$$

The new phase variable ψ is chosen to satisfy the equations

$$\frac{\partial \psi}{\partial t} = - \omega_o + \varepsilon B_1(a) + \ldots. ,$$

$$\frac{\partial\psi}{\partial x} = k_o + \varepsilon\, C_1(a) + \dots\ .$$

The as yet undetermined functions are to be chosen so as to eliminate the "secular" terms which can grow linearly in x and/or t. Substituting the "Ansatz" for f back into the differential equation, one obtains

$$\frac{\partial^2 f}{\partial t^2} - c^2 \frac{\partial^2 f}{\partial x^2} + \lambda^2 f = (-\omega_o^2 + c^2 k_o^2 + \lambda^2)\, a \cos\psi$$

$$+ \varepsilon\Big[\lambda^2 \left(\frac{\partial^2 u_1}{\partial\psi^2} + u_1\right) + 2(\omega_o A_1 + c^2 k_o D_1) \sin\psi$$

$$+ 2\, a\, (\omega_o B_1 + c^2 k_o C_1) \cos\psi\Big] + O(\varepsilon^2)$$

$$= \varepsilon\, F(f, \frac{\partial f}{\partial t}, \frac{\partial f}{\partial x})\ .$$

Replace in F the arguments by their value for the zero-order solution and expand in a Fourier series

$$F(a\cos\psi,\ \omega_o\, a \sin\psi,\ -k_o\, a \sin\psi)$$

$$= g_o(a) + \sum_{n=1}^{\infty}(g_n(a) \cos n\psi + f_n(a) \sin n\psi).$$

Equating the coefficients for ε, one obtains the equation for u_1,

$$\lambda^2\left(\frac{\partial^2}{\partial\psi^2} + 1\right) u_1 = g_o(a) + [f_1(a) - 2\omega_o A_1 - 2c^2 k_o D_1]\sin\psi$$

$$+ [g_1(a) - 2\, a\, \omega_o B_1 - 2\, a\, c^2 k_o C_1]\cos\psi$$

$$+ \sum_{n=2}^{\infty}[g_n(a) \cos n\psi + f_n(a) \sin n\psi]\ .$$

The secular perturbations are those that vary at the same frequency ψ as the unperturbed solution. Therefore, we choose A_1, B_1, C_1 and D_1 such that they satisfy

$$2(\omega_o A_1 + c^2 k_o D_1) = f_1(a),$$

$$2\ a\ (\omega_o B_1 + c^2 k_o C_1) = g_1(a).$$

Since $\frac{\partial^2 \psi}{\partial t \partial x} = \frac{\partial^2 \psi}{\partial x \partial t}$ and $\frac{\partial^2 a}{\partial t \partial x} = \frac{\partial^2 a}{\partial x \partial t}$,

we have the following two additional relations,

$$A_1 \frac{\partial C_1}{\partial a} = D_1 \frac{\partial B_1}{\partial a},$$

$$A_1 \frac{\partial D_1}{\partial a} = D_1 \frac{\partial A_1}{\partial a}.$$

These four equations for the four unknowns A_1, B_1, C_1 and D_1 must now be solved, with the aid of the boundary conditions of the physical problem. It is clear that B_1 can be interpreted as a frequency shift and C_1 as a wave number shift. In practice f is often prescribed to vary sinusoidally with t at x = 0, because a laser beam with a given frequency is incident on the nonlinear medium. In that case A_1 and B_1 may be taken equal to zero. This leaves a wave number shift

$$C_1 = g_1(a)/2\ c^2 k_o a$$

and

$$D_1 = f_1(a)/2\ c^2 k_o .$$

Conversely, in a laser oscillator, the wave vector may be prescribed to have a certain value because the mirror spacing has to be equal to a specified number of half wavelengths. Then one may take $C_1 = D_1 = 0$, etc.

If F contains only quadratic terms, say f^2 or $f \frac{\partial f}{\partial t}$, then $f_1(a)$ and $g_1(a)$ vanish in the Fourier expansion. There is no problem of secularity. A_1, B_1, C_1 and D_1 may all be taken equal to zero,and a completely straightforward perturbation theory results. If, however, F is proportional to f^3, then $f_1(a) = 0$, but $g_1(a) = 3\, a^3/4$. A wave number shift occurs.

When the secular perturbations have thus been eliminated, the solution for u_1 is

$$u_1 = \frac{1}{\lambda^2} \left[g_o(a) + \sum_{n=2}^{\infty} \frac{g_n(a)}{-n^2+1} \cos n\,\psi + a\, \frac{f_n(a)}{-n^2+1} \sin n\,\psi \right].$$

When this is substituted back into $F(a \cos \psi + \varepsilon\, u_1, \frac{\partial}{\partial t} \,.., \frac{\partial}{\partial x} \,..)$, one gets corrections which are one order higher in ε. It should be emphasized that the amplitude of higher harmonics goes as ε/λ^2. The nonlinearity must be compared with the dispersion. In most practical cases this expansion parameter is very small compared to unity.

For $\lambda = 0$, the secular terms cannot be eliminated. Then the solutions are valid only for sufficiently small time intervals and/or small distances.

When two or more monochromatic waves are incident, the general problem of secularity becomes quite complex. We shall now turn to a method which was developed independently of these theories and is specifically adapted to experimental cases of optical dispersion. It turns out that this method is capable of handling several monochromatic waves at the same time and is also capable of handling secular perturbations, if these arise. The mathematical model of dispersion used in the differential equation is somewhat artificial. The situations of experimental importance often display very different dispersion characteristics.

3. Boundary Conditions and Coupling Between Light Waves in a Dispersive Medium

Since the linear dispersion is usually large in comparision to the nonlinear coupling (the quantity $\varepsilon/\lambda^2 << 1$ in the preceding mathematical example of dispersion), it is natural to consider the free eigenwaves in the linear dispersive medium as the starting point for the discussion. The nonlinearity will provide a "weak" coupling between them. It is again advantageous to expand both the fields and the polarization in a Fourier series. The linear polarization may be lumped with the electric field by means of the linear dielectric constant.

In many practical situations, the boundary conditions are such that there are one or more incident light waves of *prescribed* frequencies. In this case the amplitudes of the waves of the linear solution may be considered as slowly varying functions of the position z. This type of solution is used in this section. Solutions where the amplitude of a geometrically fixed electromagnetic mode function is considered as a function of time can be obtained in a similar manner. Examples will be given in Chapters 4 and 5. Mixed spatial and time variations can also be treated, but are somewhat more complex.

3.1 Parametric Generation

Consider an infinite nonabsorbing isotropic medium and suppose that at z = 0 only one homogeneous, plane, monochromatic wave is present,

$$E_1 = A_1 \, e^{ik_1 z - i\omega t}$$

with

$$k_1 = \varepsilon^{1/2}(\omega) \, \omega \, c^{-1} \; .$$

Assume that the crystal is piezoelectric and only the lowest order nonlinearity is of importance. The nonlinear polarization at the second harmonic frequency is

$$P(2\omega) = \chi^{NL} \, A_1^2 \, e^{2ik_1 z - 2i\omega t} \; .$$

This nonlinear phased array of dipoles will drive a wave at the second harmonic frequency. The nonlinear wave equation for this component takes the form

$$-\frac{\partial^2 E_2}{\partial z^2} - \frac{\varepsilon(2\omega)(2\omega)^2}{c^2} E_2 = \frac{4\pi(2\omega)^2}{c^2} \chi^{NL} \, A_1^2 \, e^{2ik_1 z - 2i\omega t} \; .$$

As long as the amplitude of the harmonic waves remains small, A_1 may be considered as a constant parameter. This is the basis of the parametric approximation, which linearizes the problem. The solution of the inhomogeneous wave equation is

$$E_2 = A_2 \, e^{ik_2 z - 2i\omega t} + \frac{4\pi(2\omega/c)^2 \chi^{NL} A_1^2}{|2k_1|^2 - k_2^2} \, e^{2ik_1 z - 2i\omega t} \; .$$

The first term is the solution of the homogeneous equation for a linear medium with $k_2 = \varepsilon^{1/2}(2\omega)c^{-1}2\omega$. Due to color dispersion, $\varepsilon\,(2\omega) \neq \varepsilon(\omega)$ and $2k_1 \neq k_2$. The amplitude A_2 is chosen so that $A_2 = 0$ for $z = 0$. One thus finds

$$E_2 = \frac{4\pi(2\omega/c)^2 \chi^{NL} A_1^2}{4k_1^2 - k_2^2} \left[1 - e^{i(2k_1 - k_2)z}\right] e^{ik_2 z - 2i\omega t} \; . \qquad (3.1)$$

The second-harmonic intensity will vary periodically as

$$|E_2|^2 = 16\ \pi^2\ |\chi^{NL} A_1^2|^2\ \frac{4\ \sin^2\ [\frac{1}{2}\ \omega c^{-1}(\varepsilon^{1/2}(\omega)\ -\ \varepsilon^{1/2}(2\omega))]z}{[\varepsilon(\omega)\ -\ \varepsilon(2\omega)]^2}\ .$$

Without special precautions the harmonic production will always remain small. In many experiments, less than $1{:}10^6$ of the laser intensity is converted to harmonics. A more precise discussion of the boundary conditions shows (see Sect. 4.1 and Appendix 2 of Nonlinear Optics) that there is also a small reflected harmonic wave. A small term, equal in amplitude to the reflected wave, must be added to Eq. (3.1). The directions of the waves are fixed by the condition that the components of momentum tangential to the boundary must be conserved. The momentum mismatch and the amplitude variation is therefore restricted to the direction normal to the boundary.

3.2 Secular Harmonic Conversion

If one wants a large energy conversion to harmonic frequencies, phase matching must be achieved. If $\varepsilon(\omega) = \varepsilon(2\omega)$ and the fundamental and harmonic waves travel at the same velocity, the parametric solution breaks down. In this case A_2 increases linearly with z. Physically it is clear that conservation of energy will prevent A_2 from becoming larger than A_1 $(z = 0)$. When A_2 becomes comparable to A_1, the reaction of the harmonic wave back on the fundamental cannot be ignored.

The experimentally important case is not that of a dispersionless medium in which all harmonics, and eventually a shock wave, would be created, but in which the color dispersion between the fundamental and one harmonic is compensated by birefringence in uniaxial crystals. This technique was first employed by GIORDMAINE and TERHUNE. The phase velocity of the extraordinary ray at 2ω is matched to the velocity of the ordinary polarized ray at ω. In KDP at the ruby frequency, e.g., the beams have to travel at an angle of 49° with respect to the optic axis.

We thus encounter the situation where two Fourier components are strongly coupled but other waves with other frequencies and polarizations can be ignored. The technique for solving such a secular problem was developed by ARMSTRONG, BLOEMBERGEN, DUCUING, and PERSHAN (Phys. Rev. 127, 1918, 1962). See Sects. 4, 5, and 6 of this paper and Sect. 4.2 of Nonlinear Optics. The case of three coupled waves which satisfy the energy and momentum condition $\omega_1 + \omega_2 = \omega_3$, $\underline{k}_1 + \underline{k}_2 = \underline{k}_3$ has also been treated. The problem is described by complex coupled amplitude equations.

3.3 The Propagation of a Wave in a Medium with a Complex Intensity Dependent Index of Refraction

In a medium with inversion symmetry, the lowest order nonlinearity is a polarization cubic in the field strength. One may write down the coupled wave equations of the fundamental and third harmonic,

$$\nabla \times \nabla \times E_1 - \varepsilon(\omega)\, \omega^2 c^{-2} E_1$$

$$= 4\pi(\omega/c)^2 \chi_1^{NL} |E_1|^2 E_1 + \frac{4\pi\omega^2}{c^2} \{\chi_2 |E_3|^2 E_1 + \chi_3 E_3 E_1^{*2}\},$$

$$\nabla \times \nabla \times E_3 - \varepsilon(3\omega)(3\omega)^2 c^{-2} E_3$$

$$= 4\pi(3\omega/c)^2 \chi_3 E_1^3 + 4\pi(3\omega/c)^2 \chi_2 |E_1|^2 E_3 + 4\pi(3\omega/c)^2 \chi_4 |E_3|^2 E_3 ,$$

together with the corresponding complex conjugate equations. An exact solution for this set of equations has been given by ARMSTRONG et al. (Phys. Rev. 127, 1918, 1962). See also Nonlinear Optics p. 189.

With the boundary condition, $E_3(z = 0) = 0$, we may take $|E_1|^2$ constant. Thus far no significant percentage of power has been converted to third harmonic in this manner, even if the third harmonic was near phase matching. In the second equation we therefore ignore the term in χ_4. The term in χ_2 gives a change in index of refraction at 3ω due to $|E_1|^2$. This term may be lumped with the linear term in $\varepsilon(3\omega)$ on the left-hand side. The third harmonic intensity can thus be solved in the parametric approximation.

The fundamental Fourier component, to first order in the nonlinearity, is then described by omitting the last two terms. The complex amplitude equation for the fundamental wave $A_1 e^{ik_1 z - i\omega t}$ becomes

$$\frac{dA_1}{dz} = \frac{2\pi i\omega}{c^2 k_1} (\chi_1' + i\,\chi_1'')^{NL} |A_1|^2 A_1 .$$

Separating this nonlinear equation into real and imaginary parts, one finds immediately the amplitude and phase equations – on the assumption that there is no linear absorption and k_1 is real –

$$\frac{d|A_1|}{dz} = -\frac{2\pi\omega}{c^2 k_1} (\chi_1'')^{NL} |A_1|^3$$

$$\frac{d\phi_1}{dz} = \frac{2\pi\omega}{c^2 k_1} (\chi_1')^{NL} |A_1|^2 .$$

If there is no nonlinear dissipative process, such as two photon absorption, then $(\chi_1'')^{NL} = 0$ and the amplitude of the fundamental wave is constant. The phase equation describes a wave number shift, proportional to the intensity. It should be emphasized that this procedure is exactly equivalent to the "refined first approximation" of Bogoliubov for a cubic nonlinearity. The last two equations give the correct treatment for the secularity. If the medium is dissipative, k_1^{-1} may be complex also. One may still integrate an equation for the amplitude of the form

$$\alpha = \mathrm{Im}\, k_1, \quad \beta = \mathrm{Im}\left|\frac{2\pi\omega\, \chi^{NL}}{c^2 k_1}\right| ,$$

$$\frac{d|A_1|}{dz} = -\beta|A_1|^3 - \alpha|A_1| .$$

Multiply both sides by $|A_1|$ and one obtains a differential equation for the intensity

$$\frac{dI_1}{dz} = -\, 2\beta\, I_1^2 - 2\alpha\, I_1 ,$$

and solving

$$\frac{I_1}{1 + \frac{\beta}{\alpha} I_1} = \frac{I_1(0)}{1 + \frac{\beta}{\alpha} I_1(0)}\, e^{-2\alpha z} .$$

For $\alpha = 0$, the solution is $I_1 = \dfrac{1}{I_1^{-1}(0) + 2\beta z}$, which would apply for pure two-photon absorption processes only. The phase equation can be solved by substitution of $|A_1|^2 = I_1(z)$ in the expression for $d\phi/dz$.

3.4 Saturable Absorber

In this case it is of interest to retain all powers of the intensity of the laser field. The effect of saturation is that the population difference between the pair of levels responsible for the absorption is changed by the intense absorption rate. The transitions induced by the light occur at a rate comparable to or faster than the rate of relaxation mechanisms. It can be shown (see Eq. 5.5 of Chap. 5, or Appendix 3 of Nonlinear Optics, especially Eq. 3.6 on page 215) that the absorption coefficient varies as $\alpha(1 + \beta' I_1)^{-1}$, where $\beta' = |er|^2 T_1 T_2$, and $|er|$ is the electric dipole matrix element between the pair of energy levels involved in the absorption process, T_1 is the relaxation rate of the population difference, and T_2^{-1} a measure for the line width. This formula applies to a single resonant transition, but could readily be extended to energy band situations. The differential equation for the intensity is, accordingly,

$$\frac{dI_1}{dz} = \frac{-\alpha I_1}{1 + \beta' I_1},$$

$$\left(\frac{1}{I_1} + \beta'\right) dI_1 = -\alpha\, dz.$$

Integrating, we obtain

$$\ln I_1 + \beta' I_1 = -\alpha z + \ln I(0) + \beta' I(0).$$

When the intensity is sufficiently high, $\beta' I(0) >> 1$, the term in $\ln I_1$ and $\ln I(0)$ may be ignored in first approximation,

$$I_1 = I(0) - \frac{\alpha}{\beta'} z.$$

The intensity only decreases, as it is necessary to retain saturation in each subsequent layer dz. This leads to a very small relative power loss at high intensities.

3.5 Second Harmonic Generation in Dissipative Crystals Without Inversion Symmetry

As a last example we consider the harmonic generation in a crystal without inversion symmetry, which is absorbing at 2ω, but not at the fundamental frequency. Interesting interference effects between the harmonic generation, one- and two-photon processes may then occur. We consider only the case of exact resonance in which the χ^{NL} are pure imaginary and of exact phase matching.

$$\frac{d^* A_1}{dz} = \frac{2\pi i \omega^2}{k_1 c^2} \left[\chi^{NL*}(2\omega, \omega, \omega) A_2 A_1^* + \chi^{NL}(\omega, -\omega, \omega, -\omega) A_1^2 A_1^*\right],$$

$$\frac{d^* A_2}{dz} = + \frac{4\pi i \omega^2}{k_1 c^2} \chi^{NL}(2\omega, \omega, \omega)\, A_1^2 - \alpha_2 A_2.$$

Under the conditions mentioned, one may find a steady state solution, in which the left- and right-hand sides are zero with

$$A_2 = \frac{4\pi\omega^2}{k_1 c^2 \alpha_2} \chi''^{NL} A_1^2,$$

$$\chi''^{NL}(\omega, -\omega, \omega, -\omega) = \frac{4\pi\omega^2}{k_1 c^2} \frac{|\chi^{NL}(2\omega, \omega, \omega)|^2}{\alpha_2}.$$

This last relation between the nonlinear susceptibilities follows from an explicit calculation, if only one- and two-photon resonant processes are retained.

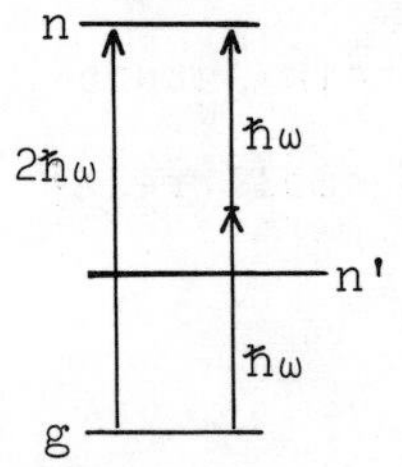

For the indicated ratio of fundamental to second harmonic fields the net transition probability between levels g and n vanishes. The material is nonabsorbing under these conditions due to exact destructive interference between one- and two-photon processes. In practice this situation is not likely to arise, because one usually has a distribution of energy levels and several laser and second harmonic modes. The exact cancellation could occur for only one pair of levels for only one pair of modes. The example indicates, however, that rather unexpected phenomena may occur and illustrates the meaning of the imaginary part of the second harmonic nonlinear susceptibility.

Photo-electric currents have been induced in materials (PbSe at 10.6 micron of a CO_2 laser) due to two-photon absorption. In piezo-electric crystals such as CdS the photo-electric current might either increase or decrease if radiation at the second harmonic is admitted in addition to the fundamental, depending on the relative phase of the harmonic with respect to the fundamental, $(\phi_2 - 2\phi_1)$.

4. Stimulated Brillouin, Raman and Rayleigh Scattering Effects

4.1 Self-Focusing of the Laser Beam

Thus far we have considered only homogeneous plane waves of infinite cross section. Actually the laser beam has a finite transverse intensity profile. For the lowest single mode from a gas laser this profile is given by a Gaussian. The transverse distribution leads to an index of refraction which is slightly higher on the axis, where the intensity is highest. As a consequence a light ray initially parallel to the axis will be bent toward the axis because of the transverse gradient in index of refraction. Consider the initial distance from the axis r. Let a be a measure of the spot size of the light. The transverse second derivative in intensity of the laser beam is on the order of $4|E_1|^2/a^2$. The radius of curvature ρ of a light ray at a distance r is

$$\frac{1}{\rho} = \frac{1}{n_o}\frac{dn}{dz} \approx \frac{n_2}{n_o} \frac{d^2|E_1|^2}{dz^2} r = 4\frac{n_2|E_1|^2 r}{n_o a^2} ,$$

where the total dielectric function $n = n_o + n_2|E_1|^2$, and n_2 is related in a straightforward manner to the nonlinear susceptibility discussed earlier. The light ray will cross the axis at an approximate distance, $(\rho r)^{1/2}$

$$z_{foc} \approx \frac{1}{2} a \left(\frac{n_o}{n_2|E_1|^2}\right)^{1/2} .$$

For an intense laser beam in a material with a large value of n_2 (CS_2 and other liquids with highly anisotropic molecules) the distance z_{foc} may be as short as $10^3 a \sim 1$ meter. We shall now give a more quantitative derivation (see, e.g., P. L. KELLEY, Phys. Rev. Letters 15, 1005, 1965). The wave equation has already been given in the preceding chapter. The third harmonic contribution is dropped, the variation along the beam in the z-direction is described by a slowly varying amplitude, but the second derivative in the transverse direction must be retained. Assume that the field remains linearly polarized and has a form

$$\underset{\sim}{E} = \underset{\sim}{e} A \exp(ikz - i\omega t) + c.c., \qquad E_o = 2|A|,$$

$$2ik \frac{\partial A}{\partial z} + \frac{\partial^2 A}{\partial x^2} + \frac{\partial^2 A}{\partial y^2} + \frac{8\pi\chi^{NL}\omega^2}{c^2} |A|^2 A = 0.$$

For a plane parallel wave $\partial_x = \partial_y = 0$ and χ^{NL} real, we have again the phase variation with intensity,

$$A = A_o \exp\left(-\frac{4\pi i\chi^{NL}\omega^2}{c^2 k} z\right) = A_o \exp\left(\frac{i n_2 k}{n_o} |E_o|^2 z\right),$$

where

$$n^2 = n_o^2 + 8\pi\chi^{NL}|A|^2,$$

$$n = n_o + \frac{4\pi\chi^{NL}|A|^2}{n_o} = n_o + n_2|E_o|^2.$$

If we separate the wave equation again into real and imaginary parts, we get the amplitude and phase equations,

$$k\frac{d|A|^2}{dz} = -\frac{\partial |A|^2}{\partial x}\frac{\partial \phi}{\partial x} - \frac{\partial |A|^2}{\partial y}\frac{\partial \phi}{\partial y} - |A|^2 \left(\frac{\partial^2 \phi}{\partial x^2} + \frac{\partial^2 \phi}{\partial y^2}\right),$$

$$2k\frac{\partial \phi}{\partial z} + \left(\frac{\partial \phi}{\partial x}\right)^2 + \left(\frac{\partial \phi}{\partial y}\right)^2 = \frac{n_2}{n_o} k^2 |A|^2 + \frac{1}{|A|}\left(\frac{\partial^2 |A|}{\partial x^2} + \frac{\partial^2 |A|}{\partial y^2}\right).$$

On the axis the intensity and the phases have extremal values. The equations reduce to

$$k\frac{\partial I_o}{\partial z} = - I_o \nabla_\perp^2 \phi \quad \text{and} \quad \phi = \frac{n_2 k}{2n_o} \int_0^z I_o \, dz.$$

In integrating the last equation, the phase change due to diffraction (the last term) has been ignored. Approximate the last equation by

$$\phi = \frac{n_2 k}{2n_o} I_o z \quad \text{and} \quad \nabla_\perp^2 I_o = -4 I_o^2/a^2 I_o(z = 0).$$

This last equation implies that the effective radius of the beam in any cross section is inversely proportional to the intensity. The beam retains its shape while it narrows. With these assumptions the first equation becomes

$$\frac{\partial I_o}{\partial z} = \frac{I_o^3}{I_o(z=0)} \frac{z n_2}{2n_o a^2} = \frac{I_o^3}{I_o^2(z=0)} \frac{z}{z_{foc}^2} .$$

The intensity on the axis increases as

$$\frac{I_o(z)}{I_o(0)} = \frac{z_{foc}}{(z_{foc}^2 - z^2)^{1/2}} .$$

The focal distance so derived must be corrected for diffraction effects. CHIAO, GARMIRE, and TOWNES have shown that a stationary solution for beam trapping exists, in which the z-dependence vanishes.

The light propagates in a self-focused wave guide, where diffraction and focusing effects are balanced. It is not clear that such a solution is stable against small perturbations in the total power carried in the beam. If one equates the focal distance to the Fresnel diffraction length of the beam, one obtains an approximate threshold for self-focusing of a cylindrical beam,

$$\frac{1}{2} a \left(\frac{n_o}{n_2|E_1|^2}\right)^{1/2}_{th} = \frac{4a^2 n_o}{(1.22\ \lambda)} .$$

Squaring both sides, this may be brought into the form

$$a^2 |E_1|^2_{th} = \frac{n_o}{4n_2} (1.22\ \lambda)^2 .$$

The total power in the beam for which self-focusing is just balanced by diffraction is

$$P_{cr} = \frac{c\pi a^2 n |E_1|^2_{th}}{8\pi} = \frac{(1.22\ \lambda)^2 c}{512\ n_2} .$$

For $n_2 \sim 10^{-11}$esu, which is a large value pertaining to CS_2, this critical intensity is only about 10 kw and is easily exceeded in Q-switched laser operation. The diffraction may be included in the net focal length formula by

$$l_{foc} = \frac{1}{2} a \left(\frac{n_o}{n_2}\right)^{1/2} \left(\frac{1}{|E_1| - |E_{cr}|}\right) = \frac{1}{4} n_o \left(\frac{c}{n_2}\right)^{1/2} a^2 \frac{1}{P^{1/2} - P_{cr}^{1/2}} .$$

This self-focusing is most pronounced in fluids with optical anisotropy. These molecules align themselves preferentially with their axis of easy polarization parallel to the electric field of the linear polarized laser beam. This minimizes the dielectric energy. This orientation effect in turn enhances the index of refraction.

The time averaged dielectric energy of the molecule with an axis of symmetry, making an angle θ with the laser field with real amplitude $|E_L|$, is

$$- \frac{1}{4} |E_L|^2 (\alpha_\parallel \cos^2\theta + \alpha_\perp \sin^2\theta)$$

$$= \frac{1}{12}|E_L|^2(\alpha_\parallel + 2\alpha_\perp) - \frac{1}{12}|E_L|^2(\alpha_\parallel - \alpha_\perp)(3\cos^2\theta - 1).$$

The energy due to permanent electric dipole moments is linear in E_L and must be excluded because the molecular orientation cannot follow the light frequency. The probability of having an angle θ is proportional to the Boltzmann factor,

$$f(\theta) = \exp[+ \frac{1}{12} |E_L|^2 (\alpha_\parallel - \alpha_\perp)(3\cos^2\theta - 1)/kT].$$

The change in polarizability from the average, $\frac{1}{3}(\alpha_\parallel + 2\alpha_\perp)$, due to the orientation is consequently

$$\Delta\alpha = \int_0^\pi \frac{1}{3} (\alpha_\parallel - \alpha_\perp)(3\cos^2 \theta - 1) f(\theta) \, d\theta \Big/ \int_0^\pi f(\theta) \, d\theta.$$

The change in index of refraction, to terms in $(kT)^{-1}$, becomes

$$n_2 |E_L|^2 = \frac{2\pi(\alpha_\parallel - \alpha_\perp)^2 N_o}{45 \; kT \; n_o} L \, |E_L|^2,$$

where L is a Lorentz local field correction factor,

$$L = [n_o^2 + 2]^4 / 81.$$

Other possible contributions to the intensity dependent index come from the electronic orbitals, as discussed in the anharmonic oscillator model, and from the electrostrictive effect.

4.2 Parametric Down Conversion

Consider the coupling between three electromagnetic waves with the energy and momentum relations at least approximately satisfied. Suppose that initially only the wave at the highest frequency, $\omega_L = \omega_3 = \omega_1 + \omega_2$, is excited. The question is whether an oscillation can be set in motion at the smaller frequencies, $\omega_1 = \omega_s$ (s for "signal") and $\omega_2 = \omega_i$ (i for "idler"). At least initially the fields E_1 and E_2 will be very small in comparison to E_L and the latter quantity may be considered as a constant parameter. In this case the problem reduces to two coupled linear equations

$$\nabla \times \nabla \times E_s + \frac{1}{c^2}\frac{\partial^2}{\partial t^2}\varepsilon(\omega_s)E_s = \frac{4\pi\omega_s^2}{c^2}\chi(-\omega_s, \omega_L, -\omega_i)E_L E_i^*,$$

$$\nabla \times \nabla \times E_i + \frac{1}{c^2}\frac{\partial^2}{\partial t^2}\varepsilon^*(\omega_i)E_i^* = \frac{4\pi\omega_i^2}{c^2}\chi^*(-\omega_s, \omega_L, -\omega_i)E_L^* E_s .$$

In Sect. 4.4 of Nonlinear Optics this problem is discussed for prescribed frequencies ω_i and ω_s. The linearized set of coupled equations leads to a determinantal problem in the propagation vectors. The quartic equation can be reduced to a quadratic one in ΔK, the deviation from unperturbed values k_s and k_i for the linear medium,

$$\Delta K = -\frac{1}{2}\Delta k \pm \left[\frac{1}{4}(\Delta k)^2 - \left(\frac{4\pi\omega_i\omega_s}{c^2}\right)^2 \frac{1}{4k_s k_i} |\chi^{NL}|^2 |E_L|^2\right]^{1/2} .$$

For sufficiently small momentum mismatch, $\Delta k = k_L - k_i - k_s$, and there is one root with an exponential gain, corresponding to an exponentially growing wave with mixed frequency character, ω_i and ω_s.

The same problem may be treated in the time domain. Suppose that the mode wave functions are prescribed, for example, by resonator boundary conditions. Write the electric field in the terms of dynamical variables $p_\lambda(t)$ which are proportional to the mode amplitudes

$$\underset{\sim}{E}(\underset{\sim}{r}, t) = -(4\pi)^{1/2} \sum_\lambda p_\lambda(t)\, \hat{a}_\lambda\, u_\lambda(\underset{\sim}{r}).$$

Introducing a damping factor in a phenomenological manner for each mode, one finds from the Hamiltonian

$$\mathcal{H}_{EM} + \mathcal{H}_{int} = \frac{1}{2} \sum_\lambda (p_\lambda^2 + \omega_\lambda^2 q_\lambda^2) - \int \underset{\sim}{P}^*(\underset{\sim}{r}, t) \cdot \underset{\sim}{E}(\underset{\sim}{r}, t)\, dV,$$

the equations of motions,

$$\ddot{p}_\lambda + \frac{\omega_\lambda}{Q_\lambda} \dot{p}_\lambda + \omega_\lambda^2 p_\lambda = + 4\pi \omega_\lambda^2 \int \underset{\sim}{P}\, \underset{\sim}{P} \cdot \underset{\sim}{a}_\lambda\, u_\lambda(\underset{\sim}{r}) dV.$$

The effect of the large linear polarization may be incorporated as a change in the resonant eigenfrequency ω_λ^o from the corresponding vacuum value. The linear eigenmodes are then driven by the nonlinear polarization projections on the mode function,

$$\ddot{p}_s + \frac{\omega_s^o}{Q_s} \dot{p}_s + (\omega_s^o)^2 p_s = (\omega_s^o)^2\, \eta\, E_L\, p_i^\dagger$$

$$\ddot{p}_i^\dagger + \frac{\omega_i^o}{Q_i} p_i^\dagger + (\omega_i^o)^2 p_i^\dagger = (\omega_i^o)^2\, \eta^* E_L^*\, p_s,$$

where

$$\eta = 4\pi \int \chi^{NL} : \underset{\sim}{a}_s^*\, \underset{\sim}{a}_i^*\, \underset{\sim}{a}_i\, u_1^*(\underset{\sim}{r})\, u_2^*(\underset{\sim}{r})\, u_3(\underset{\sim}{r})\, d^3\underset{\sim}{r} \Big/ \left[\int u_1^*(\underset{\sim}{r})\, u_1(\underset{\sim}{r})\, d^3\underset{\sim}{r}\right].$$

Find a solution in the form where p_s has a time variation $\exp(-i\omega_s t)$ and p_i a time variation $\exp[-i(\omega_L - \omega_s)t]$. This leads to the determinantal condition (if the trivial solution $E_s = E_i = 0$ is excluded)

$$0 = \begin{vmatrix} -\omega_s^2 - \dfrac{i\omega_s\omega_s^o}{Q_s} + (\omega_2^o)^2 & -\eta\, E_3\, (\omega_s^o)^2 \\ -\eta^* E_3^*\, (\omega_i^o)^2 & -(\omega_L - \omega_s)^2 + \dfrac{i\omega_i^o(\omega_L - \omega_s)}{Q_i} + (\omega_i^o)^2 \end{vmatrix}.$$

The solution is of interest only if ω_s lies close to the eigenfrequency of the linear unperturbed mode frequency, $\omega_s = \omega_s^o + \Delta\omega$. Ignore small terms in $\Delta\omega^2$, $\omega_s^o \Delta\omega/Q_s$, etc. This approximation is equivalent to ignoring terms at other frequencies. We obtain a quadratic equation for $\Delta\omega$,

$$0 = \begin{vmatrix} -2\omega_1^o \Delta\omega - \dfrac{i(\omega_s^o)^2}{Q_s} & -\eta E_3 (\omega_s^o)^2 \\ -\eta_3^* E_3^* (\omega_i^o) & +2\omega_i^o \Delta\omega + 2\omega_i^o(\omega_L - \omega_s^o - \omega_i^o) + \dfrac{i(\omega_i^o)^2}{Q_i} \end{vmatrix} .$$

For exact tuning of resonator at both signal and idler one has $\omega_L = \omega_s^o + \omega_i^o$. Then the eigenvalues are

$$-i\Delta\omega = -\frac{1}{2}\left(\frac{\omega_s^o}{2Q_s} + \frac{\omega_i^o}{2Q_i}\right) \pm \left[\frac{1}{4}\left(\frac{\omega_s^o}{2Q_s} + \frac{\omega_i^o}{2Q_i}\right)^2 + |\eta|^2 |E_L|^2 \omega_s^o \omega_i^o - \frac{\omega_s^o \omega_i^o}{Q_s Q_i}\right]^{1/2} .$$

If $|\eta|^2 |E_3|^2 > (Q_i^o Q_s^o)^{-1}$, there will be one root with exponential gain. This indicates the threshold condition for parametric down conversion. It should be noted that the same quadratic equation for $\Delta\omega$ would have been obtained from the coupled amplitude equations. In general there will be a mixed space-time variation of the "slowly varying" complex amplitudes, and the variation $\frac{\partial A_\lambda}{\partial t} + v_g \frac{\partial A_\lambda}{\partial z}$ is proportional to the nonlinear driving term. Here v_g is the group velocity of the wave. An example of the treatment with coupled amplitude equations is given in the next section.

4.3 The Stimulated Raman Scattering as a Parametric Process[1]

Start with the Lagrangian density function

$$L = L_{rad} + L_{vib} + L_{int}.$$

The vibrational optical phonon wave is derivable from a Lagrangian expressed in terms of the normal coordinate $Q_v = R(2\rho)^{1/2} e^{-i\underset{\sim}{k}_v \cdot \underset{\sim}{r}_i}$. Here R is the vibrational deviation and ρ the reduced vibrational mass density, $\underset{\sim}{k}_v$ the vibrational wave vector and $\underset{\sim}{r}_i$ the location of a vibrating molecule.

$$L_{vib} = \frac{1}{2} \dot{Q}_v^2 - \frac{1}{2} \omega_{v,o}^2 Q_v^2 + \frac{1}{2} \beta (\nabla Q_v)^2.$$

The interaction energy between the material system and the electromagnetic waves may be expressed in terms of a polarizability which is a function of the vibrational coordinates,

$$L_{int} = \frac{1}{2} \sum_N \underset{\approx}{\alpha}_o : \underset{\sim}{E}\,\underset{\sim}{E} + \sum_{k_v} \frac{1}{2} \left(\frac{\partial \underset{\approx}{\alpha}}{\partial Q_v}\right)\Big|_{Q=0} Q_v \underset{\sim}{E}\,\underset{\sim}{E} .$$

The equation for the optical phonon wave derived from this Lagrangian, with a damping term added, is

$$\ddot{Q}_v - \beta \nabla^2 Q_v + \omega_{v,o}^2 Q_v + 2\Gamma \dot{Q}_v = N \left(\frac{\partial \alpha_{ij}}{\partial Q_v}\right)_0 E_i E_j .$$

For optical phonons we have the dispersion relation, $\omega^2 = \omega_{v,o}^2 - \beta k_v^2$, where β is small and positive. The group velocity, $d\omega/dk_v = \frac{k_v}{\omega_{v,o}} \beta$, is very small for relatively small k values of interest, namely, $|k_v| = |k_L - k_s|$. The laser wave parametrically couples a light wave at the Stokes frequency, $\omega_s = \omega_L - \omega_o$, and the optical phonon

---

[1] See Y.R.SHEN and N.BLOEMBERGEN, Phys. Rev. <u>137</u>, 1787, 1965, or Nonlinear Optics, Sects. 4.5 and 4.6.

wave with wave vector $\underset{\sim}{k}_L - \underset{\sim}{k}_s$. The coupled wave equations for *prescribed* spatial functions are

$$\ddot{Q}_v^* - \beta\,|k_L - k_s|^2\,Q_v^* + \omega_{v,o}^2 Q_v^* + \frac{2i}{\tau_v}\,\dot{Q}_v^* = N\,\frac{\partial\alpha}{\partial Q} E_L^* E_s\,,$$

$$\ddot{E}_s + \omega_{s,o}^2\,E_s - \frac{2i}{\tau_s}\dot{E}_s = 4\pi\omega_s^2\,N\,\frac{\partial\alpha}{\partial Q} E_L Q_v^*.$$

The damping constant τ_s^{-1} in the Stokes wave is negligible in comparison to the damping in the vibrational wave, $\tau_v^{-1} = \Gamma_v$. With amplitudes slowly varying in time, after separation of the main time dependence $\exp(-i\omega_v t)$ and $\exp(-i\omega_{s,o}t)$, respectively, one has the coupled amplitude equations,

$$\dot{A}_v^* + \frac{A_v^*}{\tau_v} = +\,\frac{1}{2i\omega_v}\,N\left(\frac{\partial\alpha}{\partial Q}\right)^* A_L^* A_s\,,$$

$$\dot{A}_s + \frac{A_s}{\tau_s} = -\,\frac{4\pi\omega_s}{2i}\,N\frac{\partial\alpha}{\partial Q}\;A_L A_v^*.$$

Assume that the amplitudes vary as $e^{\alpha t}$. The eigenvalues for α are

$$\alpha = -\frac{1}{2}\left(\frac{1}{\tau_v} + \frac{1}{\tau_s}\right) \pm \left[\frac{1}{4}\left(\frac{1}{\tau_v} - \frac{1}{\tau_s}\right)^2 + \frac{4\pi\omega_s\,N^2\,\left|\frac{\partial\alpha}{\partial Q}\right|^2\,|E_L|^2}{4\omega_v}\right]^{1/2}.$$

There is one root with exponential gain. In the stimulated Raman effect the vibrational damping is heavy, $\tau_v \ll \tau_s$ and $\tau_v^{-1} \gg N\omega_s\,|\partial\alpha/\partial Q|\,|E_L|\,(\omega_s\omega_v)^{-1/2}$. In that case the mode with gain has almost complete Stokes light character with a gain constant,

$$\alpha_s = \frac{4\pi\omega_s\,N^2\left|\frac{\partial\alpha}{\partial Q}\right|^2 |E_L|^2\tau_v}{4\omega_v} - \frac{1}{\tau_s}\,.$$

The gain is proportional to the laser intensity and inversely proportional to the damping constant τ_v^{-1} of the vibrational wave. This same physical result is obtained from the description with the Raman susceptibility,

$$\ddot{E}_s + \omega_{s,o}^2 E_s - \frac{2i}{\tau_s}\dot{E}_s = 4\pi\omega_s^2 \, i\chi''_{Raman}|E_L|^2 E_s$$

or

$$\alpha_s = -\frac{1}{\tau_s} - \frac{4\pi\omega_s}{2}\chi''_{Raman}|E_L|^2.$$

There is exponential gain if $\chi''_{Raman} < 0$ and large enough to overcome the small losses at the Stokes frequency. Comparison of the two results for α_s shows that

$$\chi''_{Raman} = -\frac{N^2(\partial\alpha/\partial Q)^2 \tau_v}{2\omega_v}.$$

This result may also be derived by a direct quantum mechanical calculation of χ''_{Raman} (see Nonlinear Optics).

In the same manner the coupling of a laser wave, a "Stokes" light beam displaced toward lower frequencies, and either an acoustic wave, a spin wave, or a plasma wave may be treated (see papers by Y. R. SHEN and N. BLOEMBERGEN, Phys. Rev. <u>141</u>, 298 and <u>143</u>, 372, 1966).

The build-up of the Stokes intensity eventually results in a depletion of the laser intensity. If no other light waves are involved and intrinsic linear attenuation is negligible, the situation would be described by the coupled amplitude equations:

$$\dot{A}_s = +2\pi\omega_s |\chi''_{Raman}||A_L|^2 A_s,$$

$$\dot{A}_L = -2\pi\omega_L |\chi''_{Raman}||A_s|^2 A_L.$$

Multiply these equations by A_s^* and A_L^*, respectively, to obtain the coupling for the intensity,

$$\frac{dI_s}{dt} = 4\pi\omega_s |\chi''_{Raman}| I_L I_s,$$

$$\frac{dI_L}{dt} = -4\pi\omega_L |\chi''_{Raman}| \; I_L \, I_s .$$

Introduce the photon numbers in the two modes,

$$n_s = \int \frac{|E_s|^2 \varepsilon_s}{8\pi\hbar\omega_s} d^3r, \quad n_L = \int \frac{|E_L|^2 \varepsilon_L}{8\pi\hbar\omega_L} d^3r.$$

We then obtain the coupled equations,

$$\frac{dn_s}{dt} = W(n_s + 1)n_L,$$

$$\frac{dn_L}{dt} = -W(n_s + 1)n_L,$$

with

$$W = \frac{32\,\pi^2\,\hbar\,\omega_s\omega_L\,|\chi''|}{\varepsilon_s\,\varepsilon_L},$$

which may be considered as a time proportional probability for the spontaneous Raman scattering of a quantum $\hbar\omega_L$ into $\hbar\omega_s$. We have added the +1 to n_s in an *ad hoc* manner to account for these spontaneous processes, which must initiate and precede the stimulated production of Stokes radiation. Actually, this factor follows naturally from a quantized treatment of the field. The introduction of the normal coordinate $p_\lambda(t)$ as a canonical dynamical variable in the preceding section has put the theory in the form in which quantization can proceed readily (see A. YARIV, J. of Quantum Electronics 1, 28, 1965).

Note that $n_s + n_L = n_s(0) + n_L(0)$ is a constant of the motion, as follows directly from the nature of Raman scattering. One thus finds

$$\frac{d(n_s + 1)}{dt} = W(n_s + 1)\,\{1 + n_s(0) + n_L(0) - (n_s + 1)\}$$

with the solution

$$n_s + 1 = \frac{(n_s(0) + 1)\,(n_s(0) + n_L(0) + 1)}{n_s(0) + 1 + n_L(0)\exp[-\{n_s(0) + 1 + n_L(0)\}Wt]}.$$

For $t \to \infty$, $n_s \to n_s^o + n_L^o$ and the laser power is completely depleted. Long before this happens, other waves have to be taken into account. Higher-order Stokes and antiStokes waves are produced.

The experimental situation in the stimulated Raman emission was plagued for a number of years by several anomalies:

1) Anomalous Raman Gain,

2) Anomalous Frequency Broadening,

3) Anomalous Angles of AntiStokes Cones.

The first anomaly arises because the calculated Raman gain in a typical unfocused laser beam passing through a cell about 20 cm long is $e^{0.2}$, whereas the experimental value was more nearl e^{20}. The stimulated Stokes intensity is very large compared to the weak spontaneous emission, and a discrepancy of a factor 10^{10} existed between theory and several experiments.

Recently this anomaly has been correlated with self-focusing of the laser beam. In the bright filaments with diameter 10 - 80 microns, the laser intensity can be two orders of magnitude higher than the nominal input intensity.

The frequency broadening is caused by a modulation of the index of refraction, the difference frequency $\omega_L - \omega_L'$, when the incident laser beam contains two frequencies, ω_L and ω_L'. The molecular reorientation can follow frequencies up to 10^{12} cps. in certain fluids. This effect is described by a theory which considers the time dependent distribution function $f(\theta, t)$ for molecular orientation. The quantity n_2 becomes time dependent. The intensity dependent index becomes complex. This leads to the phenomenon of stimulated elastic Rayleigh scattering. The role of the Stokes wave is now taken over by a light wave at frequency $\omega_L' = \omega_L - 1/\tau_c$, where τ_c is a correlation time for molecular reorientation (see BLOEMBERGEN and LALLEMAND, Phys. Rev. Letters 16, 81, 1966).

The anomalous antiStokes angles are also explained by the fact that these frequencies, $\omega_{as} = 2\omega_L - \omega_s$, etc., are generated in narrow filiamentary regions where both the laser and the Stokes intensity are high due to self-focusing.

5. Quantum Mechanical Calculation of Nonlinear Susceptibilities. Lamb's Theory of Coupled Laser Modes

5.1 The Coupling of One Electromagnetic Mode with Independent Particles with Two Energy Levels

The coupling of the electromagnetic mode with mode function $u_\lambda(\underset{\sim}{r})$ to the polarization of the medium is described by the mode equation,

$$\ddot{p}_\lambda + \frac{\omega_\lambda}{Q_\lambda}\dot{p}_\lambda + \omega_\lambda^2 p_\lambda = -4\pi\,\omega_\lambda^2 \int (\underset{\sim}{P}(\underset{\sim}{r},\,t)\cdot\underset{\sim}{a}_\lambda)\,u_\lambda(\underset{\sim}{r})\,d^3\underset{\sim}{r}. \qquad (5.1)$$

It must be emphasized that in the equation of motion p really stands for D. This is easiest to see in the linear case. Consider the linear medium. The wave equation for D has the same form as the equation for p,

$$\frac{\partial^2 \underset{\sim}{D}}{\partial t^2} + c^2\, \nabla \times \nabla \times (\underset{\sim}{D} - 4\pi \underset{\sim}{P}) = 0\ .$$

For a cavity mode with an eigenfrequency ω_{vac}^o, when the cavity is not filled with the dielectric, this wave equation can be written in the form

$$\frac{\partial^2 D}{\partial t^2} + (\omega_{vac}^o)^2 D = 4\pi\left(\omega_{vac}^o\right)^2 P_L.$$

When the equation of motion is derived from the Hamiltonian in a strong linear dielectric, one has

$$\mathcal{H} = \mathcal{H}_{vac} - \underset{\sim}{E}_{vac} \cdot \underset{\sim}{P} + \text{Interaction between dipoles.}$$

This last quantity cannot be ignored. It is automatically taken into account, if one writes for the field Hamiltonian

$$\mathcal{H} = \frac{D^2}{8\pi} - \underset{\sim}{D} \cdot \underset{\sim}{P} = \frac{1}{2}\left[(\omega^L_{vac})^2 q^2 + p^2\right] - \underset{\sim}{p} \cdot \underset{\sim}{P}.$$

One should really go through the quantization of the linear dielectric with the Ewald-Oseen integral equation. Note that the equation of motion for a *given mode* function (given $\underset{\sim}{k}$ vector) yields the correct eigenfrequency, with

$$P_L = \chi E = \frac{\chi^L D}{1+4\pi\chi^L} \quad \text{and with D as the dynamical variable,}$$

$$\frac{\partial^2 D}{\partial t^2} + (\omega^o_{vac})^2 \left(1 - \frac{4\pi\chi^L}{1+4\pi\chi^L}\right) D = 0,$$

$$\text{or} \quad \omega^o = \omega^o_{vac} \frac{1}{(1+4\pi\chi^L)^{1/2}} = \omega^o_{vac}\, \varepsilon^{-1/2} = k\, c\, \varepsilon^{-1/2}.$$

In the equation of motion for E, P must still be treated as a dynamical quantity,

$$\nabla \times \nabla \times E + \frac{1}{c^2}\frac{\partial^2 E}{\partial t^2} = -\frac{4\pi}{c^2}\frac{\partial^2 P^L}{\partial t^2} \quad \text{or} \quad + (\omega^o_{vac})^2 - (\omega^o)^2 = 4\pi\chi(\omega^o)^2.$$

We obtain again $(\omega^o)^2 = \dfrac{(\omega^o_{vac})^2}{1+4\pi\chi}$.

The polarization of the medium is in turn determined by the field. We should use the quantum mechanical expectation value for $N(\underset{\sim}{r})$ independent two level systems per unit volume. Assume that the dipole moment operator has only nondiagonal elements,

$$\underset{\sim}{P}(\underset{\sim}{r}, t) = N(\underset{\sim}{r})\,(e\, r_{ab}\, \rho_{ba} + e\, r_{ba}\, \rho_{ab}). \qquad (5.2)$$

The only approximation in this quantum mechanical theory made here is that the trace is taken over the material system only. In other words, it is assumed that the density matrix factorizes,

$$\mathrm{Tr}\; P\rho = \mathrm{Tr}\; P\rho_{mat}\, \mathrm{Tr}\rho_{field} = \mathrm{Tr}\; P\rho_{mat}.$$

This point is discussed in detail by JAYNES and CUMMINS, Proc. IEEE 51, 89, 1963. The equations of motion for the elements of the density matrix are given by

$$i\hbar\, \dot{\rho}_{ba} = \mathcal{H}_{ba}\, (\rho_{aa} - \rho_{bb}) + \hbar\, \omega_{ba}\rho_{ba} - i\,\hbar\, T_2^{-1}\rho_{ba}. \tag{5.3}$$

$$\dot{\rho}_{bb} - \dot{\rho}_{aa} = \frac{2}{i\hbar}\, (\rho_{ab}\mathcal{H}_{ba} - \mathcal{H}_{ab}\, \rho_{ba}) - \left[\rho_{bb} - \rho_{aa} - \rho_{bb}^{(0)} + \rho_{aa}^{(0)}\right] T_1^{-1}\;. \tag{5.4}$$

Here $\mathcal{H} = -e\, \underset{\sim}{r} \cdot \underset{\sim}{E}$ is the interaction Hamiltonian acting on the material system, $\hbar\, \omega_{ba}$ is the splitting between the two energy levels. The terms with T_2 and T_1 represent transverse and longitudinal relaxation processes. $\rho_{bb}^{(0)} - \rho_{aa}^{(0)}$ is the population difference, when $p_\lambda = 0$. This need not be the thermal equilibrium value, because random pump mechanisms, due to a flashlamp or discharge collision processes, may achieve a different steady state value. In fact, in lasers, $\rho_{bb}^{(0)} - \rho_{aa}^{(0)}$ must have the opposite sign from the thermal equilibrium value, because the upper level must have a higher population to achieve laser action.

The set of equations gives a rather general description of the excitation of one laser mode, because in general only one pair of energy levels of the material system is near resonance with the EM mode. The set is nonlinear because Eqs. (5.1) and (5.4) contain $E\, \rho_{ab}$ and Eq. (5.3) contains $E(\rho_{aa} - \rho_{bb})$. If Eq. (5.2) is substituted into Eq. (5.1), we have actually five real nonlinear equations with five real unknowns. The complex variables p(t) and $\rho_{ab} = \rho_{ba}^{*}$

have both an amplitude and a phase. The diagonal elements ρ_{aa} and ρ_{bb} are real and positive.

This set of equations can be specialized to describe diverse phenomena, such as the transient behavior of Q-switched lasers, light echo phenomena, nonlinearities in magnetic resonance, etc. The set of equations is indeed equivalent to the coupling of a classical magnetization, precessing around a magnetic field H_0, in a cavity tuned close to resonance. If the cavity field is "precessing," no harmonics are generated. For a linearly polarized field, the harmonic components in $\underset{\sim}{P}$ may still be ignored because they are off resonance and cannot couple effectively to the mode. In this case one obtains the well-known saturation formula for a two-level system,

$$\rho_{bb}^{dc} - \rho_{aa}^{dc} = (\rho_{bb}^{(0)} - \rho_{aa}^{(0)}) \frac{1 + (-\omega + \omega_{ba})^2 T_2^2}{1+(-\omega + \omega_{ba})^2 T_2^2 + |e\, r_{ba}|^2 |E|^2 T_1 T_2}, \tag{5.5}$$

$$\rho_{ba}^{(\omega)} = \frac{-\hbar^{-1}\, e\, r_{ba}\, E}{\omega - \omega_{ba} + i/T_2} (\rho_{bb}^{dc} - \rho_{aa}^{dc}). \tag{5.6}$$

Actually these expressions retain the non-vanishing Fourier components to all powers in the amplitude in the rotating field approximation. If the denominator is expanded in power series in $|E|^2$, one obtains besides a linear polarization,

$$\underset{\sim}{P}^{L}(\omega) = \frac{-\hbar^{-1} e^2 |r_{ab}|^2}{\omega - \omega_{ba} + i/T_2} E\, (\rho_{bb}^{(0)} - \rho_{aa}^{(0)}) = \chi_{pump}^{L} E, \tag{5.7}$$

a polarization, cubic in the field amplitude, describing the incipient saturation effect,

$$\underset{\sim}{P}^{NL}(\omega) = \frac{-\hbar^{-1} e^2 |r_{ab}|^2}{\omega - \omega_{ba} + i/T_2} \frac{|e\, r_{ab}|^2\, T_1 T_2}{1 + (-\omega + \omega_{ba})^2 T_2^2} |E|^2 E = \chi^{NL} |E|^2 E. \tag{5.8}$$

5.2 Quantum Mechanical Calculation of Nonlinear Susceptibilities

Since for most purposes in nonlinear optics a power series expansion is adequate, and since the coupled equations between all EM modes and all elements of a general density matrix lead quickly to a "hopeless" set of equations, it is useful to have a general method of successive approximation in which the components of the density matrix are each found as an ascending power series in the field amplitudes, p_λ, $p_\lambda^\dagger$, etc. Since ρ_{mat} commutes with these field operators and p_λ, $p_\lambda^\dagger$ are the only noncommuting pairs, this procedure can be carried quite far in a scheme in which both the material system and the fields are quantized. For most applications, with the important exception of noise and quantum fluctuation problems, the fields may be treated classically.

The systematic development of the susceptibility in this semiclassical treatment has been presented in Nonlinear Optics, Chap. 2. When the expectation value of the electric dipole moment has thus been found, we substitute it into the right-hand side of Eq. (5.1) and other mode equations when more than one EM mode is retained.

LAMB (Phys. Rev. 134, A 1429, 1964) has in this manner discussed the oscillations in a gas laser. The situation is more complex than the bare outline given here. In the first place there is an "inhomogeneous broadening" mechanism because of the Doppler shift of gas molecules moving with different velocities. In the second place the motion of the molecules causes the density matrix to be a function of velocity and position as well, $\rho(\underset{\sim}{r}, \underset{\sim}{v}, t)$ instead of $\rho(t)$. Although this increases the complexity of the calculations and the phenomena, our example of a homogeneous broadened resonance of particles fixed in space will show many of the important features.

5.3 Oscillation in a Single Mode

It is well known that the amplitude of an electronic oscillator is stabilized by the nonlinearity. In lasers this stabilization is provided by the saturation mechanism.

The mode equation becomes

$$\ddot{E} + \frac{\omega_\lambda}{Q_\lambda}\dot{E} + \omega_\lambda^2 E = -4\pi\chi^L_{pump}\ddot{E} - 4\pi\chi^{NL}\frac{d^2}{dt^2}(E^2E^*),$$

where χ^L_{pump} and χ^{NL} are given by Eqs. (5.7) and (5.8). Due to the presence of an appropriate pump mechanism, the population difference, $\rho^{(0)}_{bb} - \rho^{(0)}_{aa}$, is inverted and χ^L_{pump} has a negative imaginary part. Try a solution of the form $E = A\exp(-i\,\omega_o t)$, where the oscillation frequency ω_o is close to ω_λ but not necessarily equal to it. If we ignore in the usual way small quantities of higher order, such as terms in $\ddot{A}$, $\dot{A}/Q_\lambda$, etc., the following complex amplitude equation results:

$$(-\omega_o^2 + \omega_\lambda^2)A - 2i\omega_o\dot{A} - \frac{i\omega_\lambda^2}{Q_\lambda}A = +4\pi(\chi' + i\chi'')_{pump}A\omega_o^2 + 4\pi\chi^{NL}\omega_o^2|A|^2A,$$

$$\dot{A} = \alpha A - \beta|A|^2A,$$

with

$$\alpha = \left(-\frac{1}{2}\frac{\omega_\lambda}{Q_\lambda} - 2\pi\,\omega_o\,\chi''_{pump}\right) + 2\pi i\,(\omega_o\chi'_{pump}(\omega_o) + \omega_\lambda - \omega_o),$$

$$\beta = 2\pi\,\omega_o\,\chi''^{NL} - 2\pi i\,\omega_o\,\chi'^{NL},$$

where we have replaced $-\omega_o^2 + \omega_\lambda^2$ by $2\omega_o(-\omega_o + \omega_\lambda)$ and $\omega_\lambda^2/Q_\lambda$ by $\omega_o\omega_\lambda/Q_\lambda$. In the steady state, $\dot{A} = 0$. In order to get oscillation, $\chi''_{pump} < 0$ and $4\pi\omega_o|\chi''|_{pump} > \omega_\lambda/Q_\lambda$ are required. Note that $\chi''^{NL} > 0$, because by the nature of the saturation process it makes the susceptibility less negative.

This can be shown more explicitly by writing out the real and imaginary parts. Introduce $A = |A|e^{i\phi}$. The amplitude determining equation becomes

$$|\dot{A}| = 0 = \{-\frac{1}{2}\frac{\omega_\lambda}{Q_\lambda} - 2\pi\omega_o \chi''_{pump}(\omega_o)\}|A| - 2\pi\omega_o \chi''^{NL}|A|^3.$$

The frequency determining equation becomes

$$i\dot{\phi} = 0 = -2\pi i(\omega_o \chi'_{pump}(\omega_o) + \omega_\lambda - \omega_o) + 2\pi i\omega_o \chi'^{NL}|A|^2.$$

This equation determines the oscillating frequency, $\omega_o + \dot{\phi}$, in terms of ω_λ. The last term represents a power dependent nonlinear pulling.

5.4 Two Oscillating Modes

When the pump power of a gas laser is increased sufficiently above threshold, it is possible to obtain a regime where two or more longitudinal modes of the Fabry-Perot resonator are excited simultaneously. Eventually modes with different transverse distributions will also be excited.

The index of refraction depends also on the intensity of light at other frequencies. Denote the two modes by E_1 and E_2. The coupled amplitude equations take the form

$$\dot{A}_1 = \alpha_1 A_1 - \beta_1|A_1|^2 A_1 - \theta_{12}|A_2|^2 A_1,$$

$$\dot{A}_2 = \alpha_2 A_2 - \beta_2|A_2|^2 A_2 - \theta_{21}|A_1|^2 A_2.$$

The nonlinear susceptibility to which θ_{12} is proportional corresponds to a nonlinear polarization, $P^{NL}(\omega_1) = \chi^{NL}(-\omega_1,\omega_2,-\omega_2,\omega_1)E_2 E_2^* E_1$, proportional to the field E_1 and to the intensity at the other frequency. In gas lasers this is a term caused by the fact that the same atoms may contribute to two adjacent longitudinal modes.

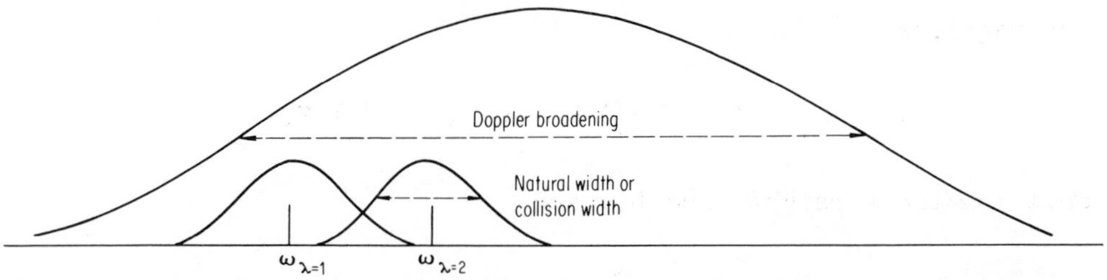

Two Fabry-Perot modes in relation to the inhomogeneous Doppler broadening and the homogeneous broadening of the atomic resonance absorption

The saturation by one mode field causes a change in the polarization at the frequency of the other mode. The two modes compete partially for the same atoms, due to the overlap of the wings of homogeneously broadened line shapes. The coefficient θ_{12} can be computed in the usual way with the density matrix formalism.

Again the complex amplitude equations can be split into two real amplitude equations and two phase or frequency determining equations. It is simpler to consider the intensity equations,

$$X = A_1 A_1^*, \qquad\qquad Y = A_2 A_2^*.$$

Multiply the equations for $\dot{A}_1$ by A_1^* and the equation for $\dot{A}_2$ by A_2^*. One obtains the coupled equations for the intensities:

$$\dot{X} = (\alpha_1 + \alpha_1^*)\, X - (\beta_1 + \beta_1^*)\, X^2 - (\theta_{12} + \theta_{12}^*)\, Y X \ ,$$

$$\dot{Y} = (\alpha_2 + \alpha_2^*)\, Y - (\beta_2 + \beta_2^*)\, Y^2 - (\theta_{21} + \theta_{21}^*)\, X Y \ .$$

These equations have solutions which can be discussed by the method of characteristics. A condition of steady state oscillation ($\dot{X}=\dot{Y}=0$) for either one or the other mode or for both modes depends on the nonlinear coupling.

For a steady state with simultaneous oscillation in both modes one requires

$$\alpha_1 + \alpha_1^* = (\beta_1 + \beta_1^*)\ X + (\theta_{12} + \theta_{12}^*)\ Y,$$

defining the straight line L_1, and

$$\alpha_2 + \alpha_2^* = (\beta_2 + \beta_2^*)\ Y + (\theta_{21} + \theta_{21})\ X,$$

defining the straight line L_2.

These straight lines have to cross in the first quadrant because X and Y are positive. The solution has been discussed by LAMB (loc. cit., Sect. 10). The transient solutions have $\dot{X} = 0$ on the first line, $\dot{Y} = 0$ on the second line. There are, of course, single mode steady state oscillations characterized by

$$X = 0 \qquad Y = \frac{\alpha_2 + \alpha_2^*}{\beta_2 + \beta_2^*}$$

or

$$Y = 0 \qquad X = \frac{\alpha_1 + \alpha_1^*}{\beta_1 + \beta_1^*}.$$

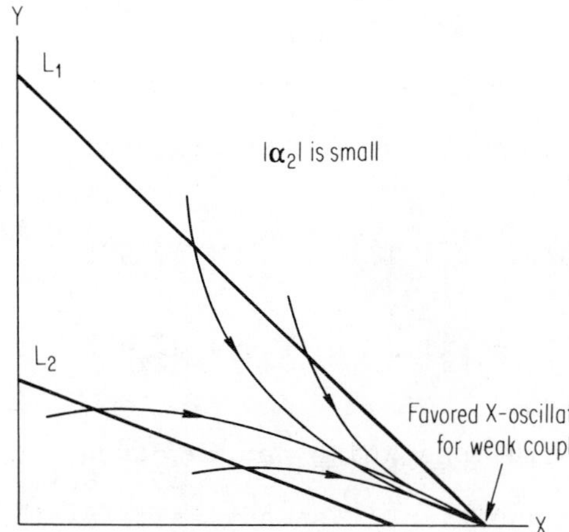

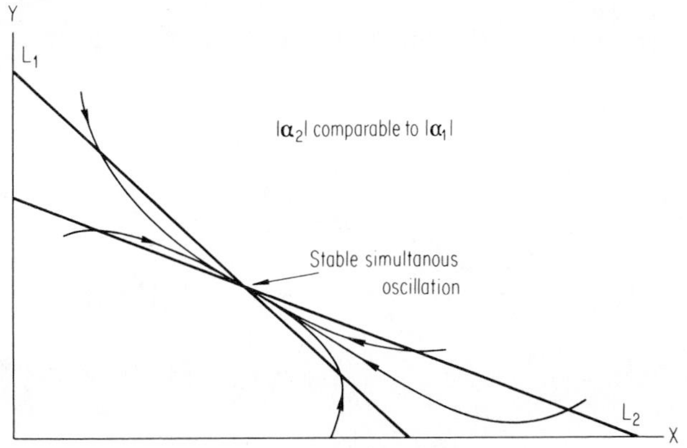

The gas laser situation is one of weak coupling, because the mode separation is usually so large that each atom is coupled strongly to not more than one mode.

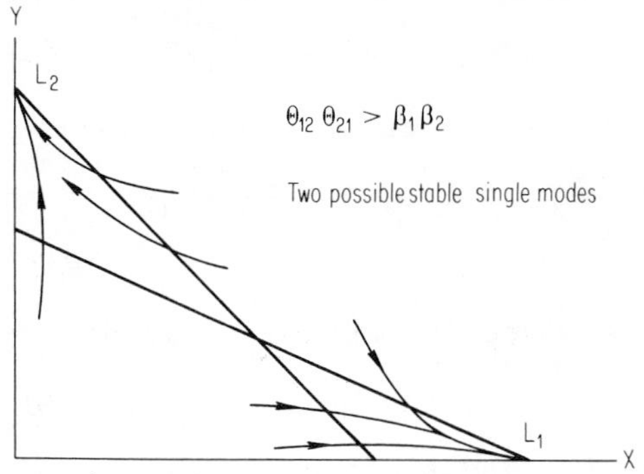

If $\theta_{12}\theta_{21} > \beta_1\beta_2$, the case of strong coupling results. The system oscillates in one mode even if $|\alpha_2|$ is comparable to $|\alpha_1|$. Which of the two stable single mode oscillations the system chooses depends on previous history (hysteresis).

5.5 Three Coupled Modes in a Gas Laser

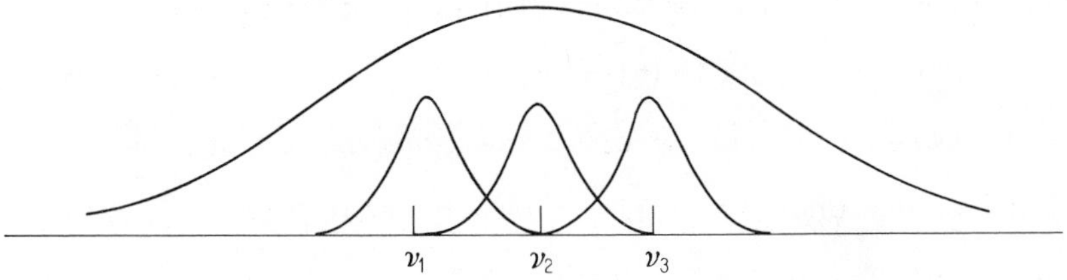

The three coupled amplitude equations are:

$$\dot{A}_1 = \alpha_1 A_1 - \beta_1 |A_1|^2 A_1 - \theta_{12}|A_2|^2 A_1 - \theta_{13}|A_3|^2 A_1$$
$$- \eta_{23} A_2^2 A_3^* e^{2\pi i(-2\nu_2 + \nu_3 + \nu_1)t},$$

$$\dot{A}_2 = \alpha_2 A_2 - \beta_2 |A_2|^2 A_2 - \theta_{21}|A_1|^2 A_2 - \theta_{23}|A_3|^2 A_2$$
$$- \eta_{13} A_3 A_1 A_2^* e^{2\pi i(+2\nu_2 - \nu_3 - \nu_1)t},$$

$$\dot{A}_3 = \alpha_3 A_3 - \beta_3 |A_3|^2 A_3 - \theta_{31}|A_1|^2 A_3 - \theta_{32}|A_2|^2 A_3$$
$$- \eta_{12} A_2^2 A_1^* e^{i(-2\nu_2 + \nu_3 + \nu_1)t}.$$

Other time dependent coupling terms are omitted because their time variation is too fast. Since ν_2 lies nearly midway between ν_3 and ν_1, the last term may become a secular perturbation. In many cases even the time variation of exp $i(\nu_3 + \nu_1 - 2\nu_2)t$ is too fast and results only in a small periodic perturbation, unless very careful tuning adjustment of the modes is made. In this situation, in the steady state, $\dot{A}_1 = \dot{A}_2 = \dot{A}_3 = 0$, one has three coupled algebraic equations. They are complex and there are amplitude and frequency determining parts. Stable oscillation may occur at one, two, or three frequencies. One can draw three-dimensional stability diagrams. The frequencies of oscillation are determined from the imaginary parts of the equations. If the system oscillates simultaneously at ν_1 and ν_2 there is of course a combination tone at $2\nu_2 - \nu_1$ which may be close to the resonant frequency ν_3. When the pumping is sufficiently strong, the system oscillates at three frequencies, ν_1, ν_2, and ν_3. When the combination tone $2\nu_2 - \nu_1$ is close enough ($\sim$1 kc/sec) to ν_3, frequency or phase locking of the oscillation will occur. This phenomenon is well known in radio technology and has been discussed by B. VAN DER POL (Proc. IRE 22, 1051, 1934) and earlier references quoted therein. This can be achieved by oscillating the laser in three modes and then carefully tuning ν_2 to the center of the Doppler profile.

Consider the imaginary or phase determining parts of the three coupled complex amplitude equations,

$$\dot{\phi}_1 = \mathrm{Im}\alpha_1 - \left[\mathrm{Im}\beta_1|A_1|^2 + \mathrm{Im}\theta_{12}|A_2|^2 + \mathrm{Im}\theta_{23}|A_3|^2\right] - \mathrm{Re}\eta_{23}\frac{|A_2|^2|A_3|}{|A_1|}\sin\psi - \mathrm{Im}\eta_{23}\frac{|A_2|^2|A_3|}{|A_1|}\cos\psi,$$

where we have introduced the relative phase,

$$\psi = 2\phi_2 - \phi_1 - \phi_3 - 2\pi\,(2\nu_2 - \nu_1 - \nu_3)\,t.$$

One has, of course, analogous expressions for $\dot{\phi}_2$ and $\dot{\phi}_3$. In this way one derives an equation for $\dot{\psi}$,

$$\dot{\psi} = 2\dot{\phi}_2 - \dot{\phi}_1 - \dot{\phi}_3 - 2\pi\,(2\nu_2 - \nu_1 - \nu_3) = \sigma + C_1\sin\psi + C_2\cos\psi.$$

Here σ is a constant, independent of ψ, which contains both the linear and nonlinear pulling. It depends somewhat on the amplitudes. The constants C_1 and C_2 are given by

$$C_1 = \mathrm{Re}\eta_{23}\frac{|A_2|^2|A_3|}{|A_1|} + \mathrm{Re}\eta_{12}\frac{|A_2|^2|A_1|}{|A_3|} + \mathrm{Re}\eta_{13}|A_1||A_3|,$$

$$C_2 = \mathrm{Im}\eta_{23}\frac{|A_2|^2|A_3|}{|A_1|} + \mathrm{Im}\eta_{12}\frac{|A_2|^2|A_1|}{|A_3|} - \mathrm{Im}\eta_{13}|A_1||A_3|.$$

The equation for $\dot{\psi}$ can be integrated implicitly with respect to time,

$$t = \int_{\psi_o}^{\psi}\frac{d\psi}{\sigma + C_1\sin\psi + C_2\cos\psi}.$$

If $(C_1^2 + C_2^2)^{1/2} < |\sigma|$, the relative phase oscillates. That is, one has three independent frequencies, ν_1, ν_2, and ν_3.

If $(C_1^2 + C_2^2)^{1/2} > |\sigma|$, the time becomes infinite for ψ = arc cos $\frac{\sigma}{(C_1^2 + C_2^2)^{1/2}}$. This means that ψ reaches an asymptotic value independent of time. This implies frequency locking, $2\nu_2 - \nu_1 - \nu_3 = 0$. Further details are given by LAMB and VAN DER POL (loc. cit.).

Lectures on Homogeneous Turbulence

P. G. Saffman

Preface

These notes contain an extended version of ten 45-minute lectures given in June-July, 1966 at the International School of Nonlinear Physics and Mathematics held in Munich at the Max-Planck-Institute. An attempt has been made to examine critically the state of knowledge about homogeneous turbulence, and if the tone is generally pessimistic and destructive, it is because our real understanding of the mechanics of turbulence appears to be practically nil. For example, the voluminous literature contains no serious discussion, with the notable exception of G.I. Taylor, of the truly remarkable fact that the rate of energy dissipation appears to be independent of the viscosity. The corresponding problem in aerodynamics that the drag of a body is both nonzero and independent of viscosity is recognized as a real problem that is not properly understood. The lectures were prepared at short notice and suffer from all the faults of hasty preparation and insufficient reflection. In an effort to be positive and not just destructive, some material not yet to be found elsewhere has been included. The work described in §3 on the structure and invariance of the large eddies is believed to be both new and correct, but is of no real importance. On the other hand, the ideas of §7 on the structure of turbulence are new and hopefully important, but are speculative and quite possibly in serious error. The lecturer owes a great debt to Professor H. W. LIEPMANN of the California Institute of Technology for many stimulating ideas; most of them have been incorporated into the lectures. The lectures are not an authorative account of the subject, but are very much a personal viewpoint. Indeed, anything else seems impossible in a subject so devoid of sound, logical theoretical arguments and definitive reliable experimental data.

1. Introduction

It is not easy to give an altogether satisfactory definition of turbulent motion. We say, however, that a fluid is in turbulent motion when the velocity field varies both temporally and spatially in an irregular, apparently random manner. A property of turbulent motion is that the boundary conditions do not suffice to determine the detailed flow field but only average or mean properties. For example, pipe flow or the flow behind a grid in a wind tunnel at large Reynolds number is such that it is impossible to determine from the equations of motion the detailed flow at any instant. The true aim of turbulence theory is to predict the mean properties and their dependence on the boundary conditions.

Turbulent flows are characterized by random vorticity. Turbulence can be defined as a field of *chaotic vorticity*. If $\underset{\sim}{\omega} \equiv 0$, $\underset{\sim}{u} = \nabla\phi$; and since div $\underset{\sim}{u} = 0$, $\nabla^2\phi = 0$; and in this case ϕ and $\underset{\sim}{u}$ are uniquely determined at every instant by the boundary conditions. Random irrotational flows do exist, e.g. the flow outside a turbulent jet, wake or boundary layer, or random sound and water waves, but the fundamental dynamical processes are not the same as in turbulent flows with chaotic vorticity.[1]

Diffusive Property of Turbulence

Turbulence has the property that fluid particles tend to drift apart. This is a matter of common experience. Another way of expressing this property is to say that material lines in the fluid are

---

[1] For example, the joint normal hypothesis which has unsatisfactory features as an approximation to turbulence dynamics can be shown to be rigorously correct for the nonlinear interaction of random waves in a dispersive medium [1].

stretched on average.

There is apparently no mathematical proof in existence of the observed fact that two fluid particles P and Q in a random velocity field $\underset{\sim}{u}(\underset{\sim}{x},t)$ move so that on average

$$\frac{d}{dt} \mid \underset{\sim}{x}(P) - \underset{\sim}{x}(Q) \mid > 0.$$

Even if P and Q are close together, relatively little can be done. Let us suppose that $\vec{PQ}$ is infinitesimal. Write $\vec{PQ} = \underset{\sim}{\ell}$. Then

$$\frac{d\ell_i}{dt} = A_{ij}(t)\ell_j, \tag{1}$$

where $A_{ij} = \frac{\partial u_i}{\partial x_j}(P)$, i.e. it is the velocity gradient tensor evaluated at the moving point P. For a field of turbulence, the tensor A_{ij} is a random function of time. Suppose for simplicity that A_{ij} is a stationary random function of t, i.e. the statistical properties are independent of t. We should like to be able to show that in general

$$\frac{d}{dt}\left(\overline{\ell_i^2}\right)^{1/2} > 0. \tag{2}$$

A partial proof exists in 1 and 2 dimensions (see Appendix 1), but not in 3 dimensions unless A_{ij} can be diagonalized, which is a special case.

It is unsatisfactory that such a basic idea as the relative diffusion of two particles should be without convincing mathematical support and be based mainly on intuition, but this is typical of turbulence theory. We stress this point because it is the diffusive properties of turbulence that make it a subject of the greatest physical importance. Turbulent transport or convection of heat, momentum, energy, and concentration dominate the molecular transport if the fluid is turbulent, and the vast majority of fluid flows of engineering, geophysical and astrophysical interest are turbulent. Also, the diffusive property as manifested by the stretching of vortex lines controls the energy dissipation of turbulent motion.

The transport of momentum by turbulent diffusion distinguishes between fluid turbulence and random irrotational velocity fields. The motion of the fluid is governed by the Navier-Stokes (NS) equation

$$\frac{\partial \underset{\sim}{u}}{\partial t} + (\underset{\sim}{u}\cdot\nabla)\underset{\sim}{u} = -\frac{\nabla p}{\rho} + \nu\nabla^2\underset{\sim}{u}, \tag{3}$$

$$\operatorname{div} \underset{\sim}{u} = 0. \tag{4}$$

Write $\underset{\sim}{u} = \underset{\sim}{U} + \underset{\sim}{u}'$, where $\underset{\sim}{U} = \overline{\underset{\sim}{u}}$ is the mean velocity. Taking the mean of Eq. (3) gives

$$\frac{\partial \underset{\sim}{U}}{\partial t} + (\underset{\sim}{U}\cdot\nabla)\underset{\sim}{U} = -\frac{\nabla P}{\rho} + \nu\nabla^2\underset{\sim}{U} - \operatorname{div}\overline{\underset{\sim}{u}'\underset{\sim}{u}'}. \tag{5}$$

Thus the turbulent fluctuations act like a stress $\overline{u_i'u_j'}$ on the mean flow. Now if the fluctuations are irrotational,

$$\frac{\partial}{\partial x_j}\overline{u_i'u_j'} = \overline{u_j'\frac{\partial u_i'}{\partial x_j}} = \overline{u_j'\frac{\partial u_j'}{\partial x_i}} = \frac{\partial}{\partial x_i}\frac{1}{2}\overline{\underset{\sim}{u}'^2}.$$

Thus the turbulent stresses (which are called the Reynolds stresses) can be absorbed into the pressure and they have no dynamical action on the mean flow.

Strong Nonlinearity

Turbulent flows occur when the Reynolds number Re is large, and turbulent flows are found experimentally to be only weakly dependent upon Re. The viscous terms in the equations of motion are generally small, the exceptions being in the neighborhood of rigid walls or on 'singular surfaces' in the fluid, and the nonlinear terms in the equations of motion are usually dominant. Thus turbulence is in general a nonlinear problem with strong coupling. All attempts to solve the turbulence problem using the ideas developed in theoretical physics for nonlinear problems with weak-coupling (and these ideas seem to form the basis of much recent analytical work) should be treated with suspicion. Note that the pressure is a nonlinear term since taking the divergence of Eq. (3) gives

$$\nabla^2 p = -\rho \frac{\partial^2 u_i u_j}{\partial x_i \partial x_j}. \tag{6}$$

Intermittency

A further complication arises because the viscous term, although linear, contains the highest order derivative. This is the recipe for a singular perturbation problem and we can expect that there will be thin regions, of volume proportional to some power of ν, in which large velocity gradients occur so that in these regions the viscous forces and the nonlinear acceleration forces will be comparable. It is therefore to be expected that many properties of the turbulent motion, in particular functions of velocity gradients, will show a strongly intermittent nature, and experimental evidence that this is so seems to be accumulating.[1]

Homogeneous Turbulence

The types of problem we really wish to solve are those under the name of shear flow or nonhomogeneous turbulence, e.g. we wish to find the profile of mean velocity and the intensity of the wall stress for flow in a pipe, or the angle at which a jet diverges, or the heat transport between horizontal plates with a cold plate above a hot plate.

Considerable progress has been made in these problems by a combination of experimental observation, dimensional analysis, and inspired physical assumptions. But it is not unfair to say that hardly any progress has been made on the problem of predicting the phenomena by mathematical arguments based on the Navier-Stokes equations.

The concept of homogeneous turbulence, in which all mean properties are independent of position, simplifies the problem of turbulent motion by separating the interaction of the turbulent fluctuations with themselves from their interaction with the mean flow. The largest amount of theoretical work has been devoted to homogeneous turbulence and further to the more restricted special case of isotropic turbulence, in which all directions are the same statistically.

---

[1] Intermittency in homogeneous turbulence was noted first by BATCHELOR & TOWNSEND [2]. Intermittency in shear flow usually refers to the existence of a sharp boundary between the turbulent and 'non-turbulent' regions of a flow [3]. In these notes, intermittency will always refer to the intermittent structure inside a region of fully developed turbulence. There is probably a lot in common between these two types of intermittency, but it is nevertheless unfortunate that the same term is applied to both.

However, homogeneous turbulence is a theoretical idealization and it has been said [4] with too much truth for comfort that the purpose of homogeneous turbulence is to provide full employment for mathematicians. Nobody has produced homogeneous turbulence or can say where it is to be found. The hope is that homogeneous turbulence may provide an approximation to certain aspects of real turbulence, namely properties with a characteristic length scale small compared with the scale of inhomogeneity. This is probably true for questions like the small scale fluctuations in concentration of a convected scalar. But time may show that as far as the mechanics of real turbulence are concerned, the problem was made harder by removing the interaction with a nonuniform mean flow. And also homogeneous turbulence has constant Reynolds stresses with zero divergence, so that a field of homogeneous turbulence can have no effect on a distribution of nonuniform mean velocity, if it stays homogeneous. These lectures will henceforth be devoted entirely to the problem of homogeneous turbulence,but the reasons are frankly academic and not because of any strong conviction that the study is of practical utility.

The closest approximation to homogeneous turbulence is provided by the flow downstream of grids in wind tunnels. After the wakes of individual rods have coalesced, the turbulence is approximately uniform across the tunnel and decays so slowly with distance that the turbulence at any station x is approximately homogeneous. Identifying time t with x/U, where U is the mean velocity in the tunnel, the grid turbulence gives an approximate representation of the decay of a field of homogeneous turbulence, and provides the major source of experimental data with which the theory has been compared. But experimental work [5] has shown that grid turbulence is not as homogeneous, and certainly not as isotropic, as earlier workers appear to have believed. How much caution, if any, should be taken in comparing the experimental results with the theories, or in basing theories on the observations, is unclear at the present time.

Appendix 1

The one-dimensional form of equation (1) is

$$\frac{d\ell}{dt} = A(t)\,\ell, \tag{A1}$$

where A(t) is a stationary random function of time with zero mean. Integration is immediate to give

$$\log \ell/\ell_0 = \int_{t_0}^{t} A(t')dt'. \tag{A2}$$

Under fairly weak conditions, the r.h.s. of (A2) is Gaussian as $t-t_0\to\infty$ with variance

$$\sigma^2 = 2t \int_0^\infty \overline{A(t)A(t+\tau)}\, d\tau. \tag{A3}$$

Thus, provided the variance is nonzero (it cannot be negative), the probability distribution of $\ell(t)$ is log-normal, and it follows that

$$\overline{\ell} \to \ell_0\, e^{\sigma^2/2} \quad \text{as} \quad t-t_0\to\infty. \tag{A4}$$

The two-dimensional form of equation (1) can be written without loss of generality as

$$\dot{\ell}_1 = \alpha\ell_1 + \beta\ell_2\,, \quad \dot{\ell}_2 = -\beta\ell_1 + \alpha\ell_2, \tag{A5}$$

where α and β are stationary random functions of time with zero mean. An incomplete but plausible proof is as follows. Write

$$\ell_1 + i\ell_2 = re^{i\theta}\,, \quad \alpha + i\beta = \rho e^{ix}. \tag{A6}$$

Then,

$$\frac{\dot{r}}{r} = \rho\cos(x-2\theta), \quad \dot{\theta} = \rho\sin(x-2\theta). \tag{A7}$$

The second of these equations shows that θ is changing in such a way as to make $|x-2\theta|$ decrease, so that on balance $x-2\theta$ will lie between $-\pi/2$ and $\pi/2$ more often than not. Thus, we expect that $\overline{\cos(x-2\theta)} > 0$, and it then follows that $\overline{\dot{r}/r} > 0$ since ρ is always positive. Hence, the length of the line tends to increase.

The argument can be made more convincing if the variations of α and β are such that either $\rho \ll \dot{x}$ or $\rho \gg \dot{x}$. In the first case, we write $\theta - \frac{1}{2}x = \phi$. Then,

$$\dot{\phi} = -\tfrac{1}{2}\dot{x} - \rho\sin 2\phi, \tag{A8}$$

and, approximately,

$$\phi = -\tfrac{1}{2}x + \int^t \rho'\sin x'\,dt'. \tag{A9}$$

Hence,

$$\rho\cos 2\phi \doteqdot \rho\cos x + \rho\sin x\int^t \rho'\sin x'\,dt'. \tag{A10}$$

The mean value of the first term is zero. The mean value of the second term is

$$\int \overline{\rho(t) \sin x(t) \; \rho(t+\tau) \sin x(t+\tau)} \, d\tau > 0. \tag{A11}$$

Thus $\overline{\rho \cos 2\phi} > 0$ and $\overline{(\dot{r}/r)} > 0$. In the second case with $\rho \gg \dot{x}$, the angle ϕ is small and (A8) gives

$$\phi \doteqdot -\tfrac{1}{4}\dot{x}/\rho \;, \qquad \cos 2\phi \doteqdot 1 \;, \tag{A12}$$

and hence

$$\overline{\left(\frac{\dot{r}}{r}\right)} \doteqdot \overline{\rho} > 0. \tag{A13}$$

So far, it has not even been possible to obtain arguments as weak as these for the three-dimensional case, which rests entirely on physical intuition, unless the matrix A_{ij} can be diagonalized.

2. The Description of Homogeneous Turbulence

Correlation Tensors

We are interested in mean values of the velocity and its derivatives, i.e. quantities of the form[1]

$$Q^{(m)}_{ij \,\ldots\, p}(\underset{\sim}{x}_1, \ldots \underset{\sim}{x}_m, t_1, \ldots t_m) = \langle u_i(\underset{\sim}{x}_1, t_1) \ldots u_p(\underset{\sim}{x}_m, t_m)\rangle,$$

[1] Mean values will be denoted by either angle brackets or an overbar, depending upon the length of the expression.

where the points $\underset{\sim}{x}_1 \ldots \underset{\sim}{x}_m$ and times $t_1 \ldots t_m$ are not necessarily different. For homogeneous turbulence, the mean value depends on the relative orientation of the m-points, and not on the absolute location.

(There are pedantic arguments about the meaning of probability or the method of taking averages, but these add nothing to the physical understanding and they will not be considered here. It will be simply assumed that ensemble averages can be taken in a meaningful way and that they are equivalent to spatial averages by virtue of the ergodic theorem. For grid turbulence, averages are time means at points fixed relative to the grid.)

In practice, we are interested mainly in the velocity correlation tensor

$$R_{ij}(\underset{\sim}{r},t) = \langle u_i(\underset{\sim}{x},t)u_j(\underset{\sim}{x}+\underset{\sim}{r},t)\rangle . \tag{7}$$

The two-point two-time velocity correlation tensor is also used sometimes,

$$R_{ij}(\underset{\sim}{r},t,t+\tau) = \langle u_i(\underset{\sim}{x},t)u_j(\underset{\sim}{x}+\underset{\sim}{r},t+\tau)\rangle ,$$

particularly in stationary homogeneous turbulence when mean values are independent of the absolute value of the time and only the time difference comes in. Stationary turbulence requires the existence of external forces to maintain the motion against viscous decay.

We take axes such that the mean velocity is zero, i.e. $\overline{\underset{\sim}{u}} = 0$. This is possible because the operations of taking averages and spatial derivatives commute. Then, the mean of the Navier-Stokes equations gives

$$\frac{\partial \bar{u}_i}{\partial t} = -\frac{1}{\rho}\frac{\partial \bar{p}}{\partial x_i} + \nu\nabla^2 \overline{u_i} - \frac{\partial}{\partial x_j}\overline{u_i u_j} = 0,$$

since all mean values are independent of position. Thus, the mean value does not change and can be taken as zero.

Various properties of $R_{ij}(\underset{\sim}{r},t)$ should be noted. (The dependence on t will not be shown explicitly henceforth unless different times are involved.)

$$R_{ii}(0) = \overline{\underset{\sim}{u}^2} = \text{twice mean kinetic energy per unit mass} \quad (8)$$

$$R_{ij}(\underset{\sim}{r}) = R_{ji}(-\underset{\sim}{r}) \quad (9)$$

$$\frac{\partial R_{ij}}{\partial r_j}(\underset{\sim}{r}) = \langle u_i(\underset{\sim}{x}) \frac{\partial}{\partial r_j} u_j(\underset{\sim}{x}+\underset{\sim}{r})\rangle$$

$$= \langle u_i(\underset{\sim}{x}) \frac{\partial}{\partial x_j'} u_j(\underset{\sim}{x}')\rangle \text{ , where } \underset{\sim}{x}' = \underset{\sim}{x}+\underset{\sim}{r}$$

$$= 0 \text{ , because of the equation of continuity.} \quad (10)$$

Similarly, $\frac{\partial R_{ij}}{\partial r_i} = 0.$

From the Schwartz inequality, $\overline{ab} \leq (\overline{a^2})^{\frac{1}{2}} (\overline{b^2})^{\frac{1}{2}}$, it follows that

$$R_{ij}(\underset{\sim}{r}) \leq \left|R_{ii}(0)R_{jj}(0)\right|^{\frac{1}{2}} \text{ (no summation).}$$

In particular,

$$R_{ii}(\underset{\sim}{r}) \leq R_{ii}(0).$$

The third-order correlation tensor

$$S_{ijk}(\underset{\sim}{r}) = \langle u_i(\underset{\sim}{x})u_j(\underset{\sim}{x})u_k(\underset{\sim}{x}+\underset{\sim}{r})\rangle, \qquad (11)$$

also comes in to a discussion of the dynamics of decay. This has the properties

$$S_{ijk}(\underset{\sim}{r}) = S_{jik}(\underset{\sim}{r}) \; ; \quad \frac{\partial}{\partial r_k} S_{ijk}(\underset{\sim}{r}) = 0;$$

but note that $\frac{\partial}{\partial r_i} S_{ijk}(\underset{\sim}{r}) \neq 0$.

An example of a velocity correlation involving velocity derivatives is the vorticity correlation tensor

$$W_{ij}(\underset{\sim}{r}) = \langle \omega_i(\underset{\sim}{x})\omega_j(\underset{\sim}{x}+\underset{\sim}{r})\rangle \; , \quad \text{where } \omega_i = \varepsilon_{ijk} \frac{\partial u_k}{\partial x_j},$$

$$= \nabla^2 R_{ji} - \delta_{ij}\nabla^2 R_{kk} + \frac{\partial^2 R_{kk}}{\partial r_i \partial r_j} \; , \qquad (12)$$

on making repeated use of results like

$$\left\langle \frac{\partial u_i}{\partial x_j} \frac{\partial u'_k}{\partial x'_\ell} \right\rangle = - \frac{\partial^2}{\partial r_j \partial r_\ell} \langle u_i u'_k \rangle.$$

For further details and more examples, reference should be made to the monograph by BATCHELOR [6].

Spectral Tensors

Many of the results for homogeneous turbulence are expressed in terms of the spectrum tensors, the Fourier transforms of the velocity correlation tensors. The most important is the *energy spectrum tensor*

$$\phi_{ij}(\underset{\sim}{k}) = \frac{1}{(2\pi)^3} \int R_{ij}(\underset{\sim}{r})\, e^{-i\underset{\sim}{k}\cdot\underset{\sim}{r}}\, d\underset{\sim}{r}. \tag{13}$$

This exists if R_{ij} is square integrable. It will be shown below that in general $R_{ij} = O(r^{-3})$ as $r \to \infty$, so that $\phi_{ij}(\underset{\sim}{k})$ exists except at $k = 0$. The inverse relation is

$$R_{ij}(\underset{\sim}{r}) = \int \phi_{ij}(\underset{\sim}{k})\, e^{i\underset{\sim}{k}\cdot\underset{\sim}{r}}\, d\underset{\sim}{k}. \tag{14}$$

In particular,

$$\frac{1}{2}\, \overline{\underset{\sim}{u}^2} = \int \frac{1}{2}\, \phi_{ii}(\underset{\sim}{k})\, d\underset{\sim}{k} \tag{15}$$

so that $\phi_{ii}(\underset{\sim}{k})$ gives the density of the contributions to the kinetic energy in wavenumber space. Integrating $\phi_{ii}(\underset{\sim}{k})$ over a sphere of radius $|\underset{\sim}{k}|$ gives

$$\frac{1}{2}\, \overline{\underset{\sim}{u}^2} = \int_0^\infty E(k)\, dk, \qquad E(k) = \int_{|\underset{\sim}{k}|=k} \frac{1}{2}\phi_{ii}(\underset{\sim}{k})\, dA(\underset{\sim}{k}). \tag{16}$$

$E(k)$ is called the energy spectrum function.

The properties of $\phi_{ij}(\underset{\sim}{k})$ corresponding to symmetry and incompressibility, (9) and (10), are

$$\phi_{ij}(\underset{\sim}{k}) = \phi_{ji}(-\underset{\sim}{k}) \;, \tag{17}$$

$$k_i\phi_{ij}(\underset{\sim}{k}) = k_j\phi_{ij}(\underset{\sim}{k}) = 0. \tag{18}$$

Also, since $R_{ij}(\underset{\sim}{r})$ is real,

$$\phi_{ij}(\underset{\sim}{k}) = \phi^{*}_{ij}(-\underset{\sim}{k}) = \phi^{*}_{ji}(\underset{\sim}{k}) \quad \text{by (17)}, \tag{19}$$

where the star denotes the complex conjugate, so that $\phi_{ij}(\underset{\sim}{k})$ is Hermitian, and by Cramer's theorem it is a positive definite matrix

$$\text{i.e.} \qquad X_i X_j^{*} \, \phi_{ij}(\underset{\sim}{k}) \geq 0 \tag{20}$$

for arbitrary vectors X_i. In particular

$$E(k) \geq 0. \tag{21}$$

The result (21) has proved to be an important check on the unphysical nature of some of the approximations or assumptions that have been made.

Spectral tensors can be defined similarly for correlation tensors of any order. However, the correlation tensors may not vanish as the multidimensional separation vector tends to infinity in certain directions. This difficulty is overcome by forming the cumulant of the correlation tensor, the cumulant being the correlation tensor less products of lower order mean values chosen so that the resulting quantity vanishes as the multidimensional separation vector goes to infinity in all directions. For example, the fourth-order tensor

$$\langle u_i \; u_j \; u_k' \; u_\ell' \rangle$$

does not vanish as $r \to \infty$, but the cumulant

$$\langle u_i u_j u_k' u_\ell' \rangle - \langle u_i u_j \rangle \langle u_k' u_\ell' \rangle - \langle u_i u_k' \rangle \langle u_j u_\ell' \rangle - \langle u_i u_\ell' \rangle \langle u_j u_k' \rangle \tag{22}$$

does vanish and presumably does have a Fourier transform.

The spectrum tensor of the vorticity is

$$\Omega_{ij}(\underset{\sim}{k}) = \frac{1}{(2\pi)^3} \int W_{ij}(\underset{\sim}{r})\, e^{-i\underset{\sim}{k}\cdot\underset{\sim}{r}}\, d\underset{\sim}{r} \, . \tag{23}$$

The relation between $\Omega_{ij}(\underset{\sim}{k})$ and $\phi_{ij}(\underset{\sim}{k})$ follows from (12) and is

$$\Omega_{ij}(\underset{\sim}{k}) = (k^2\delta_{ij} - k_i k_j)\, \phi_{kk}(\underset{\sim}{k}) - k^2\phi_{ji}(\underset{\sim}{k}) \, . \tag{24}$$

In particular

$$\Omega_{ii}(\underset{\sim}{k}) = k^2\phi_{ii}(k) \, , \tag{25}$$

and

$$\overline{\underset{\sim}{\omega}^2} = 2 \int_0^\infty k^2\, E(k)\, dk, \tag{26}$$

so that $2k^2\, E(k)$ is the density of contributions to $\overline{\underset{\sim}{\omega}^2}$ per unit wavenumber magnitude.

The Probability Distribution

Instead of the correlation or spectrum tensors, we can describe the turbulent motion statistically by means of the joint probability distribution function (j.p.d.f.)

$$P_n(\underset{\sim}{u}^{(1)}, \ldots \underset{\sim}{u}^{(n)};\ \underset{\sim}{x}_1, \ldots \underset{\sim}{x}_n;\ t_1, \ldots t_n)$$

which gives the probability density of the velocity at n points at the times $t_1, \dots t_n$ (which may be identical) being in the range $d\underset{\sim}{u}_1 \dots d\underset{\sim}{u}_n$. It is a result of probability theory that a knowledge of the n-point correlation tensor of all order determines P_n. Because the turbulence is homogeneous, P_n depends only on the relative position of the n-points.

If the probability distribution were exactly Gaussian, then we should have

$$P_n = \frac{\pi^{3n/2}}{||\lambda_{\alpha\beta}||} \exp \left(- \sum_1^{3n} \lambda_{\alpha\beta}\, u_\alpha\, u_\beta \right), \tag{27}$$

where the $\lambda_{\alpha\beta}$ are functions of the $\underset{\sim}{x}_i$ and the times t_i. Note that it can be shown that if the distribution is Gaussian, the cumulants of order higher than the second vanish. In particular, the cumulant (22) would be zero.

Experimental measurements in grid turbulence suggest strongly that the probability distribution of the velocity at a single point is Gaussian. A proof or even a plausible demonstration that a non-Gaussian single point probability distribution relaxes to a Gaussian distribution in a time short compared with the lifetime of the turbulence does not seem to have been given. It is not altogether certain that the Gaussian distribution for the velocity at a single point is not a manifestation of particular initial or generating conditions associated with the grid, rather than a general property of homogeneous turbulence.[1]

---

[1] In work as yet unpublished, T.S. LUNDGREN has derived from the Navier-Stokes equations in an ingenious manner the equations for the j.p.d.f.s. and considered the approach to a Gaussian distribution of the velocity fluctuations at a fixed point.

The two-point j.p.d.f. is not Gaussian unless the separation of the two points is comparable with length scales characteristic of the turbulent motion. As a measure of the skewness, we have, for example,

$$S(x_1) = \frac{\overline{(u_1' - u_1)^3}}{\left[\overline{(u_1' - u_1)^2}\right]^{3/2}} , \tag{28}$$

where x_1 is the difference in the 1-coordinate, the x_2 and x_3 coordinates being the same, and

$$S_0 = \overline{\left(\frac{\partial u_1}{\partial x_1}\right)^3} \Big/ \left[\overline{\left(\frac{\partial u_1}{\partial x_1}\right)^2}\right]^{3/2} . \tag{29}$$

The flatness factor of the two-point j.p.d.f. is

$$F(x_1) = \frac{\overline{(u_1' - u_1)^4}}{\left[\overline{(u_1' - u_1)^2}\right]^2} ; \quad F_0 = \frac{\overline{\left(\frac{\partial u_1}{\partial x_1}\right)^4}}{\left[\overline{\left(\frac{\partial u_1}{\partial x_1}\right)^2}\right]^2} . \tag{30}$$

Experimental measurements in grid turbulence suggest values of S_0 about -0.4 and values of F_0 about 4. (For a Gaussian distribution, $S_0 = 0$, $F_0 = 3$).

Similarly, hyperskewness and hyperflatness factors can be defined. They show increasing departures from Gaussian behavior [7].

Formulation of the Problem of Homogeneous Turbulence

Quite simply, the problem to be solved is: Given the correlation tensors or the j.p.d.f.s. for all sets of points at an initial instant t_0, calculate the quantities for later times.

There are two questions. First, is the problem well posed mathematically in the sense that a solution exists and is unique? An answer to this question does not seem to have been given. Second and much more important, is the problem of any physical significance? Here the answer is clear. The problem in full generality is of no physical significance. In order to be significant there must exist simple results which depend only on simple properties of the initial data. For instance, we should like to calculate the one-point and two-point j.p.d.f.s. given only the values of these quantities at the initial instant. Physical intuition suggests that this task should be sensible as the low-order correlations should be reasonably independent of the high-order ones.

It is perhaps appropriate to stress again the point that homogeneous turbulence may be of limited practical importance and the effort put into the theory is not likely to be commensurate with the rewards.

Isotropic Turbulence

In order to reduce the algebraic complications, the further idea of isotropy is introduced, where the statistical properties are independent of the direction of the axes. At one time it was thought that grid turbulence was approximately isotropic as well as homogeneous, but more detailed measurements showed that the assumption of isotropy is not in fact justified. Nevertheless, the simplifications of isotropy are well worth considering, particularly for theoretical arguments, and indeed much of the experimental work is described in the language of isotropic turbulence.

Consider the velocity correlation tensor $R_{ij}(\underset{\sim}{r})$. If the form of $R_{ij}(\underset{\sim}{r})$ is independent of the choice of axes, it can be shown [6] that

$$R_{ij}(\underset{\sim}{r}) = F(r)\, r_i r_j + G(r)\, \delta_{ij},$$

where F and G are functions of $r^2 = |\underset{\sim}{r}|^2$ if $R_{ij}(\underset{\sim}{r})$ is analytic. Making use of the equation of continuity gives

$$R_{ij}(\underset{\sim}{r}) = \overline{u_1^2}\left(\frac{f(r)-g(r)}{r^2}\, r_i r_j + g(r)\, \delta_{ij}\right),$$

$$\text{where} \quad g = f + \tfrac{1}{2} r f', \qquad \text{or}$$

$$R_{ij}(\underset{\sim}{r}) = \overline{u_1^2}\left[\left(f + \frac{1}{2} r f'\right) \delta_{ij} - \frac{1}{2}\frac{f'}{r}\, r_i r_j\right]. \tag{31}$$

The functions f(r) and g(r) are called the longitudinal and lateral velocity correlations.

$u_n \uparrow \cdot\!\rightarrow u_p \qquad u'_n \uparrow \cdot\!\rightarrow u'_p$

$$f(r) = \frac{\overline{u_p u'_p}}{\overline{u_1^2}}, \quad g(r) = \frac{\overline{u_n u'_n}}{\overline{u_1^2}}.$$

Thus, for isotropic turbulence, one scalar function is sufficient to describe the velocity correlation tensor.

For small r,

$$f(r) = 1 - \frac{r^2}{2\lambda^2} + O(r^4), \qquad (32)$$

where

$$\lambda^{-2} = -\frac{\partial^2 f}{\partial r^2}(0) = \overline{\left(\frac{\partial u_1}{\partial x_1}\right)^2} \Big/ \overline{u_1^2} \; .$$

(It is believed that $f(r) > 0$ for any realistic set of initial conditions.) There is an integral associated with $f(r)$

$$\int_0^\infty r^4 \, f(r) \, dr,$$

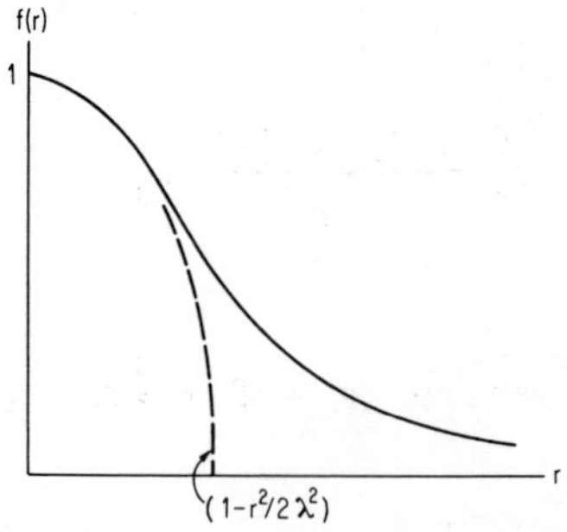

which is called the Loitsianskii invariant because it was believed that the integral remained constant during decay. However, it was shown by BATCHELOR and PROUDMAN [8] that the assumptions on which the proof of invariance rested were invalid and that the integral is not a dynamical invariant. We shall see below (§3) that for a general initial condition the integral does not in fact exist because $f(r) \sim r^{-3}$ as $r \to \infty$.

The length λ is called the dissipation-length parameter or microscale. Its importance lies in the fact that the rate of dissipation of energy per unit mass ε is given by

$$-\frac{d}{dt}\frac{1}{2}\overline{\underset{\sim}{u}^2} = \varepsilon = \nu \overline{\left(\frac{\partial u_i}{\partial x_j}\right)^2} = \nu \overline{\underset{\sim}{\omega}^2}$$

$$= 15\nu\overline{\left(\frac{\partial u_1}{\partial x_1}\right)^2} = 15\nu \frac{\overline{u_1^2}}{\lambda^2} \quad \text{for isotropic turbulence.} \tag{33}$$

Thus, a knowledge of λ gives the mean square vorticity and the rate at which energy is dissipated by the turbulence. The length λ is thus an important parameter. We shall consider later the intriguing question of whether there is any physical process in the turbulence that takes place with the length scale λ.

Note that the relation

$$\nu\overline{\left(\frac{\partial u_i}{\partial x_j}\right)^2} = \nu \overline{\underset{\sim}{\omega}^2} \tag{34}$$

holds generally for homogeneous turbulence.

The corresponding results for the energy spectrum tensor are

$$\phi_{ij}(\underset{\sim}{k}) = \frac{E(k)}{4\pi k^4}(k^2\delta_{ij} - k_ik_j). \tag{35}$$

Note that

$$\frac{1}{\lambda^2} = \frac{\int_0^\infty k^2 E(k)\, dk}{5\int_0^\infty E(k)\, dk}, \tag{36}$$

so that $5/\lambda^2$ is the "mean-square wavenumber."

For the third-order correlation tensor, we write

$$(\overline{u_1^2})^{3/2}\ k(r) = \overline{u_p^2 u_p'} = \frac{r_i r_j r_k}{r^3}\ S_{ijk}(\underset{\sim}{r}). \tag{37}$$

It can be shown [6] that

$$S_{ijk}(\underset{\sim}{r}) = (\overline{u_1^2})^{3/2}\left[\left(\frac{k-rk'}{2r^3}\right) r_i r_j r_k + \frac{(2k+rk')}{4r}(r_i\delta_{jk}+r_j\delta_{ik})-\frac{k}{2r}r_k\delta_{ij}\right]. \tag{38}$$

The triple correlation function k(r) is related to the skewness factor as defined in (28). For

$$\overline{(u_1' - u_1)^3} = 3\ \overline{u_1^2 u_1'}\ -3\ \overline{u_1 u_1'^2} = 6(\overline{u_1^2})^{3/2}\ k(r),$$

since $\overline{u_1^2 u_1'} = -\overline{u_1 u_1'^2}$ because of symmetry under reflection and change of sign on reversal of velocities. For small r, $k(r) \sim r^3$. Also

$$\overline{(u_1' - u_1)^2} = 2\ \overline{u_1^2}\ \{1-f(r)\}.$$

Therefore,

$$S(r) = \frac{6k(r)}{2^{3/2}\{1-f(r)\}^{3/2}}. \tag{39}$$

In particular,

$$S_0 = \lambda^3 k'''(0) = \frac{\overline{\left(\frac{\partial u_1}{\partial x_1}\right)^3}}{\left[\overline{\left(\frac{\partial u_1}{\partial x_1}\right)^2}\right]^{3/2}}. \tag{40}$$

The functions described above are the most common ones in the language of homogeneous turbulence. Further correlation and spectrum tensors will be introduced if and when necessary.

Do All Mean Values Exist?

It is certainly reasonable to assume that the solutions of the Navier-Stokes equations are continuous and infinitely differentiable. The reason is that viscosity smooths out discontinuities infinitely rapidly. However, the corollary that all mean values of the velocity and its derivatives exist requires the further assumption that the j.p.d.f.s. are exponentially small for large values of the velocity, or that the probability of large velocities and large velocity derivatives is exponentially small. This is undoubtedly correct because of the smoothing action of viscosity, so that even if mean values did not exist at the initial instant, they would exist immediately afterwards, but the physical mechanism by which viscosity reduces the probability of large values is by no means obvious, and a proof that the proposition is correct would perhaps be of value.

3. Structure and Invariance of the Large Eddies

It is well known that any motion of an incompressible fluid can be generated instantaneously from rest by a distribution of impulsive forces. Hence as an initial condition for the turbulent motion we may specify not the velocities but rather the impulsive force system which produces the velocities at the initial instant. Physically, this is indeed much more satisfactory as any motion is generated by the application of forces and we control the forces, not the velocities.

Thus, we suppose the turbulence is generated from a state of rest at $t = 0$ by a random, spatially homogeneous distribution of impulsive forces $\underset{\sim}{f}(\underset{\sim}{x})$ per unit mass with zero mean, i.e. $\overline{\underset{\sim}{f}(\underset{\sim}{x})} = 0$. Since viscous stresses are finite, they can have no effect on the motion generated impulsively. The velocity field at $t = 0+$ is the solution of the pair of equations,

$$\underset{\sim}{u} = \nabla\phi + \underset{\sim}{f}\ , \qquad \text{div}\ \underset{\sim}{u} = 0, \tag{41}$$

where ϕ is the impulsive pressure which is produced when the force is applied because of the incompressibility of the fluid. Clearly $\nabla^2\phi = -\text{div}\ \underset{\sim}{f}$, and hence

$$\phi(\underset{\sim}{x}) = \frac{1}{4\pi}\int \frac{\text{div}\ \underset{\sim}{f}}{|\underset{\sim}{x}-\underset{\sim}{x}'|}\, d\underset{\sim}{x}'\ . \tag{42}$$

Fourier Analysis of the Velocity Field

This result can be expressed more simply in terms of a Fourier analysis of the velocity field. We can write

$$\underset{\sim}{u}(\underset{\sim}{x}) = \int \underset{\sim}{a}(\underset{\sim}{k})\, e^{i\underset{\sim}{k}\cdot\underset{\sim}{x}}\, d\underset{\sim}{k}\ ; \quad \underset{\sim}{a}(\underset{\sim}{k}) = \frac{1}{(2\pi)^3}\int \underset{\sim}{u}(\underset{\sim}{x})\, e^{-i\underset{\sim}{k}\cdot\underset{\sim}{x}}\, d\underset{\sim}{x}, \tag{43}$$

where it is to be understood that $\underset{\sim}{a}(\underset{\sim}{k})$ is a "generalized" function of wavenumber $\underset{\sim}{k}$ because the second integral in (43) does not converge since $\underset{\sim}{u}(\underset{\sim}{x})$ is a stationary random function of position. The mean values of $\underset{\sim}{a}(\underset{\sim}{k})$ are related to the spectrum tensors as follows.

$$\overline{a_i(\underset{\sim}{k})a_j(\underset{\sim}{k}')} = \frac{1}{(2\pi)^6}\int \overline{u_i(\underset{\sim}{x})u_j(\underset{\sim}{x}')}\, e^{-i(\underset{\sim}{k}\cdot\underset{\sim}{x}+\underset{\sim}{k}'\cdot\underset{\sim}{x}')}\, d\underset{\sim}{x}\, d\underset{\sim}{x}'$$

$$= \frac{1}{(2\pi)^6}\int R_{ij}(\underset{\sim}{x}'-\underset{\sim}{x})\, e^{-i\underset{\sim}{k}'\cdot(\underset{\sim}{x}'-\underset{\sim}{x})}\, e^{-i(\underset{\sim}{k}+\underset{\sim}{k}')\cdot\underset{\sim}{x}}\, d\underset{\sim}{x}\, d(\underset{\sim}{x}'-\underset{\sim}{x})$$

$$= \phi_{ij}(\underset{\sim}{k}')\,\delta(\underset{\sim}{k}+\underset{\sim}{k}')\ , \tag{44}$$

since
$$\frac{1}{(2\pi)^3}\int e^{-i\underset{\sim}{k}\cdot\underset{\sim}{x}}\, d\underset{\sim}{x} = \delta(\underset{\sim}{k}), \tag{45}$$

where $\delta(\underset{\sim}{k})$ denotes the three-dimensional delta function, i.e.

$$\delta(\underset{\sim}{k}) = \delta(k_1)\delta(k_2)\delta(k_3). \tag{46}$$

If $\underset{\sim}{m}(\underset{\sim}{k})$ denotes the Fourier transform of the impulsive force system, then the transform of (41) is

$$\underset{\sim}{a} = i\underset{\sim}{k}\tilde{\phi} + \underset{\sim}{m}\ , \qquad \underset{\sim}{k}\cdot\underset{\sim}{a} = 0\ , \tag{47}$$

where $\tilde{\phi}$ is the generalized transform of ϕ. The solution of these equations is

$$a_i(\underset{\sim}{k}) = (\delta_{ij} - \frac{k_i k_j}{k^2})\, m_j(\underset{\sim}{k}) = \Delta_{ij}\, m_j(\underset{\sim}{k}). \tag{48}$$

Suppose now that the impulsive forces are such that the force correlations $\langle f_i(\underset{\sim}{x})f_j(\underset{\sim}{x}+\underset{\sim}{r})\rangle = F_{ij}(\underset{\sim}{r})$ are square integrable. This is an assumption, but it is a very weak one, and it certainly would seem to apply to any "real" situation. Then the force distribution has a spectrum tensor

$$M_{ij}(\underset{\sim}{k}) = \frac{1}{(2\pi)^3} \int F_{ij}(\underset{\sim}{r})\, e^{-i\underset{\sim}{k}\cdot\underset{\sim}{r}}\, d\underset{\sim}{r} , \tag{49}$$

and it follows from (48) that at the initial instant t = 0,

$$\phi_{ij}(k,0) = (\delta_{ip} - \frac{k_i k_p}{k^2})(\delta_{jq} - \frac{k_j k_q}{k^2})\, M_{pq}(\underset{\sim}{k}). \tag{50}$$

In a similar but more complicated manner the initial values of the spectrum tensors of all orders can be expressed in terms of the spectrum tensors of the impulsive force distribution.

Form of the Energy Spectrum Tensor for Small Wavenumber

Let us make the further physical assumption that $M_{pq}(0)$ exists. It would be sufficient that $F_{ij}(\underset{\sim}{r}) = O(r^{-4})$ as $r \to \infty$. Again the assumption is weak, and would seem to be realistic for any field of homogeneous turbulence which is an approximation to a real flow field, but it may well be possible to generate impulsive force fields which do not satisfy the criterion. It is necessary to consider what a nonzero value of $M_{pq}(0)$ implies.

We have

$$\begin{aligned} M_{pq}(0) &= \frac{1}{(2\pi)^3} \int \overline{f_p(\underset{\sim}{x}) f_q(\underset{\sim}{x}+\underset{\sim}{r})}\, d\underset{\sim}{r} \\ &= \frac{1}{(2\pi)^3} \lim_{V\to\infty} \left[\frac{1}{V^{1/2}} \int_V f_p(\underset{\sim}{x})dV \times \frac{1}{V^{1/2}} \int_V f_q(\underset{\sim}{x}')d\underset{\sim}{x}'\right] \end{aligned} \tag{51}$$

on interpreting the averages as spatial means. Now

$$\lim_{V\to\infty} \frac{1}{V} \int_V f_p(\underset{\sim}{x})\, dV = \overline{f}_p(\underset{\sim}{x}) = 0,$$

but
$$\frac{1}{V^{1/2}} \int_V f_p(\underset{\sim}{x})\, dV$$

does not necessarily become zero for large V. For a general force distribution, this last quantity will fluctuate in a finite manner corresponding to the fact that the sum of N random variables with zero mean fluctuates with an amplitude of order $N^{1/2}$ as $N \to \infty$, and the mean-square fluctuation of the sum exists and is proportional to N. Put roughly, it is absolutely consistent with the average momentum of the fluid being zero to have the total momentum in a large volume V of the fluid being of order $V^{1/2}$. Unless, therefore, there is a constraint on the impulsive force system so that the integral (51) vanishes, the quantity $M_{pq}(0)$ will be nonzero.

At a later time t, the velocity field will have changed and will correspond to another impulsive force system $\underset{\sim}{f}'(\underset{\sim}{x})$, say, with a spectrum tensor $M'_{pq}(\underset{\sim}{k})$. We now give an heuristic argument to show that

$$M'_{pq}(0) = M_{pq}(0). \tag{52}$$

The momentum of the fluid inside a very large volume V is initially of order $V^{1/2}$ because it is proportional to the total impulse of the forces applied to V. During the motion, the momentum changes because of the transport of momentum across the surface of V by convection and pressure forces. Because the transfer processes are random, the net transfer will be proportional to the square root of the surface area, i.e. $V^{1/3}$. Thus, the momentum changes by an amount $V^{1/3}$ and therefore the total impulse of the force system $\underset{\sim}{f}'(\underset{\sim}{x})$ will differ from that of $\underset{\sim}{f}(\underset{\sim}{x})$ by an amount $O(V^{1/3})$. Hence,

$$\frac{1}{V^{1/2}}\left[\int_V \underset{\sim}{f}'(\underset{\sim}{x})\, d\underset{\sim}{x} - \int_V \underset{\sim}{f}(\underset{\sim}{x})\, d\underset{\sim}{x}\right] = O(V^{-1/6}) \to 0$$

as $V \to \infty$, and the result (52) follows. It is interesting that this argument is based simply on the conservation of linear momentum and properties of the sum of a large number of random variables, and is independent of the precise form of the Navier-Stokes equations. (An alternative more formal but not more rigorous argument is outlined in [9]).

It follows now that for small k,

$$\phi_{ij}(\underset{\sim}{k},t) = (\delta_{ip} - \frac{k_i k_p}{k^2})\,(\delta_{jq} - \frac{k_j k_q}{k^2})\, M_{pq}(0) + o(1). \qquad (53)$$

Thus, the form of the energy spectrum tensor for small wavenumber is an invariant of the motion. Note that the energy spectrum tensor is not analytic at $k = 0$.

The energy spectrum function is easily calculated from (16). We find

$$E(k) = \frac{4\pi}{3} M_{pp}(0)\, k^2 = C\, k^2, \text{ say} \qquad (54)$$

where C is a constant. Thus, the curvature of the energy spectrum tensor at $k = 0$ should be constant during the decay of the turbulence.

The Form of the Velocity Correlation Tensor at Large Separation

The behavior of $R_{ij}(\underset{\sim}{r})$ for large r is determined by the non-analytic form of $\phi_{ij}(\underset{\sim}{k})$ for small k. Using the results (for generalized functions) that

$$\int e^{i\underset{\sim}{k}\cdot\underset{\sim}{r}}\, d\underset{\sim}{k} = 8\pi^3\,\delta(\underset{\sim}{r}),$$

$$\int \frac{k_i k_j}{k^2} e^{i\underset{\sim}{k}\cdot\underset{\sim}{r}}\, d\underset{\sim}{k} = -2\pi^2 \frac{\partial^2}{\partial r_i \partial r_j}\left(\frac{1}{r}\right),$$

$$\int \frac{e^{i\underset{\sim}{k}\cdot\underset{\sim}{r}}}{k^4}\, d\underset{\sim}{k} = -\pi^2 r,$$

it follows that

$$R_{ij}(\underset{\sim}{r}) = -\pi^2 M_{pq}(0)\left(\delta_{ip}\nabla^2 - \frac{\partial^2}{\partial r_i \partial r_p}\right)\left(\delta_{jq}\nabla^2 - \frac{\partial^2}{\partial r_j \partial r_q}\right) r. \tag{55}$$

Note that $R_{ij}(\underset{\sim}{r})$ is $O(r^{-3})$ as $r \to \infty$.

For the special case of isotropic turbulence, we can write

$$M_{pq}(0) = \frac{C}{4\pi}\,\delta_{pq}, \tag{56}$$

and comparing (55) with (31) shows that

$$\overline{u_1^2}\, f(r) \sim \frac{\pi C}{r^3} \tag{57}$$

which demonstrates that the Loitsianskii invariant does not exist if $C \neq 0$.

For homogeneous turbulence, we can define the average energy correlation for two points distance r apart by

$$R(r) = \frac{1}{4\pi r^2} \int_{|\underset{\sim}{r}|=r} R_{ii}(\underset{\sim}{r})dA(r). \tag{58}$$

Then it follows from the various relations that

$$E(k) = \frac{1}{\pi}\int_0^\infty R(r)\ kr \sin kr\ dr. \tag{59}$$

It follows that

$$\int_0^\infty r^2 R(r)\ dr = \pi C. \tag{60}$$

Thus, the velocity correlation tensor does have a nontrivial integral invariant. Note that it is an immediate consequence of (55) that $R(r) = o(r^{-3})$. (LANDAU & LIFSHITZ [10] show that the Loitsianskii invariant can be related to the conservation of angular momentum, but the argument depends on the convergence of integrals that are in fact divergent.)

The Case C = 0

If the invariant C is zero, the results are more complicated. If we assume that $M_{pq}(0) = 0$ but that the second integral moments of $F_{ij}(\underset{\sim}{r})$ exist, i.e. the integrals

$$\int r_p r_q F_{ij}(\underset{\sim}{r})\ d\underset{\sim}{r}$$

converge absolutely, then $M_{pq}(\underset{\sim}{k})$ has the expansion

$$M_{pq}(\underset{\sim}{k}) = C_{pqij}k_i k_j + \dots \tag{61}$$

and we obtain from (50) the form of the energy spectrum tensor.[1] It now follows that at the initial instant

$$E(k) = \frac{4\pi}{15}(2C_{iijj} - C_{ijij})\,k^4 + \dots . \tag{62}$$

BATCHELOR and PROUDMAN [8] argued, on the basis of the assumptions that all integral moments of the correlation tensor exist initially and that the spectrum functions are analytic functions of the time, that the forms (50) with $M_{pq}(\underset{\sim}{k})$ given by (61) and the result (62) hold for all time, but the coefficients C_{pqij} are not invariants and in general may be expected to decay with time.

The corresponding velocity correlation tensor is $O(r^{-5})$ except in isotropic turbulence when it becomes $O(r^{-6})$.

It must be stressed that all these results of the invariance and structure of the large eddies are academic because any real flow will not be homogeneous over large distances. However, they are worth giving because they are the only analytical results for homogeneous turbulence in which one can have any confidence, and they may provide a useful check for the approximate theories.

As far as comparison with experiment is concerned, this is not possible. It would however be desirable to consider whether the invariant C is likely to be zero. A sufficient condition is that

---

1 There are no terms linear in the wavenumber vector $\underset{\sim}{k}$ because of Cramer's theorem equation (20).

div $\underset{\sim}{f} = 0$, as then $k_p M_{pq}(\underset{\sim}{k}) = 0$, from which it follows that $M_{pq}(0)=0$ on expanding about $k = 0$. Physically this condition would mean that the impulsive force applied to a volume V could be expressed as surface stresses over the surface of V, however large V may be. For motion produced by stirring a fluid with a lot of paddles which are then removed, the vanishing of C would seem to require a degree of correlation between the paddles which is not likely to exist. For "homogeneous" turbulence which develops out of the instability of a laminar flow, it is not clear whether C is zero; arguments can be made each way. However, for the experimentally important case of grid turbulence, it is likely that C = 0. The reason is that the lift and drag on a rod in an air stream is periodic in time even at large Reynolds numbers. Thus the momentum communicated to the air which flows past the rod in a time t, which corresponds to a length Ut of turbulent flow downstream of a grid of rods, does not increase like $t^{1/2} \propto x^{1/2}$, but will be bounded. It follows that the equivalent force system has $M_{pq}(\underset{\sim}{k}) = 0$. This raises an important question. Since most experimental results on decaying homogeneous turbulence are based on grid turbulence, does turbulence with $C \neq 0$ differ in any fundamental way from turbulence with $C = 0$, and are the wind-tunnel results special and not typical of general decaying homogeneous turbulence? The usual belief that Fourier components of very different wavenumber are statistically independent suggests that the form of the energy spectrum function near $k = 0$ should not affect the decay process, but nothing should be taken for granted in turbulence.

Stationary Homogeneous Turbulence

It is interesting to see if we can extend these results on decaying homogeneous turbulence to stationary homogeneous turbulence where external forces are applied continuously to maintain the motion against viscous decay. It is almost obvious that for the large scale features of the motion, the convective acceleration terms and the viscous terms are negligible, and that the equation of motion is a balance between acceleration terms and applied forces. That is, the approximate form of the equation of motion is

$$\frac{\partial \underset{\sim}{u}}{\partial t} = -\nabla\phi+\underset{\sim}{f}(\underset{\sim}{x},t), \text{ div } \underset{\sim}{u} = 0, \tag{63}$$

where now $\underset{\sim}{f}$ describes an external force per unit mass that is a stationary random function of position and also of time. The Fourier transform of this equation with respect to space gives

$$\frac{\partial a_i}{\partial t} = (\delta_{ij} - \frac{k_i k_j}{k^2})\ m_j(\underset{\sim}{k},t). \tag{64}$$

The approximation of dropping the convective acceleration terms and the viscous terms corresponds in wavenumber space to making an expansion about $\underset{\sim}{k} = 0$ and retaining the leading terms, it being an immediate consequence of the equations that the nonlinear and viscous terms are $O(k)$ and $O(k^2)$, respectively.

We now introduce Fourier transforms with respect to time and define

$$\tilde{M}_{pq}(\omega) = \frac{1}{(2\pi)^4} \int \langle f_p(\underset{\sim}{x},t) f_q(\underset{\sim}{x}+\underset{\sim}{r},t+\tau)\rangle\ e^{-i\omega\tau}\ d\underset{\sim}{r}\ d\tau, \tag{65}$$

and

$$\tilde{\Phi}_{ij}(\underset{\sim}{k},\omega) = \frac{1}{(2\pi)^4} \int \langle u_i(\underset{\sim}{x},t) u_j(\underset{\sim}{x}+\underset{\sim}{r},t+\tau)\rangle\ e^{-i\underset{\sim}{k}\cdot\underset{\sim}{r}-i\omega\tau}\ d\underset{\sim}{r}\ d\tau. \tag{66}$$

It follows from (64) that

$$\tilde{\Phi}_{ij}(0,\omega) = \Delta_{ip}\Delta_{jq} \frac{\tilde{M}_{pq}(\omega)}{\omega^2} , \tag{67}$$

and $\phi_{ij}(\underset{\sim}{k})$, which is independent of time for a stationary process, is given at $\underset{\sim}{k} = 0$ by

$$\phi_{ij}(0) = \Delta_{ip}\Delta_{jq} \int_{-\infty}^{\infty} \frac{\tilde{M}_{pq}(\omega)}{\omega^2} \, d\omega \tag{68}$$

and

$$E(k) = \frac{4\pi}{3} k^2 \int_{-\infty}^{\infty} \frac{\tilde{M}_{pp}(\omega)}{\omega^2} \, d\omega . \tag{69}$$

The expression (68) shows that a truly stationary velocity field is not possible unless

$$\tilde{M}_{pq}(0) = 0, \qquad \int \langle f_p(\underset{\sim}{x},t) f_q(\underset{\sim}{x}+\underset{\sim}{r},t+\tau) \rangle \, d\underset{\sim}{r} \, d\tau = 0. \tag{70}$$

If the force field does not satisfy the restriction (70), the amount of energy in the large eddies will increase without bound and a stationary flow field does not exist. However, the result must be interpreted with caution as the region of validity of (68), i.e. the range of wavenumbers in which it is permissible to neglect the nonlinear terms which are of order k but proportional to a higher power of the velocity than the terms retained, probably shrinks as the density of energy at k = 0 increases. Nevertheless, it shows that the assumption implicit in discussions of stationary homogeneous turbulence that an arbitrary stationary random force field will produce a stationary random velocity field is not altogether correct, although it may be a reasonable approximation.

Given that (70) is satisfied, the big eddies have the same large scale structure as in decaying turbulence with the tensor $M_{pq}(0)$ replaced by

$$\int_{-\infty}^{\infty} \frac{\tilde{M}_{pq}(\omega)}{\omega^2}\, d\omega = \frac{-1}{(2\pi)^4} \int_{-\infty}^{\infty} |\tau| \langle f_p(\underset{\sim}{x},t) f_q(\underset{\sim}{x}+\underset{\sim}{r},t+\tau)\rangle \, d\underset{\sim}{r}\, d\tau. \qquad (71)$$

4. The Problem of Decay

So far we have been concerned with the description of homogeneous turbulence. Even the results of the previous section were consequences of the conservation of linear momentum rather than of the detailed structure of the Navier-Stokes equations,

$$\frac{\partial u_i}{\partial t} + \frac{\partial}{\partial x_j}(u_i u_j) = -\frac{1}{\rho}\frac{\partial p}{\partial x_i} + \nu \nabla^2 u_i, \quad \frac{\partial u_i}{\partial x_i} = 0. \qquad (72)$$

Let us consider isotropic turbulence, as the extra generality of homogeneous turbulence does not compensate at the moment for the increase in algebraic complexity. We multiply Eq. (72) by u_i' and add the equation with $\underset{\sim}{x}$ and $\underset{\sim}{x}'$ interchanged. We note from (38) that

$$\frac{\partial}{\partial r_j} \overline{u_i u_j u_i'} = -\frac{1}{2}\left(r\frac{\partial}{\partial r} + 3\right)\left(\frac{\partial}{\partial r} + \frac{4}{r}\right) k(r) (\overline{u_1^2})^{3/2}.$$

Further

$$\frac{\partial}{\partial t}\overline{u_i u_i'} = \frac{\partial}{\partial t}\left\{ \overline{u_1^2}\left(r\frac{\partial}{\partial r} + 3\right) f(r) \right\}$$

$$\nabla^2 \overline{u_i u_i'} = \overline{u_1^2} \left(r \frac{\partial}{\partial r} + 3\right)\left(\frac{\partial^2}{\partial r^2} + \frac{4}{r}\frac{\partial}{\partial r}\right) f(r),$$

and $\overline{pu_i'} = A(r) r_i$ since it is a first-order isotropic tensor, and by the equation of continuity

$$\frac{\partial A}{\partial r} + 3A = 0,$$

so that $A \equiv 0$ if it is regular at $r = 0$. It follows after a little reduction that

$$\frac{\partial}{\partial t}\left[\overline{u_1^2} f(r)\right] = (\overline{u_1^2})^{3/2}\left(\frac{\partial}{\partial r} + \frac{4}{r}\right)k(r) + 2\nu\, \overline{u_1^2}\left(\frac{\partial^2}{\partial r^2} + \frac{4}{r}\frac{\partial}{\partial r}\right) f(r). \quad (73)$$

This is the Karman-Howarth equation. It gives an expression for the second-order correlation in terms of the triple correlation. The Karman-Howarth equation expresses in a nutshell the fundamental difficulty of turbulence; namely, the equation for f involves k. As is quite obvious, if we use the Navier-Stokes equations to form an equation for f it will involve a fourth-order correlation, and so on. Thus we obtain an infinite number of equations in an infinite number of unknowns. All the quantitative theories of turbulence involve truncations of this system in order to obtain finite closed systems. None has yet shown itself to be of any value, and only the qualitative theories associated with the name of Kolmogorov have proved useful and even these are open to doubt as we shall see. (Recently theories have been proposed by EDWARDS [11] and HERRING [12] based on a Liouville equation for the probability distribution, but these seem to be equivalent in some sense to truncations of the infinite set.)

Note that the Karman-Howarth equation is consistent with the previous results on the large scale structure if $k(r)$ is $O(r^{-3})$ as $r \to \infty$. The analysis of the initial velocity distributions can be used to show that initially $k(r)$ is $O(r^{-4})$ and remains of this order if the fourth-order cumulant is $O(r^{-3})$, and so on. Multiplying the Karman-Howarth equation by r^4 and integrating gives

$$\frac{\partial}{\partial t} \int_0^r \overline{u_1^2} f(r)\, r^4\, dr = (\overline{u_1^2})^{3/2}\, r^4\, k(r) - 2\nu\, \overline{u_1^2}\, r^4 \frac{\partial f}{\partial r}. \tag{74}$$

The divergent part of the integral on the left-hand side is constant and its time derivative is therefore zero. If the invariant $C = 0$, we have

$$\frac{\partial}{\partial t} \int_0^\infty \overline{u_1^2}\, r^4\, f\, dr = (\overline{u_1^2})^{3/2} \lim_{r\to\infty} r^4\, k(r) \neq 0. \tag{75}$$

(This result was originally deduced by PROUDMAN and REID [13] on the basis of the joint normal approximation.)

It is useful to rewrite the Karman-Howarth equation in terms of the functions

$$U(r,t) = \overline{u_1^2}(1-f(r)) = \overline{\tfrac{1}{2}(u_1 - u_1')^2} \tag{76}$$

(which measures the way in which the velocity correlation decreases from unity as the separation increases) and the skewness $S(r)$ defined by (39). We have

$$\frac{\partial U}{\partial t} + \frac{2^{1/2}}{3}\left(\frac{\partial}{\partial r} + \frac{4}{r}\right) S U^{3/2} - 2\nu\left(\frac{\partial^2}{\partial r^2} + \frac{4}{r}\frac{\partial}{\partial r}\right) U = \frac{\partial \overline{u_1^2}}{\partial t} = -\frac{2\varepsilon}{3}, \tag{77}$$

where ε is the mean rate of energy dissipation per unit mass. This shows clearly that the nonlinearity enters through the skewness of the j.p.d.f. at two points.

For small r, we have

$$U = \frac{\varepsilon}{30\nu} r^2 = \frac{r^2}{2\lambda^2} \overline{u_1^2}, \tag{78}$$

which just recovers the previous results.

The main problem of turbulent motion, which has been overlooked in recent years, is the calculation of ε. How fast does turbulent energy decay and what is the mechanism by which the mechanical energy is dissipated by viscosity? For a given distribution of velocity, is ε independent of ν and if so, why? Most of the attempts to calculate ε seem to be based on dimensional considerations and essentially assume that ε is independent of ν. The only attempt to find ε for an actual velocity field seems to be that by TAYLOR and GREEN [14], and their results are inconclusive.

We can define a Reynolds number for the turbulence by introducing a length scale

$$L_p = \int_0^\infty f(r)\, dr = \frac{3\pi}{4} \int_0^\infty k^{-1} E(k)\, dk \Big/ \int_0^\infty E(k)\, dk, \tag{79}$$

and then define

$$Re = \frac{(\overline{u_1^2})^{1/2} L_p}{\nu} . \tag{80}$$

It is also customary to define a second Reynolds number based on the dissipation length parameter λ, namely

$$R_\lambda = \frac{(\overline{u_1^2})^{1/2}\lambda}{\nu} = (15\ \mathrm{Re})^{1/2}\left[\frac{\varepsilon L_p}{(\overline{u_1^2})^{3/2}}\right]^{-1/2} . \qquad (81)$$

(Although these definitions are for isotropic turbulence, they are also used for the description of experimental turbulence, with the 1-direction singled out as the direction of mean flow and λ defined either in terms of the curvature of $f(r)$ at $r = 0$ or by $\lambda = \sqrt{\frac{15\nu\overline{u_1^2}}{\varepsilon}}$. According to the universal equilibrium theory, these two definitions are identical if the Reynolds number is large.)

It is a common belief that ε is independent of Reynolds number when the Reynolds number is large compared with unity, i.e.

$$\varepsilon = A\,\frac{(\overline{u_1^2})^{3/2}}{L_p} , \qquad (82)$$

where for any particular type of isotropic turbulence A is a constant of order unity when Re >> 1. This result is fundamental to an understanding of turbulence and yet still lacks convincing theoretical support. The argument given says that L_p is a measure of the length scale of the "energy-containing eddies" of the turbulence. Or more precisely L_p^{-1} is the wavenumber near where $E(k)$ has its maximum. The nonlinear interactions degrade the energy into scales smaller than L_p until the scale of the motion becomes so small that viscosity dissipates the energy.[1] The rate at which energy is degraded, or energy is transfered along the wavenumber axis, is determined by the slowest stage in the process and this is the first

---

[1] Big whirls have little whirls that feed on their velocity; little whirls have smaller whirls and so on till viscosity (attributed to RICHARDSON).

process with time scale $L_p/(\overline{u_1^2})^{1/2}$. Hence

$$\varepsilon \propto \overline{u_1^2} \Big/ \left\{ L_p/(\overline{u_1^2})^{1/2} \right\}$$

which is (82). This argument is plausible, but a result as important as (82) deserves something better. Also objections can be raised. When two Fourier components interact nonlinearly. they produce components with sum and *difference* wavenumbers. Thus the nonlinear interactions transfer energy from small scales to large scales. What is required is that there should be a net transfer which is independent of the small scale structure and so independent of viscosity. This is indeed likely, but solutions of the Navier-Stokes equations can have complex behavior at large Reynolds number, and the possibility that A depends weakly on the Reynolds number can by no means be completely discounted. The experimental evidence is far from convincing and would not rule out a dependence of A on Re like

$$A \propto Re^{-1/4} \quad \text{or} \quad A \propto (\log Re)^{-1}. \tag{83}$$

It is not the intention to claim here that (83) holds; the purpose is to appeal for a demonstration that A is independent of Re. The arguments based on similarity hypotheses which show that $\overline{u_1^2}$ decays like some inverse power of t are not relevant, as they essentially assume that ε is independent of Re. Another difficulty may be mentioned here. The inviscid hydrodynamic equations are time reversible. Thus, for every solution in which there is degradation of energy, there is another solution (the time image) in which the energy goes backwards from small scales to large scales. This paradox is well known in kinetic theory, where it can be resolved by introducing the concept of "coarse graining" or by recognizing the existence of two different time scales, the duration of a collision and the time between collision, the system being reversible over the smaller time scale but irreversible over the larger (phase mixing). But in turbulence the time scales are the same, and it is by no means obvious how the reversible inviscid equations produce irreversible effects.

The result (82) can be given an alternative rough interpretation. If we were to think of the turbulence as composed of blobs of fluid of size L_p moving around with velocity $(\overline{u_1^2})^{1/2}$, the idea that A is of order unity says that the blobs have a constant drag coefficient of unity, which then needs to be reconciled with D'Alemberts paradox. A recent examination by ROSHKO (unpublished) of the experimental data on cylinder drag shows substantial Reynolds number dependence in the range 10^2 to 10^7 !

The Generation of Vorticity

It was long ago pointed out by TAYLOR [15] that the physical mechanism of decay is associated with the stretching of vortex lines. Thus,

$$\varepsilon = \nu\overline{\underset{\sim}{\omega}^2},$$

and the decay of kinetic energy must correspond to the production of vorticity. The vorticity equation obtained by taking the curl of the Navier-Stokes equations is

$$\frac{\partial \underset{\sim}{\omega}}{\partial t} + \underset{\sim}{u} \cdot \nabla \underset{\sim}{\omega} = \underset{\sim}{\omega} \cdot \nabla \underset{\sim}{u} + \nu \nabla^2 \underset{\sim}{\omega} \quad . (84)$$

(convection of vorticity) (amplification of vorticity by stretching of vortex lines) (diffusion of vorticity by viscosity)

Taking the scalar product of (84) with $\underset{\sim}{\omega}$, taking averages, and using the homogeneity gives

$$\overline{\frac{\partial \underset{\sim}{\omega}^2}{\partial t}} = \overline{\omega_i \omega_j \frac{\partial u_i}{\partial x_j}} - \nu \overline{\left(\frac{\partial \omega_i}{\partial x_j}\right)^2} \quad . \tag{85}$$

The first term on the right-hand side describes the generation of $\overline{\underset{\sim}{\omega}^2}$ by the mean stretching of the vortex lines. It is a manifestation of the diffusive property of turbulent motion. For if $\nu \equiv 0$, vortex lines move with the fluid and the magnitude of $\underset{\sim}{\omega}$ is proportional to the length of a vortex line. If fluid particles tend to separate, vortex lines stretch and vorticity is generated. The stretching term must be positive as the viscous term is clearly negative.

For homogeneous turbulence, there is an interesting identity that relates the stretching term to the mean value of the product of the principal rates of strain in the fluid (BATCHELOR and TOWNSEND [16], BETCHOV [17]). We have

$$\omega_i \omega_j \frac{\partial u_i}{\partial x_j} = u_{i,j}\, u_{j,k}\, u_{k,j} - u_{i,j}\, u_{i,k}\, u_{j,k}$$

on using $\omega_i = \varepsilon_{ijk}\, u_{k,j}$ and expanding. Further, if α, β, γ denote the principal values of the rate of strain tensor $e_{ij} = \frac{1}{2}(u_{i,j}+u_{j,i})$,

$$\alpha^3+\beta^3+\gamma^3 = 3\alpha\beta\gamma = e_{ij}e_{jk}e_{ki} = \frac{1}{4}\, u_{i,j}u_{j,k}u_{k,i} + \frac{3}{4}\, u_{i,j}u_{i,k}u_{j,k} .$$

Now

$$\overline{u_{i,j}\, u_{j,k}\, u_{k,i}} = \overline{\frac{\partial}{\partial x_i}(u_{i,j}\, u_{j,k}\, u_k)} - \frac{1}{2}\, \overline{\frac{\partial}{\partial x_k}(u_{i,j}\, u_{j,i}\, u_k)} \ ,$$

where the equation of continuity has been used repeatedly. The right-hand side is zero because of homogeneity. Hence

$$\overline{\omega_i \omega_j \frac{\partial u_i}{\partial x_j}} = -4\overline{\alpha\beta\gamma} \ .$$

If further, the turbulence is isotropic, one can show that

$$\overline{\alpha\beta\gamma} = \frac{35}{8}\, \overline{\left(\frac{\partial u_1}{\partial x_1}\right)^3} = \frac{35}{8}\, S_0\, \frac{(\overline{u_1^2})^{3/2}}{\lambda^3} \ .$$

For vorticity production to balance vorticity destruction, it is necessary therefore that $S_0 < 0$.

It is clear that the mechanism by which turbulent energy is dissipated is intimately connected with the production of vorticity. Since Eq. (26) the spectral density of $\overline{\omega^2}$ is $2k^2E(k)$, the production of vorticity can be associated with a transfer of spectral energy density from small wavenumbers to large wavenumbers, i.e., there is a cascade of energy from small k to large k. However, there may be definite advantages in thinking not of a general, somewhat vague, cascade but of the actual physical processes associated with vortex line stretching. TAYLOR and GREEN [14] attempted a numerical calculation in order to show how vorticity is produced (they also called this process the grinding down of large eddies to produce small ones), but since then effort in this direction has been minimal compared with the attention paid to the cascade process in wavenumber space. This may prove to have been a mistake. We shall see later that the general cascade arguments associated with the universal equilibrium theory give the *wrong* answer when applied to the Burgers equation, which is a one-dimensional model of turbulence. Thus, the number of dimensions in physical space is important. The dimensions of physical space must appear in physical models of turbulence or in considerations of vortex stretching, but are hidden in models of the cascade process in wavenumber space.

It is worth placing on record here the relative orders of magnitude of the first two terms in (85). We have for isotropic turbulence

$$\frac{\frac{\partial \overline{\underset{\sim}{\omega}^2}}{\partial t}}{\overline{\omega_i \omega_j \frac{\partial u_i}{\partial x_j}}} = \frac{8}{35 S_0} \frac{\nu^{1/2}}{\varepsilon^{3/2}} \frac{d\varepsilon}{dt} \,. \tag{86}$$

The experimental observations that ε is at most only weakly dependent on the Reynolds number and that S_0 is of order unity (the measurements give values of about -0.4) imply that (86) is $O(R^{-1/2})$ and that therefore the vorticity production and dissipation will be approximately in balance at each instant.

It is a consequence of the equations that in isotropic turbulence, a negative skewness of the j.p.d.f. and the stretching of vortex lines go together, but there doesn't seem to be a simple physical explanation of the association.

It does not seem unfair to say that beyond the accumulation of experimental data, there has been no progress in understanding the fundamental mechanics of turbulent energy dissipation or the so-called energy cascade since the work of G.I. TAYLOR.

5. The Universal Equilibrium Theory (Kolmogorov)

It is tempting to apply the ideas of statistical mechanics to the turbulence problem, despite the fact that there is little reason to believe that turbulence is independent of initial conditions because we can argue that the time for turbulence to reach a state independent of the initial conditions is $L_p/(\overline{u_1^2})^{1/2}$; and this appears to be the time characteristic of decay (if ε is independent of ν), so that the turbulence has disappeared by the time any kind of statistical equilibrium is reached.

However, it has been argued (originally by KOLMOGOROV, and independently by von WEIZSÄCKER and ONSAGER, see [6]) that although there is no statistical equilibrium of the turbulent motion as a whole, the small scale structure of the turbulence, "the small eddies," which receives its energy by a cascade process will be statistically independent of the "large or energy-containing eddies." These terms are vague, but the concept of a cascade process is also vague and the lack of precision is a flaw in the whole argument. We talk about large and small eddies but we can't define them precisely, and to talk about Fourier components only begs the question as these have no real physical structure.

Experimentally, it is found at large Reynolds number that the energy and dissipation spectra are separate. If we argue

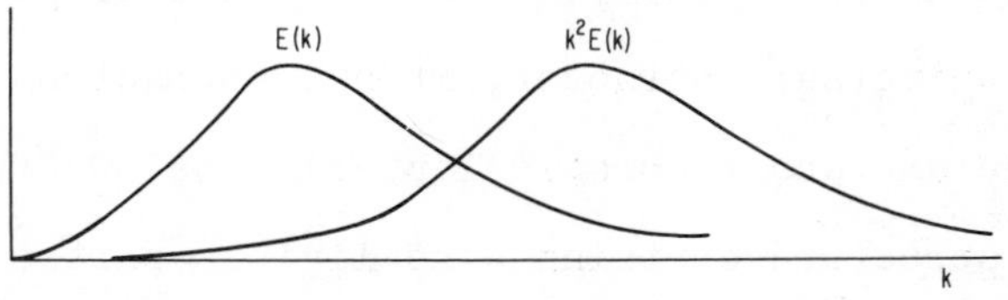

that the characteristic time scale of the small eddies is $(\overline{\underset{\sim}{\omega}^2})^{-1/2}$, the ratio of the time scales of large and small eddies is $(15\ L_p/\lambda)^{1/2} >> 1$. It is therefore plausible to argue that the small eddies are in a state of statistical equilibrium determined by the rate at which they are dissipating energy ε, and by the only other physical parameter ν. This is the hypothesis of local similarity. The consequences are

(a) the small scale motion is isotropic

(b) all statistical properties of the small scale motion depend only on ε and ν.

The necessary conditions for the hypothesis to be self-consistent are that ε is independent of ν (the energy being transmitted to the small eddies by the cascade process) and that the Reynolds number be large enough for the energy and dissipation spectra to be distinct. A sufficient condition would be $L_p/\lambda >> 1$, or according to (81) with the quantity A taken as of order unity,

$$\frac{15}{R_\lambda} << 1 \ . \tag{87}$$

The importance of the local similarity hypothesis lies in the fact that the argument should apply to all turbulence, not necessarily homogeneous or isotropic or decaying or stationary, provided the Reynolds number is large enough. For this reason it is called the universal equilibrium theory. Being the part of turbulence which seems to have a universal character, it has attracted the attention of the mathematicians and theoretical physicists. On the other hand, the small scale structure contains a negligible part of the energy, makes a negligible contribution to the Reynolds stresses in shear-flow turbulence, and is not really of great practical importance.

Deductions

From ε and ν we can form a length and velocity scale

$$\ell = \left(\frac{\nu^3}{\varepsilon}\right)^{1/4}, \quad v = (\nu\varepsilon)^{1/4}. \tag{88}$$

The hypothesis says that all statistical quantities on scale $r \ll L_p$ should be isotropic and scale in terms of ℓ and v. Thus

$$U = \overline{(u_1-u_1')^2} = (\varepsilon\nu)^{1/2} f_e(\tfrac{r}{\ell}); \quad \frac{\overline{(u_1-u_1')^3}}{\left[\overline{(u_1-u_1')^2}\right]^{3/2}} = S_e(\tfrac{r}{\ell});$$

$$\frac{\overline{(u_1-u_1')^4}}{\left[\overline{(u_1-u_1')^2}\right]^2} = Q_e(\tfrac{r}{\ell}); \quad \text{etc.} \tag{89}$$

where f_e, S_e, Q_e are universal functions. In particular S_0 and Q_0, the skewness and flatness factors of $\partial u_1/\partial x_1$, should be constants independent of Re.

ℓ and v are called Kolmogorov length and velocity, or the internal length and velocity scale.

In terms of the energy spectrum function, the hypothesis implies that

$$E(k) = v^2 \ell E_e(k\ell), \quad \text{for } kL_p \gg 1. \tag{90}$$

If the Kolmogorov hypothesis is correct, then it should be possible to deduce from the equations of motion that (89) and (90) are valid asymptotically in the limit $\nu \to 0$. This has so far not been done in a satisfactory manner although there have been many attempts.

Inertial Subrange

If $\ell <<<< L_p$, there can exist separations r such that

$$\ell << r << L_p, \quad \text{or corresponding wavenumbers} \quad L_p^{-1} << k << \ell^{-1}. \tag{91}$$

This is the inertial subrange. The further hypothesis is made that in the inertial subrange, energy is simply being transfered at the rate ε from the energy-containing eddies of scale L_p to the dissipating eddies of scale ℓ. This is the inertial subrange, the statistical quantities depend only on ε and hence

$$U(r) = C\,\varepsilon^{2/3}\,r^{2/3} \quad , \quad E(k) = K\,\varepsilon^{2/3}\,k^{-5/3}, \tag{92}$$

where C and K are universal constants.

Experimental Evidence

The experimental evidence both supports and contradicts the hypothesis. Against are measured values of the flatness factors of the velocity derivatives $\overline{(\partial^n u_1/\partial x_1^n)^4} \Big/ \left[\overline{(\partial^n u_1/\partial x_1^n)^2}\right]^2$. According to Kolmogorov, these should be constant. Measurements are limited but show a significant dependence on Reynolds number (see [6],figure 8.5).

Measurements of the skewness factor S_0 are indecisive. They show a decrease with Re but it cannot be determined whether S_0 goes to zero as Re increases or tends to a constant.

Measurements of the energy spectrum are evidence in support, but there are disagreements and difficulties. The experiments measure the Fourier transforms of f(r) and g(r), i.e.

$$\phi_1(k_1) = \frac{2}{\pi}\int_0^\infty \overline{u_1^2} f(r) \cos k_1 r \, dr$$

and
$$\phi_2(k_1) = \frac{2}{\pi}\int_0^\infty \overline{u_1^2} g(r) \cos k_1 r \, dr \quad . \tag{93}$$

These are called the longitudinal and transverse spectra, respectively. If the turbulence were isotropic

$$\phi_2 = \tfrac{1}{2}\phi_1 - \tfrac{1}{2}k_1 \frac{d\phi_1}{dk_1} , \qquad \phi_1(k_1) = \int_{k_1}^\infty \left(1 - \frac{k_1^2}{k^2}\right) \frac{E(k)}{k}\, dk \ . \tag{94}$$

If the Kolmogorov hypothesis is correct, in the inertial subrange

$$\phi_1(k_1) = \frac{18}{55} K \, \varepsilon^{2/3} \, k^{-5/3} , \tag{95}$$

$$\phi_2(k_1) = \frac{4}{3} \phi_1(k_1) \ . \tag{96}$$

From (93), we have that

$$\phi_1(k_1) = \frac{2}{\pi}\int_0^\infty \frac{\sin kr}{k} \frac{dU}{dr} \, dr$$

$$= \frac{1}{\pi\sqrt{3}} \Gamma(\tfrac{2}{3}) \, C \, \varepsilon^{2/3} \, k^{-5/3} \text{, on substituting (95).}$$

$$K = \frac{55}{18} \frac{\Gamma(\frac{2}{3})}{\pi\sqrt{3}} C. \tag{97}$$

GIBSON [18] in a round jet finds agreement with K = 1.6 and verifies the isotropy. GRANT *et. al.* [19] report measurements in a tidal channel which agree with (95) if $K = 1.45$ and also find that the transverse spectra fit the -5/3 law but were unable to check (96) for the isotropy.On the other hand, KISTLER and VREBALOVICH [20] found for grid turbulence at Reynolds numbers large enough for the inertial subrange to exist that the longitudinal spectrum satisfied the -5/3 law, but that (96) was definitely not satisfied. These authors did not give a value for K, but the data would seem to fit (95) with a K of about 2. Experiments by LAUFER in a pipe showed lack of isotropy, but these were at lower Reynolds number and it could be argued that the inertial subrange did not exist. To sum up, the experimental evidence on the inertial subrange seems to support (95), but there is conflict about the isotropy condition (96).

In the dissipation range, which can exist at Reynolds numbers too low for the inertial subrange to be present, the scaling (90) of $\phi_1(k_1)$ on v and ℓ seems to be found by all measurements, which is impressive support. However, as TOWNSEND [21] points out, the scaling works at Reynolds numbers low enough for a significant part of the energy dissipation to occur in the energy-containing eddy range, so that the "small eddies" are not receiving energy at the rate ε. Thus the Kolmogorov hypothesis works where it shouldn't, which is somewhat of a puzzle.

In view of this conflict, and the others mentioned above (in particular the trouble with the flatness factors) it is worthwhile seeing if the dependence of E(k) on ε and ν alone for large wave-numbers and the $k^{-5/3}$ dependence in the inertial subrange can be obtained without applying the Kolmogorov hypothesis or using it in a restricted form.

Alternative Explanation of the Dependence on ε and ν and the $k^{-5/3}$ Dependence

We return to the Karman-Howarth equation in the form (77) and integrate with respect to r giving

$$2\nu \frac{\partial U}{\partial r} - \frac{\sqrt{2}}{3} S\, U^{3/2} = \frac{2\varepsilon r}{15} + \frac{1}{r^4}\int_0^r r^4 \frac{\partial U}{\partial t}\, dr, \tag{98}$$

with $\varepsilon = \lim_{r\to 0} \frac{15\nu}{r}\frac{\partial U}{\partial r}$. We remember that $U = U(r,t)$ and $S = S(r,t)$, so (98) cannot be solved for U without some further hypothesis. Further $U(r) = \frac{\varepsilon r^2}{30\nu}$ as $r \to 0$ and $U \to \overline{u_1^2}$ as $r \to \infty$.

It was pointed out by KOLMOGOROV [22] that according to his hypothesis, the second term on the right-hand side is negligible because of local equilibrium for $r \ll L_p$. Moreover, according to the Kolmogorov hypothesis S is a function of r/ℓ alone and constant for $r \gg \ell$, and the viscous term is negligible. Hence the Kolmogorov hypothesis applied to the Karman-Howarth equation gives

$$U = \left[\frac{\sqrt{2}}{5}\right]^{2/3} \frac{(\varepsilon\, r)^{2/3}}{\left[-S_e(\infty)\right]^{2/3}} \tag{99}$$

which leads to the inertial subrange $k^{-5/3}$ law.

However, it is clear that (99) will follow from (98) under assumptions much less restrictive. We make no assumptions now about the dependence of ε or other quantities on Reynolds number, or that a cascade exists.

We introduce dimensionless quantities

$$r^* = \frac{r}{\ell}, \quad U^* = \frac{U}{(\varepsilon\nu)^{1/2}}, \quad S^* = -S, \tag{100}$$

i.e., we use the internal length and velocity scales. Then (98) becomes

$$\frac{\partial U^*}{\partial r^*} + \frac{1}{3\sqrt{2}} S^* U^{*3/2} = \frac{r^*}{15} + \left(\frac{\nu}{\varepsilon}\right)^{1/2} \int_0^{r^*} r^{*4} \frac{\partial U^*}{\partial t} dr^* \Big/ r^{*4}. \qquad (101)$$

We need two assumptions:

$$\text{(i)} \qquad \frac{\partial}{\partial t} = O(\varepsilon/\overline{u_1^2}) \quad ,$$

i.e., time rates of change of mean quantities are characteristic of the decay time of the turbulence. Then the second term on the right-hand side of (101) can be neglected provided

$$\frac{U^*}{R_\lambda} \ll 1 \ , \quad \text{or} \quad \frac{U}{\overline{u_1^2}} \ll 1. \qquad (102)$$

Condition (102) will be satisfied if $r \ll L_p$.

Then granted assumption (i), which is very weak, and that we restrict ourselves to values of the separation such that $r \ll L_p$, we have

$$\frac{\partial U^*}{\partial r^*} + \frac{1}{3\sqrt{2}} S^* U^{*3/2} = \frac{r^*}{15} \ , \quad U^*(0) = 0. \qquad (103)$$

For stationary isotropic turbulence, $\partial U\ /\partial t = 0$, but then the Karman-Howarth equation is not correct because forcing terms must be included. However, it is almost obvious that if the energy is put in on the scale L_p, Eq. (103) will hold under the same conditions.

Equation (103) still contains the unknown S^*. We need the second assumption:

(ii) S^* is independent of r^* for $r^* >> 1$, and is nonzero.

A more restrictive form of assumption (ii), but still plausible, is to say that the j.p.d.f. of the velocity is independent of the separation for $r << L_p$. We don't know why the j.p.d.f. is skew (except that if not skew there is no vorticity produced) so the significance of the assumption is hard to assess. However, it seems that skewness is associated with the stretching of vortex lines and that therefore the assumption would seem to be equivalent intuitively to supposing that the vortex lines are straight over scales $r << L_p$. This indeed is consistent with intuitive ideas on the way vortex lines will be stretched (see §7 below). Note that S^* can be a function of Re and that (ii) is weaker than the Kolmogorov hypothesis, but on the other hand does not have at present the intuitive support that the cascade idea has.

It follows from (ii) that for $r^* >> 1/\sqrt{S^*}$

$$U^* = \left(\frac{\sqrt{2}}{5}\right)^{2/3} \frac{r^{*\,2/3}}{S^{*\,2/3}} , \tag{104}$$

independently of whether ε or S depends on Re.

The corresponding result for the spectrum function is

$$E(k) = \frac{55}{18} \frac{\Gamma(\frac{2}{3})}{\pi\sqrt{3}} \frac{2^{1/3}}{5^{2/3}} \frac{\varepsilon^{2/3} k^{-5/3}}{|S|^{2/3}} , \quad k << \ell^{-1}\sqrt{S^*} \tag{105}$$

where S may be a function of Re, or of the particular structure of the energy-containing eddies.

To repeat, the existence of the $k^{-5/3}$ law is not proof that the full Kolmogorov hypothesis is valid. In fact, if the skewness of the j.p.d.f. is a kinematical property of three-dimensional motion, then the $k^{-5/3}$ law appears simply as a kinematical constraint on the dynamical equations in isotropic turbulence. The puzzle that needs to be explained is why for nonisotropic turbulence, the longitudinal velocity correlation may obey the $k^{-5/3}$ law but not the transverse.

The scaling of the dissipation range spectra with ε and ν would follow immediately from the assumption that S^* is independent of Reynolds number for $r^* \sim 1$.

To sum up, it cannot be said with confidence that the experimental evidence supports the Kolmogorov hypothesis. Indeed, it does not seem unduly extreme to say that the conclusion to be drawn from the experiments is that the skewness factor is a slowly varying function of Re and r/ℓ for $r << L_p$. Admittedly, we have no plausible arguments at present to explain why the skewness factor has these properties, but the absence of such arguments is not necessarily sufficient basis for believing that the Kolmogorov hypothesis must be valid.

Exact solutions of (103) can be constructed trivially by allowing S^* to vary for $r^* \leq 1$. Any function U^* which is $O(r^{*2})$ for r^* small and $O(r^{*2/3})$ for r^* large gives a solution. For example

$$U^* = \frac{1}{30} \frac{r^{*2}}{(1+\alpha r^{*2})^{2/3}} \quad \text{where} \quad \alpha = \frac{S^*(\infty)}{12\sqrt{15}} . \qquad (106)$$

GOLITSYN [23] has solved (103) numerically with $S^* = \text{const.}$, but negative spectra are found for $k << \ell^{-1}$. Whether this unphysical result is due to S = const. being a bad approximation or to the neglect of the second term on the right-hand side of (101) is not known.

6. "Burgerlence"[1]

It was suggested by Burgers that one should consider the equation

$$\frac{\partial u}{\partial t} + u \frac{\partial u}{\partial x} = \nu \frac{\partial^2 u}{\partial x^2} \tag{107}$$

as a one-dimensional model of turbulence. Eq. (107) also describes the formation and decay of weak shock waves in a compressible fluid, so that the solution is of considerable physical interest in its own right, as well as being a one-dimensional model of the Navier-Stokes equations.

There is a formal similarity between the statistical equations for the Burger's equation and the Navier-Stokes equations. We consider random solutions $u(x,t)$ which are stationary random functions of x. Then $\frac{1}{2}\overline{u^2}$ is the mean energy density which decays according to the expressions

$$\varepsilon = -\frac{\partial}{\partial t} \frac{1}{2} \overline{u^2} = \nu \overline{\left(\frac{\partial u}{\partial x}\right)^2} = \nu \frac{\overline{u^2}}{\lambda^2} = \nu \overline{\omega^2}. \tag{108}$$

The quantity $\omega = \frac{\partial u}{\partial x}$ is analogous to the vorticity, and the analogy of the other quantities is obvious. Also,

$$\overline{u(x)\, u(x+r)} = \overline{u^2} f(r) = \overline{u^2} - U(r,t), \tag{109}$$

[1] This name was coined by Dr. J. D. Cole.

$$S(r) = \overline{(u'-u)^3} \Big/ \left[\overline{(u'-u)^2}\right]^{3/2} . \tag{110}$$

The assumption of isotropy implies

$$f(r) = f(-r) \ ; \ S(r) = -S(-r). \tag{111}$$

Then we can form in a manner analogous to (73) the Karman-Howarth equation for "Burgerlence"

$$\frac{\partial U}{\partial t} + \frac{2^{3/2}}{3} \frac{\partial}{\partial r} (S\ U^{3/2}) - 2\nu \frac{\partial^2}{\partial r^2} U = -2\varepsilon \quad , \tag{112}$$

which integrates to

$$2\nu \frac{\partial U}{\partial r} - \frac{2^{3/2}}{3} S\ U^{3/2} = 2\varepsilon r + \int_0^r \frac{\partial U}{\partial t}\ dr \ . \tag{113}$$

Apart from numerical coefficients and a change in the last term on the right-hand side, this equation is the same as (98).

We can define a length scale

$$L_p = \int_0^\infty f(r)\ dr. \tag{114}$$

Under the weak assumption that the triple correlation $\overline{u^2 u'} \to 0$ as $r \to \infty$, it is clear that

$$\overline{u^2}\ L_p = \text{constant}, \tag{115}$$

is a consequence of (112).

A Reynolds number

$$Re = (\overline{u^2})^{1/2} L_p/\nu \tag{116}$$

can be introduced and we can consider the turbulence in the limit of large Re.

An energy spectrum function is

$$E(k) = \frac{\overline{u^2}}{4\pi} \int_{-\infty}^{\infty} f(r)\, e^{ikr}\, dr \ , \ \frac{1}{2}\,\overline{u^2} = \int_{-\infty}^{\infty} E(k)\, dk \tag{117}$$

and the rate at which energy is dissipated by viscosity is

$$\varepsilon = 2\nu \int_{-\infty}^{\infty} k^2 E(k). \tag{118}$$

We can now make the hypothesis equivalent to Kolmogorov's. At large Reynolds number, we suppose that ε is independent of ν and therefore

$$\varepsilon = A\, \frac{(\overline{u^2})^{3/2}}{L_p} \ , \tag{119}$$

where A is independent of ν and depends only on the large-scale structure. Further, when Re >> 1, we make the hypothesis that the spectra of $E(k)$ and $k^2E(k)$ do not overlap, and that dissipation is due to the formation of small-scale motions from large-scale motions, and the statistical properties of the small-scale motion are therefore in a state of local equilibrium. Thus, according to the Kolmogorov hypothesis, the j.p.d.f. of u'-u is universal when scaled against

$$\ell = \left(\frac{\nu^3}{\varepsilon}\right)^{1/4} , \quad v = (\nu\varepsilon)^{1/4} , \tag{120}$$

and in particular for $r << L_p$

$$U = (\nu\varepsilon)^{1/2} \, U_e\left(\frac{r}{\ell}\right)$$

and (121)

$$E(k) = \nu^{5/4} \, \varepsilon^{1/4} \, E_e(k\ell) \; .$$

Further, if there is an inertial subrange with $\ell << r << L_p$,

$$U = C' \, \varepsilon^{2/3} r^{2/3} \; , \quad E = K' \, \varepsilon^{2/3} k^{-5/3} \tag{122}$$

where C' and K' are universal constants,

$$K' = \frac{4}{\pi\sqrt{3}} \, \Gamma\left(\frac{2}{3}\right) \, C' \; .$$

Again, according to Kolmogorov, the skewness S will be constant for $r >> \ell$ and

$$S(\infty) = - \frac{3}{\sqrt{2}} \quad C'^{-3/2} \; . \tag{123}$$

There is nothing in the argument about a cascade process that appears to depend on the number of dimensions or the detailed structure of the equations. So if Kolmogorov's argument is valid in three dimensions it should apply also to Burgerlence.

We shall now present arguments to show that these results are incorrect for Burgerlence, and we conclude that the Kolmogorov hypothesis is in general unsound.

Burgers' equation has the well-known exact solution

$$u(x,t) = -2\nu \frac{\partial}{\partial x} \log \theta$$

where (124)

$$\frac{\partial \theta}{\partial t} = \nu \nabla^2 \theta \ .$$

The inviscid equation

$$\frac{\partial u}{\partial t} + u \frac{\partial u}{\partial x} = 0 \tag{125}$$

has the simple exact solution

$$u = f(x-ut). \tag{126}$$

A rigorous complete discussion of the statistical properties of an arbitrary random solution has not yet been given using the exact solution, but enough is known about the behavior of Burgers' equation to predict the behavior with confidence. The method used is to employ the inviscid equation (125) and its solution (126) and to make the solution single valued by inserting shock waves at the appropriate points.

Decay of a Single Pulse

Suppose that at $t = 0$, u has the form of a sloping roof. Then for $u_1 \ell/\nu >> 1$, the solution has the following well-known form

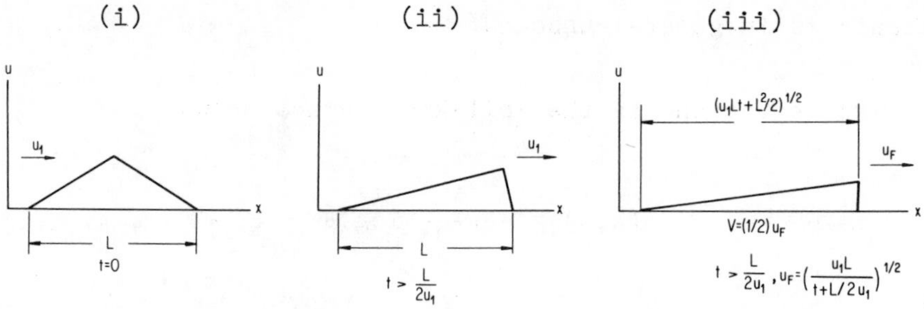

where it is well known that the velocity of the shock V has a value equal to 1/2 the jump in u. From (i) to (ii), the solution is described by the inviscid equation. For $t > \frac{L}{2u_1}$, a shock wave forms at the front which propagates into the medium at rest. In the shock, velocity gradients are very large, of order $1/\nu$ and energy is dissipated.

Note that momentum = $\int u\,d\,x$ is conserved; but energy

$$\int \frac{1}{2} u^2\, dx = \frac{1}{6}\left[\frac{u_1^3 L^3}{t+\frac{1}{2}\,\frac{L_1}{u}}\right]^{1/2} \qquad \text{for} \quad t > \frac{1}{2}\,\frac{L}{u_1}$$

decays at a rate proportional to $t^{-1/2}$ for $t > \frac{1}{2}\frac{L}{u_1}$. The rate of decay of energy is independent of ν.

The shock has the approximate profile

$$u = -\frac{u_F}{2} \tanh\left[\frac{u_F}{4\nu}\,(x-x_S)\right] + \frac{u_F}{2}\,. \tag{127}$$

Decay of a Periodic Train of Pulses

We generalize the problem of a single pulse to a periodic train of pulses,each of roof top form.

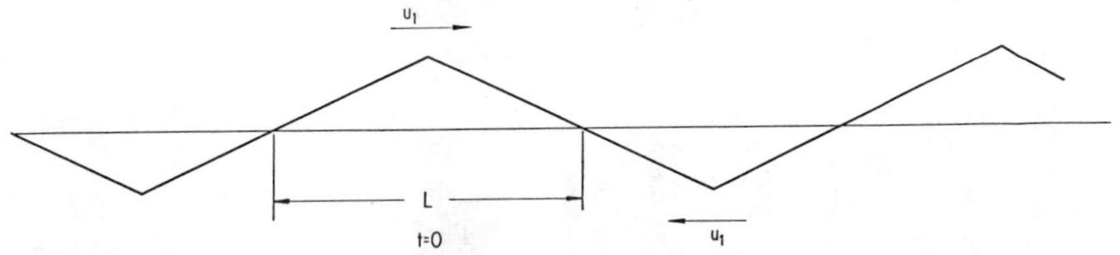

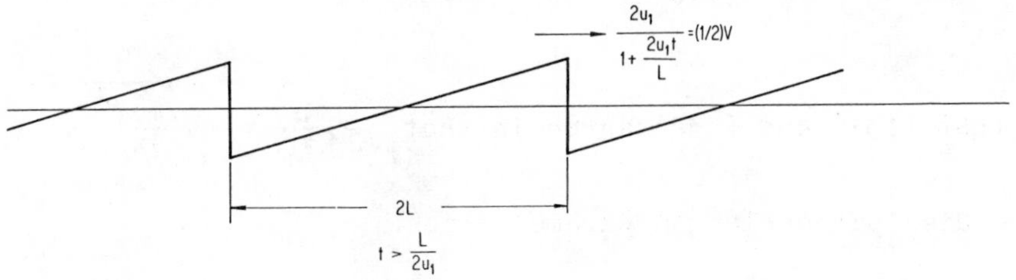

After a time $t = L/2u_1$, the waves break and a series of discrete shocks form, in which energy is dissipated. In this case the shocks are fixed in space. The velocity jump across the shock is easily seen to be

$$\frac{4u_1}{1 + \frac{2u_1 t}{L}} = V(t), \quad \text{say} . \tag{128}$$

For large values of VL/ν, the profile is given approximately by

$$u = \frac{V(t)}{2}\left[- \tanh\left(\frac{Vx}{4\nu}\right) + \frac{x}{L}\right], \quad -2L < x < 2L . \tag{129}$$

(Note that (129) is actually an exact solution of Burgers' equation.) The thickness of the shock is of order $4\nu/V$. We define mean values as averages over a period.

Thus, retaining only the leading terms in an expansion in powers of Reynolds number,

$$\overline{u} = \frac{1}{2L}\int_{-L}^{L} u\,dx \equiv 0 , \tag{130}$$

$$\overline{u^2} = \frac{V^2}{12} , \tag{131}$$

$$\overline{\left(\frac{\partial u}{\partial x}\right)^2} = \frac{V^3}{24\nu L} . \tag{132}$$

Note that (131) and (132) agree in that $\frac{1}{2}\frac{\partial \overline{u^2}}{\partial t} = -\nu\overline{\left(\frac{\partial u}{\partial x}\right)^2}$.

The dissipation-length parameter is

$$\lambda = \sqrt{\frac{2\nu L}{V}} \tag{133}$$

and clearly has no physical significance; the natural length-scale of the dissipating process being

$$\delta = \frac{4\nu}{V} . \tag{134}$$

The skewness of the velocity derivatives is

$$\overline{\left(\frac{\partial u}{\partial x}\right)^3} = -\frac{1}{240}\frac{V^5}{\nu^2 L} . \tag{135}$$

Notice that the skewness factor S_0 is

$$S_0 = -\frac{\sqrt{6}}{5}\left(\frac{VL}{\nu}\right)^{1/2} . \tag{136}$$

We now calculate the velocity correlation. From (129),

$$u(x + r) - u(x) = u' - u = \frac{V}{2}\left[\frac{r}{L} - \tanh\frac{V(x+r)}{4\nu} + \tanh\frac{Vx}{4\nu}\right] .$$

On squaring and taking mean values, we find after some reduction that for r << L,

$$\overline{(u' - u)^2} = \frac{V^2}{2L}\left[r \coth\frac{Vr}{4\nu} - \frac{4\nu}{V}\right] . \qquad (137)$$

Note that (137) agrees with (132) as $r \to 0$. Further analysis gives

$$\overline{(u' - u)^3} = -\frac{V^3}{4L}\left[3r \coth^2\frac{Vr}{4\nu} - r - \frac{12\nu}{V}\coth\frac{Vr}{4\nu}\right] , \qquad (138)$$

for r << L. Hence

$$S(r) = -\frac{L^{1/2}}{\sqrt{2}}\,\frac{\left[3r \coth^2\frac{Vr}{4\nu} - r - \frac{12}{V}\coth\frac{Vr}{4\nu}\right]}{\left[r \coth\frac{Vr}{4\nu} - \frac{4\nu}{V}\right]^{3/2}} , \qquad (139)$$

for r << L. Finally, we note that

$$\varepsilon = \frac{V^3}{24L} . \qquad (140)$$

In these formulae, it is clear that for r << L the major contribution to mean values of velocity differences has come from the shock waves.

Decay of a General Disturbance

Although no general formal analysis is available for the decay of a general wave, it is clear that the periodic train reproduces the qualitative features of the small-scale behavior of a general motion.

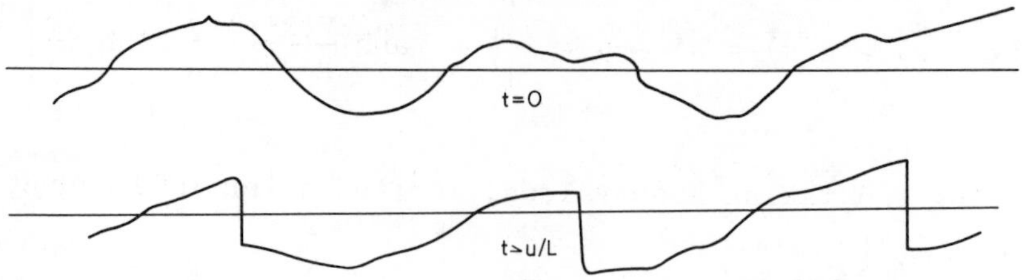

(u is a characteristic velocity and L a characteristic length)

After a certain finite time in which the dissipation is completely negligible, the wave will break into shocks, which will be spaced an average distance L apart and have across them a jump in velocity of order V. Once the shocks form, energy is dissipated at a rate independent of the viscosity and which depends only on the initial conditions which determine how many shocks there are and how they move. Thus, Burgerlence reproduces the feature that is expected of turbulence, namely that ε is independent of ν. Moreover, the motion of the shocks or the regions in which energy is dissipated is also independent of ν, so that we appear to have a situation in which the statistical properties of the small-scale motion are statistically independent of the large-scale motion and the conditions for the Kolmogorov hypothesis to be valid should be satisfied.

Now since the small-scale properties are determined by the shock structure, it is clear that (137), (138), (139) will hold with V and L some suitable average of the large-scale motion, and moreover ε will be related to V and L by (140). It is immediate that the Kolmogorov hypothesis is completely inconsistent with these results. The small-scale properties are not functions of ε and ν. Thus although the rate at which energy is dissipated is independent of ν, the general idea of a cascade process is not correct.

In particular, we see that

$$\left.\begin{aligned} \overline{(u' - u)^2} &\sim \frac{V^2 r}{L}, \\ \overline{(u' - u)^3} &\sim \frac{V^3 r}{L}, \\ S(r) &\sim \left(\frac{L}{r}\right)^{1/2}, \end{aligned}\right\} \tag{141}$$

when $r \gg \frac{\nu}{V}$ and $r \ll L$,
which are substantially different from the behavior predicted by Kolmogorov.

The Spectrum Function

With the assumption of isotropy, we have from (117) on repeated integration by parts that

$$E(k) = -\frac{1}{2\pi k^2}\,\overline{u^2}\int_0^\infty \frac{d^2 f(r)}{dr^2}\cos kr\,dr.$$

Substituting (137) gives

$$E(k) = \frac{1}{2\pi k^2}\,\frac{V^2}{4L}\int_0^\infty \frac{d^2}{dr^2}\left\{r\coth\frac{Vr}{4\nu} - \frac{4\nu}{V}\right\}\cos kr\,dr,$$

$$= \frac{2\nu^2\pi}{4L}\,\text{cosech}^2\left(\frac{\pi\nu k}{2V}\right). \tag{142}$$

For $k \ll V/\nu$,

$$E(k) = \frac{2V^2}{\pi L}\,k^{-2}. \tag{143}$$

This of course is the power spectrum of a continuous stochastic process with random discontinuities. For $k \gg V/\nu$,

$$E(k) = \frac{2\nu\pi}{L}\, e^{-\frac{\pi\nu k}{V}}\,, \tag{144}$$

which shows the exponential cut-off due to viscosity.

Note that for Burgerlence the dissipation spectra $k^2E(k)$ is constant until cut off by viscous decay. In a sense, there is equipartition of dissipation.

Simple Derivation of the Qualitative Dependence

The qualitative structure of the results in equations (137)-(140) can be determined simply as follows. It is useful to look for simple arguments, since in turbulence we do not have exact solutions and general qualitative arguments must be employed. We start from the observation that the general solution will consist of shock waves of average strength $(\overline{u^2})^{1/2}$, an average distance L apart. The thickness δ of each shock is of order $\nu/(\overline{u^2})^{1/2}$. The fraction of length occupied by shocks is of order δ/L. We have the following table.

Quantity	Order of magnitude in shock	Order of magnitude between shocks
u^2	$\overline{u^2}$	$\overline{u^2}$
$(\frac{\partial u}{\partial x})^2$	$\frac{\overline{u^2}}{\delta^2}$	$\frac{\overline{u^2}}{L^2}$
$(\frac{\partial u}{\partial x})^3$	$-\frac{(\overline{u^2})^{3/2}}{\delta^3}$ †	$\frac{(\overline{u^2})^{3/2}}{L^3}$

† $\partial u/\partial x$ is negative in the shock whether the shock is moving to the left or to the right.

Remembering that in forming mean values (which are spatial averages) we weight the quantities in the shock by δ/L, we see that

$$\overline{\left(\frac{\partial u}{\partial x}\right)^2} \sim \frac{\overline{u^2}}{\delta L} \sim \frac{(\overline{u^2})^{3/2}}{\nu L} , \tag{145}$$

$$\overline{\left(\frac{\partial u}{\partial x}\right)^3} \sim - \frac{(\overline{u^2})^{3/2}}{\delta^2 L} \sim - \frac{(\overline{u^2})^{5/2}}{\nu^2 L} . \tag{146}$$

These should be compared with (132) and (135). It follows of course that

$$S_0 \sim - \left[(\overline{u^2})^{1/2} \quad L/\nu\right]^{1/2} .$$

For separations $r << L$, it is clear on dimensional grounds that the expected values of $(u'-u)^2$ inside the shock wave will be of order $\overline{u^2}$ times a function of r/δ. Moreover, the contribution from the shocks will outweigh the contribution from the smooth regions between the shocks, even though they are weighted by a factor δ/L. Thus, for $r << L$,

$$\overline{(u'-u)^2} \sim \overline{u^2} \; h\left(\frac{r}{\delta}\right) \frac{\delta}{L} = \overline{u^2} \; \frac{r}{L} \; H(r/\delta), \tag{147}$$

and

$$\overline{(u'-u)^3} = (\overline{u^2})^{2/3} \; \tilde{h}\left(\frac{r}{\delta}\right) \frac{\delta}{L} = (\overline{u^2})^{3/2} \; \frac{r}{L} \; \tilde{H}(r/\delta), \tag{148}$$

where h, $\tilde{h}$, H, and $\tilde{H}$ are dimensionless functions of the argument. These forms are exactly those of (147) and (148).

If we assume further that for $L >> r >> \delta$, the expressions should approach forms independent of the actual thickness of the shock or equivalently independent of the viscosity, we conclude that H and $\tilde{H}$ should tend to constants when $r/\delta >> 1$, and we regain (141).

The reason the Kolmogorov law does not hold is clearly because the skewness factor cannot be approximated by a constant, being of order $((\overline{u^2})^{1/2} L/\nu)^{1/2}$ when $r << \delta = \nu/(\overline{u^2})^{1/2}$, and of order $(L/r)^{1/2}$ for $r >> \delta$. For Burgerlence, this behavior of the skewness factor is related to the shock wave structure and the fact that fluid crosses shock waves. In turbulence, the surfaces of discontinuity are vortex sheets which have the property that fluid flows along them and not across them. Thus, for simple kinematical reasons we expect that the skewness factor in turbulence will have a different behavior.

Note that an important feature of Burgerlence is the "intermittency." For this reason, the analysis of Burgerlence takes place naturally in physical space. It is a far more difficult problem to derive the above results by working with the Fourier transform of Burgers' equation. It is highly likely, as has been speculated for some time, that turbulence is intermittent and that the intermittent structure is fundamental to the process of decay or of turbulent dissipation. If so, Fourier analysis would be an awkward and unnatural way to attack the problem, and it is not surprising that such approaches have met with little real success.

The Kolmogorov Length in Burgerlence

If we scale the Karman-Howarth equation for Burgerlence in terms of the Kolmogorov length and velocity variables, we obtain the approximate equation (cf. 103) for $r << L$

$$\frac{\partial U^*}{\partial r^*} + \frac{\sqrt{2}}{3} S^* U^{*\,3/2} = r^*,$$

where

$$S^* \sim \left(\frac{L}{\ell}\right)^{1/2} \frac{1}{r^{*\,1/2}} \quad \text{for } r^* \text{ large,}$$

$$S^* \sim \left(\frac{L}{\ell}\right)^{2/3} \quad \text{for } r^* \text{ small, using } \varepsilon \sim V^3/L.$$

Thus, since S^* is not constant or independent of the parameters, the Kolmogorov scaling is not natural.

The example of Burgers' equation demonstrates unequivocally that the Kolmogorov hypothesis in isolation is untenable. This fact was known to BURGERS [24] who derived all the results of this section and more in unpublished lectures given at the California Institute of Technology in 1951. The counter argument, given by von NEUMANN [25] and others, is that Burgers' equation is too special and does not have sufficient degrees of freedom to show the statistical independence of different modes required by the Kolmogorov hypothesis. This may well be so, and perhaps the Kolmogorov hypothesis is right for three dimensions but wrong in one dimension, but how do we know that three dimensions is sufficient? There is a cascade process in the Burgers' equation, as energy has continually to be transferred to higher wavenumbers to make up for the dissipation around $k = V/\nu$; the point is that the cascade process which in physical space corresponds to shock wave formation does not uncouple statistically the small-scale motion from the large-scale motion. In Burgerlence, the formation of large gradients of $\partial u/\partial x$ is the cascade and it depends on the large-scale motion. In turbulence, we do not understand the mechanism of vorticity production which is the physical process of the cascade; but just because we do not understand it, there is not sufficient justification to assert that vorticity is generated in such a way that it is statistically independent of the large-scale structure and depends only on the rate at which the large-scale structure loses energy. An excellent criticism of the physical basis of the Kolmogorov hypothesis has been given by KRAICHNAN [26, §9.1], where it is pointed out that simple observation of the tangled sheets and filaments, seen, for example, in cigarette smoke, suggests strongly that the small-scale and large-scale components are coupled. (In his later work, Kraichnan however appears to have dropped his criticism of the Kolmogorov hypothesis and become a believer in the theory.) The fact is that there is no real theoretical support for the Kolmogorov hypothesis, and before we accept it on the grounds that its predictions are in agreement with experiment, we must be sure that there is no other theory which makes the same experimentally verified predictions and does not lead to other predictions which are in conflict with the observations. A crude attempt at such a theory will be described in the next section. The failure in two dimensions of the cascade hypothesis is so well known to be hardly worth mentioning. The Karman-Howarth equation is

$$\frac{\partial}{\partial t}(u^2 f) = u^3 \left(\frac{\partial}{\partial r} + \frac{3}{r}\right) k + 2\nu u^2 \left(\frac{\partial^2}{\partial r^2} + \frac{3}{r}\frac{\partial}{\partial r}\right) f,$$

which is formally similar to the one- and three-dimensional case. However, it is a consequence of isotropy (without any dynamics) that $k'''(0) = 0$, so that $S_0 = 0$ which makes the behavior in two dimensions very different from that in one or three.

7. The Structure of Turbulence

A theory of homogeneous turbulence needs to deal with three questions.

(i) What is the physical process by which energy is dissipated and how does ε depend on the Reynolds number?

(ii) What does the skewness factor S depend on, and if constant and negative, why?

(iii) Is there any actual physical process associated with the scale $\ell = (\nu^3/\varepsilon)^{1/4}$?

There are of course many other questions, but these would seem to be the first that should be tackled. The study of Burgerlence indicates that understanding the intermittency of turbulent flow may be fundamental to answering these questions. It also suggests strongly that Fourier analysis or the decomposition of the flow into wave modes is not the way to tackle the problem.

To see why turbulent flows are probably intermittent, consider a blob of vorticity added to a turbulent flow and suppose it so weak that the extra vorticity does not change the velocity field. If viscous effects are negligible, vortex lines move with the fluid and it is a kinematic consequence of the diffusive nature of turbulence (unproved mathematically but intuitively obvious) that the blob is pulled out into a long thin ribbon or sheet. As the blob is deformed, the vortex lines are stretched and the result is a long thin ribbon of concentrated and large vorticity. This process continues until either the gradients of vorticity become so large that viscosity starts to produce significant decay, or until the velocity fields induced by the concentration of vorticity start to act against the process by which the vorticity is produced.

Thus the diffusive nature of random velocity fields will produce sheets of concentrated vorticity. On the other hand the process is circular because in the absence of moving walls or external forces the velocity field is itself produced by the vorticity distribution. A type of self-consistent field model, in which we study the velocity field induced by a collection of random vortex sheets and tubes, each sheet or tube being maintained by the stretching due to all the other sheets, is worth searching for although the task will not be easy.

Two Exact Solutions of the Navier-Stokes Equations

Corresponding to the steady shock wave of Burgers' equation where the dissipative action of viscosity is balanced by the steepening effect of the nonlinear convective terms, there are two exact solutions of the Navier-Stokes equations which show how a vortex tube or sheet can be maintained against viscous dissipation by stretching due to an external irrotational velocity field. These solutions are due to BURGERS [24].

(a) Two-dimensional flow. Consider the velocity and vorticity fields

$$\underset{\sim}{u} = (-\alpha x, \alpha y, w(x)) \quad \text{and} \quad \underset{\sim}{\omega} = (0, \omega(x), 0), \qquad \omega = -\frac{\partial w}{\partial x} .$$

This is an exact solution of the Navier-Stokes equation if

$$-\alpha x \frac{\partial \omega}{\partial x} = \alpha \omega + \nu \frac{\partial^2 \omega}{\partial x^2} \quad . \tag{149}$$

The solution is

$$\omega = \omega_0 \, e^{-\alpha x^2/2\nu} \tag{150}$$

which demonstrates that an irrotational velocity field can maintain the vorticity against viscous decay. The value of ω_0 is arbitrary. The width of the vortex sheet is of order $\sqrt{\nu/\alpha}$.

The local rate of dissipation of energy by viscosity is

$$\frac{\nu}{2}\left(\frac{\partial u_i}{\partial x_j} + \frac{\partial u_j}{\partial x_i}\right)^2 = 4\nu\alpha^2 + \nu\omega^2. \tag{151}$$

Then the dissipation per unit volume averaged over a length L in the x-direction is

$$4\nu\alpha^2 + \frac{1}{2L}\int_{-L}^{L} \nu\omega^2 dx = 4\nu\alpha^2 + \frac{\omega_0^2}{2}\left(\frac{\nu^3\pi}{\alpha L^2}\right)^{1/2} . \tag{152}$$

If the velocity difference in the z-direction across the vortex sheet is V, this expression becomes

$$4\nu\alpha^2 + \frac{1}{4}\nu\,\frac{V^2}{L^2}\left(\frac{L^2\alpha}{\nu}\right)^{1/2} . \tag{153}$$

If α is of order V/L, this shows that ε is increased by a factor $Re^{1/2}$ by the concentration of vorticity, and that most of the dissipation takes place in the vortex sheet.

(b) Axisymmetric flow. In cylindrical polar coordinates r,θ,z, the velocity field

$$u_r = -\alpha r, \quad u_\theta = u_\theta(r), \quad u_z = 2\alpha z,$$

$$\omega_r = 0, \quad \omega_\theta = 0, \quad \omega_z = \omega(r),$$

is an exact solution if

$$\omega = \omega_0 \, e^{-\alpha r^2/2\nu}, \quad u_\theta = \frac{\nu\omega_0}{\alpha r}\left[1-e^{-\alpha r^2/2\nu}\right]. \qquad (154)$$

In this flow, we have a vortex tube of radius $\sqrt{\nu/\alpha}$ maintained by the convergence. Now the dissipation per unit mass averaged over a circle of radius L is

$$12\nu\alpha^2 + \frac{C^2\alpha}{4\pi}, \qquad (155)$$

where $C = \dfrac{2\pi\nu\omega_0}{\alpha}$ is the circulation around the vortex. The process by which the vortex is formed will conserve circulation. (It is easy to show that the solution (154) is the asymptotic solution as $t \to \infty$ of an exact unsteady solution of the Navier-Stokes equations with $u_r = -\alpha r$, $u_z = 2\alpha z$, $\omega_z = \omega(r,t)$.) If, therefore, we suppose that the circulation of the vorticity distribution from which the vortex is produced is of order VL, where V is a characteristic velocity of the turbulence and L the length scale, we see that the convergence will amplify the vorticity and produce a mean dissipation ε which is $O(V^3/L)$ and independent of viscosity, provided that the rate of strain α of the convergence is also of order V/L. The dimension of the vortex tube is of order $\sqrt{\nu/\alpha} \sim \sqrt{\nu L/V} \sim \sqrt{\nu V^2/\varepsilon} \sim \lambda$, the Taylor microscale.

On the other hand, the kinetic energy of the motion associated with the circulation is in a circle of radius L

$$\int_0^L 2\pi r \, \frac{1}{2} \, \frac{C^2}{4\pi^2 r^2} \left[1-e^{-\frac{\alpha r^2}{2\nu}}\right]^2 dr \propto C^2 \log\left(\frac{\alpha L^2}{\nu}\right) . \tag{156}$$

Then with $C \sim VL$, the kinetic energy density is proportional to $V^2 \log(VL/\nu)$, so that if the kinetic energy is normalized to be of order unity, the energy dissipation is of order $[\log(VL/\nu)]^{-1}$, and is therefore not independent of Reynolds number. Thus the Burgers line vortex does not in fact give an ε which is Reynolds number independent, but the dependence is only logarithmic.

Random Collections of Sheets and Tubes

We can try to construct a flow field by a random superposition of the exact solutions (a) and (b) in analogy with the way in which a random set of shock waves describes the dissipative processes in Burgerlence. Indeed TOWNSEND [21] has done this, but his motivation was to explain the scaling of the large wavenumber energy spectrum for $k > \ell^{-1}$ with ε and ν at small Reynolds numbers, and he envisaged the sheets and tubes as being present as a small perturbation in a general homogeneous straining field of rate of strain $(\varepsilon/\nu)^{1/2}$. The strength-times-density of the sheets is an arbitrary parameter in Townsend's theory, but it can be chosen to give good agreement with the measurements. However, as Townsend has pointed out privately, this choice of the parameter leads to practically all the vorticity and straining motion being concentrated in the sheets and tubes, and it is not a consistent picture to regard them as perturbations on a general straining motion due to the small eddies distributed uniformly through space and of characteristic magnitude $(\overline{\underset{\sim}{\omega}^2})^{1/2} = (\varepsilon/\nu)^{1/2}$. On the other hand, the fact that the model seems to work suggests that it may indeed be a reasonable representation of the small eddies. That is to say, instead of supposing that the small eddies are distributed uniformly in space and using the model to determine the fine-scale structure of the small eddies, we suppose that the small or dissipating eddies, which contain most of the vorticity, are localized in space and occupy only a small fraction of the volume of the turbulence and say that the model actually describes the structure of the small eddies themselves. We now follow Townsend and calculate the statistical properties of the velocity field induced by a random collection of vortex sheets. It is as well to point out now that the model is too simple to explain all the observed features of turbulence, but it can certainly be regarded as a first step which might be capable of development towards understanding the mechanics of turbulent flow.

We proceed by calculating the longitudinal velocity correlation $\overline{u_p u_p'}$. Let s denote distance along the line PP' joining the points. We suppose the velocity field is due to a random collection of sheets like the exact solutions (a), but we do not identify α with $(\varepsilon/\nu)^{1/2}$. We say that α is a property of the energy-containing eddies given by the not-understood mechanism which concentrates the vorticity into sheets or tubes.

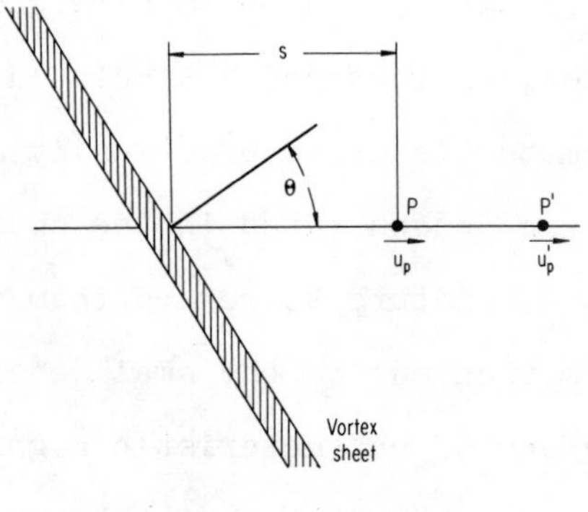

From a sheet a distance s from P, the value of u_p is

$$u_p = \omega_0 \sin\phi \sin\theta \, q(s \cos\theta) \tag{157}$$

where

$$q(\eta) = \int_0^{\eta} e^{-\alpha\eta^2/2\nu} \, d\eta,$$

θ and ϕ are the azimuthal angles that define the direction of the velocity in the vortex sheet, and ω_0 is the vorticity at the center of the sheet. The value of u_p' is given by (157) with s replaced by s + r.

To calculate the mean value of $u_p u_p'$, we can proceed in a variety of ways. We shall regard ω_0 as a stationary random function of s, θ, ϕ, and write $\omega_0 = \zeta(s,\theta,\phi)$ and regard ζ as a continuous function. Then

$$\overline{u_p u_p'} = \int\limits_{\substack{-\infty<s,s^*<\infty \\ 0<\phi,\phi^*<2\pi \\ 0<\theta,\theta^*<\pi}} \overline{\zeta(s,\theta,\phi)\sin\phi \sin\theta \, q(s\cos\theta)\zeta(s^*,\theta^*,\phi^*)\sin\phi^* \sin\theta^*} \, q([s^*+r]\cos\theta^*)\, ds\, d\theta\, d\phi\, ds^*\, d\theta^*\, d\phi^* \tag{158}$$

If the sheets are discrete, oriented at random, and statistically independent of one another, then

$$\overline{\zeta(s,\theta,\phi)\zeta(s^*,\theta^*,\phi^*)} = N\, \overline{\omega_0^2}\, \frac{\sin\theta}{4\pi}\, \delta(s-s^*)\delta(\phi-\phi^*)\delta(\theta-\theta^*), \tag{159}$$

where N is the number of sheets per unit length of line and $\overline{\omega_0^2}$ is the mean-square strength of the vorticity. Then,

$$\overline{u_p u_p'} = \frac{N\omega_0^2}{4} \int_{-\infty}^{\infty} \int_0^{\pi} \sin^3\theta \; q(s \cos\theta)\, q([s+r]\cos\theta)\, d\theta\, ds. \tag{160}$$

Now

$$q(\eta) = \sqrt{\frac{\nu}{2\pi\alpha}} \int_{-\infty}^{\infty} \frac{e^{-\frac{\nu k^2}{2\alpha}}}{ik}\, e^{ik\eta}\, dk.$$

Substituting into (160) and integrating with respect to s gives

$$\overline{u_p u_p'} = \frac{N\omega_0^2}{4} \frac{\nu}{\alpha} \int_0^{\pi} \int_{-\infty}^{\infty} \frac{\sin^3\theta}{\cos\theta} \frac{e^{-\nu k^2/\alpha}}{k^2}\, e^{ikr\cos\theta}\, dk\, d\theta. \tag{161}$$

It follows that the longitudinal spectrum function, the Fourier transform of $\overline{u_p u_p'}$ is

$$\frac{N\omega_0^2}{4} \frac{\nu}{\alpha} \int_0^{\pi} \sin^3\theta\, d\theta\, \frac{e^{-\nu k^2/\alpha \cos^2\theta}}{k^2}\,. \tag{162}$$

Thus, for wavenumbers small compared with $\sqrt{\alpha/\nu}$, the spectrum function behaves like k^{-2}. Equivalently, for separations large compared with $\sqrt{\nu/\alpha}$, the thickness of the vortex sheets, the correlation function behaves linearly with separation r, just as in Burgerlence. This similarity in behavior is to be expected, because the array of vortex sheets is similar to the random set of shock waves. However, the vortex sheet array is dynamically inconsistent because it does not agree with the Karman-Howarth equation. This is clear becuase the skewness is clearly zero for the array in the absence of any kind of dynamical interaction between the sheets.

A random collection of vortex tubes gives rise to a somewhat more complicated spectrum function [21], which however also does not have the $k^{-5/3}$ behavior and is also dynamically inconsistent because the skewness factor due to an array of tubes is zero from the symmetry. Thus, turbulence is fundamentally more complicated than Burgerlence, which is not really a model at all. The reasons for the extra complexity are the three dimensionality and the incompressibility; the singular surface in turbulence (i.e. the vortex sheets or tubes) are characteristic surfaces of the inviscid equations, whereas in Burgerlence the singular surfaces (i.e. the shock waves) are not characteristic surfaces in physical space.

As the discussion makes clear, we cannot construct a model of turbulent flow out of a random array of exact nonlinear solutions, in the way we can regard Burgerlence as a random collection of shock waves. This is not altogether surprising, as the exact solutions are rather artificial since the vortex lines are always going in the same direction and do not close — mathematically,

$$\int \underset{\sim}{\omega} \cdot d\underset{\sim}{s} \neq 0, \tag{163}$$

in the exact solutions where the integral is over an infinite plane, whereas in a real flow this integral would vanish if the surface is large enough to intersect all the vortex lines. We can look for exact solutions, two-dimensional and axisymmetric, like (a) and (b), but with the property that the total amount of vorticity is zero. It is easily seen that no steady solutions exist, but that unsteady exact solutions exist which eventually decay to zero, although there will be initial amplification if the viscosity is small.

These results are also true for general three-dimensional motions, provided we do not look for exact solutions but rather consider the behavior of a weak general distribution of vorticity superposed on a uniform straining motion of infinite extent. Thus suppose $\underset{\sim}{U}$ is an unperturbed velocity field, with velocity in suffix notation

$$U_i = A_{ij} x_j. \tag{164}$$

In general, there will be a vorticity field $\underset{\sim}{\Omega}$ associated with the field $\underset{\sim}{U}$. A weak perturbation of vorticity $\underset{\sim}{\omega}'$ and the associated velocity field $\underset{\sim}{u}'$ satisfy the equation (obtained from the Navier-Stokes equations by neglecting squares and products of $\underset{\sim}{u}'$ and $\underset{\sim}{\omega}'$)

$$\frac{\partial \underset{\sim}{\omega}'}{\partial t} + (\underset{\sim}{U}\cdot\nabla)\underset{\sim}{\omega}' = (\underset{\sim}{\omega}'\cdot\nabla)\underset{\sim}{U} + (\underset{\sim}{\Omega}\cdot\nabla)\underset{\sim}{u}' + \nu\nabla^2\underset{\sim}{\omega}' \tag{165}$$

where

$$\underset{\sim}{\omega}' = \text{curl } \underset{\sim}{u}'.$$

This linear equation can be solved without undue difficulty and it can be shown that provided $|\underset{\sim}{\Omega}|$ is not too large compared with the symmetrical part of A_{ij}, then there exist steady solutions if (163) is satisfied, but for distributions in which the flux of vorticity across a plane is zero, then the perturbation eventually decays to zero. If $|\underset{\sim}{\Omega}|$ is larger than a critical value which depends in a fairly complicated way on the relative magnitudes of the components of A_{ij}, the perturbation always decays to zero, even if (163) is satisfied.

In any event, it is not just the special nature of the exact solutions which prevents them from "explaining" turbulence, but also a more fundamental defect which is that they cannot produce the velocity field necessary to maintain themselves. Thus, considerably more subtle and complicated flow fields are involved, and clearly it will not be possible to neglect the interaction of the vorticity with itself.

A Model of Turbulent Motion

One of the stumbling blocks in the way of understanding turbulent energy dissipation is the significance of the Kolmogorov internal scale $\ell = (\nu^3/\varepsilon)^{1/4}$. If ε is independent of ν (and this although intuitively appealing is by no means certain and is, as mentioned previously, one of the main features to be explained), the internal scale depends on the 3/4 power of the viscosity. Now this is not a natural length scale for an incompressible fluid moving at large Reynolds number in the absence of nonconservative forces. As is well known, viscous effects in motions at large Reynolds number are confined in laminar flows to boundary layers and vortex sheets whose thickness is proportional to the 1/2 power of the viscosity. It is a mystery why turbulence has the $\nu^{3/4}$ rather than $\nu^{1/2}$ dependence for the length scale of the viscous effects. In Burgerlence, the width of a shock is proportional to ν and we have seen that the scales of the motion in which the energy is dissipated is this natural length scale proportional to ν.

To see how the Kolmogorov length scale may arise, consider the following heuristic model. We suppose that the kinetic energy of the turbulent motion is associated with large eddies of characteristic velocity $u = \sqrt{\frac{1}{3}(\overline{\underset{\sim}{u}^2})}$, say, and length scale L. These eddies produce a local straining field whose magnitude α is of order u/L. We assume that the convergence associated with these motions produces concentrated sheets and tubes of vorticity, and it is the enhanced dissipation in the regions of large vorticity that dissipates the energy of the turbulence. The thickness of the sheets or the radius of the tubes has a characteristic length scale

$$\delta = \sqrt{\frac{\nu}{\alpha}} = \sqrt{\frac{\nu L}{u}}. \tag{166}$$

We shall assume that the characteristic velocity associated with the vorticity in the sheets and tubes is also u. This assumption is not easy to justify and may well not be right. It should be stressed again that the aim here is to present some speculative ideas which outline an approach that it is hoped will be useful. The reason for believing that the characteristic velocity in the tubes and sheets is of order u is that the vorticity stretching process is very efficient and transfers kinetic energy from the straining motion to the vortex sheets as rapidly as possible, yet cannot produce an energy density greater than that of the convergence which gives rise to the concentrated vorticity.

The sheets and tubes will presumably not have a permanent existence. They will probably continually be formed and will continually decay. Mathematically, we can describe their formation and decay fairly easily by taking equation (165) for the special case

$$\underset{\sim}{U} = (\alpha x,\ \beta y,\ \gamma z), \quad \alpha + \beta + \gamma = 0 \tag{167}$$

and an initial weak vorticity perturbation which is confined to a sphere of radius L, say. The solution of (165) is then quite straightforward, and it is easy to see that if $\alpha > 0$, $\beta > 0$, $\gamma < 0$ (product of the principal rates of strain negative), the sphere of vorticity is pulled out into a sheet of thickness $\sqrt{\nu/(-\gamma)}$; whereas if $\alpha > 0$, $\beta < 0$, $\gamma < 0$ (product positive), it is pulled out into an elliptic cylinder with principal axes $\sqrt{\nu/(-\beta)}$, $\sqrt{\nu/(-\gamma)}$. The vorticity is amplified by a factor of order

$$L\sqrt{(-\gamma)/\nu} \sim \sqrt{\frac{uL}{\nu}},$$

so that if the initial strength of the vorticity was of order u/L, the characteristic vorticity amplitude in the sheets will be of order u/δ, which is consistent with supposing that the characteristic velocity associated with the flow in the sheets and tubes is u. (Of course the mathematics is now not consistent, because the condition that $\omega' \ll \alpha$ is violated, but the order-of-magnitude considerations should still hold generally.)

We must now ask what proportion of volume is occupied by sheets and tubes. Our belief that turbulence is of an intermittent structure means that only a fraction of the volume is occupied by the regions of intense vorticity. Now for sheets the fraction is likely to be greater than δ/L, because as the sphere is pulled out into a sheet, the volume of the region containing vorticity continually increases. The thinning of the sheet is stopped by the viscous diffusion and the thickness stays at the value δ, but the stretching of the sheets continues unabated. (In fact, although the amplitude of the perturbation tends to zero, the total energy of the perturbation continues to increase because the extent of the region in which there is a significant perturbation increases more rapidly than the level of the perturbation decays, provided of course that the perturbation remains in a region of uniform strain.) We cannot answer the question in a convincing manner, but we shall make a guess that seems to give reasonable results. First consider sheets, and think about the vorticity in a volume L^3. The convergence of the flow in this volume concentrates the vorticity (and amplifies it) in a sheet of thickness δ, but in the process the area of the sheet becomes of order L^3/δ. Thus apparently σ_S, the proportion of volume occupied by the sheets, is unity. But this ignores the bending of the sheet back on itself and the amalgamation of neighboring sheets. We can guess that the area of the sheet reaches the value L^3/δ by L/δ increments of magnitude L^2, since L^2 is the area over which the straining motion of the large eddies can be regarded as persistent. If we now say that the increments of number L/δ are not all of the same sign because of bending back and amalgamation, but are either positive or negative with equal chance, the net area of the sheet would be $(L/\delta)^{1/2} L^2 = (\frac{L^5}{\delta})^{1/2}$, and then

$$\sigma_S = \delta(\frac{L^5}{\delta})^{1/2} \frac{1}{L^3} = (\frac{\delta}{L})^{1/2} . \qquad (168)$$

This argument is no more than suggestive, but the estimate is reasonable and leads to sensible results.

We can apply the same arguments to tubes. When the radius of a tube is of order δ, the length is of order $L^3/\delta^2 = L(L^2/\delta^2)$ if the stretching is uniform. But the bending and amalgamation of the tubes decreases the effective length and makes it of order $L(L^2/\delta^2)^{1/2} = L^2/\delta$ assuming that the stretching is persistent over lengths L. Then the proportion of volume occupied by tubes is

$$\sigma_T = \delta^2 \frac{L^2}{\delta} \frac{1}{L^3} = \frac{\delta}{L} . \qquad (169)$$

The process of forming tubes and sheets with a characteristic thickness δ, a characteristic vorticity u/δ , and a typical volume density σ_T and σ_S, we shall call the PRIMARY CASCADE. It is to be emphasized that we do not understand in detail the primary cascade process, but on the basis of our present knowledge of fluid mechanics it is reasonable to suppose that it exists.

We can now estimate the rate at which energy is dissipated by the primary cascade. The local rate of energy dissipation is $\nu\omega^2$. The quantity ε is the mean rate of energy dissipation. The contribution to ε from the straining motion is ε_0, say, where

$$\varepsilon_0 = \nu \frac{u^2}{L^2} (1-\sigma_T-\sigma_S) \approx \nu \frac{u^2}{L^2} = \frac{1}{Re} \frac{u^3}{L} , \tag{170}$$

where $Re = uL/\nu$ is the Reynolds number of the turbulence.

The straining eddies therefore dissipate energy at a rate inversely proportional to the Reynolds number. However, from the primary cascade, i.e. the vorticity concentrated in sheets and tubes, we have a greater rate of energy dissipation. From the sheets, we have dissipation of order

$$\varepsilon_1^S = \nu \frac{u^2}{\delta^2} \sigma_S = \frac{u^3}{L} \sqrt{\frac{\delta}{L}} = \frac{1}{Re^{1/4}} \frac{u^3}{L} , \tag{171}$$

and from the tubes

$$\varepsilon_1^T = \nu \frac{u^2}{\delta^2} \sigma_T = \frac{u^3}{L} \frac{\delta}{L} = \frac{1}{Re^{1/2}} \frac{u^3}{L} . \tag{172}$$

Thus, the primary cascade enhances the dissipation by an order of magnitude, particularly in the sheets, which it is reasonable to assume are as common as the tubes. However, the dissipation ε_1 is still dependent on the Reynolds number, and the Kolmogorov length scale is still not relevant.

These results appear to be consistent with the Karman-Howarth equation (98), or rather the approximate form (103) which written in dimensional form is

$$\nu \frac{\partial}{\partial r} \overline{(u_1'-u_1)^2} - \frac{1}{6} \overline{(u_1'-u_1)^3} = \frac{2\varepsilon r}{15} \qquad (173)$$

for r << L. Since by dimensional considerations of the type applied to Burgers' equation we have

$$\overline{(u_1'-u_1)^2} = u^2 fn(\tfrac{r}{\delta})(\sigma_T+\sigma_S) + u^2 fn(\tfrac{r}{L})(1-\sigma_T-\sigma_S) \sim u^2 \frac{r^2}{\delta^2} \sigma_S \qquad (174)$$

for small r,

because we know that for small separations the quantity $\overline{(u_1'-u_1)^2}$ is proportional to r^2. Also, the mean cube of the velocity difference is proportional to r^3 for small r, so substituting for σ_S and δ we see that (173) is not violated for r small, on taking $\varepsilon = \varepsilon_1^S+\varepsilon_1^T \approx \varepsilon_1^S$.

Let us now consider larger values of r, which are greater than δ, and the behavior of the skewness. Again from dimensional considerations, we have that

$$\left[(u_1'-u_1)^3\right] = u^3 \; fn(\tfrac{r}{L},\tfrac{r}{\delta}) \qquad (175)$$

where [ ] here denotes an average value over the sheets and tubes. Now we can argue that the mean cube is a property of the stretching mechanism and is not dependent in a significant way on the actual structure of the vortex sheet, especially if r >> δ. We saw that a random collection of noninteracting vortex sheets gives zero skewness. The skewness therefore comes from vorticity interaction which will be a large-scale phenomenon. We can make the hypothesis, therefore, that (175) is independent of δ. On the other hand, the skewness will be large only where vorticity is being produced, and therefore

$$\overline{(u_1'-u_1)^3} = u^3 \; fn(\tfrac{r}{L})(\sigma_S+\sigma_T) \quad \text{for} \quad r >> \delta, \tag{176}$$

the bar denoting the usual spatial (or ensemble) average. A linear dependence of this function on the argument gives consistency with (173).

The spectrum function due to the sheets can be found from (162), according to which for $L^{-1} << k << \delta^{-1}$ we have

$$E_S(k) \propto \frac{N}{k^2} u^2 \quad \text{where } N\delta = \sigma_S,$$

i.e.

$$E_S(k) \propto u^{9/4} \; \nu^{-1/4} \; L^{-3/4} \; k^{-2}. \tag{177}$$

This model is of course highly speculative and we have not demonstrated in any way that the ideas can be given a firm foundation, but there are no obvious dynamical inconsistencies. The most serious objection is that the experiments are against it. The dependence of the dissipation on $Re^{-1/4}$ is not inconsistent with the observations, but the k^{-2} behavior is certainly not in agreement, and even more strongly against is the observation that the energy spectrum function scales with ε and ν alone at large wavenumbers. According to the functional dependence of (174), and the estimate of ε given by (171), $\varepsilon^{-1/4} \; \nu^{-5/4} \; Re^{-1/4} \; E(k)$ should be a function of $k\nu^{3/4} \; \varepsilon^{-1/4} \; Re^{5/16}$ for large k.

However, it is possible that the somewhat organized sheets and tubes of the primary cascade are unstable and will degenerate into smaller scale motions. It is known for instance that curved vortex sheets are unstable to Taylor-Görtler instability, and the tubes may be unstable to the classical Taylor-Couette instability. Both these instabilities have the property that a stable secondary motion is formed, with a cellular structure whose size is about δ, the thickness or radius of the original vortex sheet or tube. Provided the conditions are not too close to critical, the characteristic velocities of the cellular motion are of order u, and there are boundary layers on the edges of the cells of thickness

$$\eta = \sqrt{\frac{\nu\delta}{u}} = \left(\frac{\nu^3 L}{u^3}\right)^{1/4} . \qquad (178)$$

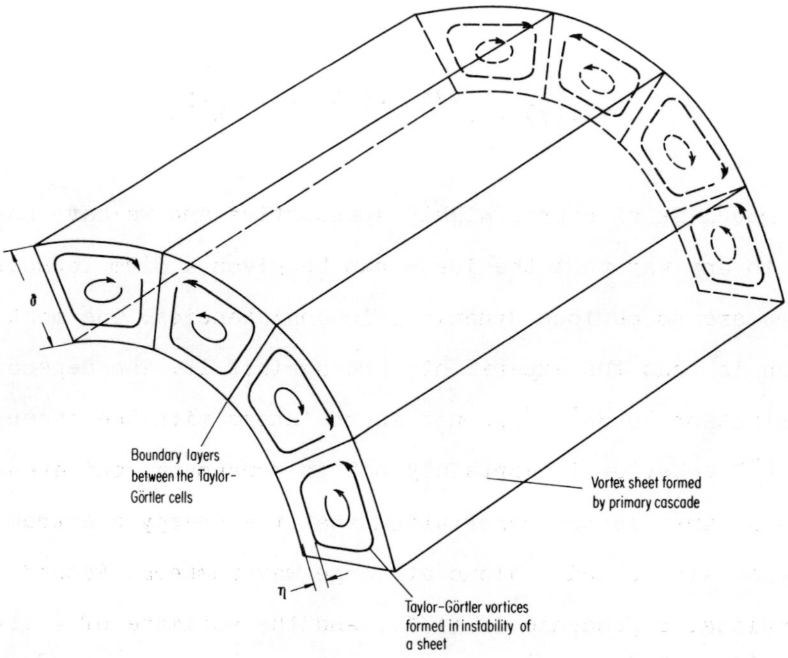

The vorticity in the cells is parallel to the velocities in the sheets and tubes and perpendicular to the vorticity produced by the primary cascade.

The important effect of the instability is to produce small regions of thickness η in which the characteristic vorticity is of order u/η. The volume of these regions is a fraction η/δ of the volume of the tubes and sheets so that the fraction of volume occupied by these regions of concentrated vorticity is

$$\frac{\eta}{\delta}(\sigma_S+\sigma_T) \approx \frac{\eta}{\delta}\sigma_S. \tag{179}$$

Since $\sigma_T << \sigma_S$, we will henceforth forget about the tubes and just consider the sheets. This process of vorticity concentration we shall call the SECONDARY CASCADE.

We now calculate the energy dissipation in the secondary cascade. It is

$$\varepsilon_2 \sim \nu \frac{u^2}{\eta^2} \frac{\eta}{\delta} \sigma_S \quad . \tag{180}$$

Using (178), (168), and (166), it follows that

$$\varepsilon_2 \sim \frac{u^3}{L}, \tag{181}$$

and moreover that

$$\eta = \left(\frac{\nu^3}{\varepsilon}\right)^{1/4} = \ell. \tag{182}$$

Thus, the dissipation in the secondary cascade, provided all the assumptions that have been made are reasonable, is such as to make the average dissipation independent of Reynolds number, and provides a physical basis for understanding the Kolmogorov length. Further, the thickness δ of the primary cascade vortex sheets is now proportional to the Taylor microscale λ, since $\varepsilon = 15\nu u^2/\lambda^2$, and (166) gives

$$\delta = \sqrt{\frac{\nu L}{u}} = \sqrt{\frac{\nu u^2}{\varepsilon}} .$$

To summarize, we can, by introducing the idea of a primary and secondary cascade, give physical significance to the energy dissipation being independent of Reynolds number, to the Kolmogorov length scale, and also to the Taylor microscale. It is hoped that further work may put these ideas on a firmer foundation. A necessary condition for the secondary cascade to exist is that the Taylor-Görtler instability should occur. Assuming that the curvature of the vortex sheets is 1/L, this condition is [27, p.503]

$$\frac{u\delta}{\nu} \sqrt{\frac{\delta}{L}} > \frac{1}{2} , \text{ i.e. } Re > \frac{1}{2^4} . \qquad (183)$$

Thus there seems to be little doubt that the primary cascade is unstable for the sheets at least. The Reynolds number of the secondary cascade motion is $u\eta/\nu$ which also is of order $Re^{1/4}$. We can ask whether there is a tertiary cascade and so on, because of an instability of the Taylor-Görtler cellular motion. This is an intriguing question, to which the answer is not clear. As we shall see, a tertiary cascade is not necessary to explain the Kolmogorov similarity, and possibly the lifetimes of the individual elements of the primary cascade are so short that the tertiary cascade never gets established, but the whole picture is still too vague for anything to be said with certainty. For the present, we shall assume that the process stops with the secondary cascade.

Predictions of the Model

First we consider the velocity correlation. We need a notation to denote expected values inside the general straining motion, the primary cascade, and the secondary cascade. We denote these by square brackets with suffices 0,1,2, respectively. Then

$$\overline{} = \Big[\quad\Big]_0 + \Big[\quad\Big]_1 \underset{(\nu^{1/4})}{\sigma_S} + \Big[\quad\Big]_2 \frac{\ell}{\lambda} \underset{(\nu^{1/2})}{\sigma_S} , \tag{184}$$

because the primary and secondary cascades occupy only a fraction of the volumes, the fractions being σ_S and $(\ell/\lambda)\sigma_S$, respectively. By dimensional considerations,

$$\Big[(u_1'-u_1)^2\Big]_0 = u^2 fn(\tfrac{r}{L}) \ ; \quad \Big[(u_1'-u_1)^2\Big]_1 = u^2 fn(\tfrac{r}{\lambda}) \ ;$$
$$\Big[(u_1'-u_1)^2\Big]_2 = u^2 fn(\tfrac{r}{\ell}) \ . \tag{185}$$

The functions are quadratic in their arguments for small values of the argument, so that for $r \leq \ell$ it is clear that the contribution from the secondary cascade dominates and we have

$$\overline{(u_1'-u_1)^2} = u^2 \frac{\ell}{\lambda} \sigma_S \, fn(\tfrac{r}{\ell}) = (\varepsilon\nu)^{1/2} \, fn(\tfrac{r}{\ell}), \tag{186}$$

which is just the Kolmogorov prediction and has been well verified by experiment.

On the other hand, if we consider the flatness factors we get substantial disagreements. Thus, by dimensional considerations

$$\left[\left(\frac{\partial u_1}{\partial x_1}\right)^4\right]_2 = \frac{u^4}{\ell^4} \quad \text{and} \quad \left[\left(\frac{\partial u_1}{\partial x_1}\right)^2\right]_2 = \frac{u^2}{\ell^2} \,. \tag{187}$$

Clearly, the contribution from the secondary cascade will dominate, and therefore

$$F_0 = \frac{\overline{\left(\frac{\partial u_1}{\partial x_1}\right)^4}}{\left\{\overline{\left(\frac{\partial u_1}{\partial x_1}\right)^2}\right\}^2} \sim \frac{\frac{1}{\ell\sigma_S}}{\lambda} = Re^{1/2} \,, \tag{188}$$

and similarly for the flatness factors of the other velocity derivatives. On the other hand, F_0 is constant according to the Kolmogorov hypothesis. The experimental evidence supports (188) (see [6, Fig. 8.5]).

Further, the model is not inconsistent with the Karman-Howarth equation. We can argue, as for (175), that

$$\left[(u_1'-u_1)^3\right]_2 = u^3 \, \text{fn}\left(\frac{r}{\lambda},\frac{r}{\ell}\right) \,, \tag{189}$$

and again that since the stretching is the interaction between the velocity field that produces the vorticity and the velocity field induced by the vorticity, the actual thickness of the second cascade should not be relevant when $r \gg \ell$. Then, noting that we need a linear dependence to satisfy the Karman-Howarth equation, we have

$$\left[(u_1'-u_1)^3\right]_2 \sim -u^3 \frac{r}{\lambda} \quad \text{for } r \gg \ell. \tag{190}$$

By a similar argument (see (176)),

$$\left[(u_1'-u_1)^3\right]_1 \sim u^3 \text{ fn}(\tfrac{r}{L}) \tag{191}$$

and is negligible in comparison. Then

$$\overline{(u_1'-u_1)^3} \sim -\frac{u^3 r}{\lambda}\frac{\ell}{\lambda}\sigma_S = \frac{u^3 r}{L}, \quad \text{for} \quad r >> \ell, \tag{192}$$

which is consistent with (173).

The model does not appear to have any gross inconsistencies, although some may show up in later work. The main trouble with it is that we cannot predict the skewness factor for $r >> \ell$ to obtain the $r^{2/3}$ law. This is a serious deficiency. We can, however, make some predictions on the basis of the assumption that (189) is (r/λ) times function of (r/ℓ). Then the skewness factor is

$$S(r) = \frac{\left[(u_1'-u_1)^3\right]_2}{\left\{\left[(u_1'-u_1)^2\right]_2\right\}^{3/2}} \quad \frac{1}{\left\{\frac{\ell\sigma_S}{\lambda}\right\}^{1/2}} = Re^{1/4}\,\frac{r}{\lambda}\,\text{fn}(\tfrac{r}{\ell}) = \text{fn}(\tfrac{r}{\ell}). \tag{193}$$

Of course, the scaling (193) follows from the Karman-Howarth equation and the dependence (186), so we are only demonstrating consistency. The result (193) does say that S_0 is independent of Re, and this is not inconsistent with some measurements [6, Fig.6.3], which however do show some decrease of S_0 with increase of Re. Again, if S(r) does become independent of r, we have that

$$S \to \text{constant} \tag{194}$$

so that the Kolmogorov constant is independent of Reynolds number.

Remarks

The purpose of the model is to provide some physical basis for the cascade process, which must be more than just the nonlinear interaction of Fourier components. The idea here is that turbulence should be thought of as a random array of laminar flows. This may prove to be completely wrong. Even if the physical model described above proves to be a reasonable first step towards a description of turbulent motion, it is entirely qualitative and needs to be complemented by a mathematical discussion based on the Navier-Stokes equations. Part of the purpose in thinking about physical models is the hope that they may provide clues to the mathematical problem. The lecturer feels that progress towards understanding turbulence is most likely to come from a synthesis of physical ideas and mathematical analysis, and that mathematical analysis by itself will be extremely difficult, but this remains to be seen.

It is perhaps appropriate to mention here a problem whose solution would be of interest, namely the pressureless Navier-Stokes equations or the three-dimensional Burgers' equation

$$\frac{\partial \underset{\sim}{u}}{\partial t} + (\underset{\sim}{u}\cdot\nabla)\underset{\sim}{u} = \nu\nabla^2\underset{\sim}{u} \qquad (195)$$

where $\underset{\sim}{u}$ is a vector with three components. Here we have both shock waves and vortex sheets, and it would be of interest to see if the statistical quantities in this problem are given at all by the Kolmogorov hypothesis. As COLE [28] has pointed out, this equation has the exact solution $\underset{\sim}{u} = -2\nu\ \mathrm{grad}(\log\theta)$ where $\frac{\partial\theta}{\partial t} = \nu\nabla^2\theta$, but only the class of irrotational solutions can be expressed in this way.

An additional comment about compressible turbulence is in order here. It seems to be believed that supersonic turbulence, with velocity fluctuations greater than sonic speed, cannot exist because the energy would be dissipated by shock waves very rapidly until the motion became subsonic. But the property of shock waves, that they dissipate energy at a rate independent of viscosity, is shared by the turbulent dissipation process, and if the average spacing of the shocks is L, the shocks and the cascade processes will dissipate at the same rate u^3/L. There seems no reason why supersonic turbulence could not exist if sufficient energy were put into the flow.

It may also be mentioned that the strengths of the vortex sheets produced by the primary cascade are likely to be distributed according to a log normal probability law, because the strength will be proportional to the extension of the vortex lines and Appendix 1 suggests strongly that the extension of material lines will obey the log normal distribution. Since the vorticity in the secondary cascade will be proportional to the strength of the primary cascade, it is to be expected that the energy dissipation is distributed according to a log normal probability distribution.

8. Quasianalytical Theories of Turbulence

Attempts to construct mathematical solutions of the Navier-Stokes equations to describe turbulent motion are proliferating at an increasing rate. Unfortunately, as the comparatively more elementary theories prove inadequate, the newer theories tend to be increasingly complicated and more difficult to assess. The discussion here will be confined to the theories that have excited the most interest or appear to have the greatest potentiality.

The theories to be considered are attempts to close the hierarchy of equations for the correlation tensors or their Fourier transforms in order to obtain a finite number of equations in a finite number of unknowns. There are two distinct types: those which employ arbitrary physical assumptions or those which employ arbitrary mathematical assumptions. None of the theories (and this includes those that will not be discussed) seems to be a rational approximation in the sense that there are reasonable grounds for believing that the errors are small or that the equations derived provide a satisfactory description of turbulence.

Theories Based on Physical Approximations

For simplicity, we restrict the discussion to isotropic turbulence. If we take the Fourier transform of the Karman-Howarth equation, we obtain an equation for the energy spectrum function

$$\frac{\partial E(k)}{\partial t} = T(k) - 2\nu k^2 E(k) \qquad (196)$$

where

$$T(k) = \frac{u^3}{\pi} \int_0^\infty (r \frac{\partial}{\partial r} + 3)(\frac{\partial}{\partial r} + \frac{4}{r})\ kr\ k(r) \sin kr\ dr \qquad (197)$$

is related to the Fourier coefficient of the triple correlation k(r). The function T(k) describes the net transfer of energy to Fourier components of wavenumber magnitude k by the nonlinear interactions. It can be shown by manipulation of the Fourier transform of the Navier-Stokes equation that (for homogeneous as well as isotropic turbulence)

$$T(k) = \int\int Q(\underset{\sim}{k},\underset{\sim}{k}')\, d\underset{\sim}{k}'\, dA(\underset{\sim}{k}) , \tag{198}$$

where

$$Q(\underset{\sim}{k},\underset{\sim}{k}') = -\frac{i}{2}\, k_j \left[\Gamma_{jii}(-\underset{\sim}{k},\underset{\sim}{k}') - \Gamma_{jii}(\underset{\sim}{k},-\underset{\sim}{k}')\right] , \tag{199}$$

and

$$\overline{a_i(\underset{\sim}{k}''-\underset{\sim}{k}')a_j(\underset{\sim}{k})a_k(\underset{\sim}{k}')} = \delta(\underset{\sim}{k}+\underset{\sim}{k}'')\Gamma_{ijk}(\underset{\sim}{k},\underset{\sim}{k}') . \tag{200}$$

These complicated expressions show that the transfer of energy to the Fourier components of wavenumber vector $\underset{\sim}{k}$ comes about by the interaction of the Fourier component $\underset{\sim}{k}$ with all other pairs of Fourier components whose wavenumber vectors add up to $-\underset{\sim}{k}$, i.e., the interactions are through triads which form closed triangles.

We define

$$S(k) = \int_k^\infty T(k)\, dk. \tag{201}$$

Then S(k) describes the flow of energy across wavenumber k, i.e.,it gives the rate at which Fourier components with wavenumber less than k transfer energy to Fourier components with wavenumber greater than k. It can be deduced from (200) that

$$Q(\underset{\sim}{k},\underset{\sim}{k}') + Q(\underset{\sim}{k}',\underset{\sim}{k}) = 0 , \tag{202}$$

so that

$$S(0) = \int_0^\infty T(k)\, dk = 0 . \tag{203}$$

(Alternatively, (203) can be deduced directly from (197).) This demonstrates, as is obvious, that the nonlinear interactions transfer energy between Fourier components without creating or destroying energy. Integrating (196), we obtain

$$\frac{\partial}{\partial t} \int_k^\infty E(k')\, dk' = S(k) + \int_0^k 2\nu k'^2 E(k')dk' - \varepsilon \,. \qquad (204)$$

The phenomenological theories assume that S(k) is some functional of E(k). There are four main theories, although an infinite number clearly exist. For further details and comparison with experiment, see [29].

(i) Heisenberg's theory

We take

$$S(k) = N(k) \int_0^k 2k'^2\, E(k')\, dk' \,,$$

where (205)

$$N(k) = \gamma \int_k^\infty k'^{-3/2} \left[E(k')\right]^{1/2} dk' \,,$$

γ being an arbitrary constant. This theory is based on the idea that the small eddies act like an eddy viscosity on the big eddies, the energy transfer across wavenumber k then being some function of k multiplied by the mean square vorticity of the Fourier components with wavenumber less than k.

(ii) Obukhov's theory

$$S = \gamma \int_k^\infty E(k')\, dk' \left[\int_0^k k'^2\, E(k')\, dk'\right]^{1/2} . \qquad (206)$$

The physical idea here is that the small eddies act on the large eddies like a Reynolds stress.

(iii) Kovasznay's theory

$$S = \gamma k^{5/2}\Big[E(k)\Big]^{3/2}. \tag{207}$$

(iv) Modified Obukhov

$$S = \gamma k\, E(k)\left[\int_0^k k'^2 E(k')\, dk'\right]^{1/2}. \tag{208}$$

The last two have no simple physical analogue.

For wavenumbers large compared with L^{-1}, the left-hand side of (204) may be assumed negligible compared with the right-hand side, if the energy transferred across k is passed on to the dissipating scales as soon as it arrives. This assumption is consistent with the ideas of the universal equilibrium theory, and leads to the approximate equation

$$\varepsilon = S(k) + \int_0^k 2\nu k'^2 E(k')\, dk'. \tag{209}$$

Solutions of the equation (209) for the 4 different approximations have been obtained, either numerically or analytically, and have been discussed by Ellison. The first three theories give physically unacceptable results for large wavenumber. For instance, the Heisenberg theory gives $E(k) \propto k^{-7}$, while (ii) and (iii) require E(k) to fall discontinuously to zero at some wavenumber. Theory (iv) gives an exponential decay as $k \to \infty$ and is on these grounds less objectionable.

Note that all the theories predict that

$$E(k) = \varepsilon^{2/3} k^{-5/3} fn(k\ell), \tag{210}$$

in accordance with the universal equilibrium theory, because equation (209) contains only ε and ν as parameters and (210) must therefore follow on dimensional grounds. This does not confirm the universal equilibrium theory, because the Kolmogorov assumption of the statistical independence of large- and small-scale motions, the coupling being due to the transfer of energy at rate ε, is essentially built in to all the assumptions. In fact, the theories are properly interpreted as just giving semiquantitative forms to the Kolmogorov hypothesis.

A strong criticism of the theories is that they would apply equally well to Burgerlence, since nowhere in the formulation of the hypothesis is the actual form of the Navier-Stokes equations used, and the Fourier transform of the Karman-Howarth equation for Burgerlence is qualitatively the same. And as we have seen the structure of Burgerlence is not consistent with the Kolmogorov hypothesis.

There would therefore seem to be no point in pursuing these or similar theories further, as their interest seems to be now entirely historical. Conceptually, the theories are also objectionable as they ignore the phases of the Fourier components and almost regard Fourier components as having a real physical existence, rather than being a mathematical representation of the motion. As BURGERS [24] pointed out, it is doubtful if Fourier analysis is the natural way to treat problems with strong nonlinear interactions, and the intermittent structure of Burgerlence and the possible intermittent structure of turbulence support the belief that theories based on assumptions about the statistical independence of Fourier coefficients with different wavenumbers are unlikely to be correct. We turn now to the theories based on mathematical approximations without any real physical justification.

The Joint Normal Approximation or Zero Fourth-Order Cumulant Hypothesis

The equations in physical space for the velocity correlations are closed by assuming that

$$\overline{u_i\, u_j\, u_k'\, u_\ell''} = \overline{u_i\, u_j} \cdot \overline{u_k'\, u_\ell''} + \overline{u_i\, u_k'} \cdot \overline{u_j\, u_\ell''} + \overline{u_i\, u_\ell''} \cdot \overline{u_j\, u_k'} \qquad (211)$$

where the ' denotes a value measured at $\underset{\sim}{x} + \underset{\sim}{r}$ and " a value measured at $\underset{\sim}{x} + \underset{\sim}{r}'$, all the velocities being measured at the same point of time. This assumption was first suggested by MILLIONSHTCHIKOV, but was developed by PROUDMAN and REID [13], and TATSUMI [30]. The algebraic complexities are formidable, and reference should be made to [13] and [30] for the details. The assumption lacks any real physical content. It essentially asserts that in calculating the triple correlation, one can ssume that the probability distribution of the products of four velocities at three points is Gaussian. If the points are widely separated, the assumption would seem to be good, as is indeed indicated by the available experimental evidence, but if the points are close together the assumption is of more doubtful validity. The attraction of the hypothesis is its relative simplicity, but if the process of energy dissipation is due to small-scale intermittency, it is to be expected that the predictions will need to be interpreted with caution.

KRAICHNAN [31] has objected to a more general form of the hypothesis for nonsimultaneous velocities on the grounds that it violates energy conservation, but MEECHAM [32] has shown that Kraichnan's objection is invalid, or more precisely that the violation of energy conservation is due to an unsymmetrical application of the hypothesis and that the theory can be constructed so that energy is conserved.

A more fundamental objection is that numerical calculations by OGURA [33] of the equations show that negative values of E(k) appear if the initial Reynolds number is sufficiently large. However, MEECHAM and colleagues have shown in unpublished work that for Burgerlence the joint normal hypothesis gives answers that are good approximations and that do not have negative spectra if the initial conditions are chosen to be close to numerical solutions of the Burgers' equation that develop from random initial conditions; but that negative spectra do appear if the initial conditions are not close to a solution. Thus the joint normal hypothesis may be a reasonable approximation provided the initial values of the correlation functions are close to real values such as would exist in a real flow field sometime after the initial instant.

PROUDMAN and REID [13] derived the interesting result for the inviscid equations in isotropic turbulence that

$$\frac{d^2}{dt^2}\overline{\omega_1^2} = \left[\overline{\omega_1^2}\right]^2, \tag{212}$$

the solution of which in terms of elliptic functions shows that $\overline{\omega_1^2} = \infty$ when $t \doteqdot 5.9/\left[\overline{\omega_1^2}(0)\right]^{1/2}$. Thus, according to the joint normal hypothesis, a random distribution of vorticity is violently unstable and the extension of the vortex lines becomes infinitely large in a finite time. This of course is correct for Burgerlence, because compression waves "break" in a finite time, but it is hard to believe that this result is correct for the motion of a perfectly inviscid fluid, and it indicates that small-scale phenomena are probably not well described by the hypothesis. (Nevertheless, the result is interesting and supports the hypothesis that the primary cascade is very efficient in producing vorticity.)

It may be possible to construct more general closure schemes than (211) which have the property that negative spectra are not permitted. ORSZAG and KRUSKAL [34] have given a scheme of this type, but it remains doubtful how well any such closure will describe the small-scale motion. A test of any such scheme is Burgerlence, but unfortunately turbulence is so much more complicated that there is no guarantee that if a theory is right for Burgerlence, it is right for turbulence.

The Wiener-Hermite Expansion

A method of some promise has been pointed out by MEECHAM and SIEGEL [35] and applied by them to Burgers' equation. We shall describe this method now as applied to the Burgers equation. The application to the Navier-Stokes equations has not yet been carried out, but is awaited with interest. There will doubtless be a flood of papers in the next decade.

The method is an expansion of the random variable $u(x,t)$ in a Wiener-Hermite functional expansion, based on the "ideal random function" or white noise function $a(x)$. These functions have the property that they are distributed according to a Gaussian probability distribution, with the correlation given by

$$\overline{a(x_1)\ a(x_2)} = \delta(x_1-x_2). \tag{213}$$

It is known that any stationary random Gaussian function f(x) can be expressed as

$$f(x) = \int K(x-\xi)\ a(\xi)\ d\xi\ , \tag{214}$$

where K is an ordinary nonrandom function. To describe nonGaussian stationary random functions, we construct statistically orthogonal polynomial functions of a(x). We take

$$\left.\begin{aligned} &H^{(0)}(x) = 1,\quad H^{(1)}(x) = a(x),\quad H^{(2)}(x_1,x_2) = a(x_1)a(x_2)-\delta(x_1-x_2),\\ &H^{(3)}(x_1,x_2,x_3) = a(x_1)a(x_2)a(x_3)-a(x_1)\delta(x_2-x_3)-a(x_2)\delta(x_3-x_1)\\ &\qquad\qquad -a(x_3)\delta(x_1-x_2), \end{aligned}\right\} \tag{215}$$

etc.

It is clear that

$$\overline{H^{(i)}\ H^{(j)}} = 0 \quad \text{for} \quad i \neq j, \tag{216}$$

and

$$\left.\begin{aligned} &\overline{H^{(0)}\ H^{(0)}} = 1,\quad \overline{H^{(1)}(x_1)\ H^{(1)}(x_2)} = \delta(x_2-x_1),\\ &\overline{H^{(2)}(x_1,x_2)\ H^{(2)}(x_3,x_4)} = \delta(x_1-x_3)\delta(x_2-x_4) + \delta(x_1-x_4)\delta(x_2-x_3) \end{aligned}\right\} \tag{217}$$

etc.

Any product of the Wiener-Hermite polynomials can be expressed as a sum of such polynomials. Thus,

$$\left.\begin{aligned} H^{(1)}(x_1)\, H^{(1)}(x_2) &= H^{(2)}(x_1,x_2) + \delta(x_2-x_1) \\ H^{(1)}(x_1)\, H^{(2)}(x_2,x_3) &= H^{(3)}(x_1,x_2,x_3) + H^{(1)}(x_3)\delta(x_2-x_1) \\ &\quad + H^{(1)}(x_2)\delta(x_3-x_1), \end{aligned}\right\} \quad (218)$$

etc.

It can be shown that the Wiener-Hermite polynomials are complete so that an arbitrary stationary random function u(x) with zero mean can be expanded in the form

$$u(x) = \int K^{(1)}(x-\xi)H^{(1)}(\xi)\, d\xi + \int K^{(2)}(x-\xi_1,x-\xi_2)H^{(2)}(\xi_1,\xi_2)d\xi_1 d\xi_2$$
$$+ \int K^{(3)}(x-\xi_1,x-\xi_2,x-\xi_3)H^{(3)}(\xi_1,\xi_2,\xi_3)d\xi_1 d\xi_2 d\xi_3 + \ldots \, . \quad (219)$$

The Burgers equation then gives an infinite set of equations for the rate of change of the infinite number of nonrandom kernels $K^{(i)}$. The advantage of the method is that it gives a realizable velocity field and so cannot by construction produce negative spectra like the joint normal approximation. On the other hand, the method suffers from the same objection as the joint normal hypothesis (or all the other truncation schemes) that closure is effected by an arbitrary truncation of the hierarchy of equations for the kernels, and it is not clear that the error is small.

There is a variety of ways to obtain the equations. One method is to express $u^2(x)$ as an expansion in Wiener-Hermite functions by squaring (219) and using the relations (218). Thus from the kernels

$K^{(1)}$ and $K^{(2)}$ we obtain after some reduction, for the case of symmetrical kernels,

$$u^2(x) = \int \left[K^{(1)}(\xi)\right]^2 d\xi + 2\int \left[K^{(2)}(\xi,\eta)\right]^2 d\xi\, d\eta$$

$$+ \int H^{(1)}(\xi)\, K^{(1)}(x-\eta)\, K^{(2)}(x-\xi,x-\eta)\, d\xi\, d\eta$$

$$+ \int H^{(2)}(\xi,\eta)\, K^{(1)}(x-\xi)\, K^{(1)}(x-\eta)\, d\xi\, d\eta$$

$$+ 4\int K^{(2)}(x-\xi,x-\zeta)\, K^{(2)}(x-\eta,x-\zeta)\, H^{(2)}(\xi,\eta)\, d\xi\, d\eta\, d\zeta$$

$$+ \dots , \tag{220}$$

where the remaining terms contain either higher-order kernels or are proportional to the higher-order Wiener-Hermite functions. Substituting into Burgers' equation and equating the coefficients of $H^{(1)}(\xi)$, $H^{(2)}(\xi,\eta)$, etc. or using the statistical orthogonality, gives

$$\frac{\partial}{\partial t} K^{(1)}(x) + 2\frac{\partial}{\partial x}\int K^{(1)}(x-\eta)\, K^{(2)}(x,x-\eta)\, d\eta - \nu \frac{\partial^2}{\partial x^2} K^{(1)}(x) = \dots \tag{221}$$

$$\frac{\partial}{\partial t} K^{(2)}(x,x-\xi) + \frac{1}{2}\frac{\partial}{\partial x}\left[K^{(1)}(x)\, K^{(1)}(x-\xi)\right]$$

$$+ 2\frac{\partial}{\partial x}\int K^{(2)}(x,x-\zeta)\, K^{(2)}(x-\xi,x-\zeta)\, d\zeta - \nu\frac{\partial^2}{\partial x^2} K^{(2)}(x,x-\xi) = \dots \tag{222}$$

where the ... are terms which all involve higher-order kernels, but also contain $K^{(1)}$ and $K^{(2)}$. Thus, the right-hand side of (222) contains a term $K^{(1)}K^{(3)}$.

The Wiener-Hermite expansion method thus clearly leads to a hierarchy of equations. However, truncation of these equations cannot lead to negative spectra since the expansion is of the random function itself and not of the probability distribution. The hierarchy can be closed by neglecting the right-hand sides of (221) and (222) and the higher-order equations for the higher-order kernels. It can be verified that these equations conserve energy to the order of the truncation through the nonlinear interactions, as the constancy of the first two terms in (220), which give $\overline{u^2}$, can be deduced from the truncated equations.

MEECHAM and SIEGEL [35] have solved these equations by approximate analytical means and also numerically. Actually, they solved for the Fourier transforms of the kernel functions. Also, their equations are not quite identical with (221) and (222), as they neglect the square of the second-order kernel in (222) on the reasonable grounds that the main nonGaussian part of the variable u(x) is determined by the $K^{(1)}K^{(1)}$ interaction. The error involved in the truncation clearly cannot be assessed, but the method is certainly attractive. It is conceptually very simple and has a sound physical basis, being consistent with the experimental observation that the probability distribution of u(x) is Gaussian to a good approximation. Moreover, it predicted a spectrum function like k^{-2} for Burgerlence, and therefore it is apparently a satisfactory approximation for Burgers' equation. The results for turbulence, i.e. the three-dimensional Navier-Stokes equations, should therefore be of great interest and it is hoped that results will be available in the near future. All told, the Wiener-Hermite expansion method would appear to be the most promising and least objectionable of all the truncation schemes, and although optimism may be unfounded, it gives hope that a satisfactory analytical description of homogeneous turbulence may be obtained fairly soon.

It should be noted that the equations for the kernels when $\nu = 0$ possess the time reversal properties of the original inviscid equation. This property is also possessed by the joint normal equations as is exemplified by (212). However, it is plausible, and very readily seen for (212), that although reversible, the solutions of the equations for given initial conditions will behave similarly as $t \to +\infty$ or $t \to -\infty$. In other words, for arbitrary initial conditions the "cascade" may for a finite time send energy to small wavenumbers but eventually the transfer will be to large wavenumbers. There appears to be nothing inconsistent about such behavior, and indeed since, as pointed out earlier, there are not the two time-scales needed for "coarse graining" or "phase mixing", approximate equations that are irreversible at any instant would be very suspicious.

The Kraichnan Direct Interaction Approximation

Over the last eight years, there has been a large number of papers by R. H. KRAICHNAN and co-workers, expounding and analyzing a method which purports to give an analytical description of homogeneous turbulence, the so-called direct interaction hypothesis. However, despite all the papers, the foundations of the method are still obscure to most workers in the field. This is perhaps due to the absence of any clear physical basis for the approximation, and the fact that the theory is apparently derived by a series of mathematical assumptions whose accuracy cannot be assessed. It must also be said that the paper [26] which attempts to describe in detail the foundations of the theory is obscure at an important state in the argument, at least to the lecturer. An attempt will now be made to outline the main steps in the direct interaction approximation.

However, it should first be noted that this task is probably of no value, since not only has the theory recently been modified and replaced by a new form, which KRAICHNAN calls the Lagrangian History Direct Interaction Approximation, but in a recent review article [36] KRAICHNAN says that his equations are not rational approximations to the Navier-Stokes equations, in the sense that presumably there is no guarantee whatsoever that the terms neglected have a small effect but are model equations constructed to satisfy some of the invariance properties that the Navier-Stokes equations possess, and with the further property that they are a valid approximation in the limit $Re \to 0$. (It is difficult to believe the relevance of this last criterion as turbulence is a large Reynolds number phenomenon, and it is well known that the Navier-Stokes equations form a singular perturbation problem in the limit $Re \to \infty$, at least for flows with solid boundaries or free surfaces but presumably also for turbulence.) It should be noted that WYLD [37] has shown that the direct interaction approximation is mathematically equivalent to truncating a perturbation solution of the Navier-Stokes equations in powers of the Reynolds number. As KRAICHNAN [36] says, it will be an accident or miracle if any such truncation scheme of a low Reynolds number expansion yields quantitative results of value, and any qualitative results may also be of doubtful significance.

We turn now to the direct interaction approximation as expounded in [26] . In order to reduce the algebraic complexity, we outline the theory for Burger's equation. The extension to the Navier-Stokes equations is straightforward. The Fourier transform of the Burger equation in terms of the generalized Fourier components a(k) [see Eq. (43)] is

$$\frac{\partial a(k)}{\partial t} + i\,k \int a(k')a(k-k')dk' = -\nu k^2 a(k). \qquad (223)$$

For reference, it may be noted that the Fourier transform of the Navier-Stokes equation is

$$\frac{\partial a_i(\underset{\sim}{k})}{\partial t} + i\,k_j \left(\delta_{ik} - \frac{k_i k_k}{k^2}\right) \int a_j(\underset{\sim}{k}')a_k(\underset{\sim}{k}-\underset{\sim}{k}')d\underset{\sim}{k}' = -\nu k^2 a_i(\underset{\sim}{k}). \qquad (224)$$

Thus Burgers' equations and the Navier-Stokes equations differ only through the three dimensionality and the presence of the projection operator which arises from the elimination of the pressure through the equation of continuity. (If therefore the Kolmogorov hypothesis that small-scale and large-scale components are statistically independent is true, it is necessary to explain why the three dimensionality and the projection operator produce this independence which is not true for Burgers' equation.)

To formulate the direct interaction approximation, we must use a discrete representation of the equations. That is, we assume the motion is periodic with period L and write

$$u(x) = \sum_k A(k)e^{ikx}\ , \quad k = \frac{2\pi n}{L}\ , \quad n = 0, 1, 2, \ldots \qquad (225)^1$$

The relation between the continuous and discrete representations is summarized by the relations

---

[1] The symbol L no longer refers to the scale of turbulence, but is the size of the interval. The theory is supposed valid in the limit $L \to \infty$.

$$\frac{2\pi}{L}\sum_k = \int dk \quad , \quad \frac{L}{2\pi}\delta_{kk'} = \delta(k-k') \ . \tag{226}$$

Thus, if $\phi(k)$ is the spectrum function of $u(x)$,

$$\langle A(k)A^*(k')\rangle = \frac{2\pi}{L}\phi(k)\delta_{kk'} \ ; \ \langle a(k)a^*(k')\rangle = \phi(k)\delta(k-k') , \tag{227}$$

where we now use angle brackets to denote ensemble averages, and the star denotes a complex conjugate. The equation of motion for $A(k)$ is

$$\mathcal{B}\,A(k) \equiv \frac{\partial}{\partial t}A(k) + ik\sum_{k'}A(k')A(k-k') + \nu k^2 A(k) = 0 \ , \tag{228}$$

from which it follows, on using the property that $A(k) = A^*(-k)$ and changing variables, that

$$\frac{\partial}{\partial t}A(k)A^*(k) + 2\nu k^2 A(k)A^*(k) - ik\sum_{k+k'+k''=0}\left[A(k)A(k')A(k'') - A^*(k)A^*(k')A^*(k'')\right] = 0. \tag{229}$$

Thus, the triple interactions are required to determine the rate of change of $A(k)A^*(k)$.

Let us consider a particular triple interaction

$$T(\alpha,\beta,\gamma) = A(\alpha)A(\beta)A(\gamma) \ , \quad \alpha+\beta+\gamma = 0 \ . \tag{230}$$

Consider now the equation for a new set of Fourier coefficients $B(k)$ which satisfy the equation

$$\mathcal{B}\,B(k) = ik\left[\delta_{k\alpha}A(-\beta)A(-\gamma) + \delta_{k\beta}A(-\alpha)A(-\gamma) + \delta_{k\gamma}A(-\alpha)A(-\beta)\right] , \tag{231}$$

$$B(k) = A(k) \quad \text{at } t = t_0 ,$$

where t_0 is an initial instant. As PROUDMAN [38] pointed out, the direct interaction hypothesis asserts that

Assumption (i)

$$\langle B(\alpha)B(\beta)B(\gamma)\rangle = 0 . \tag{232}$$

Note that (231) has taken out from the equation for $A(\alpha)$ say, the Fourier components $A^*(\beta)A^*(\gamma)$ and therefore it gives an $A(\alpha)$, which we denote by $B(\alpha)$, that has part of the interaction with wavenumbers β and γ reduced. This hypothesis is intuitively quite attractive, and although one has no real idea of its accuracy, it is not particularly objectionable. The main criticism is that (231) cannot be given a sensible form in terms of the continuous representation, and one feels that a correct theory should be capable of expression in either the continuous or discrete representations, and also directly in physical space. The theory does not satisfy this criterion.

We now define

$$\Delta A(k) = B(k) - A(k), \text{etc.} \tag{233}$$

Then, we make the assumption that the difference between $A(\alpha)$ and $B(\alpha)$ is infinitesimal in some sense; i.e.,

Assumption (ii)

$$|\Delta A(k)| << A(k), \text{etc.} \tag{234}$$

and therefore

$$\langle T(\alpha,\beta,\gamma)\rangle = -\langle \Delta A(\alpha)A(\beta)A(\gamma)\rangle - \langle A(\alpha)\Delta A(\beta)A(\gamma)\rangle - \langle A(\alpha)A(\beta)\Delta A(\gamma)\rangle . \quad (235)$$

Assumption (ii) seems very sensible, as we can think of the typical Fourier component as having an order of magnitude like $L^{-1/2}$, in which case the terms on the r.h.s. of (231) are of order L^{-1} and smaller than the terms on the left which are of order $L^{-1/2}$. Thus the difference between $B(\alpha)$ and $A(\alpha)$ should be smaller by a factor $L^{-1/2}$. Note also that the mean value of the triple interaction has to be of order L^{-2}, in order that the triple correlation in physical space be finite, as may easily be seen by expressing the triple product in terms of the Fourier transform of the 3-point triple velocity correlation. From (230) and (231) it follows that $\Delta A(k)$ satisfies the equation

$$\mathcal{L}\psi(k) \equiv \frac{\partial\psi(k)}{\partial t} - 2\, i\, k \sum_{k'} \psi(k')A(k-k') - \nu k^2 \psi(k) = F(k,\alpha,\beta,\gamma) \quad (236)$$

where $F(k,\alpha,\beta,\gamma)$ is the right-hand side of (231), with the initial condition $\psi(k) = 0$ when $t = t_0$. Thus, the direct interaction approximation leads us now to consider the linear equation (236) for $\psi(k)$. However, because the equation is linear, it does not mean that it is necessarily easier to solve than the original nonlinear equation.

We proceed further by introducing a Green's function for the linear equation (236). Let $g(k|\rho;t|t_1)$ be the solution of

$$\mathcal{L}\, g(k|\rho;t|t_1) = \delta_{k\rho}\; \delta(t-t_1) \; . \quad (237)$$

In the linear operator, the Fourier coefficients $A(k)$ are supposed to have the exact (but unknown) values corresponding to the real solution. We can call this Green's function an impulse response function since it describes the response of the system to an infinitesimal perturbation applied to the wavenumber ρ at time t_1. The mean value of g has interesting properties that should be noted.

We return to physical space and consider a perturbation δu to the solution of Burger's equation. It satisfies the equation

$$\frac{\partial}{\partial t}\delta u + u\frac{\partial}{\partial x}\delta u + \delta u\frac{\partial u}{\partial x} - \nu\frac{\partial^2}{\partial x^2}\delta u = 0. \tag{238}$$

Let $G(x,t \mid x_0,t_1)$ be the Green's function of the linear equation for δu, and $\overline{G}(x-x_0;t,t_1)$ the ensemble average of the Green's function. It must be remembered that the ensemble average is over the ensemble of exact solutions $u(x,t)$. The average Green's function depends only on $x-x_0$ because of the statistical homogeneity. If initially at $t = t_1$, we have that $\delta u = e^{i\rho x}$, then the solution of (238) is

$$\delta u = \int e^{i\rho x_0} G(x,t \mid x_0,t_1)dx_0 \tag{239}$$

and

$$\langle\delta u\rangle = \int e^{i\rho x_0}\overline{G}(x-x_0;t|t_1)dx_0 = e^{i\rho x}\zeta(\rho;t|t_1) \tag{240}$$

when ζ is the Fourier transform of $\overline{G}$. Note that $\zeta(\rho;t|t_1)$ is an ordinary bounded function, quite independent of L, and that

$$\zeta(\rho;t_1|t_1) = 1. \tag{241}$$

Now the initial condition that gives (239) is exactly the initial condition for (237); i.e., δu and $g(k|\rho;t|t_1)$ are Fourier transforms. In particular,

$$\langle g(k|\rho;t|t_1)\rangle = \delta_{k\rho}\zeta(\rho;t|t_1), \tag{242}$$

which shows that the mean value of $g(k|\rho)$ is zero except when $k = \rho$. The function $\zeta(\rho;t|t_1)$ is the *impulse response function* of KRAICHNAN.

A further assumption that KRAICHNAN makes is

Assumption (iii)

$$|g(k|\rho)| << |g(\rho|\rho)|. \qquad (243)$$

This is again very reasonable, as the original contribution at wavenumber ρ is spread out over all wavenumbers, and the gain at any one wavenumber will be small compared with the original amplitude at wavenumber ρ. However, we must not deduce from (243) that any sum over k will remain small compared with the amplitude at ρ. Thus,

$$\sum_{k\neq\rho} |g(k|\rho)| \not\ll |g(\rho|\rho)|. \qquad (244)$$

The solution of (236) can be expressed in terms of the impulse response function. Thus, from the expression of a general solution in terms of a Green's function,

$$\Delta A(k,t) = \int_{t_0}^{t'} i\alpha A(-\beta,t')A(-\gamma,t')g(k|\alpha;t|t')dt' +\cdots +\cdots \qquad (245)$$

where the two other terms are obtained by permuting α,β,γ. The time dependence has now been put in because the Fourier components are measured at different times.

Now we use the third assumption and form the triple interaction (230). It follows that

$$\langle \Delta A(\alpha)A(\beta)A(\gamma)\rangle = i\ \alpha \int_{t_0}^{t'} \langle A(\beta)A'(-\beta)A(\gamma)A'(-\gamma)g(\alpha|\alpha;t|t')\rangle dt' \quad (246)$$

and similarly for the other contributions to (235). The primed quantities are measured at time t'. (The application of assumption (iii) to (245) may be violating (244) because the triple interaction is summed over many wavenumbers when it goes back into (229).

KRAICHNAN now makes an assumption that, as PROUDMAN [38] points out, is likely to introduce an error comparable to that in the joint normal approximation.

Assumption (iv)

$$\langle A(\beta)A'(-\beta)A(\gamma)A'(-\gamma)g(\alpha|\alpha;t|t')\rangle$$

$$= \langle A(\beta)A'(-\beta)\rangle\langle A(\gamma)A'(-\gamma)\rangle\langle g(\alpha|\alpha;t|t')\rangle. \quad (247)$$

It does not appear to be possible to justify this assumption [1], as the terms neglected seem in the absence of further argument to be as large as those retained. The resulting expression for $\langle A(\alpha)A(\beta)A(\gamma)\rangle$ is now a functional of $A(k)A'(-k)$, and substitution into (229) gives an expression for the rate of change of $\langle A(k)A(-k)\rangle$ in terms of integrals over t' and over wavenumber, of products of $\langle A(k)A'(-k)\rangle$ and $\zeta(k;t|t')$. We shall not write down the complicated equation here.

We can ask what is the purpose of this analysis since the impulse response function $\zeta(k;t|t')$ is still an unknown, and the analysis

[1] See p. 607 for a discussion of the "weak dependence principle" which was originally claimed as the justification of the assumption.

is still not closed. Even if we are prepared to accept the first four assumptions, we still have the problem of calculating the impulse response function, and there is no reason to believe that this task is any easier than solving the original equations. Unfortunately, it is at this stage that Kraichnan's argument becomes particularly obscure. We shall now present an argument that leads to Kraichnan's equations, but there may be more convincing ways of presenting the argument, which are not obvious to the lecturer.

The unaveraged impulse response function for an impulse at wavenumber ρ satisfies the equation

$$\mathcal{L}\,\psi(k) = \frac{\partial\psi(k)}{\partial t} + \nu\, k^2\psi(k) - 2\,i\,k \sum_{k'} \psi(k')A(k-k') = 0 \tag{248}$$

where

$$\psi(k) = \delta_{k\rho} \text{ at } t = t_0,$$

the solution of this equation being denoted by $g(k|\rho;t|t_0)$. By virtue of assumption (iii), the dominant term in the nonlinear sum is $\psi(\rho)A(k-\rho)$. We consider therefore the equation with the dominant term moved over to the right-hand side, but with the nonlinear operator retaining the same form, i.e.,

$$\mathcal{L}\,F(k) = 2\,i\,k\,\psi(\rho)A(k-\rho), \tag{249}$$
$$F(k) = 0 \quad\text{when}\quad t = t_0\ ,\ \text{all } k.$$

We now make an assumption about the relation between the solutions of (248) and(249). We assert

Assumption (v)

$$F(k) = \psi(k) \quad \text{for } k \neq \rho. \tag{250}$$

The difference between $F(k)$ and $g(k|\rho)$ lies in the extra term $F(\rho)A(k-\rho)$ present in the equation for $F(k)$. This one term should not make too much difference, and assumption (v) should not lead to excessive error. In terms of the unaveraged response function, we have that

$$F(k) = \sum_{k_1} \int 2ik_1 \psi'(\rho)A'(k_1-\rho)g(k|k_1;t|t')dt', \tag{251}$$

where again the prime denotes values measured at time t'.

We now write equation (248) for $k = \rho$. That is,

$$\frac{\partial\psi(\rho)}{\partial t} + \nu\rho^2\psi(\rho) = 2i\rho \sum_k \psi(k)A(\rho-k)$$

$$= 2i\rho \sum_{k\neq\rho} \psi(k)A(\rho-k) \tag{252}$$

since $A(0) = 0$ because $\bar{u} = 0$. Then with assumption (v) and substituting for (251), we have on taking mean values and remembering that $\langle\psi(\rho)\rangle = \zeta(\rho)$,

$$\frac{\partial}{\partial t}\zeta(\rho) + \nu\rho^2\zeta(\rho) \tag{253}$$

$$= -4\rho \sum_k \sum_{k_1} \int k_1 \langle A(\rho-k)g(\rho|\rho;t'|t_0)A'(k_1-\rho)g(k|k_1;t|t')\rangle dt'.$$

Finally, we make again an assumption of statistical independence like assumption (iv). We assume that the Fourier components A and the response functions g are statistically independent. Thus,

Assumption (vi)

$$\langle A(\rho-k)g(\rho|\rho;t'|t_0)A'(k_1-\rho)g(k|k_1;t|t')\rangle$$

$$= \langle A(\rho-k)A'(k_1-\rho)\rangle\langle g(\rho|\rho;t'|t_0)\rangle\langle g(k|k_1;t|t')\rangle. \qquad (254)$$

This assumption is difficult to accept, as one expects that there is likely to be a substantial contribution from the terms neglected, especially when it is remembered that this assumption is employed in (253) where there is a sum over k and k_1, so that even if the error in (254) is small, the cumulative error might be significant because of (244). Be this as it may, if we substitute (254) into (253), and use (242), we obtain

$$\frac{\partial}{\partial t}\zeta(\rho) + \nu\rho^2\zeta(\rho)$$

$$= -4\rho\sum_k\int k\langle A(\rho-k)A'(k-\rho)\rangle\zeta(\rho;t'|t_0)\ \zeta(k;t|t')dt' \qquad (255)$$

which is Kraichnan's equation for the impulse response function. We now have a complicated pair of integro-differential equations for the spectrum function at two times $A(k)A'(-k)$ and the impulse response function $\zeta(k;t|t')$.

This derivation shows, as PROUDMAN [38] stressed, that Kraichnan's equations rely heavily on assumptions of statistical independence, which cannot be assessed and are unlikely to be correct. Thus in no sense can Kraichnan's equations be regarded as a rational approximation. Since the resulting equations are moreover so complex, the method would seem to be greatly inferior to the joint normal approximation, which derives a closed set of equations in a much simpler manner. The only advantage of the method, which cannot really be called a direct interaction approximation but is a more complex joint normal type approximation, because closure is due not to assumption (i) but to assumptions (iv) and (vi), is the fact that the energy spectrum defined by the equations is positive, according to KRAICHNAN. But the Wiener-Hermite expansion also possesses these properties and would seem to be much superior.

For solutions to the equations, the reader is referred to the literature. The main point is that the equations predict an inertial subrange with a $k^{-3/2}$ behavior for turbulence. It is not known what they predict for Burgerlence, but it does not seem to be a k^{-2} law.

We have presented the derivation of Kraichnan's equations in an axiomatic way in order to clarify the mathematical steps or procedures.[1] However, some of the steps listed as assumptions, in particular numbers (iv) and (vi), are claimed in [26] to be consequences of a "maximal randomness" or "weak dependence" principle, and not assumptions. This claim is difficult to substantiate, as PROUDMAN [38] showed conclusively that the ordering implied by the "maximal randomness" principle is nothing more than a statement that the velocity field is statistically homogeneous in space, and the statistical independence of Fourier coefficients implied by assumptions (iv) and (vi) will not follow simply from spatial homogeneity. Consider for example assumption (vi). The left-hand side of (254) is, on expressing the Fourier coefficients as integrals of the physical

---

[1] It should be stressed again that the sequence of steps in deriving the equations is not unique, and there may be better ways with assumptions that seem less arbitrary. We could, for instance, replace (iii) by another assumption like (iv). In recent work, KRAICHNAN derives the equations not by the hypotheses of [26], but by an algorithm operating on a low Reynolds number expansion about a Gaussian initial state.

variables,

$$\frac{1}{L^4}\int_0^L\int_0^L\int_0^L\int_0^L \langle u(x_1)u(x_2)\delta u_\rho(x_3)\delta u_{k_1}(x_4)\rangle$$

$$e^{-ik(x_4-x_1)}\, e^{-i\rho(x_1+x_3-x_2)}\, e^{-ik_1x_2}\, dx_1dx_2dx_3dx_4 \; . \tag{256}$$

The right-hand side is the same expression with the average broken up into

$$\langle u(x_1)\; u(x_2)\rangle\langle\delta u_\rho(x_3)\rangle\langle\delta u_{k_1}(x_4)\rangle .$$

Here δu_ρ is the solution of (238) with the initial condition $\delta u = e^{i\rho x}$, and similarly for δu_{k_1}. To reduce the complexity slightly, the time dependence in the quantities has been omitted. The equality of these integrals does not follow from the fact that the statistical properties of u(x) are independent of x, and indeed the assumption that they are the same would seem to be even more sweeping than a simple joint normal hypothesis, because it ignores the interaction between u and δu which gives in general nonzero values for $\langle u(x_1)\; \delta u_\rho(x_3)\rangle$. It is important to realize that these assumptions apply to products of the impulse response function and the Fourier coefficients of the velocity field. The ordering implied by spatial homogeneity would justify them if the expressions did <u>not</u> contain the impulse response function (as shown by KRAICHNAN), but they do!

It is perhaps worth placing on record the first step necessary to derive the equations if a continuous representation is used throughout to justify the statement made before that the theory cannot be expressed sensibly in a continuous representation. Analogous to the discrete Fourier coefficients B(k) of equation (231) which have zero triple interaction, we introduce the continuous (generalized) Fourier coefficients b(k) which satisfy

$$\frac{\partial b(k)}{\partial t} - \nu k^2 b(k) + i k \int b(k') b(k-k') dk'$$

$$= i k \Big[\delta(k-\alpha) a(-\beta) a(-\gamma) + \delta(k-\beta) a(-\gamma) a(-\alpha)$$

$$+ \delta(k-\gamma) a(-\alpha) a(-\beta) \Big] \frac{\delta(\alpha+\beta+\gamma)}{[\delta(0)]^3} . \qquad (257)$$

Equation (257) and assumption (i) is the direct interaction hypothesis in the continuous representation. Introduction of an impulse response function for a continuous representation and application of the assumptions (i) - (iv), and the relations (227), lead to an equation relating the energy spectrum and the impulse response function, as may be verified on carrying out the algebra. Note that it is necessary to have in the denominator the cube of a delta-function of zero argument, which is mathematically meaningless. The series of assumptions leads formally to the delta-function of zero argument appearing three times in the numerator, which on cancelling formally with the denominator gives a sensible answer. Another mathematically meaningless equation like (257) is required for an equation akin to (249) for the impulse response function, but our point has already been made and further details will serve no useful purpose.

To sum up, it appears from the above discussion that the theory rests on a number of unjustified and probably unjustifiable assumptions of a mathematical form with no physical basis. At best, it seems that the equations can be no more than a crude approximation or model equations, but they are so complicated that their value as approximations seems minimal. However, KRAICHNAN has recently abandoned the direct interaction approximation equations and replaced them by a set of equations called the Lagrangian History Direct Interaction Approximation [39]. The details are very complicated and hard to follow, but the gist of the method is to use the algorithm on a low Reynolds number expansion (which leads to the direct interaction approximation equations) in terms of Lagrangian variables, i.e. velocities at a point moving with the fluid. These equations are not Galilean invariant, and indeed cannot be because low Reynolds approximations of the Navier-Stokes equations are not Galilean invariant, but they are made Galilean invariant by altering them until they satisfy this property. The resulting, very complicated, equations are then solved numerically with further *ad hoc* assumptions, however, and agreement with the experimental results for the inertial subrange is obtained. It is hard to see how these equations can be more than a guess, as the method of derivation appears in no sense to be a rational logical approximation.

The reason for dropping the original equations was that they do not possess invariance under Galilean transformation. Since the Navier-Stokes equations are invariant, equations which do not possess this property are certainly open to grave doubt. However, one must be careful to ensure that in the process by which the equations were obtained, it was not specifically supposed at some step that the mean velocity is zero. If so, the final equations can clearly not possess Galilean invariance, but there is no reason why they should. Thus, the joint normal hypothesis equations are not Galilean invariant, but the reason is that (211) is sensible only if the mean velocity is zero. If one uses a frame of reference in which the mean velocity is not zero, then (211) will hold only if the velocities are measured relative to the mean; it is not valid as it stands for the absolute velocities. Similar remarks apply to the Wiener-Hermite expansion. The truncated equations are not Galilean invariant, but the expansion (219) omitted the leading term on the supposition that the leading term was zero. That is, the equations are for the velocities relative to the mean, and do not apply to the absolute velocities if the mean is not zero. The point is that Galilean invariance is built into the equations of motion, and one must take drastic steps to lose it. An assumption of statistical independence or the vanishing of a cumulant cannot violate Galilean invariance if phrased properly. The same is true of energy conservation. In the derivation of the direct interaction approximation equations, it will be remembered that it was supposed at one step, see equation (252), that $A(0) = 0$. Now $A(0) = \bar{u}$, so the final equations cannot be expected to be Galilean invariant. Also the forms (247) and (254) of the assumptions of statistical independence will require modification if there is a mean velocity, because they involve spatial correlations at different times, but this could be done and Galilean invariant equations could be obtained which would give the same answers.

9. Concluding Remarks

There are many more theories of turbulent decay than have been described in these notes, but they have been omitted either because they seem to the lecturer to be irrelevant and of little real interest, or because they are difficult to understand and bitter experience has shown that the truth or significance in them tends to be inversely proportional to the effort required to understand the theory. We have omitted the theory of the so-called final period of decay for two reasons. Firstly, it is a linear theory because the nonlinear terms are put equal to zero, and the solution of the equations in terms of arbitrary initial conditions can be written down immediately (the generalized Fourier components are given by $a_i(\underset{\sim}{k},t) = a_i(\underset{\sim}{k},t_0)e^{-\nu k^2(t-t_0)}$). But secondly, there is no proof that the neglect of the nonlinear terms is a uniformly valid approximation as $t \to \infty$, and the theory is therefore not beyond a shadow of doubt.

The notes have concentrated on the problem of turbulence decay because this seems to be the central problem of homogeneous turbulence, but the problem of stationary homogeneous turbulence is of interest. Here where the turbulence is maintained against decay by the action of external forces, one would like to know how the equilibrium level of turbulent intensity depends on the external forces. The situation of turbulence theory is somewhat depressing, because as the notes have tried to point out, our understanding of turbulence is practically nil, despite the last 30 years of effort. But undue pessimism is out of order, as the position could change overnight. One aspect of the decay problem which would be of interest and has only been mentioned in passing is the tendency of homogeneous turbulence to become isotropic, particularly for the small-scale components. The experimental results indicate that the tendency towards isotropy may be much weaker than anticipated, and investigation of this aspect is needed.

Finally, it would seem that turbulence theory to date can be summarized by the quotation from *Macbeth*, "full of sound and fury, signifying nothing." It is to be hoped that future work will render this quotation inappropriate.

References[1]

[1] D. J. BENNEY and P. G. SAFFMAN: Nonlinear interactions of random waves in a dispersive medium. Proc. Roy. Soc. A 289, 301 (1966).

[2] G. K. BATCHELOR and A. A. TOWNSEND: The nature of turbulent motion at large wave-numbers. Proc. Roy. Soc. A 199, 238 (1949).

[3] H. W. LIEPMANN: Free turbulent flows. Mecanique de la Turbulence. C. N. R. S. p. 211 (1962).

[4] A. A. TOWNSEND: Review of Turbulent Flows and Heat Transfer. J. Fluid Mech. 10, 635 (1961).

[5] H. L. GRANT and I. C. T. NISBET: The inhomogeneity of grid turbulence. J. Fluid Mech. 2, 263 (1957).

[6] G. K. BATCHELOR: Theory of Homogeneous Turbulence. Cambridge University Press (1953).

[7] F. N. FRENKIEL and P. S. KLEBANOFF: Two dimensional probability distribution in a turbulent field. Physics of Fluids 8, 2291 (1965).

[8] G. K. BATCHELOR and I. PROUDMAN: Phil. Trans Roy. Soc. A 248, 369 (1956).

[9] P. G. SAFFMAN: The large scale structure of homogeneous turbulence. J. Fluid Mech. (to appear).

[10] L. D. LANDAU and E. M. LIFSHITZ: Fluid Mechanics. Pergamon Press (1959).

[11] S. F. EDWARDS: The statistical dynamics of homogeneous turbulence. J. Fluid Mech. 18, 239 (1964).

[12] J. R. HERRING: Self-consistent field approach to turbulence theory. Physics of Fluids 8, 2219 (1965).

[13] I. PROUDMAN and W. H. REID: On the decay of a normally distributed and homogeneous turbulent velocity field. Phil. Trans. Roy. Soc. A 247, 163 (1954).

[14] G. I. TAYLOR and A. E. GREEN: Mechanism of the production of small eddies from large ones. Proc. Roy. Soc. A 158, 499 (1937).

[1] No attempt has been made to provide a complete bibliography.

[15] G. I. TAYLOR: Production and dissipation of vorticity in a turbulent fluid. Proc. Roy. Soc. A 164, 15 (1938).

[16] G. K. BATCHELOR and A. A. TOWNSEND: Turbulent diffusion. Article in: Surveys in Mechanics. Cambridge Univ. Press (1956).

[17] R. BETCHOV: An inequality concerning the production of vorticity in isotropic turbulence. J. Fluid Mech. 1, 497 (1956).

[18] M. M. GIBSON: Spectra of turbulence in a round jet. J. Fluid Mech. 15, 161 (1963).

[19] H. L. GRANT, R. W. STEWART and A. MOILLIET: Turbulence spectra from a tidal channel. J. Fluid Mech. 12, 241 (1962). The spectrum of a cross-stream component of turbulence in a tidal stream. J. Fluid Mech. 13, 237 (1962).

[20] A. L. KISTLER and T. VREBALOVICH: Grid turbulence at large Reynolds numbers. J. Fluid Mech. 26, 37 (1966).

[21] A. A. TOWNSEND: On the fine-scale structure of turbulence. Proc. Roy. Soc. A 208, 534 (1951).

[22] A. N. KOLMOGOROV:Dissipation of energy in locally isotropic turbulence. Compt. Rend. (Dokl.) Acad. Sci. U. R. S. S. 32, 16 (1941).

[23] G. S. GOLITSYN: On the structure of turbulence in the small scale range. Prikl. Mat. Mekt. 24, 1124 (1960).

[24] J. M. BURGERS: Unpublished lectures on turbulence given at the California Institute of Technology (1951).

[25] J. von NEUMANN: Recent theories of turbulence. Collected Works, Vol. 6. Pergamon Press (1962).

[26] R. H. KRAICHNAN: The structure of isotropic turbulence at very high Reynolds numbers. J. Fluid Mech. 5, 497 (1959).

[27] L. ROSENHEAD (ed.): Laminar Boundary Layers. Oxford Univ. Press (1963).

[28] J. D. COLE: On a quasi-linear parabolic equation occurring in aerodynamics. Quart. App. Math. 9, 225 (1951).

[29] T. H. ELLISON: The universal small scale spectrum of turbulence at high Reynolds number. Mecanique de la turbulence. C. N. R. S. p. 113 (1962).

[30] T. TATSUMI: The theory of decay process of incompressible isotropic turbulence. Proc. Roy. Soc. A 239, 16 (1957).

[31] R. H. KRAICHNAN: Relation of fourth-order to second-order moments in stationary isotropic turbulence. Phys. Rev. 107, 1485 (1957).

[32] W. C. MEECHAM: Turbulence energy principles for quasi-normal and Wiener-Hermite expansions. Physics of Fluids 8, 1738 (1965).

[33] Y. OGURA: A consequence of the zero fourth-order cumulant approximation in the decay of isotropic turbulence. J. Fluid Mech. 16, 33 (1963).

[34] S. A. ORSZAG and M. D. KRUSKAL: Theory of turbulence. Phys. Rev. Letters 16, 441 (1966).

[35] W. C. MEECHAM and A. SIEGEL: Wiener-Hermite expansion in model turbulence at large Reynolds numbers. Physics of Fluids 7, 1178 (1964).

[36] R. H. KRAICHNAN: Invariance principles and approximation in turbulence dynamics. Symposium on the dynamics of fluids and plasmas, 1965. Academic Press (1967).

[37] H. W. WYLD: Formulation of the theory of turbulence in an incompressible fluid. Annals of Physics 14, 143 (1961).

[38] I. PROUDMAN: On Kraichnan's theory of turbulence . Mecanique de la turbulence. C. N. R. S. p. 107, (1962).

[39] R. H. KRAICHNAN: Lagrangian history closure approximation for turbulence. Physics of Fluids 8, 575 (1965).

Superspace and the Nature of Quantum Geometrodynamics

J. A. Wheeler

Allowable History Selected out of Arena of Dynamics by Constructive Interference

Particle dynamics takes place in the arena of spacetime. Geometrodynamics takes place in the arena of superspace. One needs only to mount to this point of view to have the whole content of Einstein's theory spread out before his eyes, as in the outlook from a mountain peak, and as well quantum geometrodynamics and classical geometrodynamics. What is the nature of the landscape that we see from this height? What structures can we hope to build upon this landscape? And what kinds of mysteries are hidden in the mists beyond?

In looking at the terrain of dynamics, happily, everyone by now is long accustomed to the Hamilton-Jacobi theory. It belongs entirely to the world of classical physics. Yet it carries one in an instant into the world of the quantum of action.

When better has one ever seen the transition from quantum to classical than in the theory of a particle in motion? Call the potential $V = V(x)$. Call the energy of the particle E. Then there is not the slightest hope of discussing the movement of the particle in space and time. Complementarity forbids! The wave function is spread out all over space. That one sees in no way more easily than through the semiclassical approximation for the probability amplitude function,

$$\psi_E(x,t) = \begin{pmatrix}\text{slowly varying}\\ \text{amplitude function}\end{pmatrix} \exp(i/\hbar) S_E(x,t). \qquad (1)$$

It is of no help in localizing the probability distribution that the Hamilton-Jacobi function S has in many applications a value large in comparison with the quantum of angular momentum $\hbar = 1.02 \times 10^{-27}$ gcm^2/sec. It is of no help that this "dynamical phase" (to give S another name) obeys the simple Hamilton-Jacobi law of propagation,

$$\begin{aligned}\partial S/\partial t &= H(\partial S/\partial x,\ x)\\ &= (1/2m)\ (\partial S/\partial x)^2 + V(x). \qquad (2)\end{aligned}$$

And finally, it is of no help that the solution of this equation for a particle of energy E is extraordinarily simple,

$$S(x,t) = -Et + \int^x \left[2m(E-V(x))\right]^{1/2} dx + \delta_E. \qquad (3)$$

The probability is still spread all over everywhere! There is not the slightest trace of anything like a localized worldline, $x = x(t)$!

How old the idea of building wave packets out of monofrequency waves! And how easy. The probability amplitude is now a superposition of terms, qualitatively of the form

$$\psi(x,t) = \psi_E(x,t) + \psi_{E+\Delta E}(x,t) + \ldots \qquad (4)$$

Destructive interference takes place almost everywhere. The wave packet is concentrated in the region of constructive interference. There the phases of the various waves agree; thus

$$S_E(x,t) = S_{E+\Delta E}(x,t). \qquad (5)$$

At last a worldline! And how easy to find the Newtonian motion from this condition of constructive interference:

$$0 = S_{E+\Delta E} - S_E$$

$$0 = -t\Delta E + \int^x \Delta p_E(x)dx + (\delta_{E+\Delta E} - \delta_E)$$

$$t = \int^x dx/v_E(x) + t_0 \text{ (NEWTON)}. \tag{6}$$

Here $v_E(x)$ denotes the velocity at the location x,

$$\frac{\Delta\left[2m(E - V)\right]^{1/2}}{\Delta E} = \frac{\Delta p_E}{\Delta E} \rightarrow \frac{\partial \text{ (momentum)}}{\partial E}$$

$$= \frac{1}{v_E(x)} = \left(\begin{matrix}\text{time to cover} \\ \text{a unit distance}\end{matrix}\right); \tag{7}$$

and the quantity t_0 is an abbreviation for

$$\frac{\delta_{E+\Delta E} - \delta_E}{\Delta E} \rightarrow \frac{d\delta_E}{dE} \equiv t_0 . \tag{8}$$

Marvelously, not one trace of the quantum of action appears in the final solution for the motion. Yet the quantum principle supplies the whole rationale and motivation for talking about "constructive interference". The quantum comes in only when one recognizes the finite spread of the wave packet (Fig. 1). Then the idea of a worldline has to be renounced. To the propagation of the particle from start to finish, a whole range of histories contributes. This is the way the real world of quantum physics operates!

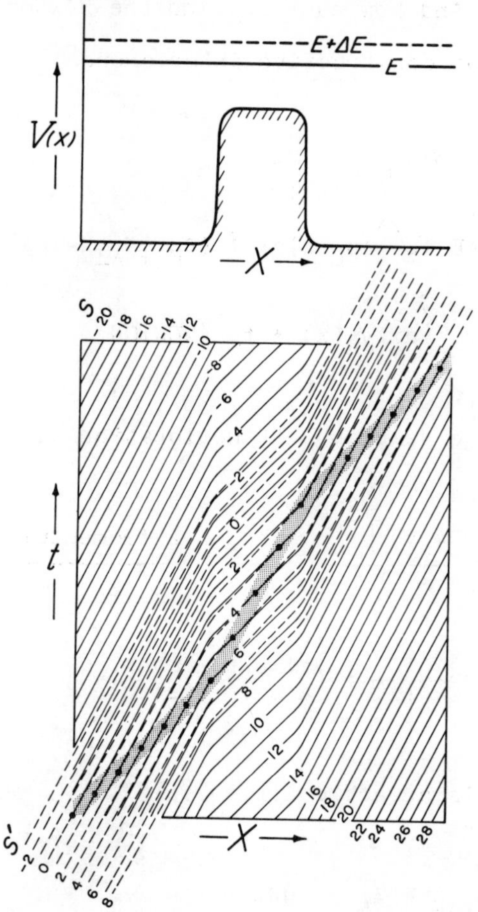

Figure 1. "Motion" and "worldline" of a particle appear in quantum mechanics as the consequence of interference between wave trains that extend over all space. Above, potential energy as a function of distance for a model problem.

Below, smooth lines numbered -20, -18,...., 28 are wave crests of probability amplitude function $\psi_E(x,t) \sim$ (slowly varying amplitude factor) exp $(i/\hbar)$. $S(x,t)$ for energy E. Dashed lines, same for energy E + ΔE. Shaded area, region of constructive interference ("wave packet"). Black dots mark locus of classical worldline $(S_{E+\Delta E} = S_E)$.

(Figure 1)

Three Dimensions, Not Four

Similarly in geometrodynamics. Here the dynamic object is not spacetime. It is space. The geometrical configuration of space changes with time. But it is space, 3-dimensional space, that does the changing. No surprise! In particle dynamics, the dynamical object is not x *and* t, but only x. How do we tell this to our friends in the world of mathematics? For so long they have heard us say that it was in default of the fourth dimension that RIEMANN could not have discovered general relativity. First there had to come special relativity and spacetime and the fourth dimension. Otherwise, how could one have had any possibility to connect gravitation with the curvature of spacetime (Fig. 2)? This understood, how can physicists change their mind and "take back" one dimension? The answer is simple. A decade and more of work by DIRAC, BERGMANN, SCHILD, PIRANI, ANDERSON, HIGGS, ARNOWITT, DESER, MISNER, DeWITT and others has taught us through many a hard knock that Einstein's geometrodynamics deals with the dynamics of geometry: of 3-geometry, not 4-geometry. [1,2]

What is a 3-geometry? To simplify the question, rephrase it in an everyday context: what is a 2-geometry? Nothing illustrates a 2-geometry more clearly than an automobile fender. In whatever way coordinates are painted on its surface, in whatever way the points of that surface are named or renamed, the fender keeps the same 2-geometry. Similarly for a 3-geometry. In mathematical terms, a $^{(3)}\mathcal{G}$ is not a positive definite 3 X 3 metric; instead, it is an *equivalence class* of such metrics which are transformable, one into another, by diffeomorphisms.

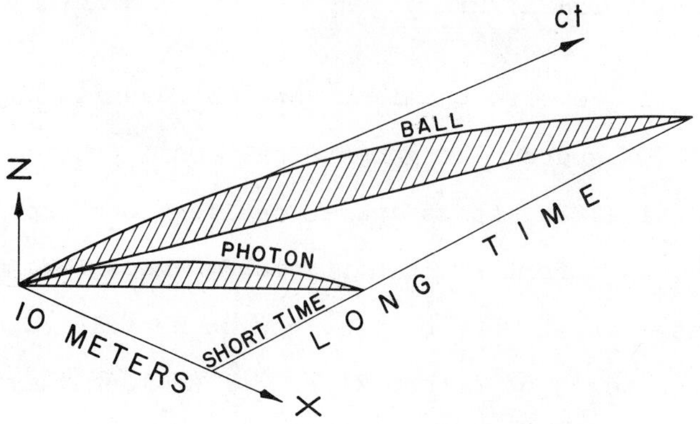

Figure 2. The track of the ball and the track of the photon through space (x, y plane) have very different curvatures, but in spacetime (x, z, ct space) the curvatures are comparable.

Mere baggage is the right term not only for the points of the space, but also for the coordinates employed to label these points, and for the metric that tells the distance from each point to all its near neighbors. Behind all of that paraphernalia lies the real idea, the concept of a 3-geometry, as solid and substantial as the 2-geometry of the fender. Down with "points"; up with "geometry"!

Superspace

One climbs up to the concept of geometry only to find a new height beyond: superspace. [3] Superspace is the manifold, a single point which stands for an entire 3-geometry (Fig. 3). A 3-geometry stands at the halfway mark between point and superspace; risen though it is above the concept of point by abstractification, any one geometry of space counts as only a single point of superspace.

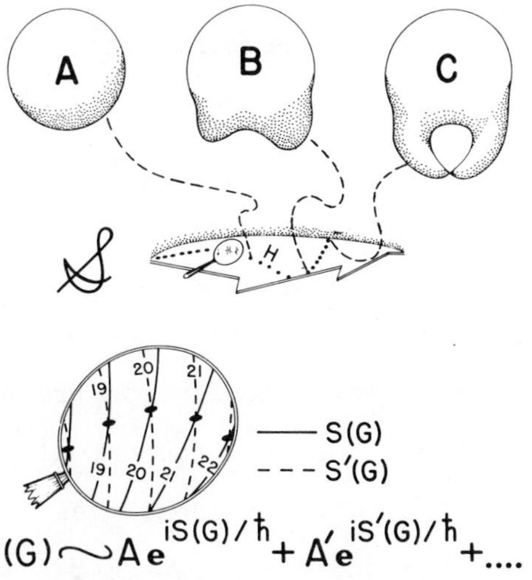

Figure 3. Superspace $\mathcal{S}$ is the manifold, each of whose "points" A,B,C,.... is an abbreviation for one 3-geometry. A submanifold H of $\mathcal{S}$ is the "classical history of the geometry of space" when space has been started off under some particular set of dynamical initial conditions. In other words, H consists of all those spacelike 3-geometries which can be obtained as spacelike sections through one particular 4-geometry (that satisfies Einstein's classical field equations). The 3-geometries of the classical history H may alternatively but equivalently be distinguished from other 3-geometries by the fact that they satisfy the classical "condition of constructive interference" $S({}^{(3)}\mathcal{G}) = S'({}^{(3)}\mathcal{G}) = S''({}^{(3)}\mathcal{G}) = \ldots$ ("coincidence of wave crests" in lower magnified view of a region of superspace). In quantum theory, the wave packet is not localized with unlimited sharpness. The probability amplitude ψ has significant values for 3-geometries at some small "distance" in superspace on each "side" of the classical history H; hence the "quantum fluctuations in the geometry of space" (depicted symbolically in more detail in Fig. 5).

(Figure 3)

Superspace is the arena for geometrodynamics as Lorentz-Minkowski spacetime is the arena for particle dynamics (Table I). The momentary configuration of the particle is an event, a single point in spacetime. The momentary configuration of space is a 3-geometry, a single point in superspace.

Is superspace a proper manifold? Its construction is simple enough: call a 3-geometry a "point", and put all such points together. Or why limit attention to 3-geometries with positive definite metric? Build "extended superspace"; call every 3-geometry a "point" even if its signature is not +++; and put all these "points" together. Does the resulting mathematical object possess a reasonable topology? Is it a true manifold? No! In a proper manifold each point has a neighborhood homeomorphic to an open set in a Banach space; and two distinct points have disjoint neighborhoods. Not so here. In an important investigation MICHAEL STERN has shown [4] that for each point of extended superspace there exists another point, distinct from it, which nevertheless cannot be isolated from it by open sets. The topology is not HAUSDORF. Extended superspace is not a manifold. It can hardly be regarded as an acceptable arena for dynamics.

Superspace, in contrast, STERN has shown, does have Hausdorf topology and does constitute a manifold. According to these mathematical considerations, *superspace is the proper arena for geometrodynamics.*

Physical considerations point to the same conclusion. To insist that a 3-geometry shall have positive definite metric is to guarantee that no light ray can traverse this 3-geometry. No physical effect can propagate from one point of the ${}^{(3)}\mathcal{G}$ to another.

Table I. Geometrodynamics Compared with Particle Dynamics

	Particle	Geometrodynamics
Dynamical entity	Particle	Space
Descriptors of momentary configuration	x,t ("event")	$^{(3)}\mathcal{G}$ ("3-geometry")
History	x = x(t)	$^{(4)}\mathcal{G}$ ("4-geometry")
History is a stockpile of configurations?	Yes. Every point on worldline gives a momentary configuration of particle.	Yes. Every spacelike slice through $^{(4)}\mathcal{G}$ gives a momentary configuration of space.
Dynamic arena	Spacetime (totality of all points x,t)	Superspace (totality of all $^{(3)}\mathcal{G}$'s)

A physical quantity local to the one point and a physical quantity local to the other have zero reciprocal coupling. They commute. Such quantities lend themselves to simultaneous specification over the entire 3-geometry. No simpler example exists of a "complete observation" in quantum geometrodynamics. Closely related is the initial value problem of classical geometrodynamics.

"Observation?" Does not observation imply an observer? If there is an observer, does he not respond to every event on his past light cone? And consequently does not that past light cone with its ++0 metric supply the appropriate geometry on which to specify physical conditions? No, and for two reasons. First, to know the state of the geometry on a past light cone, however completely, is still to be powerless to predict from Einstein's field equations the future of the geometry. Predict the 4-geometry within the past light cone? Yes. [5] Outside - to the future? No. Into the domain of the future, influences flow from afar without ever once impinging upon the light cone. "Demidynamics" - prediction into the past - one can do; full dynamics, no. The fault is in the choice of initial value hypersurface. The cone creates an unsymmetrical divide between past and future. How different from a spacelike initial value hypersurface! The initial value data on it allow a prediction of the complete geometrodynamical history, past and future.

Second, the original argument was mistaken that one should consider "a light cone converging upon an observer". The "observer" of dynamical theory is not and cannot be a single detector of events either in special relativity or in general relativity. Instead he "collects the printout" [6] from a multitude of detectors dotted densely about. Each of his detectors is sensitive only for an instant. These instants of sensitivity bear a spacelike relationship each to the other. [7] Collectively they define a spacelike hypersurface, a "simultaneity", a common moment of a rudimentary "time" variable that perhaps is not and certainly needs not be any further specified. In its moment of sensitivity each detector responds to the appropriate influence: to particle proximity in particle physics; to local field strength in electrodynamics and to local geometry in geometrodynamics. Whichever the dynamic entity, the measurements of it do not and cannot fully serve dynamic theory unless they span a spacelike hypersurface.

Distill this discussion! Where specify dynamically complete initial value data? On a spacelike 3-geometry, yes; on a null 3-geometry, no. What "points" belong to a topologically acceptable arena for geometrodynamics? Spacelike 3-geometries, yes; null 3-geometries, no. What a remarkable correspondence between dynamics and topology!

Wave Packet in Superspace and its Propagation

So much for the dynamical entity, space; and so much for the arena in which the dynamics takes place, superspace; now for the dynamics itself. Not one trace of dynamics does one see when he examines the typical probability amplitude function, $\psi = \psi({}^{(3)}\mathcal{G})$, for it is spread all over superspace. No surprise! Already in classical theory the Hamilton-Jacobi function, $S = S({}^{(3)}\mathcal{G})$, is spread out over the manifold. Moreover, this "dynamical phase function" of classical geometrodynamics gives at once in the semiclassical approximation the actual phase of ψ, according to the formula

$$\psi\ ({}^{(3)}\mathcal{G}) = \begin{matrix}\text{slowly varying}\\ \text{amplitude function}\end{matrix}\ \exp\ (i/\hbar)S({}^{(3)}\mathcal{G}) \tag{9}$$

- indication enough that ψ and S are both unlocalized! Dynamics first clearly becomes recognizable when sufficiently many such spread out probability amplitude functions are superposed to build up a localized wave packet: [8,9]

$$\psi = c_1\psi_1 + c_2\psi_2 + \ldots\ldots \tag{10}$$

Constructive interference occurs where the phases of the several individual waves agree: [10]

$$S_1\ ({}^{(3)}\mathcal{G}) = S_2\ ({}^{(3)}\mathcal{G}) = \ldots\ldots \tag{11}$$

The ${}^{(3)}\mathcal{G}$'s compatible with these conditions of constructive interference constitute the classical geometrodynamical history of space (Fig. 3): Marvellously, every 3-geometry that satisfies the conditions of constructive interference (11) can be obtained as a spacelike slice through a certain 4-geometry. More marvellously, this ${}^{(4)}\mathcal{G}$ satisfies the ten field equations of Einstein. In other words, *in addition to the principle of constructive interference one needs only the single equation of Hamilton and Jacobi for the "dynamical phase" $S\ ({}^{(3)}\mathcal{G})$ to obtain all of classical geometrodynamics. The proof of this important point has been announced by GERLACH.* [11]

The Hamilton-Jacobi equation itself was first given explicitly in the literature by PERES [12] on the foundation of earlier work by himself and others on the Hamiltonian formulation of geometrodynamics: [13]

$$g^{-1}\left(g_{ik}g_{jl} - \frac{1}{2}\, g_{ij}g_{kl}\right)(\delta S/\delta g_{ij})\ (\delta S/\delta g_{kl}) + {}^{(3)}R = 0. \quad (12)$$

Here the g_{ij} are the coefficients in the metric on ${}^{(3)}\mathcal{G}$ and g is the determinant of these metric coefficients. The quantity ${}^{(3)}R$ is the local value of the scalar curvature invariant of the geometry intrinsic to ${}^{(3)}\mathcal{G}$. The Hamilton-Jacobi function depends only upon the 3-geometry, and not upon how that 3-geometry is expressed in terms of metric coefficients in a particular coordinate patch. However, S is treated in (12) as if it depended upon the metric coefficients individually, $S = S(g_{11}, g_{12}, \ldots\ldots, g_{33})$. With this understanding, $\delta S/\delta g_{ij}$ denotes the functional derivative of S with respect to alterations in the function g_{ij} (x, y, z). The fact that coordinates in the end have nothing to do with the matter can be stressed by rewriting (12) symbolically in the form

$$\boxed{\left(\underset{\sim}{\nabla} S/\delta\, {}^{(3)}\mathcal{G}\right)^2 + {}^{(3)}R = 0} \quad . \quad (13)$$

This "Einstein-Hamilton-Jacobi equation" contains all of classical geometrodynamics in regions where no "real" sources of mass-energy are present. This one equation carries the entire content of Einsteins's ten field equations. Consequently, this equation has been checked and is subject to further check in the same sense and to the same degree that the predictions of Einstein's field equations have been verified, and are subject to further verification.

In the simplest version of particle physics a worldline has a simple meaning. It is a stockpile consisting of all those points (x, t) that satisfy the condition of constructive interference. There are ∞^1 of these points on the worldline, if we use an obvious though loose way of counting. Dynamics marks out these points and gives preference to them over all the ∞^2 points in the arena of particle dynamics. In geometrodynamics a 4-geometry has a similar significance. It is a stockpile consisting of all those 3-geometries that satisfy the condition of constructive interference. There are ∞^{∞^3} of these 3-geometries, obtainable by making a spacelike slice through the 4-geometry in one or another way:

$$t = t\,(x, y, z); \tag{14}$$

thus,

∞^3 points (x, y, z); and
∞^1 choices for t at each of these points; and hence
∞^{∞^3} choices of 3-geometry altogether. (15)

Classical geometrodynamics marks out these ${}^{(3)}\mathcal{G}$'s and gives preference to them over the infinitely more numerous totality of ${}^{(3)}\mathcal{G}$'s to be found in the entire arena of geometrodynamics. That arena, superspace, contains ${}^{(3)}\mathcal{G}$'s to the number $(\infty^3)^{\infty^3}$, calculated most simply as follows:

∞^3 points; and, in a coordinate system that makes the metric diagonal,
3 diagonal components of the metric specifiable per space point; and hence
∞^3 choices of the metric per space point; and therefore
$(\infty^3)^{\infty^3}$ choices of ${}^{(3)}\mathcal{G}$ altogether. (16)

From this totality of conceivable 3-geometries, one has to pick out and exhibit all the dynamically allowed 3-geometries before he has told in all fullness how space evolves with time. No new lesson! One has long known that time in general relativity is a many-fingered entity. The hypersurface drawn through spacetime to give one ${}^{(3)}\mathcal{G}$ can be pushed forward in time a little here or a little there or a little somewhere else to give one or another or another new ${}^{(3)}\mathcal{G}$. *"Time"*, conceived in these terms, *means* nothing more or less than *the location of the* ${}^{(3)}\mathcal{G}$ *in the* ${}^{(4)}\mathcal{G}$. In this sense "3-geometry is a carrier of information about time." [14]

"Spacetime" a Concept of Limited Validity

The child's toy can be removed from its box only to reveal another box and, that taken away, another box, and so on, until eventually there are dozens of boxes scattered over the floor. Or conversely the boxes can be put back together, nested one inside the other, to reconstitute the original package. The packaging of ${}^{(3)}\mathcal{G}$'s into a ${}^{(4)}\mathcal{G}$ is much more sophisticated. Nature provides no monotonic ordering of the ${}^{(3)}\mathcal{G}$'s. Two of the dynamically allowed ${}^{(3)}\mathcal{G}$'s taken at random will often cross each other one or more times. When one shakes the ${}^{(4)}\mathcal{G}$ apart, he therefore gets enormously more ${}^{(3)}\mathcal{G}$'s "spread out over the floor" than he might otherwise have imagined. Conversely, when one puts back together all of the ${}^{(3)}\mathcal{G}$'s allowed by the condition of constructive interference, he gets a structure with a rigidity that he might not otherwise have foreseen. This rigidity arises from the infinitely rich interleaving and intercrossing of clear cut, well defined

$^{(3)}\mathcal{G}$'s one with another. In summary, (1) the $^{(3)}\mathcal{G}$'s allowed by (11) are the basic building blocks; (2) their interconnections give $^{(4)}\mathcal{G}$ its existence, its dimensionality [15] and its "magic structure"; and (3) in this structure every $^{(3)}\mathcal{G}$ has a rigidly fixed location of its own.

How different from the textbook concept of spacetime! There the geometry of spacetime is conceived as constructed out of elementary objects, or points, known as "events". Here, by contrast, the primary concept is 3-geometry, and the event is secondary: (1) The event lies at the "intersection" of such and such $^{(3)}\mathcal{G}$'s; (2) Its timelike relation to some other $^{(3)}\mathcal{G}$ is determined by the structure of the $^{(4)}\mathcal{G}$ - which in turn derives from the inter-crossings of all the other $^{(3)}\mathcal{G}$'s.

Whether one starts with $^{(3)}\mathcal{G}$'s as primary and regards the "event" as a derived concept, or vice versa, might make little difference if one were to remain in the domain of classical geometrodynamics. It makes all the difference when one turns to quantum geometrodynamics.

There is no such thing as a 4-geometry in quantum geometrodynamics, and for a simple reason. No probability amplitude function $\psi\,(^{(3)}\mathcal{G})$ can propagate through superspace as an indefinitely sharp wave packet. It spreads (Fig. 3). It has a finite probability amplitude in a domain of superspace of finite measure. This domain encompasses a set of $^{(3)}\mathcal{G}$'s far too numerous to accommodate in any one $^{(4)}\mathcal{G}$. One can express this situation in various terms. One can say that propagation takes place in superspace, not by following any one classical history of space, not by following any one $^{(4)}\mathcal{G}$, but by summation of contributions from an infinite variety of such histories. This extension of Feynman's concept of "sum over histories" has received special attention from MISNER [16]. In whatever way one states the matter, however, the facts are clear. The $^{(3)}\mathcal{G}$'s that occur with significant probability amplitude do not fit and cannot be fitted into any single $^{(4)}\mathcal{G}$. That "magic structure" of classical geometrodynamics simply does not exist. Without that building plan to organize the $^{(3)}\mathcal{G}$'s of significance into a definite relationship, one to another, even the "time ordering of events" is a notion devoid of all meaning.

These considerations reveal that the concepts of "spacetime" and "time" itself are not primary but secondary ideas in the structure of physical theory. These concepts are valid in the classical approximation. However, they have neither meaning nor application under circumstances when quantum geometrodynamic effects become important. Then one has to forego that view of nature in which every event, past, present or future occupies its preordained position in a grand catalog called "spacetime". There is no spacetime, there is no time, there is no before, there is no after. The question what happens "next" is without meaning.

The Planck Length and Gravitational Collapse

Under everyday circumstances, these unexpected consequences of the quantum principle never come into evidence. The characteristic dimension of quantum geometrodynamics is the Planck length, [17] $(\hbar G/c^3)^{1/2} = 1.6 \times 10^{-33}$ cm. By comparison the normally relevant scale of any geometry of interest is stupendous. Negligible on the scale of the geometry is the quantum mechanical spread of the wave packet in superspace. Consequently the dynamical evolution of the geometry can be treated in the context of classical geometrodynamics. Thus the geometries that occur with significant probability amplitude can be idealized to a good approximation as if confined to a region in superspace of zero thickness. The ${}^{(3)}\mathcal{G}$'s of this limited set are sufficiently small in number to fit together into a single ${}^{(4)}\mathcal{G}$. They are sufficiently large in number to reproduce every conceivable spacelike slice of that ${}^{(4)}\mathcal{G}$. In this approximation it makes good sense to speak of the "classical geometrodynamical history of space."

If the dynamics of geometry thus normally lends itself to classical analysis, there are two contexts where it does not. One is the final stage of gravitational collapse. The other is analysis of the microscopic quantum mechanical fluctuations in the geometry of space and their consequences for physics generally.

Of all the applications of quantum geometrodynamics, none would seem more immediate than gravitational collapse. [18] Here, according to classical general relativity, the dimensions of the collapsing system in a finite proper time are driven down to indefinitely small values. The phenomenon is not limited to the space occupied by matter. It occurs also in the space surrounding the matter. In a finite proper time, the calculated curvature rises to infinity. At this point, classical theory becomes incapable of further prediction. In actuality, classical considerations go wrong before this point. A prediction that is infinity is not a prediction. The wave packet in superspace does not and cannot follow the classical history when the geometry becomes smaller in scale than the quantum mechanical spread of the wave packet. Not a new phenomenon! Throughout physics one sees examples of a wave impinging upon a region of interaction of dimensions small compared to a wavelength. The outcome is scattering or diffraction. One speaks of a probability for this, that, or the other outcome of the interaction. The photon or phonon or other entity entering from one direction emerges in another. The concept of a deterministic worldline may serve adequately during the phase of approach to the zone of interaction and during the phase of regression. It is completely out of place during the phase of scattering. So here. The concept of a deterministic history of geometry, a well defined $^{(4)}\mathcal{G}$, makes sense in the early phase of gravitational collapse, but has absolutely no application in the decisive phase. There "spacetime" is non-existent, "events" and the "time ordering of events" are without meaning, and the question "what happens after the final phase of gravitational collapse" is a mistaken way of speaking.

The correct way of speaking deals with the propagation of the probability amplitude $\psi\,(^{(3)}\mathcal{G})$ in superspace. The semiclassical treatment of the propagation (Eqs. 9, 13) is appropriate in most of the domain of superspace of interest for gravitational collapse. Not so in the decisive region. There, as in elementary problems of scattering, the mathematical analysis has to go straight back to the full and accurate wave equation for its foundation. What is the appropriate question to ask of the mathematics? One has learned how to formulate the right question, in the case of scattering, through long experience with the physics. However, if one had not had that physical experience, he would have learned out of the mathematical formalism itself to speak about incoming plane waves and outgoing spherical waves, and scattering amplitudes and about all the other relevant concepts. Similarly in quantum geometrodynamics, where one has so much less experience. The mathematical formalism itself must serve as the final arbiter on how to pose the central question about gravitational collapse as well as how to answer it!

Quantum Fluctuations in Geometry of Space

"Quantum fluctuations in the geometry of space" is the other pressing field of application of quantum geometrodynamics. What a strange combination of words! Fluctuations are well known. The term "quantum fluctuations" carries a deeper meaning. It stands for a movement that can never be frozen out, however low the temperature. Such fluctuations are universal. In the hydrogen molecule both the separation of the two atoms and their relative momentum continually fluctuate. Fixity of both would violate the uncertainty principle.. In the frozen vacuum of quantum electrodynamics, the electric and magnetic fields both fluctuate. Were both of these dynamically conjugate field variables to vanish, the uncertainty principle would likewise fail. The same is true of quantum geometrodynamics. There, the conjugate variables are the "intrinsic curvature" of three-dimensional space and the "extrinsic curvature", telling how this space is bent relative to the geometry of any four-dimensional spacetime that might envelop it. For both dynamic quantities to be stilled would equally contradict Heisenberg's uncertainty relation. Thus all space at the quantum scale of distances is the seat of the liveliest geometrodynamics, as it is also everywhere the scene of the most violent small-scale fluctuations in the electromagnetic field.

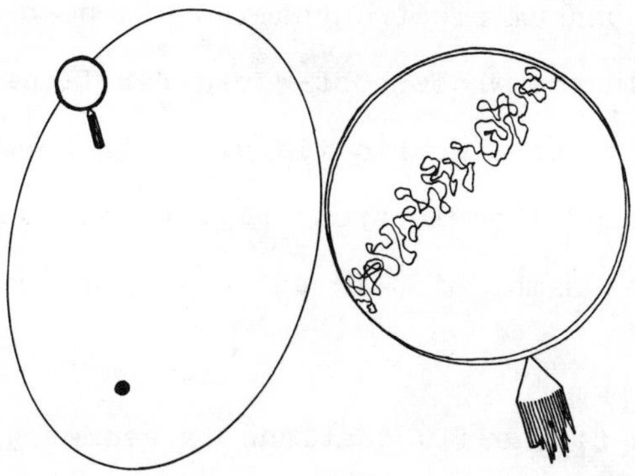

Figure 4. Symbolic representation of the motion of the electron in the hydrogen atom as affected by fluctuation in the electric field in the vacuum ("vacuum" or "ground state" or "zero point" fluctuations). The electric field associated with the fluctuation, $E_x(t) = \int E_x(\omega) e^{-i\omega t} d\omega$, brings about in the most elementary approximation the displacement $\Delta x = \int (e/m\omega^2) E_x(\omega) e^{-i\omega t} d\omega$. The average vanishes but the root mean square $<(\Delta x)^2>$ does not. In consequence the electron feels an effective atomic potential altered from the expected value V (x,y,z) by the amount

$$\Delta V\ (x,y,z) = 1/2\ <(\Delta x)^2> \nabla^2 V(x,y,z)\ .$$

The average of this perturbation over the unperturbed motion accounts for the major part of the observed Lamb-Retherford shift, $\Delta E = <\Delta V\ (x,y,z)>$, in the energy level. Conversely, the observation of the expected shift makes the reality of the vacuum fluctuations inescapably evident.

(Figure 4)

No prediction of quantum electrodynamics has been more impressively verified in the whole post-World War II era than these vacuum fluctuations in the electric field. Their perturbing influence on the motion of the electron (Fig. 4) accounts for the major component in the Lamb shift in the energy levels of the hydrogen atom. [19]

In putting numbers to the fluctuations in geometry, one is guided by the example of electromagnetism and, at a still earlier stage, by the example of the harmonic oscillator. For the oscillator the expression for the energy is

$$E = \left\langle \binom{\text{kinetic}}{\text{energy}} + \binom{\text{potential}}{\text{energy}} \right\rangle = p^2/2m + m\omega^2 x^2/2$$

$$= \int \psi^*(x) \left[-(\hbar^2/2m) d^2/dx^2 + m\omega^2 x^2/2 \right] \psi(x)dx. \qquad (17)$$

For a probability amplitude function $\psi(x)$ of Gaussian form and range a,

$$\psi(x) = \pi^{-1/4} a^{-1/2} \exp - (x^2/2a^2), \qquad (18)$$

the expectation value of the energy is

$$E = \tfrac{1}{4}\hbar^2/ma^2 + \tfrac{1}{4}m\omega^2 a^2. \qquad (19)$$

If quantum effects were absent, the first term would disappear. The minimum energy would be obtained by putting the oscillator at rest at the origin. However, in the real world of quantum physics, such a sharp localization in position ($a = 0$) would make the effective wavelength zero and make the momentum and kinetic energy arbitrarily large (divergent first term in Eq. 19). The minimum energy, $E = \hbar\omega/2$ ("half quantum" or "zero point energy" or "fluctuation energy") is obtained for a range $a = (\hbar/m\omega)^{1/2}$ ("range of zero point oscillations"). In other words, the oscillator in its ground state,

$$\psi(x) = (m\omega/\pi\hbar)^{1/4} \exp - (m\omega/2\hbar)x^2, \qquad (20)$$

can be said to "resonate" between locations in space ranging over a region of extent $\sim(\hbar/m\omega)^{1/2}$.

The electromagnetic field can be treated as an infinite collection of independent "field oscillators", with amplitudes $\xi_1, \xi_2, \ldots$. When the Maxwell field is in its state of lowest energy, the probability amplitude for the first oscillator to have amplitude ξ_1, and simultaneously the second oscillator to have amplitude ξ_2, the third ξ_3, etc., is the product of functions of the form (20), one for each oscillator. When the scale of amplitudes for each oscillator is suitably normalized, the resulting infinite product takes the form

$$\psi(\xi_1, \xi_2, \ldots) = N \exp - (\xi_1^2 + \xi_2^2 + \ldots). \qquad (21)$$

This expression gives the probability amplitude ψ for a configuration $\underset{\sim}{B}(x,y,z)$ of the magnetic field that is described by the Fourier coefficients $\xi_1, \xi_2, \ldots$. One can forego any mention of these Fourier coefficients if he so desires, however, and rewrite (21) directly in terms of the magnetic field configuration itself [20]:

$$\psi(\underset{\sim}{B}(x,y,z)) = \mathcal{N} \exp - \iint \frac{\underset{\sim}{B}(x_1) \cdot \underset{\sim}{B}(x_2)}{16\pi^3 \hbar c r_{12}^2} d^3x_1 d^3x_2 . \qquad (22)$$

No longer does one speak of "the" magnetic field; he talks instead of the probability of this, that or the other configuration of the magnetic field, and this even under circumstances, as here, where the electromagnetic field is in its ground state.

It is reasonable enough under these circumstances that the configuration of greatest probability is $\underset{\sim}{B}(x,y,z) = 0$. Consider for comparison a configuration where the magnetic field is again everywhere zero except in a region of dimension L. There let the field, subject as always to the condition div $\underset{\sim}{B} = 0$, be of the order of magnitude ΔB. The probability amplitude for this configuration will be reduced relative to the nil configuration by a factor exp(-I). Here the quantity I in the exponent is of the order $(\Delta B)^2 L^4/\hbar c$. Configurations for which I is large compared to one occur with negligible probability. Configurations for which I is small compared to one occur with practically the same probability as the nil configuration. In this sense, one can say that the fluctuations in the magnetic field in a region of extension L are of the order of magnitude

$$\Delta B \sim (\hbar c)^{1/2}/L^2. \qquad (23)$$

In other words, the field "resonates" between one configuration and another and another with the range of configurations of significance given by (23). Moreover, the smaller is the region of space under consideration, the larger is the field magnitude that occurs with appreciable probability.

Still another familiar way of speaking about electromagnetic field fluctuations gives additional insight relevant to geometrodynamics. One considers a measuring device responsive in comparable measure to the magnetic field at all points in a region of dimension L. One asks for the effect on this device of electromagnetic disturbances of various wavelengths. A disturbance of wavelength short compared to L will cause forces to act one way in some parts of the detector and nearly compensating forces in other parts of it. In contrast, a disturbance of a long wavelength λ produces forces everywhere in the same direction, but of a magnitude too low to have much effect. Thus the field, estimated from the equation

$$\begin{pmatrix}\text{energy of electromagnetic} \\ \text{wave of wavelength } \lambda \\ \text{in a domain of volume } \lambda^3\end{pmatrix} \sim \begin{pmatrix}\text{energy of one quantum} \\ \text{of wavelength } \lambda\end{pmatrix}$$

or

$$B^2 \lambda^3 \sim \hbar c/\lambda$$

or

$$B \sim (\hbar c)^{1/2}/\lambda^2 \qquad (24)$$

is very small if λ is large compared to the domain size L. The biggest effect is caused by a disturbance of wavelength λ comparable to L itself. This line of reasoning leads directly from (24) to the standard fluctuation formula (23).

Fluctuations Superposed on Classical Background

Nothing says that the electromagnetic field has to be in its ground state. It can be excited by a distant wireless antenna so that locally it is oscillating up and down at 10^6 cycles/sec in the range $B = \pm 3.3 \times 10^{-8}$ gauss, for example (accompanying electric fields 1 millivolt/meter). By comparison with this deterministic classical history of the local magnetic field,

$$B_z = 3.3 \times 10^{-8} \text{ gauss} \cos 10^6 t, \qquad (25)$$

the quantum mechanical fluctuation field of the same frequency (Eq. 24) is quite negligible:

$$B \sim 6 \times 10^{-9}\ (\text{erg cm})^{1/2} \ / \ (3 \times 10^{4} \text{cm})^{2}$$

$$\sim 10^{-17} \text{ gauss.} \qquad (26)$$

Even the total effect of all the independent quantum fluctuations in the magnetic field is small as sensed by a detector of dimension $L \sim 1$cm; thus, from Eq. 23,

$$\Delta B \sim 6 \times 10^{-9}\ (\text{erg cm})^{1/2} \ /(1 \text{ cm})^{2}$$

$$\sim 6 \times 10^{-9} \text{ gauss.} \qquad (27)$$

However, when attention is fixed on the field in a still smaller domain, say $L \sim 0.1$ cm or less, then the quantum fluctuations in the magnetic field dominate over the deterministic classical field of (25):

$$\Delta B > 6 \times 10^{-7} \text{ gauss.} \qquad (28)$$

So much for the coexistence of classical fields and quantum fluctuations in electrodynamics!

Similar considerations apply in geometrodynamics. [21] Quantum fluctuations in the geometry are superposed on and coexist with the large scale, slowly varying curvature predicted by classical deterministic general relativity. Thus, in a region of dimension L, where in a local Lorentz frame the normal values of the metric coefficients will be -1, 1, 1, 1, there will occur fluctuations in these coefficients of the order

$$\Delta g \sim L^*/L, \tag{29}$$

fluctuations in the first derivatives of the g_{ik}'s of the order

$$\Delta\Gamma \sim \Delta g/L \sim L^*/L^2, \tag{30}$$

and fluctuations in the curvature of space of the order

$$\Delta R \sim \Delta g/L^2 \sim L^*/L^3. \tag{31}$$

Here

$$L^* = (\hbar G/c^3)^{\frac{1}{2}} = 1.6 \times 10^{-33} \text{ cm} \tag{32}$$

is the so-called[22] Planck length. It is appropriate to look at orders of magnitude. The curvature of space within and near the earth, according to classical Einstein theory, is of the order

$$R \sim (G/c^2)\rho \sim (0.7 \times 10^{-28} \text{cm/g})\,(5\text{g/cm}^3)$$
$$\sim 4 \times 10^{-28} \text{ cm}^{-2}. \tag{33}$$

This quantity has a very direct physical significance. It measures the "tide producing component of the gravitational field" as sensed, for example, in a freely falling elevator or in a space ship in free orbit around the earth. [23] By comparison the quantum fluctuations in the curvature of space are only

$$\Delta R \sim 10^{-33} \text{ cm}^{-2}, \tag{34}$$

even in a domain of observation as small as 1 cm in extent. Thus the quantum fluctuations in the geometry of space are completely negligible under everyday circumstances.

Even in atomic and nuclear physics the fluctuations in the metric,

$$\Delta g \sim 10^{-33}\ \mathrm{cm}/10^{-8}\ \mathrm{cm} \sim 10^{-25}$$

and

$$\Delta g \sim 10^{-33}\ \mathrm{cm}/10^{-13}\ \mathrm{cm} \sim 10^{-20}, \tag{35}$$

are so small that it is completely in order to idealize the physics as taking place in a flat Lorentzian spacetime manifold.

The quantum fluctuations in the geometry are nevertheless inescapable if we are to believe the quantum principle and Einstein's theory. They coexist with the geometrodynamical development predicted by classical general relativity. The fluctuations widen the narrow swath cut through superspace by the classical history of the geometry. In other words, the geometry is not deterministic, even though it looks so at the everyday scale of observation. Instead, at a submicroscopic scale it "resonates" between one configuration and another and another. This terminology means no more and no less than the following: (1) Each configuration ${}^{(3)}\mathcal{G}$ has its own probability amplitude $\psi = \psi({}^{(3)}\mathcal{G})$. (2) These probability amplitudes have comparable magnitudes for a whole range of 3-geometries included within the limits (29) on either side of the classical swath through superspace. (3) This range of 3-geometries is far too variegated on the submicroscopic scale to fit into any one 4-geometry, or any one classical geometrodynamical history. (4) Only when one overlooks these small-scale fluctuations ($\sim 10^{-33}$ cm) and examines the larger scale features of the 3-geometries do they appear to fit into a single spacetime manifold, such as comports with the classical field equations.

Extrapolate Geometrodynamics to the Planck Scales of Distances?

Is it not preposterous to apply existing theory in a realm of dimensions smaller than nuclear sizes by twenty powers of ten? What a fantastic extrapolation! Yet it is the tradition of theoretical physics to adopt what might be called a "strong bargaining posture." It is not the custom to give up any long established principle without pushing it to the limit and finding out where, if anywhere, it goes wrong. A direct contradiction between a prediction and an observation, or between two points of principle, is ordinarily necessary if one is to have any solid ground for change — or even any indication where to make a change! The physicist does not have the habit of giving up something unless he gets something better in return.

To pursue systematically the consequences of quantum geometrodynamics is recommended not merely by the absence of any contradiction or by the absence of any more comprehensive theory. It is made attractive also by an example out of the past. Who in the 1850's, measuring the attraction between electric charges, and testing the Coulomb law at distances from meters to millimeters, could have predicted that it would be proved valid in 1911 to 10^{-12} cm, in 1933 to 10^{-13} cm, and in 1954 to 10^{-14} cm? The fantastic extrapolatory power of basic physical theory always seems a miracle! [24]

Electrodynamics contains no natural length. Neither do general relativity (G, c) or the quantum principle ($\hbar$) individually. The union of geometrodynamics and the quantum does: $L^* = (\hbar G/c^3)$. The importance of what is essentially this length was stressed by PLANCK as early as 1899.[25] He had taken up the study of blackbody radiation not least because it is universal: independent of the shape and size of the container, independent of the properties of its walls, and independent of the complexities of atomic, molecular and solid state physics. In keeping with this search for the universal, he asked for standards of length, mass and time which are independent of such special circumstances as the size of the planet we happen to inhabit, its period of rotation, and the density of the fluid which covers it. In excluding reference to special substances, he found it natural also to exclude reference to special particles: both the electron, with its then known mass and charge, and all heavier entities -- objects which even today pose unsolved structural problems. Excluding all else, he was left with the speed of light, the Newtonian constant of gravitation, and the constant newly discovered from the analysis of blackbody radiation as the quantities he was willing to accept as truly fundamental. Out of these three quantities there is but one way to construct a length. Planck's length, introduced to science before either special or general relativity, first acquired an understandable role in the context of quantum geometrodynamics, as measure of the fluctuations in the geometry of space.

Accept seriously a length as small as 10^{-33} cm? Try to assess in any detail at all the physics that goes on at a scale of distances shorter by twenty powers of ten than the 10^{-13} cm of elementary particle physics? What could be more preposterous? Only three numbers are still more preposterous than 10^{20}: the factor of 10^{40} that distinguishes electric forces from gravitational forces;

the 10^{40} from elementary particle dimensions to the estimated radius of the universe at the phase of maximum expansion; and the 10^{80} that furnishes an order of magnitude estimate as good as any that one knows for the number of particles in the universe. EDDINGTON [26], DIRAC [27], JORDAN [28], DICKE [29] and HAYAKAWA [30] argue that it is unreasonable to think of such enormous numbers as having independent roles in physics. The correspondence between these numbers cannot be purely accidental, they stress; [31] there could hardly be a "regularity of the large numbers" if there were not a deep connection between cosmology, general relativity and elementary particle physics. But where to begin in looking for this connection? Begin with asking why there are so and so many particles in the universe? Hardly. Physics can elucidate laws of motion, but it has proved powerless to explain initial conditions. Begin with asking why the universe has such and such dimensions? Again a matter of initial conditions, outside of the present scope of physics. Begin with trying to explain the charge structure of elementary particles, or the characteristic dimension of 10^{-13} cm? More hopeful, perhaps, but still beyond present power. No, of all the quantities coupled by the large numbers, one alone has a clear status within existing theory: the Planck length. Where else than here can one begin?

It could seem plausible to stop considering physics a little below 10^{-13} cm because accelerator budgets stop a little above $ 100 million a year. What good is it, one sometimes says, to analyze what goes on if one has no way to observe it? Happily experience has taught views less biased by mankind's temporary limitations. Wait until one had mastered the work hardening of metals before one took up the microscope and saw dislocations?

Wait until one had explained dislocations before one started the study of atoms? Not so. The route to understanding did not go down the ladder, 1cm → 10^{-4}cm → 10^{-8}cm, but up: 10^{-8}cm → 10^{-4}cm → 1cm. One had to understand something about atoms before he could explain dislocations, and something about dislocations before he could uncover the rationale of work hardening. Is it possible that one similarly must have some perspective on what happens at 10^{-33}cm before one can find the rationale of particles and 10^{-13}cm? [32] Right or wrong, [33] quantum geometrodynamics alone has any suggestions to offer on this point. What then does it say?

Every new perspective offered by this long established theory radiates out from the central prediction: *geometry fluctuates violently at small distances*. This concept opens new views on the nature of electric charge, on the nature of the vacuum, and on the nature of particles.

Electricity as Lines of Force Trapped in the Topology of Space

To arrive at a new vision of electricity it is enough to question an old view of topology: "Space is Euclidean in character at small distances". The view is reasonable enough for everyday purposes. Equally reasonable is the conception of the surface of the ocean as endowed with Euclidean topology -- reasonable to one flying miles above it. To one in a small boat, the opposite impression is inescapable. He sees the breaking waves and the foam. He knows that the surface is multiply connected at the scale of millimeters and centimeters. If the ocean is violent, the geometry of space on the Planck scale of distances is even more violent. Nowhere is there any region of calm. Moreover, if the equations of hydrodynamics are non-linear, so are the equations of geometrodynamics. What a contrast to the linearity of electrodynamics! There the predicted fluctuations in potential,

$$\Delta A \sim (\hbar c)^{1/2}/L, \qquad (36)$$

and in field

$$\Delta F \sim (\hbar c)^{1/2}/L^2, \qquad (37)$$

preserve always the same character, regardless of the smallness of the distances L to which one goes in his probing. There is no natural magnitude to mark off large fluctuations as different in nature from small ones. The contrary is the case for fluctuations in the metric, governed by the formula

$$\Delta g \sim L^*/L . \qquad (38)$$

Values of Δg comparable to unity and larger indicate changes in geometry so drastic that the word "curved space" is hardly adequate to describe them. "Changes in topology" seems a more reasonable description.

It is not so natural in mathematics as in physics to consider a transformation which alters one topology to another. An oscillating drop of water undergoes fission. The topology changes. A point marks the place of separation of the two masses of liquid. That point lacks the full neighborhood of points that characterizes a normal point. Such a critical point is ruled out from any proper manifold by the very definition of the term "manifold" in mathematics. Before the division, the surface of the drop constituted a manifold. After the division, it is again a manifold, consisting of two disparate pieces. At the instant of division, it is not a manifold. But little attention does the drop pay to this distinction. It divides, despite all definitions. No more reason does one see in the definition of "manifold" against *space* changing *its* topology.

No principle is at hand that would give one topology perpetual preference over all others. On the contrary, the field equations of relativity are purely local in character. They make no statements at all about global topology, as EINSTEIN himself emphasized more than once. Moreover, the whole character of physics speaks for the theme that "everything that can happen will happen". An alpha particle penetrates through a region classically forbidden to it; the side group on a chain molecule undergoes "hindered rotation"; and the umbrella structure of an ammonia molecule turns inside out despite the apparent contradiction to the law of conservation of energy. It is difficult to resist the conclusion that likewise the topology of space can change and does change.

If these general considerations are relevant, and if fluctuations alter the topology of space as well as its curvature, [34] the consequences are decisive for the nature of the physics that goes on at small distances, and even for the nature of superspace itself. Superspace has to be broadened from the totality of positive definite 3-geometries built on one topology to the totality of positive definite 3-geometries built on the totality of all topologies. It has new implications to say that the probability amplitude $\psi({}^{(3)}\mathcal{G})$ is appreciable for a swath of points in superspace, with a finite spread about the deterministic history of classical geometrodynamics (Fig. 5):

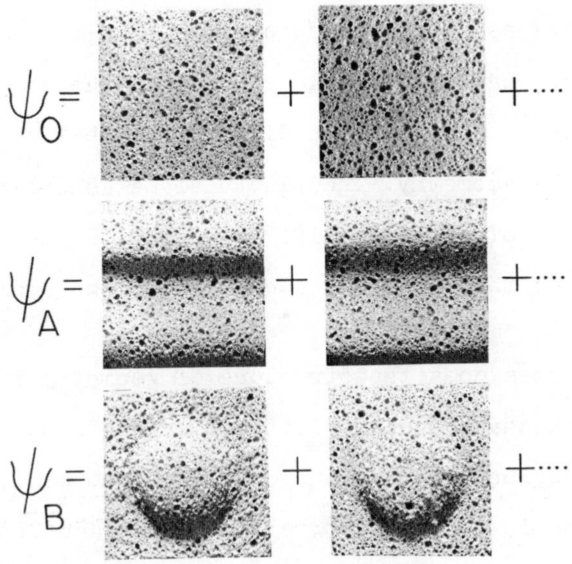

Figure 5. Symbolic representations of three alternative probability functions $\psi({}^{(3)}\mathcal{G})$. Above, ψ_0, normal fluctuations alone; middle, ψ_A, macroscopic classical gravitational wave plus superposed fluctuations; below, ψ_B, localized excitation plus superposed fluctuations. The complex number ψ gives the probability amplitude for the occurrence of the 3-geometry ${}^{(3)}\mathcal{G}$. Those 3-geometries which contribute most to the totalized probability are highly multiply connected at the scale of the Planck length ("foam-like structure of space in the small").

Geometry in the small fluctuates not only from one microscopic pattern of curvature to another but much more, from one microscopic topology to another. Moreover, those structures which are everywhere full of submicroscopic "handles" or "wormholes" are overwhelmingly more numerous than 3-geometries of simpler topology. In other words, space "resonates" between one foamlike structure and another. [35] The space of quantum geometrodynamics can be compared to a carpet of foam spread over a slowly undulating landscape. The undulations symbolize deterministic classical geometrodynamics. The continual microscopic changes in the carpet of foam as new bubbles appear and old ones disappear symbolize the quantum fluctuations in the geometry. The fluctuations change the microscopic connectivity of space itself. No longer is one entitled to take it for granted that space is Euclidean in the small.

Nowhere in physics does the structure of geometry in the small play a larger part than in electricity. Electric lines of force converge onto a region of space and none come out of it. Something strange must go on in that region; either Maxwell's equations break down, or the region is filled with a special substance, an electric jelly, a magic fluid beyond further explanation. From one picture or the other, there has never been an escape. The reason is simple. The region is tacitly assumed to have Euclidean topology. Give up this assumption,[36] but hold to Maxwell's field equations for empty space. Then the conclusion changes. The region in question must contain the mouth of at least one "handle" or "wormhole" (Fig. 6). Electric lines of force converge upon this mouth only to emerge from the other mouth, located somewhere else in space. One comes in this way to a new picture of electricity: *A classical geometrodynamical electric charge is a set of lines of force trapped in the topology of space.*

Someone can be imagined who first studies topology, next takes to heart Maxwell's equations for empty space, and then for the first time sees an electric charge. He takes it as experimental evidence that space must be multiply connected in the small (Table II). Nothing prevents us from adopting the same point of view. On this view, *the occurrence of electric charges in nature is*

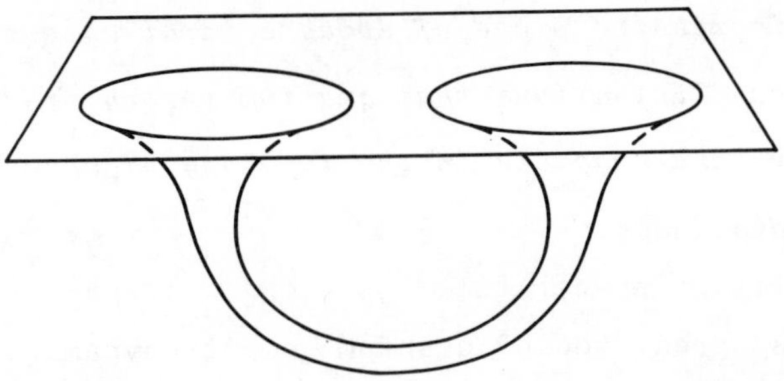

Figure 6. Classical geometrodynamical concept of charge as "electric lines of force trapped in the topology of space." The "wormhole" connects two regions in one otherwise nearly Euclidean space, not two different Euclidean spaces. The distance between the two mouths of the wormhole (1) via the nearly Euclidean space and (2) via the route through the wormhole are permitted by Einstein's field equations to be quite different, even in order of magnitude, contrary to what one might assume from the figure. There the geometry (one dimension suppressed!) is depicted for simplicity as if embedded in flat Euclidean three-space. The third dimension (distance "off" of the geometry) is to be considered as unattainable as the stratosphere is unattainable to an ant crawling on the surface of the earth. An observer endowed with an instrument of inadequate resolving power sees one wormhole mouth as a positive charge, the other as a negative charge. To be contrasted with this classical picture of single identifiable wormholes is the quantum mechanical picture (Fig. 4) of a submicroscopic foamlike wormhole structure constantly fluctuating throughout all space.

(Figure 6)

the single most impressive piece of evidence available today for the reality of the fluctuations that quantum theory predicts in the geometry of space, and suggests in the topology of space, at the Planck scale of distances.

The "wormholes" predicted by quantum geometrodynamics are a property of all space; are submicroscopic; and they and the fluxes through them arise spontaneously, through quantum fluctuations. Nothing prevents one from considering also a single wormhole, of macroscopic dimensions, created *ab initio*, with a prescribed flux threading through it, and evolving deterministically in time in accordance with the classical field equations. However this classical electric charge has not the slightest direct connection with the charges of the real world of quantum physics and requires no consideration here.

The Energy of the Vacuum

If one insight out of fluctuations in geometry has to do with the nature of electricity, a second has to do with the energy of the vacuum. Already in quantum electrodynamics one has long known from both observation and theory that when one examines a region of the vacuum of dimension L

(1) the fluctuation energy is found to be of the order $\hbar c/L$ and

(2) the effective density of energy of the fluctuation is found to be of the order $\hbar c/L^4$.

Not known out of electrodynamics is any natural lower limit for the L values which come into consideration. In quantum geometrodynamics, formulas of the same type hold, but there is now a natural cutoff:

the Planck length. It is unreasonable to apply linear theory when L is of the order of $L^* = (\hbar G/c^3)^{1/2} = 1.6 \times 10^{-33}$ or less. In other words, there is a certain sense in which one can say: (1) Elementary fluctuations measure in energy up to $\hbar c/L^*$. The magnitude of this characteristic mass-energy is $\sim 10^{-5}$g or $\sim 10^{28}$eV; that is, about twenty powers of ten greater than the mass of an elementary particle and nine powers of ten greater than the energy of the most energetic cosmic ray that has ever been found. (2) These fluctuations take place throughout all space. (3) The density of the electromagnetic energy associated with these fluctuations [37] is of the characteristic order of magnitude $\hbar c/L^4 = c^5/\hbar G^2 \sim 10^{95}$g/cm^3, stupendous in comparison with the 10^{14}g/cm^3 of nuclear matter. (4) The effective density of the "gravitational" or geometrodynamical wave energy [38] associated with these fluctuations is of the same order of magnitude.

Every observation shows that the net density of energy in space is negligible by comparison with these huge figures. The enormous positive energy must be compensated in some way. How the compensation comes about was a major concern of NIELS BOHR over the years. A new approach to the problem shows up when one considers the fluctuations at the Planck scale of distances. Two such fluctuations, each of mass-energy $\sim (\hbar c/G)^{1/2} \sim 10^{-5}$g, interacting gravitationally at the distance L^*, have a coupling energy which is negative,

$$E_{grav} = -Gm_1m_2/r_{12} \sim -G\,[(\hbar c/G)^{1/2}]^2/L^* \sim -\hbar c/L^* \qquad (39)$$

and of the order -10^{-5}g. This coupling between neighboring "fluctuons" thus has such a sign and such a magnitude as to be appropriate for compensating the energies of the individual

Table II. <u>The Concepts of Electric Charge in Quantum and Classical Geometrodynamics, Compared and Contrasted</u>

	Classical	Quantum
This type of electric charge interpretable as electric lines of force trapped in the topology of space?	Yes	Yes
Any "real" charge considered to be present?	No	No
Nature of topological trap?	Wormhole in 3-geometry	Wormhole in 3-geometry
Where are wormholes located?	Connecting opposite charges	Throughout all space
Dimension of wormhole?	Enormous compared to 10^{-33}cm	Comparable to 10^{-33}cm
Status of wormhole?	Classical deterministic development with time; macroscopic size; this size determined by initial conditions; no observational evidence that any	"Fluctuation"; consequence of fact geometry is not deterministic; that is, a consequence of the nonzero probability that exists for

	such macroscopic wormhole ever occurred, although by definition it is one of the solutions of Einstein's field equations	quite diverse 3-geometries; 3-geometries with wormholes almost everywhere (scale of 10^{-33}cm) are the most numerous ("foamlike 3-geometry"); wormholes do not have to be initiated; they occur naturally ("zero point disturbance of the vacuum") and cannot be avoided
Wormhole pinches off?	Yes. Throat of radius a undergoes gravitational collapse in time of order a/c (10^{-10}sec for a = 3cm)	Question strictly speaking is undefined; no meaning to deterministic small-scale geometrodynamics in context of quantum geometrodynamics; but in the loose way of speaking that is so useful in certain parts of quantum field

		theory one can say that "virtual wormholes" are continually being created and annihilated as "space resonates between one foamlike 3-geometry and another.
Charge as identified by flux through wormhole?	Does not change with time; a constant of the motion; not quantized	Flux through any one wormhole, like shape and dimensions of wormhole, subject to quantum mechanical fluctuations; order of magnitude of this "fluctuation charge", $(\hbar c)^{1/2} \sim 12e$; not "constant" and not "quantized")
Relation of charge on an elementary particle to this kind of geometrodynamical charge?	Not the slightest direct connection with classical geometrodynamical electric charge	Not the slightest direct connection with the charge on an elementary "fluctuation wormhole". Particle viewed as a quantum state of collective exci-

tation of the entire geometrical continuum ("geometrodynamical exciton"). Charge associated with this exciton to be calculated when one understands how to do the calculation!

fluctuons. It would seem surprising if this mechanism does not play a dominant part in bringing about compensation of vacuum energies.

Despite all the mysteries that enshroud the compensation process, one conclusion stands out clear: *Individually the components of the vacuum energy are enormous; and collectively they compensate.*

A Particle as a Geometrodynamical Exciton

Quantum fluctuations in geometry offer not only a new picture of electricity and a new view of the violence of the vacuum, but also a third vista: the concept of a particle as a quantum state of excitation of the geometry of space.

A bit of nuclear matter with its density of $\sim 10^{14}$g/cm^3 is completely unimportant compared to the calculated density of $\sim 10^{95}$g/cm^3 of the fluctuation energy of the vacuum. A particle means less to the physics of the vacuum than a cloud (10^{-6}g/cm^3) means to the physics of the sky (10^{-3}g/cm^3). No single fact points more powerfully than this to the conclusion that a "particle" is not the right starting point for the description of nature.

From the standpoint of geometrodynamics, the primordial entity is not one particle, nor an intercoupled family of particle fields, but the geometry of empty space itself. On this view, a particle is not itself a 10^{-33}cm fluctuation in the geometry; instead, it is a fantastically weak alteration in the pattern of these fluctuations, extending over a zone containing very many such 10^{-33} regions. In brief, a particle is a quantum state of excitation of the geometry; it is a *geometrodynamical exciton*. In mathematical terms, the vacuum is described by one probability amplitude $\psi_0 = \psi_0({}^{(3)}\mathcal{G})$; and states where one or more particles are present are described by other functionals $\psi = \psi({}^{(3)}\mathcal{G})$.

On this interpretation, elementary particle physics ranks as a new and beautiful kind of chemistry. First came the chemistry of atoms and molecules, marvelous in its complexities and also in its regularities, all built on one single simple dynamical entity, the electron. Then came "nuclear chemistry", with nuclear shapes and energies and reaction rates all going back for their explanation to the dynamics of another elementary dynamical entity, the nucleon. Today we deal, in effect, with a chemistry of the elementary particles themselves. Wonderful advances in the subject classify the particles into families, tie their masses together into mathematical regularities, and systematize their rates of transformation - without however revealing the identity of the elementary dynamical entity beneath it all. That entity, on the present view, is geometry itself.

When in the first half of the nineteenth century BERZELIUS proposed that chemical forces are a manifestation of electrical forces,[39] he excited investigations by many workers which eventually discredited his hypothesis. The homopolar bond: how could the observed affinity be reconciled with the known repulsion between like electric charges? Homopolar forces, ionic forces, Van der Waal's forces, valence forces; how could all this variety of magnitudes and particularities possibly be compatible with electrical forces, pure and simple? The tide against the electrical interpretation of chemical forces turned only with the discovery of the electron by J. J. THOMSON in 1897. The dynamical entity once identified, the unravelling of the mystery eventually had to follow. Still it was not easy for the imagination to grasp what organizing power the quantum principle possesses. In encounters in the mid 1920's more than one physicist told his colleague from the laboratory across the way, "Your chemistry is now passé. All that jumble can now be explained in terms of electrons and quantum numbers." In more than one case the then justified reply came back, "What makes you think your circular and elliptic orbits have anything to do with chemistry? Have you ever heard of the valence angles of ammonia or the tetrahedral bonds of carbon? Don't ever forget that electrical forces are electrical forces and chemical forces are chemical forces." The Coulomb law had to be supplemented by the concept of probability amplitude before HEITLER and LONDON could explain valence forces. Today no one doubts that the Schroedinger equation accounts in principle for all of chemistry. Yet no surer way could be found to stop the advance of chemistry than to require everyone to calculate the wave function of his new compound before making it. Not the contemplation of 600-dimensional configuration space, but the analysis of the regularities between molecule and molecule, proves to be the fruitful way to make progress. It can hardly be otherwise when the energy of binding is the very small difference between the very much larger total energies of the associated and the dissociated states.

In "elementary particle chemistry" the art of analyzing regularities is already highly advanced, [40] thanks not least to applications of group theory even more far reaching than the applications made in the chemistry of molecules. On the other hand, the possibility to derive particle masses from first principles would seem even more remote (from the standpoint of geometrodynamics) than the possibility to calculate the binding energy of a complex molecule from first principles. The energy with which one hopes to end up, 10^{-27}g to 10^{-24}g, is smaller by twenty powers of ten than the characteristic energy of the theory with which one starts. Still it would seem unwise to discount in advance the ingenuity of available methods to calculate, reliably, small effects against enormously larger backgrounds, as for example in the case of superconductivity [41] ($\sim 10^{-4}$eV versus $\sim$ 10eV).

How can the geometrodynamical interpretation of particles (Table III) be tested? Sooner to be expected than quantitative calculations are qualitative predictions and conceptual developments. Of such developments none gives more incentive than gravitational collapse(18) to believe in a tie between particles and geometry; and none gives more encouragement to believe in the relevance of the Planck length than the concept of charge as electric lines of force trapped in the topology of space.

Widening vistas open out for further investigation. [42] (1) What can one give in the way of simple principles to shortcircuit all the usual derivations of Einstein's field equations and pass in one leap from postulates to the Hamilton-Jacobi equation itself? (2) What deeper insights can one win into the structure of superspace? And (3) at what new point in geometrodynamics does one draw that old line between dynamic law and initial conditions which one sees throughout all of physics?

Table III. <u>Quantum-Geometrodynamical Interpretation of Particles and Forces</u>

Ultimate dynamical object:	Not an electron or any other kind of particle, but geometry itself.
Geometry of space:	Not unique and classical, but everywhere resonating at the scale of the Planck length between configurations of varied submicroscopic curvature and varied topology.
Topology of space:	Those geometries which occur with appreciable probability amplitude are highly multiply connected throughout all space ("foamlike structure").
Particle:	Not a foreign and physical entity moving about within the geometry of space, but a quantum state of excitation of that geometry itself; as unimportant for the physics of the vacuum as a cloud is unimportant for the physics of the sky. Not a localized ripple in the geometry, not a submicroscopic "wormhole" in the geometry of space at the Planck scale of distances, but an exciton-like change in the phase relations in the

probability amplitude for a very large number of such wormholes.

Charge:	Not a place where Maxwell's equations fail, not a mysterious "foreign and physical" jelly introduced into geometry from outside, but "lines of electric force trapped in the topology of space" (Table II and Problem 2).
Spin:	Not a dynamical object added to geometry but the non-classical two-valuedness associated with geometry itself -- because distinct probability amplitudes attach to a multiply connected 3-geometry endowed with alternative triad-fields or "spin structure" (Problem 2).
Force:	Strong forces, weak forces and intermediate forces no more distinct in their character than Van der Waal's forces, ionic forces and valence forces; not themselves primordial, but the residual effect of percentage-wise negligible changes in the enormous density of the energy of the zero point fluctuations taking place in the geometry of space at small distances.

These issues are the foothills. The mountain looms above them: Is a particle a state of excitation of the geometry of space?

EINSTEIN, above his work and writing, held a long term vision: There is nothing in the world except curved empty space. [43] Geometry bent one way here describes gravitation. Rippled another way somewhere else it manifests all the qualities of an electromagnetic wave. Excited at still another place, the magic material that is space shows itself as a particle. There is nothing that is foreign and "physical" immersed in space. Everything that is, is constructed out of geometry. This is the dream. Is this dream coming to life?

Note: This manuscript is the revised and extended version of reports made at the Academie Internationale de Philosophie des Sciences, Oberwolfach/Freiburg im Breisgau, July 1966; the International School of Nonlinear Mathematics and Physics, Munich, July 1966; the Society of Engineering Science, Raleigh, North Carolina, October 1966; the Colloque Internationale sur Fluides et Champ Gravitationel en Relativité Générale, Paris, June 1967; and the Battelle Rencontres in Mathematics and Physics, Seattle, July 1967. It appears in the proceedings of the second, third, and fifth organizations in English and in the proceedings of the fourth in French. Appreciation is expressed to the organizers of these conferences for their hospitality and to many colleagues for discussions and advice, among them especially Y. CHOQUET, BRYCE DeWITT and CECILE DeWITT, L. EHRENPREIS, U. GERLACH, H. LEUTWYLER, C.W. MISNER and M. STERN.

Problem 1. "Derivation" of "Einstein-Hamilton-Jacobi Equation"

If one did not know the Einstein-Hamilton-Jacobi Eq.12 (or 13), how might one hope to derive it straight off from plausible first principles, without ever going through the formulation of the Einstein field equations themselves? To find one such direct derivation of the EHJ equation would not seem rash to hope for when one already knows five ways to derive the field equations in their traditional form: (1) Einstein's original derivation based upon the correspondence with Newtonian gravitational theory; (2) Weyl's derivation based upon enumeration of all covariant differential operations on the metric tensor which are linear in the second derivatives and which contain no higher derivatives; (3) Hilbert's derivation from a variational principle; (4) Cartan's capture [44] of the geometrical content of the field equations [45]; and (5) the derivation by GUPTA, THIRRING and FEYNMAN starting with the theory of a field of spin two and mass zero in flat space.

The central starting point in the proposed derivation would necessarily seem to be "imbeddability". On the basis of experience one can reasonably ask that the desired Hamilton-Jacobi equation -- of course combined as always with the conditions of constructive interference -- should pick out ${}^{(3)}\mathcal{G}$'s that will fit together into a ${}^{(4)}\mathcal{G}$.

In what way would one violate the "condition of imbeddability" if, for example, one left the differential operator unchanged in (13) but replaced the term ${}^{(3)}R$ by the square or by some other function of this curvature scalar? Without directly answering this question, one can say that the special form of the equation is governed in the most direct possible way by the 4-dimensional character of spacetime. When geometrodynamics is put into simplest terms, it comes out as the statement

$$(\text{curvature}) = (\text{density of mass-energy}).$$

In the present considerations, one looks apart from situations where there is any "real" mass-energy present. To write the equation as if there were such a term on the right is, however, a reminder that one is speaking about a tensorial quantity dependent, therefore, not only upon one's choice of point, but also upon the choice of direction, or unit 4-vector, at that point. This circumstance illuminates the meaning of the curvature term on the left. [44, 45] It has to do with curvature of the 4-geometry in a tangent plane normal to the 4-vector in question. Good! But we are considering conditions, not merely at one point, but at a three-fold infinity of points that form a spacelike 3-geometry. This 3-geometry is not necessarily "free of extrinsic curvature" (vanishing "tensor, K_{ij}, of extrinsic curvature" or vanishing "second fundamental form") at the point under study. If it is, excellent! Then the desired curvature is given directly by the scalar curvature invariant, ${}^{(3)}R$, of the geometry intrinsic to ${}^{(3)}\mathcal{G}$ at the point in question. However, any non-vanishing "extrinsic curvature of the 3-geometry relative to the enveloping 4-geometry" makes an additional contribution to the scalar curvature intrinsic to ${}^{(3)}\mathcal{G}$. One has to correct for this contribution before one secures a proper measure,

$$(\operatorname{Tr} \underset{\sim}{K})^2 - \operatorname{Tr} \underset{\sim}{K}^2 + {}^{(3)}R$$

of the curvature of the 4-geometry itself in the tangent plane in question (Gauss-Codazzi formula). Why this special bilinear expression in the tensor K_{ij} of extrinsic curvature and not some other one? [46] The rationale shows most directly when one considers the special case where the 4-geometry itself is flat. In this case the "correction terms" in the expression for the desired curvature must exactly compensate ${}^{(3)}R$. Let the 3-geometry be imbedded in the 4-geometry with "principal radii of curvature" at the point in question equal to ρ_1, ρ_2 and ρ_3. Then the scalar curvature invariant of the intrinsic geometry is ${}^{(3)}R = -2/\rho_2\rho_3 - 2/\rho_3\rho_1 - 2/\rho_1\rho_2$. The difference in sign here compared to the familiar ${}^{(3)}R = 6/a^2$ for a 3-sphere of radius a arises from the fact that the "radius of curvature" is being measured in a 4-geometry of signature - + + +. The tensor of extrinsic curvature is

$$\underset{\sim}{K} = \begin{Vmatrix} 1/\rho_1 & 0 & 0 \\ 0 & 1/\rho_2 & 0 \\ 0 & 0 & 1/\rho_3 \end{Vmatrix} .$$

It is desired to build up out of this tensor an expression which (1) is bilinear in the reciprocals of the ρ_i,(2) contains no terms in $1/\rho_1{}^2$, etc., and (3) compensates ${}^{(3)}R$. These three requirements lead uniquely to the stated formula. The considerations outlined here by no means come close to providing a derivation of the Einstein-Hamilton-Jacobi equation, but they perhaps suggest some of the factors that may play a natural part in such a derivation.

Subscript on Relation of Hamilton-Jacobi Method to Conventional Analytic Solutions of Field Equations

To look towards the Hamilton-Jacobi equation as an illuminating way to found general relativity is not to look away from the field equations as a way to solve problems in general relativity. The Hamilton-Jacobi equation is divorced from the equations of motion no more in geometrodynamics than in particle mechanics. Even in the most elementary problem, the connection between the two methods is inescapable. Yes, one can solve Lagrange's equation

$$md^2x/dt^2 + \partial V/\partial x = 0,$$

by direct numerical integration. Yes, one can translate the problem into Hamilton-Jacobi formalism, write down the equation

$$-\partial S/\partial t = (1/2m)\,(\partial S/\partial x)^2 + V\,(x,t),$$

and integrate *it* by the most elementary numerical methods (replacement of (x,t) continuum by lattice space, $t = m\tau$, $x = n\delta$, with m and n integers; replacement of partial differential equation by difference equation). However, the most penetrating method to integrate the partial differential equation has long been known to go straight back to the Lagrange equation itself for its start. Not everywhere does one at first evaluate $S(x,t)$, but only along a classical worldline, or history, H,

$$x = x_H(t),$$

$$\dot{x} = dx_H(t)/dt;$$

thus

$$s(t) = S(x_H(t),t) = \int_{t_o}^{t} \left[\frac{1}{2}\, m\dot{x}_H^2 - V(x_H(t),t)\right] dt;$$

or, in a more general problem, with Lagrange function L,

$$s(t) = S(x_H(t),t) = \int_{t_o}^{t} L(\dot{x}_H, x_H, t)dt.$$

From a knowledge of S along the worldline one goes to a knowledge of S in a narrow band on either side of the worldline by using the relations

$$\begin{pmatrix}\text{rate of change}\\ \text{of action S}\\ \text{with position x}\end{pmatrix} = (\text{momentum}) = \partial L/\partial \dot{x}\ (\text{general}) = m\dot{x}_H(t)\ (\text{here})$$

and

$$-\begin{pmatrix}\text{rate of change of}\\ \text{action S with}\\ \text{time t}\end{pmatrix} = (\text{energy}) = \dot{x}\,\partial L/\partial \dot{x} - L\ (\text{general})$$

$$= \frac{1}{2} m\dot{x}_H^2 + V(x_H(t),t) \quad (\text{here}).$$

Thus at the point

$$x = x_H(t^*) + \delta x$$

$$t = t^* + \delta t,$$

a little way off the worldline the Hamilton-Jacobi function ("classical phase"; $\hbar$ times phase of quantum mechanical wave function) is

$$S(x,t) = s(t^*) + (\partial S/\partial x)\ \delta x + (\partial S/\partial t)\ \delta t.$$

An identical procedure gives $S_{new}(x,t)$, a new solution with infinitesimally different initial conditions, throughout a band extending on either side of an infinitesimally different worldline, $S_{new}(x,t)$. The "condition of constructive interference" between the new "wave" and the old "wave",

$$S(x,t) = S_{new}(x,t),$$

gives back at once the original solution of the equation of motion -- and all the history H that goes with that solution. So too in geometrodynamics!

One can use a known solution of the Einstein field equations to find the Hamilton-Jacobi function $S({}^{(3)}\mathcal{G})$ throughout a certain narrow swath through superspace. For this purpose one will find it easiest to start with one of the well known analytic solutions of Einstein's field equations. Let the 4-geometry be written in the form

$$ds^2 = g_{\alpha\beta}\ dx^\alpha dx^\beta$$

where the metric coefficients $g_{\alpha\beta}$ are certain known functions of the four coordinates x^μ. The 4-geometry expressed in this way summarizes the dynamical history H of the 3-geometry of space. It also defines a submanifold -- call it H -- of the superspace $\mathcal{S}$. How now to determine the values taken on by the Hamilton-Jacobi function throughout the submanifold? First define the equivalent of the starting time t_o, in the one particle problem. This equivalent is not a single number t_o, but in general a different number $t_o(x,y,z)$ for each point in 3-space; or better stated, it is a spacelike initial value hypersurface σ_o slicing through the given ${}^{(4)}\mathcal{G}$. Next give the equivalent of the point $x_H(t_o), t_o$ along the classical history. It is the "momentary state of the geometry of space" on the hypersurface σ_o; that is, it is the 3-geometry ${}^{(3)}\mathcal{G}_o$ defined by the metric

$$ds^2(\sigma_o) = \left[g_{oo}(\partial x^o/\partial x^m)(\partial x^o/\partial x^n) + 2g_{om}(\partial x^o/\partial x^n) + g_{mn}\right]dx^m dx^n.$$

Similarly with the equivalent of the running time t and the state $x_H(t)$ of the particle at this time. Thus, give a spacelike hypersurface σ by giving $t = t(x,y,z)$; and calculate the corresponding metric $ds^2(\sigma)$, which defines a 3-geometry ${}^{(3)}\mathcal{G}$ "on the classical history H." The value of the classical phase function S on this classical history is given by a four-fold integral. This integral is extended over the region of spacetime bounded by the two hypersurfaces. It has the form,

$$s(\sigma) = \int_{\sigma_0}^{\sigma} \mathcal{L}(x^\alpha) d^4x .$$

Here the Lagrange density $\mathcal{L}$ is given for example by ARNOWITT, DESER and MISNER (reference 2; see also GMD). Consider now a "point" $^{(3)}\mathcal{G}$ in superspace a little ways removed from some "point" $^{(3)}\mathcal{G}$ which lies on the classical history H. What is the value of S for this new 3-geometry? The difference between the two 3-geometries is expressed most conveniently for the present expository purpose in terms of the difference of the metric coefficients at corresponding points:

$$\delta g_{mn}(x,y,z) = g_{mn}(x,y,z;\sigma) - g_{mn}(x,y,z;\sigma^*).$$

This difference in 3-geometries has to be multiplied by the value of the conjugate geometrodynamical momentum $\pi^{mn}(x,y,z)$ at the point $^{(3)}\mathcal{G}^*$ on the classical history H (details of definition and calculation of π^{mn} given for example in GMD) and integrated to give the change in the Hamilton-Jacobi function. Thus, throughout a thin swath in superspace one can write

$$S(^{(3)}\mathcal{G}) = s(\sigma^*) + \int \pi^{mn} \delta g_{mn} d^3x .$$

Problem 2. Structure of Superspace

Spacetime is the arena of particle dynamics. Superspace is the arena of geometrodynamics. How different our knowledge of these two arenas! In particle dynamics, the structure of spacetime is taken to be given as from on high: an everywhere flat ideal Minkowski-Lorentz manifold. In geometrodynamics the structure of superspace is to be considered as defined entirely internally; that is to say, by the very form of the "Einstein-Schroedinger equation" itself. We write this equation symbolically as

$$-\nabla^2\psi / (\delta^{(3)} \mathcal{G})^2 + {}^{(3)}R\psi = 0.$$

To ask about the structure of this equation is therefore nothing more or less than to ask, what is the structure of superspace? It is conceivable that a given geometry may be more appropriately identified with a whole class of points in superspace than with a single point, when that 3-geometry has less than maximal symmetry. In this way, Professor STEPHEN SMALE kindly points out, one may keep from introducing "conical singularities" in superspace. To avoid such singularities would seem to be essential if superspace is to rise from a mere topological manifold to the status of a differential manifold.

Quantum theory once known, classical theory follows easily and uniquely by way of the correspondence principle,

$$\psi \simeq \begin{pmatrix}\text{slowly varying}\\ \text{amplitude factor}\end{pmatrix} \exp(iS/\hbar).$$

However, classical theory alone being known, it is ordinarily difficult or impossible uniquely to determine the form of the quantum wave equation. [47] No system better illustrates this point than a particle moving in a prescribed external electromagnetic field. To know the Hamilton-Jacobi equation alone

$$g_{\alpha\beta}(\partial S/\partial x^\alpha + eA_\alpha/c)(\partial S/\partial x^\beta + eA_\beta/c) + m^2c^2 = 0$$

is to have no way to decide between the Schroedinger-Klein-Gordon equation, the Dirac equation, or one or another wave equation of higher spin; they all reduce to the same Hamilton-

Jacobi equation in the appropriate semiclassical limit. Similarly in geometrodynamics. From the well-defined "Einstein-Hamilton-Jacobi" equation (12, 13)

$$(\underset{\sim}{\nabla} S/\delta^{(3)} \mathcal{G})^2 + {}^{(3)}R = 0,$$

no one knows a satisfying way to go to a unique "Einstein-Schroedinger" equation.[48]

In the example of the particle, most of the ambiguity about the form of the wave equation is resolved as soon as observation or other evidence reveals the spin[49]--or more directly--how many possible orientations there are for any spin degree of freedom over and above the three degrees of freedom that tell location in space. In other words, in addition to the Hamilton-Jacobi equation, one must know the "structure of configuration space" in order to end up with a well-defined wave equation. Hence our question here: What is the structure of superspace?

Unravel the structure of superspace? Hardly all in one jump, and hardly today! Instead, a step by step penetration of the issues is more to be anticipated if the history of electromagnetism or atomic structure or other branches of physics is any guide. At least three levels of analysis are to be perceived. First, one already knows the structure relevant to classical geometrodynamics: (a) Superspace has Hausdorf topology.(4) (b) Superspace is endowed with an indefinite metric, defined by the expression[50]

$$(1/2g)\ (g_{ik}g_{jl} + g_{il}g_{jk} - g_{ij}g_{kl})$$

that occurs in the Hamilton-Jacobi Eq. (12). Second, one searches for those features in the structure of superspace which come into play when space changes its topology. Third, one hopes in the longer term to uncover those deeper properties of superspace that give ordinary space its dimensionality, its metric structure, and its ability to propagate electromagnetic fields and neutrinos with the speed of light. It is appropriate to discuss the structure of superspace a little further at all three of these levels.

Level 1. Classical Geometrodynamics; Topology Does Not Change

In classical geometrodynamics, space does not change its topology. Space may be "preparing" to change its topology, but it cannot actually make the change within the context of classical theory. [51] It can at most signal its "intention" to change topology by developing somewhere a curvature that increases without limit ("gravitational collapse" [18]). To go further with the analysis of the collapse phenomenon and treat changes in topology forces one to go outside the framework of classical theory. So long as one stays inside classical theory, he must restrict his attention in one dynamical problem to one topology. Which topologies are acceptable?

Of all topologies none is more familiar than that associated in Einstein's theory with the geometry in and around a dilute center of attraction such as the sun. Curved at small distances, the geometry becomes asymptotically flat at large distances. The topology as distinguished from the geometry is Euclidean: E_3 or R x R x R. In a broader context, however, EINSTEIN thought of the space around one star as part of the space engulfing all stars, [52] and of this total space as closed and endowed with the topology of the 3-sphere, S_3. No one has stated more strongly than he the arguments for considering space to be closed. [53]

One acquires a new reason for considering space to be closed when he considers how hard it is to define "an open and asymptotically flat space" in the context of quantum geometrodynamics: Those $^{(3)}\mathcal{G}$'s which occur with overwhelming probability are everywhere endowed with all kinds of ripples and other geometrical structures at the scale of the Planck length. There is no direction that one can take and there is no distance that one can travel which will erase this structure. Under these circumstances it is difficult to attribute any well-defined meaning whatsoever to the term "asymptotically flat." On the other hand, one knows more than one example of a space which is open and *not* asymptotically flat, and which becomes wilder and wilder at great distances. [54] Not having any means to distinguish one open space from another as "good", one would seem justified at this stage in the development of the subject to exclude from attention all open spaces. This approach is the more attractive in that it appears possible [55] to distinguish between one type of closed 3-geometry and another by straightforward classificatory integers similar to the Betti numbers. In contrast, the concept of "asymptotically flat"--even in contexts where it is relevant--is much more complex to formulate. [56]

Out of a manifold with the topology E_3 it is possible to cut out a block with the shape of a cube and obtain the topology of the 3-torus, $S_1 \times S_1 \times S_1$. On a manifold with this topology the initial value problem of classical geometrodynamics presents difficulties in certain cases, according to unpublished considerations of BRILL and AVEZ. It is conceivable that these difficulties indicate that the topology $S_1XS_1XS_1$ is not acceptable. In that event one might almost say that this topology has inherited a "defective gene" from its parent topology, E_3.

Another "black gene" that one can perhaps reasonably exclude is non-orientability. We assume, in effect, that "transport" of a right handed glove shall never bring it back to its starting point left handed.

Compatible with these very tentative principles for selecting "acceptable topologies" are the 3-sphere (S_3); the 3-sphere with addition of one handle or "wormhole" ($S_2 \times S_1 = W_1$); and the 3-sphere with n wormholes (W_n). There may or may not be further acceptable topologies not included in this list. [55]

Fixed Topology Excludes GMD Account of Particles and Fields

Taking the most familiar of these "acceptable topologies", S_3, as case example, one can discuss and treat quantitatively at the classical level a rich variety of physical processes, including gravitational radiation; [38] gravitational geons; [20] the planetary motion, collisions and breakup of geons; [20] the Taub model of an expanding and recontracting universe; [57] and more complex model universes. [38] Even so, one is limited in the physics that he can include in any of these models. One has no place for particles. Nor can one treat classically the final stages of gravitational collapse. These limitations seem unrelated to each other. In classical theory they are unrelated. In that view, the geometry of space remains forever tied to a unique topology. Not so in the purely quantum geometrodynamical model of physics. There (1) a particle is pictured in terms of space resonating from one topology to another; [58] and there (2) the final stages of gravitational collapse are viewed as a coupling of macroscopic motion and microscopic topology ("waterfall and foam"). In excluding from consideration all changes in topology, classical theory, on these views, also excludes any account of the rationale of particles. Particles have to be introduced as foreign and physical entities and space has to be viewed as arena, not as structural material. In this limited conceptual framework, the particles and other fields are counted as having degrees of freedom over and above those of the geometry. The electromagnetic field, in particular, is viewed as an entity additional to geometry.

Attempts have been made to regard electromagnetism as an aspect of geometry. This enterprise limits itself to that classical framework of ideas where one looks apart from submicroscopic fluctuations in the geometry and the topology of space. One attempt has had some minor success. The second order equations of MAXWELL and the second order equations of EINSTEIN have been combined into one set of equations of the fourth order that make no reference to any but geometric magnitudes. [59] The Maxwell field is recognized by the "footprints" it leaves on the geometry of space. In a certain sense, those "footprints" *are* the electromagnetic field; hence the name, "already unified theory" of gravitation and electromagnetism. Faithfully though this theory reproduces in other respects the dynamic content of the equations of MAXWELL and EINSTEIN, it turns out not to be adapted to treat the initial value problem. Geometrical measurements alone on an initial spacelike hypersurface do not always suffice completely to determine the future time evolution of the geometry.[60] In other words, one cannot uphold in "already unified field theory" a strict division of dynamics into "equations of motion" and "initial value data for these equations." Yet on that division one had learned to insist, not least by reason of hard-won lessons from the Hamilton-Jacobi theory and from quantum theory. Therefore the electromagnetic field, like a particle, can hardly be treated as anything but as a foreign or "physical" entity immersed in space so long as one looks away from the microscopic geometry of space.

Formalism of Field When Treated as "Foreign and Physical"

In this non-geometric or "arena" approach to physics one considers the degrees of freedom of the "foreign and physical" entity as additional to those of the geometry. In the most elementary example, the case where only one such entity is contemplated, the pure source-free electromagnetic field, one writes the Hamilton-Jacobi function or the Schroedinger probability amplitude, as the case may be, in the form

$$S = S(g_{ik}(x,y,z), A_m(x,y,z))$$

or

$$\psi = \psi(g_{ik}, A_m);$$

similarly when more fields are involved. Even at this level of analysis the geometrical approach has a contribution to make. The Hamilton-Jacobi function, ostensibly dependent upon the individual components of the metric and of the electromagnetic vector potential, actually has to be understood to depend only on the 3-geometry ${}^{(3)}\mathcal{G}$ and on the 2-form $\underset{\sim}{B} = \underset{\sim\sim}{dA}$. Consequently, it cannot change S to "push the rubber sheet on which the coordinates are painted" over space, in such a way that the point P, which formerly had the coordinates x^i, now acquires the coordinates $x^i - \xi^i$. Similarly the point P + dP, which formerly had the coordinates $x^i + dx^i$, now acquires the coordinates

$$x^i + dx^i - \xi^i - (\partial\xi^i/\partial x^j)dx^j.$$

The separation between the two points of course remains unchanged by this alteration in coordinates:

$$\begin{aligned} ds^2 &= g_{ij}(x^s)dx^i dx^j \\ &= g_{ij}^{new}(x^s - \xi^s)\,(dx^i - \xi^i{}_{,m}dx^m)\,(dx^j - \xi^j{}_{,n}dx^n). \end{aligned}$$

To terms of the first order in the displacement ξ^s the alteration in the metric coefficients is

$$\delta g_{ij} = g_{ij}^{new} - g_{ij} = (\partial g_{ij}/\partial x^s)\xi^s + \xi_{i,j} + \xi_{j,i} = \xi_{i|j} + \xi_{j|i},$$

where the subscript $|j$ denotes covariant differentiation, in the space of the 3-geometry, with respect to the jth coordinate. In a similar way one finds the alteration in the electromagnetic vector potential,

$$\delta A_i = A_i^{new} - A_i = A_{i,s}\xi^s + A_j\xi^j{}_{,i}.$$

These changes in the metric coefficients, taking place throughout all space, ostensibly produce a change in the Hamilton-Jacobi function determined by the functional derivatives of S; thus,

$$\delta S = \int \left[(\delta S/\delta g_{ij})\delta g_{ij} + (\delta S/\delta A_i)\delta A_i \right] d^3x.$$

However, the mere shift in coordinates cannot produce any real physical change. Consequently the quantity δS must vanish, independent of the choice of the coordinate displacements ξ^i. The consequences of this coordinate invariance are easily traced out. The functional derivatives that come into consideration have well-defined physical interpretations. Thus

$$\pi^{ij} = \delta S/\delta g_{ij}$$

is the geometrodynamical momentum (2,3) conjugate to the geometrodynamical field coordinate g_{ij}. It is closely connected with the "extrinsic curvature" K^{ij} of the 3-geometry with respect to the yet-to-be-constructed 4-manifold. The other derivative,

$$\mathcal{E}^i = \delta S/\delta A_i,$$

the electrodynamic momentum conjugate to the vector potential A_i, is nothing other than the electric field. This quantity satisfies the divergence condition [61]

$$\mathcal{E}^i{}_{,i} = 0.$$

The condition that S be invariant with respect to coordinate changes takes the form

$$0 = \delta S = \int \left[\pi^{ij}(\xi_{i|j} + \xi_{j|i}) + \mathcal{E}^i \, (A_{i,j}\xi^j + A_j\xi^j{}_{,i}) \right] d^3x \, .$$

Integrating by parts, readjusting the positions of indices as appropriate, and making use of the divergence relation, one finds

$$0 = \int \left[-2\pi_{js}{}^{|s} + \mathcal{E}^i (A_{i,j} - A_{j,i}) \right] \xi^j d^3x.$$

This must vanish for arbitrary choices of the field of coordinate displacements ξ^j. Consequently the quantity in square brackets must vanish. One finds in this way those three of Einstein's field equations which link the curvature of space with the Poynting density of flow of electromagnetic field energy: [62]

$$2\pi_{js}{}^{|s} = (\underset{\sim}{\mathcal{E}} \times \underset{\sim}{B})_j .$$

How remarkable that one gets so much--including the Poynting vector itself--from the elementary condition that S should depend, not upon the components g_{ij} and A_i individually, but only upon the coordinate-independent geometrical quantities that are "dressed up" in these components! [63]

Express electrodynamics in geometric language, yes. See the footprints of the electromagnetic field on the geometry of space, yes. Conceive of classical geometrodynamic electric charge as lines of force trapped in the topology of space, yes. But explain geometrically the "necessity" for electromagnetism and other "physical" entities, no. Not within the context of classical geometrodynamics and its never-changing topology.

Level 2. Space Resonating Between 3-Geometries of Varied Topology

A new world opens out for analysis in quantum geometrodynamics. The new central concept is space resonating between one foamlike structure and another. For this multiple connectedness of space at submicroscopic distances no single feature of nature speaks more powerfully than electric charge. Yet at least as impressive as charge is the prevalence of spin 1/2 throughout the world of elementary particle physics. "It is impossible to accept any description of elementary particles that does not have a place for spin 1/2." This quotation from the book *Geometrodynamics* [64] of 1962 goes on to say, "What then has any purely geometrical description to offer in explanation of spin 1/2 in general? More particularly and more importantly, what possible place is there in quantum geometrodynamics for the neutrino--the only entity of half integral spin which is a pure field in its own right, in the sense that it has zero rest mass and moves with the speed of light? No clear or satisfactory answer is known to this question today. Unless and until an answer is forthcoming, *pure quantum geometrodynamics must be judged deficient as a basis for elementary particle physics*." Happily the concept of spin manifold [65] has subsequently come to light, not least through the work of JOHN MILNOR. This concept suggests a new and interesting *interpretation of a spinor field* within the context of the resonating microtopology of quantum geometrodynamics, *as the non-classical two-valuedness that attaches to the probability amplitude for otherwise identical 3-geometries endowed with alternative "spin structures."*

The Orientation Entanglement Relation or "Version"

Spin and topology: what is the connection? Take a cube (Fig.7). To its upper northeast corner attach one end of a long elastic string. Run the other end to the upper northeast corner of the room and fasten it there. With seven other elastic strings attach the other corners of the cube to the corresponding corners of the room. Now select any axis running through the center of the cube and rotate the figure about that axis through 360°. The cube resumes its original configuration. Not so the strings. They are in a tangle. Moreover, rejecting cuts, one has no way to untangle the strings. Consequently it needs more than orientation to tell the relation between the cube and its surroundings. The necessary bookkeeping is provided by a spinor,

$$s = \begin{pmatrix} \xi \\ \eta \end{pmatrix}.$$

Under a rotation through the angle θ about an axis making angles α, β, γ with the x, y, and z axes this spinor is transformed to the new spinor $s' = \underset{\sim}{q}\underset{\sim}{s}$. Here the quaternion or "versor" $\underset{\sim}{q}$ has a value indicated in various ways in various well known systems of nomenclature:

$$\underset{\sim}{q} = \cos \theta/2 + \sin \theta/2 \, (\underset{\sim}{i} \cos\alpha + \underset{\sim}{j} \cos\beta + \underset{\sim}{k} \cos\gamma)$$

$$\text{(with } \underset{\sim}{i}\underset{\sim}{j} = -\underset{\sim}{j}\underset{\sim}{i} = \underset{\sim}{k}\text{, etc.)}$$

$$= \cos \theta/2 - i \sin \theta/2 \, (\sigma_x \cos\alpha + \sigma_y \cos\beta + \sigma_z \cos\gamma)$$

$$\text{(with } i = (-1)^{1/2} \text{ and } \sigma_x\sigma_y = -\sigma_y\sigma_x = i\sigma_z\text{, etc.)}$$

$$= \left\| \begin{matrix} [\cos(\theta/2) + i \sin(\theta/2) \cos\gamma] & \sin(\theta/2) [\cos\beta - i \cos\alpha] \\ \sin(\theta/2) [-\cos\beta - i \cos\alpha] & [\cos(\theta/2) - i \sin(\theta/2) \cos\gamma] \end{matrix} \right\|$$

The important point is the change of sign under a 360° rotation: $\underset{\sim}{q}(360^{\circ}) \cdot s = -s$. The 360° rotation alters what one may most appropriately call the "orientation entanglement relation" between the cube and its surroundings - or, more briefly, the "version" of the cube. The spinor keeps account of this "orientation entanglement relation." Two successive rotations by 360° restore the cube to its original "orientation entanglement relation" with its surroundings. The strings at first appear to be tangled up with twice the twist they had before. Nevertheless, they can now be untangled completely, as one confirms by direct trial or by elementary reasoning.[66,67]

Arbitrarily pick out one way of placing the cube, call it the "standard orientation entanglement relation" between the cube and its surroundings, or the standard "version" of the cube, and associate with it the spinor $\binom{0}{1}$. Proceed similarly at other points of space, taking care only that any changes in the standard version from point to point shall take place smoothly.

A special situation develops when an orientable 3-geometry [68] is endowed with a handle or wormhole. The cube can be transported in imagination from A to B "through the surrounding nearby space" or "through the wormhole." With one cube possessing a certain pattern of colored faces follow one route and with another identically colored cube follow the other route. It makes physical sense

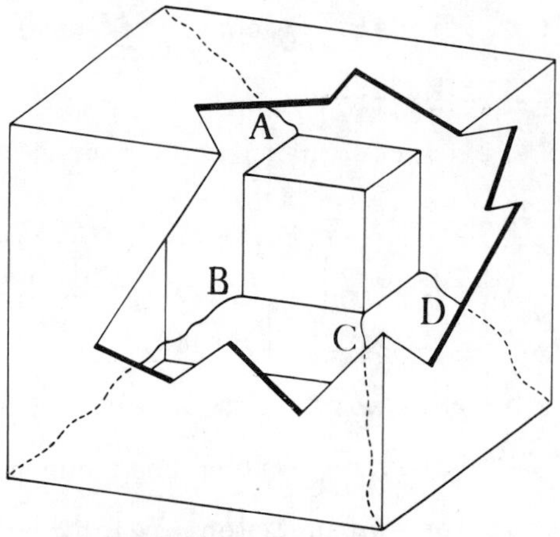

Figure 7. The elastic strings attached to the central cube keep account of its "orientation entanglement relation" with its surroundings.

to ask if the two cubes have at B identical orientation entanglement relations to their surroundings. [67] If they do not, alter the definition of the standard orientation within the wormhole by a rotation. Let this rotation increase continuously from zero at the mouth of the handle near A to 360° at the mouth near B. At each point outside the wormhole a triad of axes, a, b, c, defines the direction of the axes of the cube when the cube is located at that point in its standard orientation. These triads are not affected by the alterations made inside the wormhole. They vary in direction as smoothly as ever from point to point. At the two mouths of the wormhole they join on as smoothly to the new field of triads inside as they did to the old field of triads. Now, however, one has at last achieved an everywhere continuous field of standard orientation entanglement relations. The associated spinor field is likewise continuous, having everywhere the standard value $\begin{pmatrix} 0 \\ 1 \end{pmatrix}$, whereas before it underwent somewhere a discontinuous change from this value to its negative. One can restate in mathematical language what has been accomplished. First, one has laid down a "spin structure" upon the manifold. This structure is defined by the field of triads, or by "the class of such fields that are equivalent under homotopy." Second, one has laid down a "spinor field" upon the manifold. This spinor field is defined with respect to the given "spin structure." Other spinor fields can of course be laid down, with components also varying continuously from place to place.

Alternative Spin Structures in a Multiply Connected Space

Start again with another closed orientable 3-manifold endowed with the original topology and metric. One might hope to obtain an acceptable spin structure on this new manifold by taking over to it the identical field of triads which serves for the original manifold. It may be that this supposition is correct about the particular new manifold that one happens to have picked out. In this event identical colored cubes "taken from A to B by inequivalent routes" will preserve their orientation entanglement relation, one to the other. However, the supposition can equally well be incorrect. If so, the consequences are direct. The two cubes, carried by different routes from A to B, always in alignment with the canonical field of triads, will end up at B inequivalent to the extent of a 360° rotation. Moreover, this inequivalence in the orientation entanglement relation with the surroundings can be detected in principle by direct physical measurement.[67] In other words, the "spin structure" of the original manifold does not apply to the new manifold. To serve in the new manifold, the field of triads has to be modified to the extent of a 360° rotation within the wormhole or by an equivalent change. The difference between the new manifold and the original manifold expresses itself in these terms: the two manifolds have the same topology and metric, but they have inequivalent "spin structures." *The difference between the two manifolds is not only mathematical. It is physical.*

The Multisheeted Character of Superspace

One does not classify the closed orientable 3-manifold of physics completely when he gives its topology, its differential structure and its metric. He must tell in addition which spin structure it has. The spin structure, like the metric, lends itself in principle to observation. Consequently, it is not enough for the purposes of quantum geometrodynamics to give the probability amplitude for a "3-geometry" as the term 3-geometry was previously understood. One must introduce a new two-valued descriptor, w_k, for each (k = 1,2,...n) of the n wormholes of the manifold, to distinguish the two inequivalent ways to lay down a spin structure in the interior of that wormhole. The new, enlarged concept of a 3-geometry, $^{(3)}\mathcal{G}$, adjoins these descriptors to the continuous infinity of parameters which alone served previously to distinguish one 3-geometry, $^{(3)}\mathcal{G}^{\text{old}}$, from another; thus,

$$^{(3)}\mathcal{G} = (^{(3)}\mathcal{G}^{\text{old}};\ w_1, w_2, \ldots, w_n)$$

and

$$\psi(^{(3)}\mathcal{G}) = \psi(^{(3)}\mathcal{G}^{\text{old}};\ w_1, w_2, \ldots, w_n).$$

In other words, *superspace acquires a multisheeted character* (Fig. 8) with 2^n distinct sheets in that region of superspace where the 3-geometry is endowed with n wormholes.

For the two values to be assigned to the descriptor w_n it is natural to pick +1 and -1. However, there is nothing anomalous about the one "spin structure" as compared to the other. No canonical way has ever been proposed to give preference to one as compared to the other. Therefore it is a matter of arbitrary choice to which "spin structure" to assign the descriptor +1; and to which, -1.

The word "spin structure"can mislead. It suggests that there is something special about "laying down a spinor field" upon the mani-

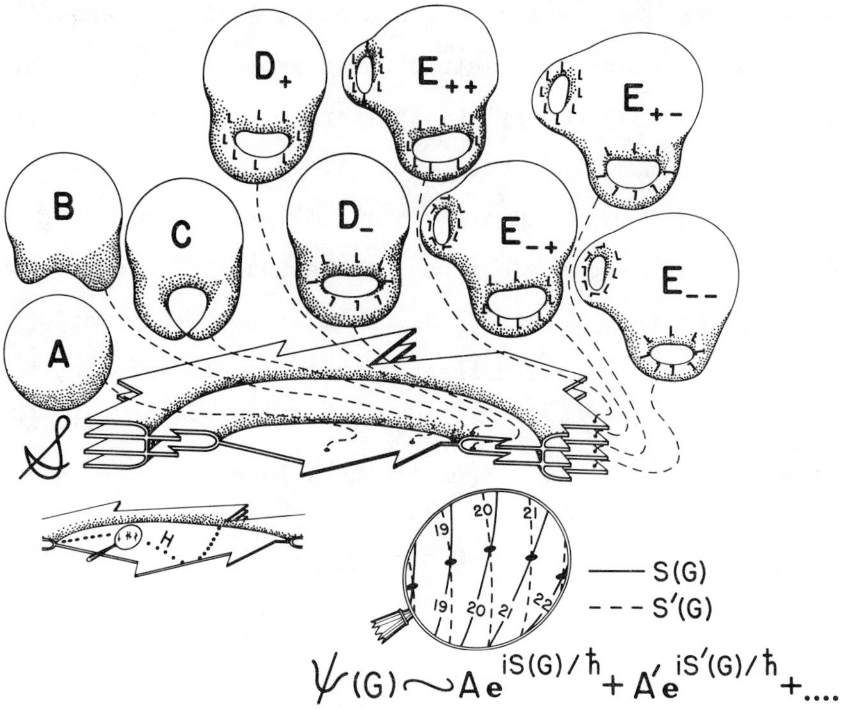

Figure 8. The multisheeted character of superspace.

fold. It conjures up visions of laying down upon the 3-geometry other kinds of fields which transform according to groups other than the spinor group SU (2), for example SU (n) or SL (n). However, the relevant point in the whole analysis is not the field of spinors, but the field of triads and their orientation-entanglement relations. There is nothing in the concept of "spin structure" that one could not have said with less chance of misunderstanding in the phrase "triad structure". Moreover, there is not the slightest indication that there is any other structure of a closed orientable 3-manifold that remains to be brought to light. Consequently, we take ${}^{(3)}\mathcal{G}$ in its new and enlarged sense to be the full indicator of the configuration of space, and as containing the full set of variables upon which ψ depends. In other words, we take the new ${}^{(3)}\mathcal{G}$ to comprise a set of commuting observables, complete in the sense of quantum mechanics, and therefore suitable for analysis of the probability amplitude ψ.

We do not add a spinor field to geometry. Quite the contrary. We take a spinor field away from geometry. A 3-geometry, augmented by a "spin structure" (as might, for example, be indicated by the descriptor $(w_1, w_2, \ldots, w_5) = (+1,+1,+1,-1,+1)$) is a possible habitation for a spinor field--but we have thrown out the inhabitant. [69]

Spin 1/2, if it occurs naturally in the context of quantum geometrodynamics, can hardly occur in any other sense than the sense in which PAULI spoke of spin from the very beginning, as a "non-classical two-valuedness." Moreover, a non-classical two-valuedness is already inescapable in the formalism. There are separate probability amplitudes for a 3-geometry with descriptor $w_k = +1$ and for an otherwise identical 3-geometry with descriptor $w_k = -1$. Does this circumstance imply that quantum geometrodynamics supplies all the machinery one needs to describe fields of spin 1/2 in general and the neutrino field in particular? That is the proposal. That is the only way that has ever turned up within the framework of Einstein's general relativity and Planck's quantum principle to account for spin. Is this the right path? It is difficult to name any question more decisive than this in one's assessment of "everything as geometry."

Electromagnetism as a Statistical Aspect of Geometry? Other Questions

It would be tempting to stop with this major issue if it did not open the door to so many tributary questions. (1) When a new handle develops and the number of descriptors rises by one, what boundary condition in superspace connects the probability amplitude, ψ , for 3-geometries of the original topology with the probability amplitudes, ψ_+ and ψ_-, for the two "spin structures" of the new topology? (2) One may wish to describe for each newly developing wormhole something like "the fractional amplitude going from ψ into ψ_+." However, with the fantastic number of wormholes that typically come into consideration ($10^{99}/\mathrm{cm}^3$) any such individual bookkeeping would seem for many purposes to be out of the question. It is equally impractical to keep account of the orientation of each of the 10^{23} spins in a ferromagnet. It is much more appropriate to speak of the "density of magnetization" and of perturbations in that density carried by "magnons." [71] What are the analogous quantities and concepts in geometrodynamics? What statistical approach is best suited to keep account of the many "matching ratios" ψ_+/ψ, associated with all the nascent wormholes in the geometry? (3) One speaks of "magnetization" knowing well that the term "magnetization" has not the slightest real meaning at subatomic distances. Is the term "electromagnetic field" equally without submicroscopic physical significance? In other words, among the statistical parameters most appropriate for keeping account of the 10^{99} "matching ratios" per cm^3, *is there one set of statistical parameters that one can identify with the electromagnetic field?*

The Example of 2-Geometries

No one can ask about the physics of changes in the topology of 3-space without at least a look at the mathematics of changes in the topology of 2-space. The superspace built on all 3-geometries is like no mathematical object so much as the superspace "built on all 2-geometries." No one did so much to bring this mathematical object to light as RIEMANN, the same BERNHARD RIEMANN who taught that the curvature of space is a branch of physics, and who provided the mathematical machinery to describe not only curvature (the Riemann curvature tensor, $R_{\alpha\beta\gamma\delta}$) but also topology (the Betti numbers, R_n).

In his 1857 paper RIEMANN noted that all algebraic 2-geometries endowed with the topology of a 2-sphere are equivalent to each other under conformal transformation (multiplication of all three metric coefficients by a common position dependent factor λ). In other words, the equivalence class of conformally equivalent 2-geometries of the given topology (S_2; or "genus g = 0") consists of a single object. It therefore constitutes a single "point" in what is not really a superspace itself, as we have been using that term, but a "reduced superspace": reduced in the sense that the "$(\infty^2)^\infty$" degrees of freedom in λ have here been "strained out" of superspace.

Two-geometries with the topology of the torus (T_2; or one wormhole, W_1; or "genus g = 1"), RIEMANN showed, are not all equivalent to one another under conformal transformation. Instead, when conformally equivalent 2-geomtries of the topology T_2 are identified, the family of objects that results is a complex continuum of dimension 1 (two real dimensions). [72] Two-geometries endowed with a larger number of wormholes (topology W_g; genus $g \geqq 2$), after extraction of the conformal degrees of freedom, reduce to a family of objects described by 3g-3 complex parameters (6g-6 real parameters). "Reduced superspace," built on the totality of conformally equivalent closed orientable 2-geometries of all topologies, thus appears to consist of a series of disjoint spaces, the first of dimension 0, the second of dimension 1, the next of dimension 3, the next of dimension 6, and so on. However, great developments in the analysis have taken place since the days of RIEMANN, through the efforts of many investigators.[73] Today, thanks not least to the works of LIPMAN BERS, one knows how to define one single infinite dimensional reduced superspace in which all these apparently disparate parts fit smoothly together.[74] What a model for the mathematical treatment of the superspace of general relativity! Yet, at the risk of seeming overdemanding, one has to ask for more. The superspace of physics is not to have any "conformal factor strained out of it"; rather, it is if anything to be enlarged, so as to include all the descriptors w_k of the "spin structure." How fit all the pieces of *this* superspace smoothly together? Challenging problem, at the very heart of quantum geometrodynamics!

Other Aspects of Superspace

Why all this emphasis on the structure of superspace? Why not simply spell out explicitly the form of the "Einstein-Schroedinger equation"? For the more elementary problem of a particle moving in flat 3-space it was straightforward for SCHROEDINGER to derive his wave equation. Simple considerations of invariance with respect to translation and rotation show that ∇^2 is the only simple differential operator that can come into play. That clear, the principle of correspondence with classical physics gives all the rest. One can hope that equally compelling considerations will fix the detailed mathematical form of the expression that we write down so far only symbolically,

$$\nabla^2\psi/(\delta^{(3)}\mathcal{G})^2.$$

However, a precise formulation of such considerations would seem to be out of reach until one has a knowledge of the transformations of superspace comparable to one's knowledge of the transformations of 3-space. Hence the emphasis on the structure of superspace.

A problem of such depth can hardly be examined from too many points of view. Six more aspects of superspace seem worthy of mention. One can summarize them under the names, "metric," "residual causality," "initial value," "conjugate momentum," "collapse," and "pregeometry." (1) There could hardly be a more helpful guide to the structure of superspace than the metric which obtains in it,

$$(1/2g)\ (g_{ik}g_{jl} + g_{il}g_{jk} - g_{ij}g_{kl}).$$

Reference is made to B. DeWITT[2] for the most illuminating discussion of this metric given to date. (2) This metric has a "light cone" associated with it. This "light cone" makes propagation proceed anisotropically in superspace. This anisotropy differs in character, however, from place to place according as $^{(3)}R$ is positive or negative. This anisotropy imposes a kind of "residual causality" upon superspace. CHARLES W. MISNER has pointed out in a conversation that one cannot forget this residual causality when one says that the customary ideas of "before" and "after" lose their meaning at the scale of the Planck length. Perhaps one can say more: If the principle of causality has been of service in analyzing the structure of flat spacetime, it can hardly fail to help in studying the structure of superspace.

Tangent Vectors on Superspace and the Classical Initial Value Problem

(3) In the classical mechanics of a particle one is accustomed to specifying freely x_0 and $(dx/dt)_0$. These initial conditions determine the whole future history of the particle. What are the analogous freely disposable initial value data of classical geometrodynamics? One is tempted to say: conceive of a continuous one-parameter family of 3-geometries, specified for example in one coordinate patch by the 6 metric coefficients $g_{ik}(x,y,z;\lambda)$; and use this one-parameter family to define data analogous to x_0 and $(ds/dt)_0$ in the one-particle problem; thus, ${}^{(3)}\mathcal{G}_0$ stands for the class of metrics equivalent to $g_{ik}(x,y,z;0)$; and $(d^{(3)}\mathcal{G}/d\lambda)_0$ is the "tangent vector in superspace" defined by $[\partial g_{ik}(x,y,z,\lambda)/\partial\lambda]_{\lambda=0}$, *modulo* the group of coordinate transformations.

It is easy for a mere coordinate shift to mock up the appearance of a change in geometry. Let the coordinates be shifted so that the point P, formerly characterized by the coordinates x^i, is now characterized by $x^i - \lambda\xi^i$, with the vector field ξ^i a continuous function of position. The metric g_{ik} is altered by this shift to

$$g_{ik} + \lambda(\xi_{i|k} + \xi_{k|i}).$$

The derivative $(d^{(3)}\mathcal{G}/d\lambda)_0$ is $\xi_{i|k}+\xi_{k|i}$, *modulo* the group of coordinate transformations. But this quantity, by reason of its very origin, is obviously annullable by a coordinate transformation. Consequently, there is in this case no real change in the geometry. In other words, one cannot admit any otherwise reasonable looking field of values for $(\partial g_{ik}/\partial\lambda)_0$ without running the risk of deception. One wants what has been called in GMD & IFS[2] a 3-geometry (of "acceptable" topology) and another "nearby" 3-geometry in order--one trusts (central hypothesis of the subject!)--to be able to determine the entire past and future of the space, and thereby a complete 4-geometry. But if one has been "deceived," in describing what is ostensibly a second and "nearby" 3-geometry, he may merely be repeating all over again the previously given 3-geometry. In that event, he has only half the amount of initial value data needed to predict the dynamics. One can restate the situation in the following terms in the context of the 4-geometry (regarded temporarily as known!). A spacelike slice is made through the 4-geometry. That gives the one 3-geometry demanded as one of the essential ingredients of the initial value data. However, that one slice is not adequate to distinguish the given 4-geometry from any number of other, different, 4-geometries which admit as slice the same 3-geometry. To complete the selection of the given 4-geometry from these alternative 4-geometries, erect vectors, λn^α, at each of the points of the 3-geometry, with n^α a continuous function of position. Their tips define a new hypersurface, the coordinates in which are connected continuously with the coordinates in the original hypersurface. Evaluate the metric coefficients $g_{ik}(x,y,z,\lambda)$ on this hypersurface. This hypersurface can be said to have been "pushed forward" with respect to the original hypersurface. But has it? Yes, if the normal component of λn^α nowhere vanishes. However, HANS OHANIAN and ELLIOT BELASCO have emphasized in unpublished remarks that it may happen that there are whole regions of the hypersurface where the normal component of λn^{α} vanishes. In that case one has not really pushed the hypersurface ahead at all in ${}^{(4)}\mathcal{G}$. In this event the second component of the initial value data, the derivatives $(\partial g_{ik}/\partial\lambda)_0$, will simply be inadequate for the purposes of the elliptic initial value equations. [75] This situation will be signalled by the fact that $(\partial g_{ik}/\partial\lambda)_0$ can be written in the form $\xi_{i|k} + \xi_{k|i}$. In other regions the 3-geometry *will* have been pushed forward in time. This situation will be signalled, except in special circumstances (time-symmetric initial value problem; change in g_{ik} proportional to λ^2 rather than λ; situation covered by appropriate care in the formulation), by the fact that $(\partial g_{ik}/\partial\lambda)_0$ (or $(\partial g_{ik}/\partial\lambda^2)_0$ in special cases like the time-symmetric initial value problem) is *not* representable in the form $\xi_{i|k} + \xi_{k|i}$.

It is natural to try to summarize the whole situation in the following form. *Give a point in superspace and give a "fully developed direction" at this point in superspace. Then (hypothesis!) this information is sufficient together with Einstein's equations, uniquely to determine the entire 4-geometry.* Here the term "point in superspace" implies, as earlier, the demand that the 3-geometry in question has acceptable topology. The term "fully developed direction" implies that there is no point on the 3-geometry where the quantity $(\partial g_{ik}/\partial\lambda)_0$ (or if it vanishes, the quantity $(\partial^2 g_{ik}/\partial\lambda^2)_0$) can be expressed in the form $\xi_{i|k} + \xi_{k|i}$. In brief, does superspace provide a new approach to the classical initial value problem? And in turn, does that initial value problem throw new light on the concept of "direction" in superspace?

(4) Superspace is built on the concept of 3-geometry; but dynamically conjugate to 3-geometry is the geometrodynamical momentum, with components π_{ij}. Out of these objects, with all the varied topologies that *they* can have, one can build a "conjugate superspace." What are *its* properties?

(5) No crisis stands out more insistently in all of physics than gravitational collapse. No topic connects so immediately the world of the very large and the very small. What insights can one gain from the concept of superspace into the cause and the consequences of gravitational collapse?

(6) How far can one go in analyzing the properties of superspace without getting into the problems of "pregeometry"?

Level 3. Pregeometry

WEYL remarks, "...a more detailed scrutiny of a surface might disclose that, what we had considered an elementary piece, in reality has tiny handles attached to it which change the connectivity character of the piece, and that a microscope of even greater magnification would reveal ever new topological complications of this type, *ad infinitum*." [76] Under such circumstances it would seem difficult to uphold the concept of dimensionality at the smallest distances. General arguments emphasize the same point. [77] Moreover, if electromagnetism and other fields have to do with the quantum mechanical resonance of space between one topology and another, why should not the concept of metric itself be likewise a derived concept, going back for its foundation to topological or pretopological--and at any rate to pregeometric--ideas ("distance between A and B" being defined in the last analysis, for example, by "the ramification of the connections between A and B")? One cannot even mention these topics without recalling the universal sway of the quantum principle, and without stressing the "order of creation" as one thinks of it from physical evidence: Not first geometry, and then the quantum principle; but first the quantum principle and then geometry!

It is enough to raise these issues, with all their depth, to see into what difficulties one can get with quantum geometrodynamics if one tries to think of it as an "ultimate" theory. However, physics has never depended for its progress on having an "ultimate" theory. There is no reason to think that the situation is different today. While one can raise ultimate issues of all kinds, there is no reason to believe that they all have to be settled now! Nor that one *can* resolve them now! The subject presents an ever widening list of issues which have lively physical interest and lend themselves to well-known methods of analysis. [78]

Problem 3. Initial Conditions

The classical initial value problem has already been discussed. What can one say about the corresponding problem in quantum geometrodynamics? In other words, how much information must one give about $\psi({}^{(3)}\mathcal{G})$ on an appropriate submanifold of superspace in order to be able to predict this probability amplitude everywhere in superspace? And what is the character of this submanifold? In this connection one recalls that the "Einstein-Schroedinger wave equation" is of the second order. This second order character raises a question of principle: In order to be able to calculate ψ everywhere must one know on a hypersurface of superspace not only ψ but also its normal derivative? No, LEUTWYLER suggests in a most interesting paper.[79] He points out in the context of a simplified model that the natural features of superspace itself impose certain natural boundary conditions. They reduce the effective order of the equation from second to first.

Wider questions of principle are also posed by the very structure of quantum geometrodynamics. The arena of the dynamics is not space, but superspace. At first this development seems preposterous. How can one speak sensibly of any physical predictions when the outcome depends on what is taking place in unreachable regions of superspace? Nothing could seem more at variance with the spirit of science as dealing only with the knowable. However, a closer look shows that one has broken not at all with the traditional spirit of dynamics, but only with the details. In classical dynamics a clean distinction has always been maintained between (1) the equations of motion, which one can hope to know and understand, and (2) the origin of the initial conditions for those equations of motion - which is beyond one's power to investigate.[80] Quantum geometrodynamics maintains a similar cut between the knowable and the unknowable, but the cut comes in a new place.[81] Nothing seems to exclude the possibility ultimately to know (1) the detailed form of the Einstein-Schroedinger equation, and the concomitant structure of superspace; but as for (2) the source of the initial conditions on ψ, that would seem as far as ever beyond one's power to know. Happily, neither in classical dynamics nor in quantum geometrodynamics does one have to know all initial conditions to make useful predictions! On the contrary, as WIGNER has so often stressed, [81] the role of physics is to predict the *correlations* between observations.

References

[1] L. ROSENFELD: Annalen der Physik 5, 113 (1930) and Z. Physik 65, 589 (1930) and Annales de L'Institut HENRI POINCARÉ 2, 25 (1932); P. G. BERGMANN: Phys. Rev. 75, 680 (1949); P. G. BERGMANN and J. H. M. BRUNINGS: Rev. Mod. Phys. 21, 480 (1949); BERGMANN, PENFIELD, SCHILLER and ZATZKIS: Phys. Rev. 78, 329 (1950); P. A. M. DIRAC: Can. J. Math. 2, 129 (1950); F. A. E. PIRANI and A. SCHILD: Phys. Rev. 79, 986 (1950); P. BERGMANN: Helv. Phys. Acta, Suppl. IV, 79 (1956), Nuovo Cimento 3, 1177 (1956) and Rev. Mod. Phys. 29, 352 (1957); C. W. MISNER: Rev. Mod. Phys. 29, 497 (1957); B. S. DeWITT: Rev. Mod. Phys. 29, 377 (1957); P. A. M. DIRAC: Proc. Roy. Soc. (London) A246, 326 and 333 (1958) and Phys. Rev. 114, 924 (1959); B. S. DeWITT: The Quantization of Geometry, a chapter in *Gravitation: An Introduction to Current Research*, L. WITTEN, ed. (John Wiley and Sons, New York, 1962); J. SCHWINGER: Phys. Rev. 130, 1253 (1963) and 132, 1317 (1963); R. P. FEYNMAN: Mimeographed letter to V. F. WEISSKOPF, dated 4 January to 11 February, 1961; Acta Physica Polonica 24, 697 (1963); *Lectures on Gravitation* (notes mimeographed by F. B. MORINIGO and W. G. WAGNER, California Institute of Technology, 1963); report in *Proceedings of the 1962 Warsaw Conference on the Theory of Gravitation* (PWN-Editions Scientifiques de Pologne, Warszawa, 1964); S. N. GUPTA: report in *Recent Developments in General Relativity* (Pergamon Press, New York, 1962); S. MANDELSTAM: Proc. Roy. Soc. (London) A270, 346 (1962) and Annals of Physics 19, 25 (1962); J. L. ANDERSON: In *Proceedings of the 1962 Eastern Theoretical Conference*, M. E. ROSE, ed.

(Gordon and Breach, New York, 1963) p. 387; I. B. KHRIPLOVICH: Gravitation and Finite Renormalization in Quantum Electrodynamics (mimeographed report. Siberian Section Academy of Science, U.S.S.R., Novosibirsk, 1965); H. LEUTWYLER: Phys. Rev. 134, B1155 (1964); B. S. DeWITT: Dynamical Theory of Groups and Fields, in *Relativity, Groups and Topology*, C. DeWITT and B. DeWITT, eds. (Gordon and Breach, New York, 1964); S. WEINBERG: Phys. Rev. 135, B1049 (1964) and 138, B988 (1965) and 140, B516 (1965); M. A. MARKOV: Progr. Theor. Phys., Yukawa Supplement, 1965, p. 85.

[2] P. W. HIGGS: Phys. Rev. Letters 1, 373 (1958) and 3, 66 (1959) R. ARNOWITT, S. DESER and C. W. MISNER: a series of papers summarized in The Dynamics of General Relativity in *Gravitation: An Introduction to Current Research*, L. WITTEN, ed. (John Wiley and Sons, New York, 1962); A. PERES: Nuovo Cimento 26, 53 (1962); R. F. BAIERLEIN, D. H. SHARP and J. A. WHEELER: Phys. Rev. 126, 1864 (1962); cf. also the Princeton A. B. Senior Thesis of D. H. SHARP, May 1960 (unpublished); J. A. WHEELER: *Geometrodynamics* (Academic Press, New York, 1962), cited hereafter as G M D, and Geometrodynamics and the Issue of the Final State, cited hereafter as G M D & I F S , a chapter in *Relativity, Groups and Topology*, C. DeWITT and B. DeWITT, eds. (Gordon and Breach, New York, 1964); B. DeWITT: Phys. Rev. 160, 1113 (1967), 162,1195 (1967), and 162, 1239 (1967), together cited hereafter as Q T G.

[3] J. A. WHEELER: G M D & I F S ; B. DeWITT: Q T G.

[4] MICHAEL D. STERN: *Investigations of the Topology of Superspace*, Princeton A. B. Senior Thesis, May 1967 (unpublished) and Proc.

Nat. Acad. Sci. U.S.A. (submitted for publication).

[5] R. PENROSE: *An Analysis of the Structure of Spacetime*, Adams Prize Essay, December 1966 (mimeographed for limited distribution, Princeton University, Princeton, New Jersey), gives a beautiful procedure to prescribe on the past light cone exactly enough geometrical information to determine out of Einstein's field equations the complete 4-geometry everywhere within the past light cone. The treatment is given only in the analytic case (in which case the difference between "inside" and "outside" the light cone does not make itself felt) but from general considerations one must expect in the non-analytic case that the data in question determine the 4-geometry only *within* the past light cone. When the light cone points into the future instead of the past, similar considerations of course apply, obtained by the interchange of the words "past" and "future" in what is said in the text. Problems arise with such formulations of the initial value problem "on the light cone" when the propagation proceeds far in a space of variable curvature. Then the light cone develops more than one sheet. Compare the several claps of thunder often heard from a single localized explosion!

[6] For a discussion of the "observer" as a "collector of printout" see for example E. F. TAYLOR and J. A. WHEELER: *Spacetime Physics* (W. H. Freeman and Co., San Francisco, 1966).

[7] For the idea of a general spacelike hypersurface as the manifold on which the magnitudes of quantum field theory are to be measured, see especially S. TOMONAGA: Progr. Theor. Phys. 1, 34 (1946) and J. SCHWINGER: Phys. Rev. 74, 1449 (1948).

[8] J. A. WHEELER: G M D & I F S .

[9] B. DeWITT: Q T G.

[10] This way of writing the conditions for constructive interference is symbolic only. In actuality the Hamilton-Jacobi function S depends not only upon the $^{(3)}\mathcal{G}$, but upon an infinity of parameters which distinguish one solution of the Hamilton-Jacobi equation from another. Thus, in a problem with one degree of freedom we write $S = S_o(x,E) + \delta(E)$ and in a problem with n degrees of freedom $S = S_o(x_1,\ldots,x_n; \alpha_1,\ldots,\alpha_n) + \delta(\alpha_1,\ldots,\alpha_n)$. In geometrodynamics there are two degrees of freedom per space point, the magnitudes associated with which may be designated by α and β. Thus the infinitude of freely disposable parameters may be indicated by two freely disposable functions, $\alpha(u, v, w)$ and $\beta(u, v, w)$. The ∞^3 points are given by the ∞^3 possible choices of u, v and w. The u, v, w manifold may be but is not required to be the same as the manifold, x, y, z of points in the 3-geometry ("alternative choices of parameterization of Hamilton-Jacobi function"). In any case we write S as a functional of α and β; thus, $S = S_o(^{(3)}\mathcal{G}; (\alpha(u, v, w), \beta(u, v, w)) + \delta(\alpha(u, v, w), \beta(u, v, w)))$. Then the conditions of constructive interference become statements about functional derivatives; thus,

$$\delta S/\delta\alpha = 0$$

and

$$\delta S/\delta\beta = 0.$$

This is an explicit form of the symbolic Eq. 11 of the text. Other ways of writing the equations of constructive interference also exist.

[11] ULRICH GERLACH: Bull. Am. Phys. Soc. for the Washington meeting of April 1966, paper DE7, p. 340.

[12] A. PERES: Nuovo Cimento 26, 53 (1962). Here the unit of length is $(16\pi)^{1/2}L^* = (16\pi\hbar G/c^3)^{1/2}$.

[13] References 1 and 2.

[14] R. F. BAIERLEIN, D. H. SHARP and J. A. WHEELER: Phys. Rev. 126, 1864 (1962).

[15] The phrase "dimensionality" can be translated as the requirement of "imbeddability" of all the $^{(3)}\mathcal{G}$'s in a $^{(4)}\mathcal{G}$ -- a requirement that would seem the natural starting point for a derivation of the Einstein-Hamilton-Jacobi equation straight from first principles (see Problem 1).

[16] C. W. MISNER: Rev. Mod. Phys. 29, 497 (1957); see also H. LEUTWYLER: Phys. Rev. 134, B1155 (1964) and B. S. DeWITT: Q T G.

[17] M. PLANCK: Sitzungsber. Preußische Akad. Wiss. Berlin, Math.-Phys. Klasse, 1899, p. 440; J. A. WHEELER: G M D.

[18] For a review of the subject of gravitational collapse, see for example B. K. HARRISON, K. THORNE, M. WAKANO, and J. A. WHEELER: *Gravitation Theory and Gravitational Collapse* (University of Chicago Press, 1965); also A. G. DOROSCHKEVICH, YA. B. ZEL'DOVICH and I. D. NOVIKOV: J. Exptl. Theor. Phys. 49, 170 (1965), English translation in Soviet Physics JETP 22, 122 (1966) and YA. B. ZEL'DOVICH and I. D. NOVIKOV: Usp. Fiz. Nauk. 84, 377 (1964) and 86, 447 (1965), English translations in

Soviet Physics Uspekhi 7, 763 (1965) and 8, 522 (1965).

[19] For a presentation of the quantum electrodynamical calculation of the major part of the Lamb shift of hydrogen from this point of view, see T. A. WELTON: Phys. Rev. 74, 1157 (1948) and F. J. DYSON: *Advanced Quantum Mechanics* (Cornell University, Ithaca, 1954, mimeographed), p. 54.

[20] J. A. WHEELER: G M D.

[21] J. A. WHEELER: G M D. and G M D & I F S ; B. S. DeWITT: Q T G and The Quantization of Geometry in *Gravitation: An Introduction to Current Research*, L. WITTEN, ed. (John Wiley, New York, 1962), pp. 342 ff.

[22] In G M D.

[23] For an elementary discussion of the identity between the tide-producing component of the gravitational force and the Riemann curvature, see for example E. F. TAYLOR and J. A. WHEELER: *Spacetime Physics* (W. H. Freeman and Co., San Francisco, 1966).

[24] E. P. WIGNER: "The Unreasonable Effectiveness of Mathematics in the Natural Sciences," in his book *Symmetries and Reflections* (Indiana University Press, Bloomington, 1967), reprinted from Comm. Pure App. Math. 13, No. 1 (February 1960).

[25] See Reference 17.

[26] A. S. EDDINGTON: *Relativity Theory of Protons and Electrons* (Cambridge University Press, 1936) and *Fundamental Theory* (Cambridge University Press, 1946), also Proc. Camb. Phil. Soc. 27, 15 (1931).

[27] P. A. M. DIRAC: Nature 139, 323 (1937), Proc. Roy. Soc. (London) A165, 199 (1938).

[28] P. JORDAN: *Schwerkraft und Weltall* (Vieweg und Sohn, Braunschweig, 1955) and Zeits. f. Physik 157, 112 (1959).

[29] R. H. DICKE: Science 129, 3349 (1959) and *The Theoretical Significance of Experimental Relativity* (Gordon and Breach, New York, 1964), p. 72.

[30] S. HAYAKAWA: Progr. Theor. Phys. 33, 538 (1965) and Progr. Theor. Phys. Supplement, 532 (1965).

[31] It is permissible to take at full force the argument of EDDINGTON, DIRAC, JORDAN, DICKE and HAYAKAWA, that a physical correlation exists between 10^{20}, 10^{40} and 10^{80}, without accepting the suggestion, sometimes made in the same context, that the physical constants may "change with time." Against such changes there is increasing observational evidence, and for them no incontrovertible evidence has ever been found. Among the relevant observations one can cite as examples R. H. DICKE: Nature 183, 170 (1959) and Nature 192, 440 (1961); also R. H. DICKE and P. J. E. PEEBLES: J. Geophys. Res. 67, 10 and 4063 (1962) and Phys. Rev. 128, 5 and 2006 (1962), showing no detectable change with time in the relative rates of selected

processes of radioactive decay. To search for changes with time in the reciprocal fine structure constant, $\alpha^{-1} = \hbar c/e^2 = 137.03$, is simple in principle. One has only to compare the wavelength of the 21cm line of hydrogen (red-shifted because it was given out by a rapidly receding galaxy, far away and long ago) with the wavelength of a line in the optical spectrum (which has undergone the same red shift). The ratio, R, of the two wavelengths is α^{-1} multiplied by a known function of the atomic number of the source (an integer) and of the relevant quantum numbers (also integers):

$$R = \alpha^{-1} \text{ times function of integers.}$$

The value of R for a source 1.4×10^9 light years away (recession rate $\beta = v/c = 0.1$) should differ by several percent from the value of R for a laboratory source if any of the suggestions are correct that physical constants might change in proportion to the time (or some significant power of the time) measured from the start of the expansion of the universe. The writer is indebted to the kindness of Professor R. MINKOWSKI of Berkeley for the following 28 July, 1967 summary of the observational situation: (1) Hopes have been dashed to observe the 21cm line in the spectrum of galaxies anywhere near as far away as would correspond to a recession velocity of $\beta = 0.1$. (2) Observations have been made on 30 nearer objects by DIETER, EPSTEIN, LILLEY and ROBERTS: Astrophys. J. **67**, 270 (1962) as supplemented by ROBERTS: Astrophys. J. **142**, 148 (1965). The red shift is the same for the 21cm line and for the optical lines within the limits of error of the observations. It is difficult to evaluate the accuracy because individual motions amount to as much as

20 to 30 percent of the average recession velocity of $v = 1600$ km/sec ($\beta = 0.005$). It is probably safe to say that any change in α^{-1} must be less than a few percent to be compatible with the observations. However, a change anyway smaller than this limit would be expected on almost any of the varied theories of the change of α^{-1} with time, since the time lapse in this case is only one two hundredth of the HUBBLE time. (3) Instead of comparing the wavelength of the 21cm line with the wavelength of an optical transition, one can measure the fine structure separation of a related pair of lines in the optical spectrum itself. The fractional splitting, $\Delta\lambda/\lambda$, should be independent of the red shift of the source if α^{-1} is constant. From the observations of R. MINKOWSKI: Astrophys. J. 123, 373 (1956), on Cygnus A ($v = 16830$ km/sec, $\beta = 0.056$); W. BAADE and R. MINKOWSKI: Astrophys. J. 119, 206 (1954) it is again probably safe to say that any change in the fine structure constant must be less than a few percent. BAHCALL, SARGENT and SCHMIDT: Ap.J. Lett. 149, 11 (1967) give $|\Delta\alpha/\alpha| < 0.05$ for $z = 2$.

[32] The quark, so useful in doing bookkeeping on the beautiful regularities of elementary particle physics (summarized for example in M. GELL-MANN and Y. NE'EMAN, eds.: *The Eightfold Way* (Benjamin, New York, 1964) and F. DYSON, ed.: *Symmetry Groups in Nuclear and Particle Physics* (Benjamin, New York, 1966), has sometimes been taken much more seriously, as if it were an actual "primordial building block" of matter. That view may or may not be correct. If it is, and if one still continues to take quantum geometrodynamics as the only available indicator of what goes on at very small distances, then it would still seem reasonable to expect that one must

have some perspective on what happens at 10^{-33} cm before one can find the rationale of quarks and particles. That there is no such thing as a quark in the literal sense is however a point of view accepted by many investigators, and stressed especially by HEISENBERG and DÜRR: W. HEISENBERG, *Introduction to the Unified Field Theory of Elementary Particles* (Wiley, New York, 1966) and H. P. DÜRR: "On the non-linear spinor theory of elementary particles", Acta Physica Austriaca, Supp. 3 (1966). DÜRR has made the same point even more vividly (kind personal communication of June, 1967) by considering in effect what one would conclude out of the first several dozen atomic energy levels of an atom, such for example as carbon or iron, if one had (1) good measurements of the energies and transition probabilities, and (2) today's aptitude for searching for symmetries, but (3) not the slightest idea of the actual internal machinery of an atom. He shows how groups of high symmetry will make their appearance. His discussion leads one to ask whether the innocent investigator will not conclude that the atom is made out of quarks!

[33] The strongest statement easily available against taking general relativity seriously at small distances appears to be that made by ROBERT OPPENHEIMER in his articles "On ALBERT EINSTEIN" (New York Review, 17 March, 1966, pp. 4, 5): "He also worked with a very ambitious program, to combine the understanding of electricity and gravitation in such a way as to explain what he regarded as the semblance -- the illusion -- of discreteness, of particles in nature. I think that it was clear then, and believe it to be obviously clear today, that the things that this theory worked with were too meager,

left out too much that was known to physicists but had not been known much in Einstein's student days. Thus it looked like a hopelessly limited and historically rather accidentally conditioned approach."

[34] For further discussion of the rationale of changes in topology see Problem 2 in the appendix.

[35] G M D .

[36] To take it as self-evident that space is Euclidean in character at small distances became impossible after RIEMANN. His Göttingen inaugural lecture of 10 June, 1854 pointed out that space can be highly rippled at submicroscopic distances and yet look smooth to all ordinary means of observation: "Über die Hypothesen welche der Geometrie zugrunde liegen" in his *Gesammelte Mathematische Werke* (H. WEBER, ed., 2nd ed., reprinted by Dover Publications, New York, 1953), also in a translation in Nature 8, 14 (1873), by W. K. CLIFFORD. CLIFFORD himself went further in his lecture before the Cambridge Philosophical Society, 21 Feb., 1870, "On the Space-Theory of Matter", reprinted in his *Mathematical Papers*, R. TUCKER, ed. (London, 1882), also in his *Lectures and Essays*, L. STEPHEN and F. POLLOCK, eds., Vol. 1 (London, 1879). He proposed to consider a particle as made up of nothing but curved empty space, differing from the surrounding space precisely in this localized curvature -- and perhaps also in its connectivity or local topology. In *Was ist Materie* (Springer, Berlin, 1924), esp. pp. 57, 58, HERMANN WEYL again pointed out that space may be multiply connected in the small,

and consequently: "The argument that the charge of the electron must be spread over a finite region, because otherwise it would possess infinite inertial mass, has thus lost its force. One can not at all say, there is charge, but only, this closed surface encloses charge." He went on to comment that the enormous value of the ratio of electric to gravitation forces "seems to indicate that the total number of electrons in the universe is important for the constitution of the individual electron". ALBERT EINSTEIN and NATHAN ROSEN: Phys. Rev. 48, 73 (1935), proposed the concept of two nearly Euclidean spaces, connected here and there by thin bridges or tubes, through which electric lines of force thread, to give the appearance of charges of variegated signs in the "upper" space and corresponding charges of the opposite sign in the "lower" space. J. A. WHEELER: Phys. Rev. 97, 511 (1955), reprinted in GMD, proposed instead the concept of a tube or handle or "wormhole" reaching between two different localities in one and the same Euclidean space. It is an automatic consequence of this picture that the universe should contain equal amounts of positive and negative electricity. It is another consequence, proved by MISNER in 1957 straight from Maxwell's equations for empty space, that the charge, or flux of lines through the wormhole, must stay constant with time. The proof, C. W. MISNER and J. A. WHEELER: Annals of Phys. 2, 525 (1957), reprinted in GMD, holds no matter how tortuously the lines of force may be twisted, no matter how wanting in symmetry the geometry of the wormhole may be, and no matter how violently the field and the geometry may subsequently change with time. MISNER also showed here the beautiful ties that exist between the MAXWELL theory in a multiply connected empty space and the mathematics of differential forms and homology groups. In his analysis the field and the geometry were assumed to evolve deterministically in time, in accordance with the classical equations of electrodynamics and geometrodynamics. Reasons out of fluctuation theory to consider "wormholes" a property, not of particles, but of all space, were first given by J. A. WHEELER in Annals of Phys. 2, 604 (1957), expanded in GMD.

[37] No one has pointed out a more direct tie between the energy of vacuum fluctuations and macroscopic physics than H. B. G. CASIMIR: Proc. Nederland Akad. Wetenschappen, Amsterdam, 60, 793 (1948), who predicted a force between two parallel metal plates. No attempt is made here to cite the extensive literature that verifies the existence and the predicted magnitude of this force. The same kind of fluctuations which are verified by this force at macroscopic distances are also checked at distances $\sim 10^{-12}$ cm by the Lamb shift, the most impressive single development in quantum electrodynamics in the post World War II period (Fig. 4 and reference 19).

[38] D. R. BRILL and J. B. HARTLE: Phys. Rev. 135, B271 (1964).

[39] For a historical survey which treats chemistry and atomic physics as the two parts of a single development, see for example: W. G. PALMER: *A History of the Concept of Valency to* 1930 (Cambridge University Press, 1965) and especially J. J. LAGOWSKI: *The Chemical Bond* (Houghton Mifflin, Boston, 1966).

[40] On these regularities see for example the books cited in reference 32.

[41] J. BARDEEN, L. N. COOPER and J. R. SCHRIEFFER: Phys. Rev. 108, 1175 (1957).

[42] The three issues listed here are taken up in more detail in the appendix.

[43] A. EINSTEIN in P. A. SCHILPP, ed. *ALBERT EINSTEIN: Philosopher Scientist* (Library of Living Philosophers, Evanston, Illinois, 1949, p. 81) remarks: "If one had the field-equation of the total field, one would be compelled to demand that the particles themselves would *everywhere* be describable as singularity-free solutions of the completed field-equations. Only then would the general theory of relativity be a *complete* theory."

[44] E. CARTAN: *Lęcons sur la géométrie des espaces de RIEMANN* (Gauthier-Villars, Paris, 2nd ed., 1959), chap. 8.

[45] J. A. WHEELER, chapter 4 on Cartan's geometrical interpretation of Einstein's field equations in *Gravitation and Relativity*, H. Y. CHIU and W. F. HOFFMANN, eds. (W. A. Benjamin, New York, 1964).

[46] Arguments that the propagator appropriate for any particle of spin two and mass zero necessarily has a leading term of the form

$$k^{-2}(g^{\mu\alpha} g^{\nu\beta} + g^{\mu\beta} g^{\nu\alpha} - g^{\alpha\beta} g^{\mu\nu}),$$

are given by S. WEINBERG, Phys. Rev. 138, B 988 (1965); also in expanded form in his contribution to: S. DESER and K. W. FORD, eds.: Brandeis Summer Institute in Theoretical Physics, 1964, Vol. 2, *Lectures on Particle and Field Theory* (Prentice-Hall, Englewood Cliffs, New Jersey, 1965).

[47] See for example W. PAULI in "Die allgemeinen Prinzipien der Wellenmechanik" in *Handbuch der Physik*, Vol. 24, part. 1, GEIGER and SCHEEL, eds., (Springer, Berlin, 1933); reprinted in revised form in the new *Handbuch der Physik*, Vol. 5, part 1, ed. by S. FLÜGGE (Springer, Berlin, 1958).

[48] See the discussion of the problem of factor ordering in B. DeWITT, QTG; also the references to earlier discussions of this issue cited by him there.

[49] For a determination of the wave equation from (1) the principle of LORENTZ covariance and (2) a selection of one or another set of spin quantum numbers see E. P. WIGNER: Ann. of Math. 40, 149 (1939) and V. BARGMANN and E. P. WIGNER: Proc. Nat'l. Acad. Sci. U. S. 34, 211 (1946), both reprinted in the collection *Symmetry Groups in Nuclear and Particle Physics*, ed. by F. J. DYSON (Benjamin, New York, 1966).

[50] For an analysis of the metric of superspace see B. DeWITT, QTG, reference 2, and other work cited by DeWITT; see also S. WEINBERG, reference 46.

[51] For the proof that the topology of space cannot change within the context of classical geometrodynamics, see R. P. GEROCH: J. Math. Phys. 8, 782 (1967).

[52] For a model of a closed universe with the topology S_3 put together out of 720 identical pieces each endowed with the SCHWARZSCHILD geometry ("lattice universe") see R. W. LINDQUIST and J. A. WHEELER: Rev. Mod. Phys. 29, 432 (1957) and further treatment in GMD & IFS, pp. 370-379.

[53] A. EINSTEIN, end of chapter dealing with Mach's principle in *The Meaning of Relativity* (Princeton University Press, Princeton, New Jersey, 3rd ed., 1950).

[54] See for example some of the solutions of Einstein's equations given by B. K. HARRISON: Phys. Rev. 116, 1285 (1959) and his fuller Princeton University Ph. D. thesis, *Exact Three-Variable Solutions of the Field Equations of General Relativity*, 1959 (unpublished).

[55] Here it is assumed that the conjecture of H. POINCARÉ is correct, that every simply connected compact differentiable three-dimensional manifold is homeomorphic to the three-sphere. See C. D. PAPAKYRIAKOPOULOS: "The theory of differentiable manifolds since 1950," *Proceedings International Congress of Mathematicians*, 1958 (Cambridge University Press, 1960) pp. 433-440; also J. MILNOR: *Topology from the Differentiable Viewpoint* (University of Virginia Press, Charlottesville, 1965), and the bibliography cited by MILNOR; also J. DERWENT: "Handle decomposition of manifolds," Jour. of Math. and Mech. 15, 329 (1966).

[56] Particularly to be emphasized is the distinction between "asymptotically flat" as that concept is so often understood in

the classical context of a 4-geometry, and the concept of flatness as it is applied to a 3-geometry in the context of HAMILTON-JACOBI theory or quantum geometrodynamics. No example illustrates this distinction more clearly than the SCHWARZSCHILD geometry. There the rate of approach of the 4-geometry to flatness at infinity determines always a unique value for the mass of the center of attraction; but the analogous calculation for a spacelike 3-geometry slicing through the 4-geometry gives quite different values for the apparent mass, depending upon the choice of slice. Thus, in the 4-geometry

$$ds^2 = -(1 - 2m/r)\, dt^2 + (1 - 2m/r)^{-1} dr^2 + r^2(d\theta^2 + \sin^2\theta\, d\phi^2),$$

take at large distances the spacelike slice

$$t = t_0 + (8\alpha r)^{1/2},$$

so that

$$dt = (2\alpha/r)^{1/2} dr.$$

On this slice one finds a 3-geometry in which the coefficient of dr^2, also at large distances, is

$$1 + 2\,(m - \alpha)/r.$$

The dependence of the "effective mass," $(m - \alpha)$, upon the choice of slice, through the parameter α, suggests some of the many dangers that seem to lurk in the concept of "asymptotic flatness" as applied to *three*-geometries.

[57] In the TAUB universe the effective mass-energy arises entirely from excitation of that mode of gravitational radiation which

has the longest wavelength capable of fitting into this universe. For the metric of this model see A. TAUB: Ann. of Math. 53, 472 (1959) and C. W. MISNER: J. Math. Phys. 4, 924 (1963).

[58] To say that a particle is "pictured in terms of space resonating from one topology to another" means more precisely that it is "pictured as a geometrodynamical exciton--a state of excitation in which space resonates from one topology and geometry to another according to a probability amplitude function slightly different from, and orthogonal to, the probability amplitude function $\psi({}^{(3)}\mathcal{G})$ that describes the vacuum."

[59] For a systematic development of "already unified field theory" see C. W. MISNER and J. A. WHEELER: Annals of Phys. 2, 525 (1957) (reprinted in GMD) where also reference is made to the earlier work of G. Y. RAINICH.

[60] In "already unified field theory" the electromagnetic field tensor is expressed, in accordance with Einstein's field equations, in terms of the "MAXWELL square root" of the RICCI curvature tensor and its dual

$$F_{\mu\nu} = (\text{"}R^{1/2}\text{"})_{\mu\nu} \cos\alpha + *(\text{"}R^{1/2}\text{"})_{\mu\nu} \sin\alpha .$$

The change of the "complexion" α of the electromagnetic field from place to place is fully determined by Maxwell's equations in places where there is a field. However, consider a spacelike initial value hypersurface. On this hypersurface consider two regions, I, II, endowed with field and separated by a region III, free of field. Within each region, individually, the relative complexion is well determined, which is all that matters momentarily for the electrodynamics. However, the complexion of region II relative to region I can never be found from purely geometrical measurements limited to this initial spacelike hypersurface. Moreover, this relative complexion is all important for the dynamic development of the electromagnetic field at those later points in spacetime which can be reached by disturbances both from I and from II. In this sense, the initial value problem of already unified field theory does not lend itself to purely geometrical formulation. For more on this topic see the chapter by L. WITTEN in the book of which he is also the editor, *Gravitation: An Introduction to Current Research* (John Wiley and Sons, New York, 1962).

[61] The divergence condition is well known to follow from the invariance of the Hamilton-Jacobi function with respect to the gauge transformation, $A_i^{new} = A_i + \partial\lambda/\partial x^i$; thus,

$$\begin{aligned} 0 = \delta S &= \int(\delta S/\delta A_i)\delta A_i d^3x \\ &= \int(\mathcal{E}^i \partial\lambda/\partial x^i)d^3x \\ &= -\int \mathcal{E}^i{}_{,i}\lambda d^3x. \end{aligned}$$

The vanishing of this expression for arbitrary λ gives the desired relation. It should be emphasized that the quantity $\mathcal{E}^i$ as employed here is not a contravariant vector, but $(g)^{1/2}$ times a contravariant vector ("vector density"). Were the contravariant vector itself employed, the divergence relation would have to be expressed in terms of covariant derivatives rather than ordinary derivatives, complicating the derivation in the text.

[62] Here the symbol $(\mathcal{E} \times B)_j$ stands for the covariant vector density $\mathcal{E}^i B_{ij}$.

[63] As HERMANN WEYL emphasized long ago, Math. Zeits. 23, 271 (1925), "In den geometrischen und physikalischen Anwendungen zeigte sich stets, dass eine Grössart nicht allein durch Angabe der Tensorstufe, sondern durch Symmetriebedingungen charakterisiert ist." In other words, every physical quantity is represented by an irreducible tensorial quantity; that is to say, by what S. S. CHERN terms "a geometrical object." WEYL conceived of these geometrical entities as local. However, it is a natural extension of his line of thought to speak of a *functional* S or ψ which depends *globally* upon a 3-geometry, and upon a 2-form imbedded in that 3-geometry.

[64] GMD, reference 2, p. 88.

[65] JOHN MILNOR: "A survey of cobordism theory," L'enseignement mathematique 8, 16 (1962); "Spin structures on manifolds," *ibid.* 9, 198 (1963); "On the Stiefel-Whitney numbers of complex manifolds and of spin manifolds," Topology 3, 223 (1965); "Remarks concerning spin manifolds" in S. S. CAIRNS, ed.: *Differential and Combinatorial Topology* (Princeton University Press, Princeton, New Jersey, 1965), p. 55; ANDRÉ LICHNEROWICZ: Acad. Sci. Paris, Comptes Rend. 252, 3742 (1961) and 253, 940 (1961) and 253, 983 (1961) and summary of these results in the third part of the chapter by LICHNEROWICZ, "Propagateurs, Commutateurs et Anticommutateurs en Relativité Générale," in C. and B. DeWITT, eds.: *Relativity, Groups and Topology* (Gordon and Breach, New York, 1964); D. W. ANDERSON,

E.H. BROWN, Jr., and F. P. PETERSON: "Spin cobordism," Bull. Am. Math. Soc. 72, 256 (1966); "SU-cobordism, KO-characteristic numbers and the Kervaire invariant," Ann. of Math. 83, 54 (1966); W. C. HSIANG and B. J. SANDERSON: "Twist-spinning spheres in spheres," Illinois J. Math. 9, 651 (1965). Appreciation is expressed to JOHN MILNOR, ROGER PENROSE and ROBERT GEROCH for discussions clarifying the concept of "spin structure."

[66] Take a belt. Stretch it out flat and taut. Keeping the left hand end A fixed in the left hand, twist the right hand end B through 720°. Maintaining A and B all the time parallel to their present orientations, move B in a complete circle about A (releasing for an instant one's hold on B). The belt straightens out. Not so when there is only a 360° twist in it. The belt is relevant to the cube, the room, and the eight elastic strings. Before the cube is rotated at all it can be pulled out through a window to some distance from the room. The eight elastic strings then take on the configuration of the belt. The distinction between 360° and 720° rotation for the belt applies equally to the "pseudo-belt" made up of the eight strings. Another way of seeing that a 720° rotation restores the "orientation entanglement relation" between the cube and its surroundings (picture of one cone rolling on another) is presented by R. PENROSE and W. RINDLER in a preprint of an appendix to a book that they have in preparation. Appreciation is expressed to Professor PENROSE for the privilege of seeing this preprint.

[67] The possibility has been suggested that one may eventually be able to detect what is here called the "orientation entanglement relation" between an object and its surroundings by measuring the contact potential between one metallic object (subject to rotation) and another (held fixed): Y. AHARONOV and L. SUSSKIND: Phys. Rev. 158, 1237 (1967).

[68] Cf. paragraph on orientability, Problem 2, level 1.

[69] It is not new to abstractify geometry. Einstein's curved spacetime was in the beginning nothing if it was not a home for geodesics. How else, one asked, could he predict a planetary motion. Later EINSTEIN, GROMMER, INFELD and HOFFMAN threw out the geodesics. The field equations themselves, they showed, predict the evolution of geometry with time, and hence the motion of concentrations of mass-energy.

[70] For a history of the concept of spin, see W. Pauli's 1945 Nobel prize lecture, *Exclusion Principle and Quantum Mechanics* (Editions Grisson, Neuchatel, 1947), also the relevant discussion in M. FIERZ and V. F. WEISSKOPF: *Theoretical Physics in the Twentieth Century: A Memorial Volume to WOLFGANG PAULI* (Interscience, New York, 1960).

[71] See for example C. KITTEL: *Introduction to Solid-State Physics*, 3rd ed. (John Wiley and Sons, New York, 1966).

[72] The torus can be converted into a single sheet by two cuts, and can then be conceived as laid out on the complex plane , with one corner at the origin. One adjacent corner can be

identified arbitrarily with the number 1 + 0i, by appropriate choice of scale ("conformal transformation"). The location of the other adjacent corner, $\tau = \tau_1 + i\tau_2$, is then completely determined. Also completely determined is the concomitant 2-geometry, *modulo* the group of conformal transformations. The quantity τ can be identified with the complex parameter mentioned in the text.

[73] For a survey, with many references to the literature, see H. E. RAUCH: "A transcendental view of the space of algebraic Riemann surfaces," Bull. Amer. Math. Soc. 71, 1 (1965). Appreciation is expressed to Professor LEON EHRENPREIS for elucidation of the subject and for this and the following reference.

[74] LIPMAN BERS: *On the moduli of RIEMANN surfaces*; lectures at the Forschunginstitut für Mathematik, Eidgenössische Technische Hochschule, Zürich, 1964. Notes by L. M. and R. J. SIBNER (mimeographed).

[75] For the initial value equations of classical geometrodynamics see G. DARMOIS: *Les equations de la gravitation einsteinienne* (Gauthier-Villars, Paris, 1927); K. STELLMACHER: Math. Ann. 115, 136 (1937); A. LICHERNOWICZ: J. Math. Pure Appl. 23, 37 (1944); Helv. Phys. Acta Supp. 4, 176 (1956); *Théories relativistes de la gravitation et de l'électromagnétisme*, (Masson, Paris, 1955); YVONNE FOURÈS-BRUHAT: Acta Math. 88, 141 (1952); J.Rational Mech. Anal. 5, 951 (1956); and the chapter by Y. FOURÈS (now Y. CHOQUET) in LOUIS WITTEN, ed.: *Gravitation: an Introduction to Current Research* (John Wiley and Son, New York, 1962).

[76] H. WEYL: *Philosophy of Mathematics and Natural Science* (original German in 1927); translation by O. HELMER, p. 91 (Princeton University Press, Princeton, New Jersey, 1949).

[77] GMD & IFS, pp. 495-499.

[78] An extensive list of problems open for further investigation is to be found in GMD & IFS.

[79] H. LEUTWYLER in *Battelle Rencontres: 1967 Lectures in Mathematics and Physics* (William Benjamin, New York, in publication).

[80] One is reminded in this connection of the statement of WILLIAM JAMES over a half a century ago, that "Actualities seem to float in a wider sea of possibilities from out of which they were chosen; and *somewhere*, indeterminism says, such possibilities exist, and form a part of truth." Appreciation is expressed to PAUL VAN DE WATER for this quotation.

[81] E. P. WIGNER: *Symmetries and Reflections* (Indiana University Press, Bloomington, Indiana, 1967); see J. M. JAUCH, E. P. WIGNER, and M. M. YANASE: Nuovo Cimento 48, 144 (1967), also B. DeWITT: QTG; also H. EVERETT, III: Rev. Mod. Phys. 29, 454 (1957); J. A. WHEELER: Rev. Mod. Phys. 29, 463 (1957) and GMD, p. 75.